Preparation

FOR ALGEBRA

J. Louis Nanney

John L. Cable

Book Team

Editor *Earl McPeek*
Developmental Editor *Theresa Grutz*
Production Editor *Eugenia M. Collins*
Designer *K. Wayne Harms*
Art Editor *Joseph P. O'Connell*
Photo Editor *Judi David*

Hawkes Publishing
A Division of Quant Systems, Inc.
Cover photo Kindra Clineff, Winchester, MA, © 1990

Photo credits
Chapter openers: 2: © Jim Shaffer; 3: © Jim Shaffer; 4: © Eric Meola/ The Image Bank-Chicago; 5: (*top*) © Peter Miller/The Image Bank-Chicago; (*bottom*) © H. Wendler/The Image Bank-Chicago; 7: © Bill Varie/The Image Bank-Chicago; 8: © Superstock; 9: © Jim Shaffer; 11: © Andy Caulfield/The Image Bank-Chicago. Page 79: © Jim Shaffer

Credits
Chapter openers 1, 6, 10: Diphrent Strokes, Inc. Page 79: (*left*) Dinty Moore Beef Stew Universal Product Code used courtesy of Geo. A. Hormel & Company. (*right*) JELL-O is a registered trademark of Kraft General Foods, Inc. Reproduced with permission.

Library of Congress Catalog Card Number: 92-53383

ISBN 0-697-12801-6

Printed in the United States of America by Quebecor World, Dubuque, IA 52001

10 9 8 7 6 5 4

Contents

CHAPTER 5

Concepts and Formulas from Geometry

CHAPTER 6

Operations on Signed Numbers

CHAPTER 7

Using Signed Numbers

CHAPTER 8

Solving Equations

CHAPTER 9

Ratio, Proportion, and Percent

CHAPTER 10

Interpreting Data from Statistics and Graphs

CHAPTER 11

Analyzing Word Problems

Preface

Purpose

Recent studies as well as personal experience and observation indicate that incoming college students have changed in at least two ways. First, on the average they are older and second, many are not sufficiently prepared in the basic skills of arithmetic to attempt a course in beginning algebra. The most common remedy to this problem has been the offering of a course in arithmetic to prepare the students for beginning algebra. But too many times we find that this arithmetic course merely exposes these students to the same topics they have failed to learn since elementary school with the result that they again fail to sufficiently master them.

Preparation for Algebra—An Integrated Foundation has been produced to offer an alternate solution to the current problem. Algebraic concepts are introduced as the basic skills of arithmetic are presented. The reasons for this integration are threefold. First, the student will have more interest in learning the skills of arithmetic when their purpose in algebra is evident. Second, the fact that algebra is an extension of arithmetic will become apparent to the student. Third, the concepts of algebra that are introduced will enhance the students' possibility of success in a subsequent course in elementary algebra.

Approach

All of the traditional topics of arithmetic are included in this text with concepts of algebra integrated where appropriate. The material is integrated in such a way that algebraic concepts are being studied at the same time that arithmetic skills are learned. The integration of topics from arithmetic and algebra is never forced but rather results from a natural flow. One new topic at a time is introduced and then reinforced, resulting in a text that is easily understood and easy to use in the classroom.

Format

Each section begins with an introductory paragraph that gives a purpose for the section and, where possible, relates the new material to previous sections or chapters. One new idea at a time is presented. Clear, concise, mathematically correct explanations and definitions are given and are followed by many detailed examples and an abundance of exercises.

Topics

Arithmetic with Integrated Algebraic Concepts

A very thorough presentation of the four basic operations on whole numbers, fractions, and decimals is presented. Appropriate algebraic concepts are integrated within this presentation thus enhancing the transition from arithmetic to algebra.

Equations

Signed numbers are introduced prior to the chapter on solving equations. The strategy for solving equations utilizes the properties of equality. Techniques for solving equations are then used in the presentation of the topics of ratio, proportion, and percent.

Geometry

National attention is being given to the increasing importance of topics from geometry in the prealgebra curriculum. In response to this demand a fairly extensive treatment of these topics is presented in chapter 5. Properties of lines and angles as well as the topics of perimeter, area, and volume are introduced and subsequently used in various sections of the text. In addition to these basic topics, coverage includes other concepts, formulas, and applications as well as an intuitive but thought-provoking approach to some simple proofs. This chapter addresses all of the minimal geometry competencies included in the preliminary report of the Developmental Mathematics Committee of the American Mathematical Association of Two Year Colleges.

Statistics

Chapter 10 introduces the concepts of mean, median, mode, and range as statistics. It also addresses the interpretation of pictographs, tables, bar graphs, line graphs, and circle graphs.

Applications

Whenever possible, exercise sets throughout the text include problem-solving situations involving the topics discussed. These applications progress from one step to multistep problems. Students are gradually introduced to more complex applications as their confidence builds in seeing how to use mathematics to solve real world problems. Problem-solving strategies are thus emphasized from the earliest sections. An entire chapter is devoted to the techniques of analyzing, translating, and solving word problems.

Text Features

Chapter Pretest

The Chapter Pretest helps students and instructors identify concepts the student has already mastered as well as areas the student should concentrate on. Instruction can be tailored to the particular strengths identified by the Chapter Pretest results. All answers are provided in the answer appendix and each is coded to the appropriate section.

Chapter Opening

Special care has been taken to begin each chapter with several special features to motivate students. Pertinent historical information or a current application of a concept from the chapter provides an interesting beginning to each chapter. Study Helps give specific study suggestions at the beginning of each chapter to help students achieve success in mathematics.

Introductory Paragraph

The introductory paragraph overviews the topics to be covered, explains why it will be useful to master the particular skills and concepts, and ties the new material to previously learned topics where appropriate.

Objectives

Each section of the text begins with a boxed statement of the objectives. Each objective is keyed to that portion of the section that addresses it. Exercise sets within the section are also keyed to the objectives. Thus the focus of the student and instructor is properly directed.

Numbered Examples

Each concept and skill is slowly developed using numerous examples that build on each other and increase slowly in level. The examples model the complete range of exercises in the text.

Skill Checks

A numbered example is usually immediately followed by a similar type problem to be worked by the student. By actively solving this skill check, the student is provided with immediate reinforcement of the concept presented. Thus any misunderstanding is corrected before proceeding.

Margins

A unique feature of this text is the judicious use of margins. Rather than just an answer column or an expensive scratch pad, the margins are designed to be useful to the student. They contain the objectives, helpful notes, hints, warnings against common errors, references to properties, restatements of appropriate definitions, and answers to Skill Checks. The language in the margins is less formal than that of the text. In short, the margins are designed to compliment the text with a self-contained student study guide.

Warnings

The warnings contained in the margins are a result of many years of teaching experience. Most common errors that students are likely to make have been identified. These warnings will be a special help to the student.

Section Exercises

Careful attention has been given to each set of exercises so that they properly reinforce the material presented, provide review, and where possible lead into the next concept to be presented. Exercise sets are graded in difficulty and odd–even problems are similar. Problems within the exercise set are keyed to the appropriate objective.

Concept Enrichment

Many exercise sets contain a set of questions that amplify and enrich the concepts discussed in that section. They include questions that are thought provoking and generate discussion.

Chapter Summary

Each chapter is followed by a summary that lists key words and procedures covered in the chapter. These are referenced to the sections that discuss them to help students effectively and efficiently review the chapter.

Chapter Review

After the chapter summary each chapter contains an exhaustive review of all of the skills and concepts presented in the chapter. This gives the student ample practice with the topics. Odd-numbered answers are given in the answer appendix. Each answer is coded to the section number so students can easily locate the appropriate area for review if they encounter difficulties.

Chapter Test

A Chapter Test, with answers keyed to the sections within the chapter, is provided at the end of each chapter. This test is designed to help the student review and to prepare for examinations. Each test covers all of the key concepts presented in the chapter and is similar in content and length to the Pretest so that the student may compare results and evaluate progress.

Cumulative Tests

Cumulative tests are provided after each chapter beginning with chapter 2. They are designed to help students maintain mastery of previously learned concepts. This also provides the opportunity to tie ideas together and recognize the continuity of the subject matter. All of the answers are given in the answer appendix and are coded to the appropriate chapter and section so that concepts needing reinforcement can be easily located.

Answer Appendix

Answers to the odd-numbered exercises and all answers to the pretests, chapter tests, and cumulative tests are provided in the appendix at the end of the book. Since the odd and even questions in the exercise sets and chapter reviews are similar, students can be assigned exercises with answers or without answers, depending on the particular class situation.

Supplementary Materials

The ***Instructor's Resource Manual*** includes all of the answers for the even-numbered section exercises, two reproducible supplemental worksheets for each objective (151 objectives) in the textbook, two reproducible forms of chapter tests, and two reproducible forms of a final exam. These supplemental problems are unique and different from those in the Test Item File, TestPak, and Diagnostic Masters. A complete answer key for all of the reproducibles is included in the Instructor's Resource Manual.

The ***Test Item File/Quiz Item File*** is a printed version of the computerized *TestPak* and *QuizPak* that allows you to choose test items based on chapter, section, or objective. The objectives are taken directly from each section of *Preparation for Algebra.*

HP *TestPak,* our computerized testing service, provides you with a mail-in/call-in testing program and the complete *Test Item File* on diskette for use with IBM PC, Apple, or Macintosh computers. HP *TestPak* requires no programming experience. Tests can be generated randomly, by selecting specific test items or the based-on-section objectives. In addition, new test items can be added and existing test items can be edited.

HP *QuizPak,* a part of *TestPak 3.0,* provides students with true/false, multiple choice, and matching questions from the *Quiz Item File* for each chapter in the text. Using this portion of the program will help your students prepare for examinations. Also included with the *QuizPak* is an on-line testing option that allows professors to prepare tests for students to take using the computer. The computer will automatically grade the test and update the gradebook file.

HP *GradePak,* also a part of the *TestPak 3.0,* is a computerized grade management system for instructors. This program tracks student performance on examinations and assignments. It will compute each student's percentage and corresponding letter grade, as well as the class average. Printouts can be made utilizing both text and graphics.

Diagnostic Masters to Accompany Preparation for Algebra are additional reproducible masters keyed to each section objective in the text, and reproducible chapter tests. *Diagnostic Masters* can be used for placement, additional assignments, instructional examples, quizzes, reviews, or exams. A complete answer key for all masters is included.

Acknowledgments

A text of the quality cannot be produced without the assistance of many people. The authors wish to express their gratitude to the initial editors and staff at Wm. C. Brown Publishers for their constant support. Special thanks go to Theresa Grutz, developmental editor; Gene Collins, production editor; K. Wayne Harms, designer; Joseph O'Connell, art editor; and Judi David, photo editor. Acknowledgment is also due to our new publisher, Hawkes Publishing for their continued support of this text. We also wish to thank the following reviewers for their constructive comments.

Beverly R. Broomell
Suffolk Community College

Sharon J. Edgmon
Bakersfield College

John R. Garlow
Tarrant County Junior College NW

Lynne Hensel
Henry Ford Community College

W. Arlene Jeskey
Rose State College

Linda J. Murphy
Northern Essex Community College

Nancy K. Nickerson
Northern Essex Community College

Carol O'Laughlin
Northern Essex Community College

Lazella Lawson
The Master's College

Sally Ann Low
Angelo State University

James Magliano
Union County College

John R. Martin
Tarrant County Junior College NE

Joyce Milmed
Miami Dade Community College

John A. Scoubis
Chicago City College

Fran Smith
Oakland Community College

Don W. Williams
Brazoport College

CHAPTER

1 Pretest

Answer as many of the following problems as you can before starting this chapter. When you finish the chapter, take the test at the end and compare your scores to see how much you have learned.

1. Name the operation indicated in the expression $5a$.

2. Name the operations indicated in the expression $\frac{a}{b} + 2$.

3. What is the place value of the digit 2 in the number 20,158?

4. Write the number 38,026 in words.

5. Replace the question mark in 24 ? 8 with the inequality symbol positioned to indicate a true statement.

6. Write the number 2,010 in expanded form.

7. Add: $\begin{array}{r} 315 \\ +687 \\ \hline \end{array}$

8. Add: $\begin{array}{r} 241 \\ 356 \\ +519 \\ \hline \end{array}$

9. Find the sum of 563; 4,207; 73; and 2,064.

10. Subtract: $\begin{array}{r} 693 \\ -495 \\ \hline \end{array}$

11. Subtract: $\begin{array}{r} 702 \\ -594 \\ \hline \end{array}$

12. Find the difference of 12,350 and 7,341.

13. State the number of terms in the expression $3x^2 + 1$.

14. Name the base in the expression $12x^3$.

15. Simplify: $3a + 6a$

16. Simplify: $8x - x$

17. Simplify: $2ab + 3a^2b - ab + 5a^2b - 2a^2b$

18. Round the number 624 to the nearest ten.

19. Round the number 14,602 to the nearest thousand.

20. The price of a new home was listed as one hundred eighteen thousand, nine hundred dollars. Write the number using digits.

21. On a trip a man drove 208 miles the first day, 364 miles the second day, and 179 miles the third day. What was the total distance for the three days?

22. A television set regularly priced at $450 is reduced by $95. What is the new price?

CHAPTER

1 Whole Numbers and the Language of Algebra

Counting traces back to Prehistoric civilizations. After learning to count, words were invented for numbers. Still later, symbolic numerals were developed to represent these numbers.

Hindu (300 B.C.)	Modern
	1
	2
	3
	4
	5
	6
	7
	8
	9

STUDY HELP

As you prepare to study, remember that mathematics is learned by doing. You must be an active participant. So have paper and pencil at hand as you work through the examples, skill checks, and exercises.

Algebra is an extension of arithmetic. The same concepts and techniques learned in arithmetic are included in algebra, which utilizes letters and numbers rather than just numbers. Therefore, each concept we study in arithmetic will be expanded to its algebraic generalization. This will provide a solid foundation for future work in algebra.

1–1 Words and Symbols that Identify Operations

OBJECTIVE

Upon completing this section you should be able to:

1 Identify the operation symbols of arithmetic.

Whether we are using only numbers in arithmetic or letters and numbers in algebra, the four basic operations of addition, subtraction, multiplication, and division are fundamental to our study. In this section we will examine the ways of indicating these four basic operations. Performing the actual operations will be discussed in subsequent sections.

1 Operation Symbols

The symbol + is used to indicate that two numbers are added. The expression $3 + 5$ (read "3 plus 5") indicates that the number 3 is to be added to the number 5. Instead of a specific example such as $3 + 5$, if we use letters to generalize addition of two numbers then $a + b$ indicates the addition of *any* two numbers a and b. The word most commonly used to indicate that numbers are added is **sum.** We say that $a + b$ indicates the *sum* of a and b.

Other expressions for sum are "result" and "total."

Numbers being added or subtracted are sometimes referred to as **terms.** In the expression $a + b$, a and b are *terms.*

Since in the expression $a + b$, a and b can take on various values, they are referred to as **variables.**

5 and 8 are terms.

Example 1 $5 + 8$ indicates the sum of 5 and 8.

x and y represent any two numbers.

Example 2 $x + y$ indicates the sum of x and y.

x is a variable.

Example 3 $5 + x$ indicates the sum of 5 and some number x.

Example 4 $3 + 2 + 5$ indicates the sum of three numbers.

⦿ ***SKILL CHECK 1*** What does $a + b$ indicate?

The symbol − is used to indicate the operation of subtraction. $a - b$ (read "a minus b") indicates the **difference** of the numbers a and b.

We can also say "a less b" or "b subtracted from a."

Example 5 $12 - 5$ indicates the difference of 12 and 5.

The terms x and y stand for any two numbers.

Example 6 $x - y$ indicates the difference of the numbers x and y.

Note that the number after the word "and" is the one being subtracted.

Example 7 $2 - x$ indicates the difference of 2 and some number x.

⦿ ***SKILL CHECK 2*** What does $5 - 2$ indicate?

The operation of multiplication (the **product** of two numbers) can be represented in several ways. $a \times b$ and $a \cdot b$ both indicate the product of a and b. But the times sign × and the dot · are not generally used in algebra because they can be confused with letters or other symbols. The two most commonly used methods of indicating multiplication are as follows.

The × might be confused with the letter X, and the dot might be confused with a period or decimal point.

1. If two letters or a number and a letter are written together without an operation sign, multiplication is indicated.

ab means "a times b."

Example 8 ab indicates the product of the two numbers a and b.

Numbers being multiplied together are called **factors.** In the expression ab, a and b are called factors of ab.

⦿ ***Skill Check Answers***

1. the sum of a and b
2. the difference of 5 and 2

Example 9 $5x$ indicates the product of 5 and a number x.

In the expression $5x$, 5 and x are factors.

◉ ***SKILL CHECK 3*** What does $3a$ indicate?

2. When two sets of parentheses have no operation sign between them, multiplication is indicated. Also, a letter or number preceding a parenthesis with no sign in between indicates multiplication.

(2) + (3) means add the numbers 2 and 3, but (2)(3) means multiply the numbers 2 and 3.

Example 10 $(3)(4)$ indicates the product of 3 and 4.

Example 11 $(a)(b)$ indicates the product of the two numbers a and b.

There is no sign between the parentheses.

Example 12 $3(x)$ indicates the product of 3 and a number x.

Example 13 $x(y)$ indicates the product of the two numbers x and y.

◉ ***SKILL CHECK 4*** What does $5(y)$ indicate?

The **quotient** of two numbers a and b can be written in two ways. $a \div b$ and $\frac{a}{b}$ both indicate "the number a is divided by the number b."

Example 14 $10 \div 2$ indicates the quotient of 10 and 2.

Note that the number after the word "and" is the one that does the dividing.

Example 15 $\frac{10}{2}$ also indicates the quotient of 10 and 2.

Example 16 $x \div y$ indicates the quotient of the two numbers x and y.

◉ ***SKILL CHECK 5*** What does $x \div 3$ indicate?

Example 17 $\frac{x}{y}$ also indicates the quotient of the two numbers x and y.

x is divided by y.

◉ ***SKILL CHECK 6*** What does $\frac{a}{4}$ indicate?

Many times more than one operation is indicated in a single expression.

Example 18 In the expression $ab + 5$ the operations of multiplication and addition are indicated.

◉ ***SKILL CHECK 7*** What operations are indicated in the expression $3x + y$?

Example 19 In the expression $a\left(\frac{x}{y}\right) - 4$ the operations of multiplication, division, and subtraction are indicated.

◉ ***SKILL CHECK 8*** What operations are indicated in the expression $a\left(\frac{2}{3}\right) - b$?

◉ ***Skill Check Answers***

3. the product of 3 and a number a
4. the product of 5 and some number y
5. the quotient of a number x and 3
6. the quotient of a number a and 4
7. multiplication and addition
8. multiplication, division, and subtraction

Exercise 1-1-1

A Name the operation or operations indicated in each expression.

1. $2 + x$ **2.** $5 + y$ **3.** $a - 6$ **4.** $8 - b$ **5.** $3x$

6. $3y$

7. $7(a)$

8. $(x)(y)$

9. $x \div 5$

10. $\frac{a}{2}$

11. $3a + 5$

12. $5 - 2x$

13. $\frac{3}{x} - 6$

14. $3\left(\frac{a}{b}\right) + 7$

Write an expression for each statement.

15. The sum of 2 and 6

16. The sum of a and 4

17. The difference of 5 and y

18. The difference of 9 and 5

19. The product of 8 and a

20. The product of c and d

21. The quotient of 4 and x

22. The quotient of q and r

23. The sum of a and b

24. The sum of 9 and x

25. The difference of m and n

26. The difference of a and 3

27. The product of 8 and 5

28. The product of 4 and x

29. The quotient of 9 and 3

30. The quotient of a and 5

• Concept Enrichment

31. Explain what terms are.

32. Explain what variables are.

33. What are the names of the four basic operations on the numbers of arithmetic?

34. What operations are indicated in the expression $\frac{xy + 4}{x - 2}$?

1–2 Reading and Writing Whole Numbers

The words **numeral** and **number** are generally used as if they have the same meaning. Actually *number* is an abstract idea and *numeral* is a symbol used to represent the abstract idea. In this text we will not attempt to distinguish between these meanings and will most often refer to the word number.

The set of **whole numbers** includes the number zero and all numbers used for counting. We can write the set as {0,1,2,3,4, . . .}. (The dots indicate the set continues in the pattern shown.)

At this time it may mean little to say that all numbers used in arithmetic and algebra are defined in terms of the whole numbers, but this is true. This should indicate to you that the whole numbers are of such importance that if you are not proficient in their use it is impossible to be proficient in the use of other numbers we will introduce later.

In this section you will learn to read and write whole numbers.

OBJECTIVES

Upon completing this section you should be able to:

1. Determine the place value of a digit.
2. Write a whole number in words.
3. Use the inequality symbol to compare two whole numbers.

Without a solid understanding of whole numbers you will have trouble with fractions, decimals, and so on.

1 Place Value

The **digits** 0, 1, 2, 3, 4, 5, 6, 7, 8, 9 are used to write all whole numbers. The reason only ten digits are necessary is that each digit has two values, a **digital** value and a **place** value.

For example, observe the following three numbers.

7
75
756

In each of these numbers the 7 has the same digital value but a different place value.

In the first number the 7 represents seven ones. In the second number it represents seven tens. In the third it represents seven hundreds. This **place value** is the basis of our system of writing numbers and allows us to use only ten digits to write all whole numbers.

Reading whole numbers is simplified if we note that each group of three numbers is given the place value names shown in this table.

Trillions	Billions	Millions	Thousands	Units
Hundreds, Tens, Ones	Hundreds, Tens, Ones	Hundreds, Tens, Ones	Hundreds, Tens, Ones	Hundreds, Tens, Ones

Groups above trillions are rarely used.

2 Writing Whole Numbers in Words

Example 1 Consider the number 5,814,263.

5,	8	1	4,	2	6	3
millions	hundred thousands	ten thousands	thousands	hundreds	tens	ones

a. The place value of the digit 2 is hundreds.
b. The place value of the digit 1 is ten thousands.
c. The place value of the digit 8 is hundred thousands.
d. The place value of the digit 5 is millions.

If we have only one, two, or three digits, we do not usually use the name of the group. For instance, 325 would usually be read "three hundred twenty-five" rather than "three hundred twenty-five units." However, 325,000 is read "three hundred twenty-five thousand." All group names other than the "units" group must be read.

Notice that the word "and" is not used in reading whole numbers.

Example 2 216,358,234 is read "two hundred sixteen million, three hundred fifty-eight thousand, two hundred thirty-four."

⊙ ***SKILL CHECK 1*** Write 423,156,347 in words.

Notice in examples 3 and 4 that 561 is read as "five hundred sixty-one" even though the group name changes.

Example 3 561,000 is read "five hundred sixty-one thousand."

Example 4 561,000,000 is read "five hundred sixty-one million."

Example 5 561,561,561 is read "five hundred sixty-one million, five hundred sixty-one thousand, five hundred sixty-one."

⊙ ***SKILL CHECK 2*** Write 207,207,207 in words.

Example 6 "Three hundred eighty-two thousand, twenty" written as a number is 382,020.

⊙ ***SKILL CHECK 3*** Write "five hundred twenty-four thousand, eighty" as a number.

Example 7 "Two hundred million, seventy-five thousand, six" written as a number is 200,075,006.

⊙ ***SKILL CHECK 4*** Write "four hundred million, sixty three thousand, one" as a number.

⊙ ***Skill Check Answers***

1. four hundred twenty-three million, one hundred fifty-six thousand, three hundred forty-seven
2. two hundred seven million, two hundred seven thousand, two hundred seven
3. 524,080
4. 400,063,001

Exercise 1–2–1

1 Refer to the number 714,385,602 in each question.

1. What is the place value of the digit 8?

2. What is the place value of the digit 5?

3. What is the place value of the digit 6?

4. What is the place value of the digit 3?

5. What is the place value of the digit 4?

6. What is the place value of the digit 2?

7. What is the place value of the digit 1?

8. What is the place value of the digit 0?

2 Write the number in words.

9. 321

10. 438

11. 2,129

12. 5,280

13. 10,124

14. 11,432

15. 320,004

16. 619,200

17. 2,503,100

18. 6,020,327

19. 15,200,621

20. 352,352,352

21. 763,763,763

22. 349,216,123

23. 22,519,054,111

24. 83,000,500,009

Write the words in numbers.

25. Three hundred twenty-six

26. Four hundred eight

27. One thousand, eight hundred one

28. Thirteen thousand, four hundred nineteen

29. Twenty-six thousand, forty-two

30. One hundred twenty-five thousand, four

31. Two hundred six thousand, two hundred one

32. Thirty one million, four hundred eleven thousand, nine hundred sixteen

33. Sixty-eight million, forty-five thousand, three hundred eight

34. One hundred twenty-two million, six hundred ten thousand, five

35. Five hundred eleven million, four hundred twelve thousand, one hundred

36. Fifty-eight billion, twenty million, fourteen thousand, three hundred twelve

Write the number in each sentence in words.

37. The price of a new Cadillac was $26,798.

38. The height of Mt. Everest is 29,028 feet.

39. The reading on the odometer of a car was 31,045 miles.

40. In one season quarterback Dan Marino of the Miami Dolphins football team passed for a total of 5,084 yards.

41. The play *A Chorus Line* was performed 4,552 times on Broadway.

42. The film *E. T. The Extra-Terrestrial* grossed $227,960,804.

Write the number in each sentence using digits.

43. The distance from Boston to Los Angeles is two thousand, five hundred ninety-six miles.

44. The depth of Great Slave Lake in Canada is two thousand, fifteen feet.

45. A man wrote a check for fifty-two thousand, nine hundred four dollars.

46. The football stadium at Ohio State University has a capacity of eighty-three thousand, one hundred people.

47. The State Russian Museum in St. Petersburg has two hundred fifty thousand objects of art on display.

48. In 1980 the population of metropolitan Dallas–Fort Worth was estimated at two million, nine hundred seventy-four thousand, eight hundred seventy-eight.

• Concept Enrichment

49. How many digits do we have in our number system?

50. Explain the distinction between digital value and place value.

3 Comparing Whole Numbers

What do we mean when we say that a certain number is *greater* than or *larger* than another number? For instance, we say that 6 is greater than 2 and 800 is larger than 200. This is due to the *order* of the numbers. We can represent the whole numbers on a **number line.** We start by placing the number zero at a point on the line. We will agree that all units (0 to 1, 1 to 2, 2 to 3, and so on) will be of equal length. The placement of two numbers on the line determines their relative value. The greater (or larger) of the two numbers is to the right of the lesser (or smaller) number.

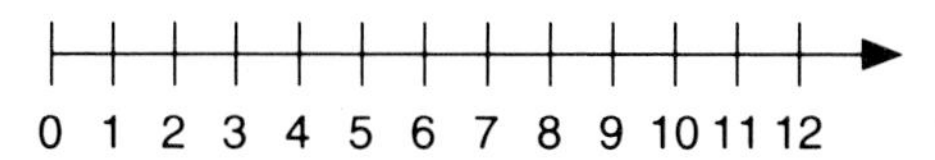

The arrow indicates that the line extends indefinitely. The number 6 is *greater* than 2 because 6 is to the right of 2 on the number line. In fact any number to the right of 2 is greater than 2.

The symbols $<$ and $>$ are **inequality symbols** or **order relations** and are used to indicate the relative position of one number with respect to another on the number line.

We usually read the symbol < "less than." For instance, $2 < 6$ is read "2 is less than 6" and indicates that 2 is to the left of 6 on the number line. We usually read the symbol > "greater than." For instance, $6 > 2$ is read "6 is greater than 2" and indicates that 6 is to the right of 2. Notice that saying "2 is less than 6" is the same as saying "6 is greater than 2." Actually then we have only one symbol that is written two ways just for convenience of reading.

Notice that the pointed end of the symbol is always toward the *lesser* of the two numbers.

Example 8 Replace the question mark in 5 ? 3 with the inequality symbol positioned to indicate a true statement.

Solution Since 5 is to the right of 3 on the number line we write $5 > 3$.

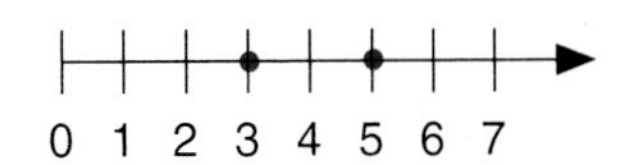

⊙ ***SKILL CHECK 5*** Replace the question mark in 8 ? 2 with the inequality symbol positioned to indicate a true statement.

Example 9 Replace the question mark in 4 ? 7 with the inequality symbol positioned to indicate a true statement.

Solution Since 4 is to the left of 7 on the number line we write $4 < 7$.

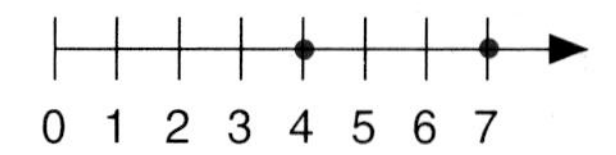

⊙ ***SKILL CHECK 6*** Replace the question mark in 5 ? 9 with the inequality symbol positioned to indicate a true statement.

⊙ ***Skill Check Answers***

5. $8 > 2$ 6. $5 < 9$

Exercise 1-2-2

3 Replace the question mark with the inequality symbol positioned to indicate a true statement.

1. 5 ? 3 **2.** 7 ? 2 **3.** 4 ? 1 **4.** 8 ? 3 **5.** 2 ? 5

6. 3 ? 9 **7.** 1 ? 2 **8.** 4 ? 6 **9.** 12 ? 27 **10.** 11 ? 18

11. 24 ? 43 **12.** 26 ? 10 **13.** 15 ? 0 **14.** 13 ? 14 **15.** 200 ? 400

16. 231 ? 198 **17.** 2,125 ? 2,115 **18.** 5,602 ? 5,712

19. 1,972,823 ? 1,973,000 **20.** 3,456,900 ? 3,456,899

• Concept Enrichment

21. How many whole numbers are less than 5?

22. How many whole numbers are greater than 5?

1–3 Exponents and Expanded Form

OBJECTIVES

Upon completing this section you should be able to:

1. Use exponent notation.
2. Write any whole number in expanded form.

In your study of arithmetic and algebra you will encounter many symbols that have been developed by mathematicians to simplify certain types of number expressions. One such symbol is the **exponent,** which we will define in this section and use to write whole numbers in expanded form.

1 Exponent Notation

Suppose you wish to indicate that the number 10 is used as a factor three times. You could write (10)(10)(10). This does not take up much space but suppose you wish to indicate 10 is used as a factor six, eight, twenty or more times. This could become quite a task. For this purpose mathematicians use a notation called an *exponent.*

WARNING

Notice that 10^3 does *not* mean "ten times three."

Using this notation (10)(10)(10) is written as 10^3. The small superscript 3 indicates that 10 is used as a factor three times. 3 is called the *exponent.* 10 is called the **base.**

> An **exponent** indicates how many times a base is used as a factor.

Example 1 $10^4 = (10)(10)(10)(10)$

Example 2 $3^5 = (3)(3)(3)(3)(3)$

Example 3 $x^3 = (x)(x)(x)$

y is used as a factor six times.

Example 4 $y^6 = (y)(y)(y)(y)(y)(y)$

⦿ ***SKILL CHECK 1*** $a^7 = ?$

The exponent is written smaller and a half-space above the number to be multiplied. 10^4 is read "ten to the fourth power" or "ten exponent four." The second power and the third power have special names that are sometimes used. x^2 can be read "x squared." In other words, to "square" a number we multiply it by itself. x^3 can be read "x cubed." Any power above the third has no special name.

Example 5 Use an exponent to write "seven cubed."

"Seven to the third power"

Answer 7^3

Example 6 Use an exponent to write "y to the sixth power."

Answer y^6

⦿ ***SKILL CHECK 2*** Use an exponent to write "five to the eighth power."

To write whole numbers in expanded form we need a definition for a special exponent.

0^0 has no meaning.

> Any number (except zero) raised to the zero power is equal to 1.

The definition for a zero exponent will be useful in later work with exponents.

Example 7 $3^0 = 1$

Example 8 $10^0 = 1$

Example 9 $x^0 = 1$ (if $x \neq 0$)

⦿ ***SKILL CHECK 3*** $6^0 = ?$

⦿ ***Skill Check Answers***

1. (a)(a)(a)(a)(a)(a)(a)
2. 5^8
3. 1

2 Expanded Form

We will now use exponents to write whole numbers in **expanded form.** In section 1–2 we noted that each digit has a place value and a digital value. The knowledge of place value is necessary as we work with whole numbers.

You may wish to refer to section 1-2.

The following table shows an important relationship.

Ten Thousands	Thousands	Hundreds	Tens	Ones
(10)(10)(10)(10)	(10)(10)(10)	(10)(10)	(10)(1)	1

Notice the number of 10s in each place value.

Notice that each place value is ten times the place value to its right. For this reason we say we have a **base ten** number system.

The following table in which we use exponents will be helpful as we write numbers in expanded form.

Millions	One Hundred Thousands	Ten Thousands	Thousands	Hundreds	Tens	Ones
10^6	10^5	10^4	10^3	10^2	10^1	10^0

Compare this table with the previous one.

Example 10 Write 536 in expanded form.

Solution

$$536 = 5 \text{ hundreds} + 3 \text{ tens} + 6 \text{ ones}$$
$$= 5(10^2) + 3(10^1) + 6(10^0)$$

Note the place value of each digit.

⊙ ***SKILL CHECK 4*** Write 346 in expanded form.

Example 11 Write 1,784 in expanded form.

Solution

$$1{,}784 = 1 \text{ thousand} + 7 \text{ hundreds} + 8 \text{ tens} + 4 \text{ ones}$$
$$= 1(10^3) + 7(10^2) + 8(10^1) + 4(10^0)$$

Example 12 Write 5,964 in expanded form.

Solution

$$5{,}964 = 5(10^3) + 9(10^2) + 6(10^1) + 4(10^0)$$

⊙ ***SKILL CHECK 5*** Write 2,435 in expanded form.

Example 13 Write 1,780,263 in expanded form.

Solution

$$1{,}780{,}263 = 1(10^6) + 7(10^5) + 8(10^4)$$
$$+ 0(10^3) + 2(10^2) + 6(10^1)$$
$$+ 3(10^0)$$

⊙ ***SKILL CHECK 6*** Write 7,230,614 in expanded form.

⊙ ***Skill Check Answers***

4. $3(10^2) + 4(10^1) + 6(10^0)$
5. $2(10^3) + 4(10^2) + 3(10^1) + 5(10^0)$
6. $7(10^6) + 2(10^5) + 3(10^4) + 0(10^3) + 6(10^2) + 1(10^1) + 4(10^0)$

Exercise 1–3–1

1 Write as a product of factors.

1. 2^3 **2.** 3^2 **3.** 5^4 **4.** 6^3 **5.** x^5 **6.** a^6

7. Use an exponent to write "five cubed."

8. Use an exponent to write "x to the ninth power."

9. Find the value of 4^0.

10. Find the value of 12^0.

2 Write each number in expanded form using exponents.

11. 251

12. 341

13. 619

14. 514

15. 2,458

16. 4,362

17. 30

18. 80

19. 500

20. 900

21. 3,041

22. 4,029

23. 6,000

24. 9,000

25. 23,426

26. 39,582

27. 4,502

28. 7,803

29. 81,294

30. 63,514

31. 15,804

32. 19,603

33. 23,038

34. 44,016

35. 172,254

36. 132,693

37. 350,400

38. 520,500

39. 5,461,006

40. 8,319,001

• Concept Enrichment

41. What is the main benefit of using exponents?

42. Why is our number system called a base ten system?

1–4 Adding Whole Numbers

OBJECTIVES

Upon completing this section you should be able to:

1. Add any group of whole numbers using the addition algorithm.
2. Recognize the additive identity, commutative, and associative properties of addition.
3. Use addition in solving applications.

The first operation on whole numbers that we will examine is addition. The facts you have learned so far about whole numbers will be of great value when performing operations on them.

One basic fact holds true throughout all of mathematics. *Only like quantities can be added or subtracted.* When applying this principle to the addition of whole numbers, we will keep in mind that ones can only be added to ones, tens to tens, hundreds to hundreds, and so on.

To prepare to add any group of numbers rapidly and correctly we must first make sure we know all the basic addition facts. We know that all numbers are written with the digits 0, 1, 2, 3, 4, 5, 6, 7, 8, 9. If we add any of these digits two at a time in all possible ways, we have the basic addition facts. It is extremely important that you know all of these basic facts. The authors have found that students not proficient in the basic facts of arithmetic have difficulty in all aspects of future studies in mathematics.

We will now look at an example of adding whole numbers. Let's examine closely what happens as we add 574 and 299. Recalling our basic principle that only like quantities can be combined, we will first write the numbers in expanded form.

$$574 = 5 \text{ hundreds} + 7 \text{ tens} + 4 \text{ ones}$$
$$299 = 2 \text{ hundreds} + 9 \text{ tens} + 9 \text{ ones}$$

Recall that this is the same as:

$$574 = 5(10^2) + 7(10^1) + 4(10^0)$$
$$299 = 2(10^2) + 9(10^1) + 9(10^0)$$

Using our addition facts and first adding hundreds to hundreds, we obtain

$$5 \text{ hundreds} + 2 \text{ hundreds} = (5 + 2) \text{ hundreds} = 7 \text{ hundreds}.$$

This illustrates an important property called the **distributive property,** which will be discussed in greater detail later. This property allows us to add (or subtract) like quantities.

Using the distributive property again, we add tens to tens, obtaining

$$7 \text{ tens} + 9 \text{ tens} = (7 + 9) \text{ tens} = 16 \text{ tens}.$$

Also

$$4 \text{ ones} + 9 \text{ ones} = (4 + 9) \text{ ones} = 13 \text{ ones}.$$

Thus we have

$$7 \text{ hundreds} + 16 \text{ tens} + 13 \text{ ones}.$$

We see, however, that only a single digit can be written in a place value. That is, we cannot write 13 in the ones place or 16 in the tens place.

We note that

$$16 \text{ tens} = 1 \text{ hundred} + 6 \text{ tens}.$$

Only a single digit is allowed for each place value.

Also,

$$13 \text{ ones} = 1 \text{ ten} + 3 \text{ ones}.$$

$$\begin{aligned} \text{So } 7 \text{ hundreds} + 16 \text{ tens} + 13 \text{ ones} &= 7 \text{ hundreds} + 1 \text{ hundred} + 6 \text{ tens} \\ &\quad + 1 \text{ ten} + 3 \text{ ones} \\ &= 8 \text{ hundreds} + 7 \text{ tens} + 3 \text{ ones}, \end{aligned}$$

Combine the 7 hundreds and 1 hundred to get 8 hundreds. Also combine 6 tens and 1 ten to get 7 tens.

which is written as 873. Thus the sum of 574 and 299 is 873.

1 Addition Algorithm

This method may explain how addition actually takes place, but you would agree that we are fortunate to have shorter methods. Let's look at the more common (and shorter) way of adding 574 and 299. This short method is called the **addition algorithm.** An **algorithm** is a step-by-step procedure for obtaining an answer. An algorithm should reduce computation time.

Example 1 Add 574 and 299 using the addition algorithm.

Solution First the numbers are written in a column, making sure the ones place is under the ones, tens under tens, and so on.

$$\begin{array}{r} 574 \\ +299 \\ \hline \end{array}$$

Now we add the $9 + 4$ obtaining 13. We recognize this as 1 ten and 3 ones, so we place the 3 in the ones column and the 1 in the tens column.

Note the importance of understanding place value.

$$\begin{array}{rl} 1 & \leftarrow \text{1 in the tens column} \\ 574 & \\ +299 & \\ \hline 3 & \leftarrow \text{3 in the ones column} \end{array}$$

This process is referred to as *carrying,* and we say "write down the three and carry the one." Of course you could "carry" mentally, but when checking it is helpful to have written down the number carried.

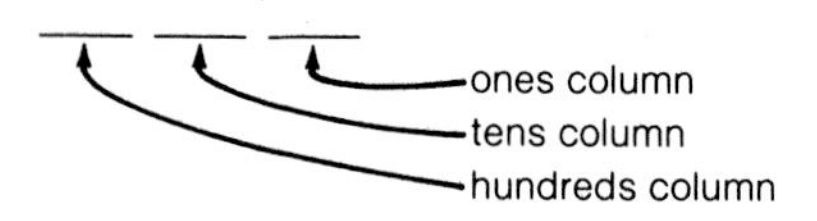

To proceed, we add the $9 + 7 + 1$ of the tens column obtaining 17 tens. We record the 7 in the tens column and carry the 1 to the hundreds column. We finally add the $2 + 5 + 1$ in the hundreds column to obtain 8 hundreds and record this in the hundreds column.

$$\begin{array}{lr} \text{1 in the hundreds column} \longrightarrow & 11 \\ & 574 \\ & +299 \\ \hline \text{8 in the hundreds column} \longrightarrow & 873 \\ \text{7 in the tens column} \longrightarrow & \uparrow \end{array}$$

⊙ ***SKILL CHECK 1*** Add 354 and 578 using the addition algorithm.

Example 2 Find the sum of 17; 325; and 983.

Solution Again, be careful to write ones, tens, and so on, in their proper places so that we will be adding like quantities.

Again, it is helpful to write down the numbers you are carrying.

$$\begin{array}{r} 11 \\ 17 \\ 325 \\ +983 \\ \hline 1{,}325 \end{array}$$

⊙ ***SKILL CHECK 2*** Find the sum of 23; 156; and 975.

2 Properties

We now consider three important properties of addition. First we note that zero added to any number will result in that number as the sum. That is $3 + 0 = 3$, $5 + 0 = 5$, $14 + 0 = 14$, and so on. Since the sum is identically the same number as that which was added to zero, zero is said to be the *additive identity.*

⊙ ***Skill Check Answers***

1. 932
2. 1,154

> **Additive Identity**
> If a represents any number, then $a + 0 = a$.

◉ ***SKILL CHECK 3*** What property is illustrated by $6 + 0 = 6$?

To establish the second property we make the following observation. If we add $5 + 4$ we obtain 9 as the sum. If we add $4 + 5$ we also obtain 9 as the sum. In other words, we will obtain the same result (sum) if we add any two numbers in any order. This illustrates the second property.

> **Commutative Property of Addition**
> If a and b represent any two numbers, then $a + b = b + a$.

Example 3 $3 + 7 = 7 + 3$

What property does this illustrate?

It does not matter whether we write $3 + 7$ or $7 + 3$. The sum in each case is 10.

◉ ***SKILL CHECK 4*** What property is illustrated by $4 + 5 = 5 + 4$?

You probably have realized by now that addition only applies to two numbers at a time. The proper mathematical term is that addition is a **binary** (that is two number) operation. So when we desire the sum of three or more numbers we must add them two at a time.

To add $2 + 3 + 6$ we can only add two of these numbers at a time. We use parentheses to indicate which two numbers are to be added first. If we write $(2 + 3) + 6$, we would first add 2 and 3 obtaining $(2 + 3) + 6 = 5 + 6 = 11$. If we write $2 + (3 + 6)$, we would first add 3 and 6 obtaining $2 + (3 + 6) = 2 + 9 = 11$. In either case we obtain the same result of 11. This leads us to the third important property.

> **Associative Property of Addition**
> If a, b, and c represent any three numbers, then $(a + b) + c = a + (b + c)$.

This means it does not matter which two numbers are added first.

◉ ***SKILL CHECK 5*** What property is illustrated by $(5 + 3) + 7 = 5 + (3 + 7)$?

The associative property of addition is sometimes referred to as the **grouping principle.** We can often use this property along with the commutative property to increase our speed in adding numbers.

> **WARNING**
> If you skip around, don't overlook a number. This process must be practiced to be effective.

Example 4 Add $5 + 3 + 7 + 6 + 5 + 4 + 9$.

Solution We first place the numbers in column form.

$$\begin{array}{r} 5 \\ 3 \\ 7 \\ 6 \\ 5 \\ 4 \\ \underline{9} \\ 39 \end{array}$$

As we add from bottom to top we might "pick out" combinations that add to ten, in this case, the 6 and 4, the two 5s, and the 7 and 3.

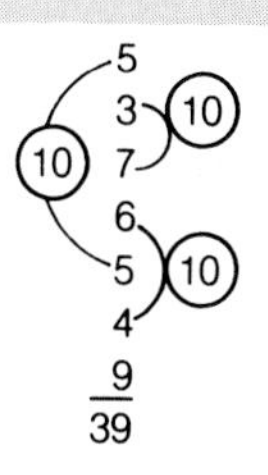

◉ ***Skill Check Answers***
3. additive identity
4. commutative property of addition
5. associative property of addition

You may even wish to rewrite the numbers in pairs such as

Example 5 Add $8 + 4 + 9 + 2 + 1 + 7 + 6$.

Solution

$$\begin{array}{r} 8 \\ 4 \\ 9 \\ 2 \\ 1 \\ 7 \\ \underline{6} \\ 37 \end{array}$$

(10) (10) (10)

◉ SKILL CHECK 6 Add $7 + 9 + 3 + 4 + 5 + 1 + 5$.

3 Applications

Example 6 Alex purchased 5 pounds of peanuts, 3 pounds of cashews, 2 pounds of almonds, and 4 pounds of pecans. What is the total number of pounds of nuts that he purchased?

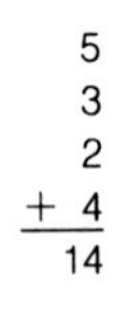

Solution When you encounter a problem that is stated in words, be certain that you read it very carefully and answer the question that is asked. In this case we are to find a *total*. That tells us we must use addition.

To find the total we need to add the number of pounds of each type of nut purchased.

Thus he purchased a total of 14 pounds.

◉ SKILL CHECK 7 What is the total in pounds of 8 pounds of peanuts, 4 pounds of cashews, 3 pounds of almonds, and 6 pounds of pecans?

◉ Skill Check Answers

6. 34
7. 21 pounds

Exercise 1–4–1

2 Name the property illustrated by the statement.

1. $4 + 0 = 4$

2. $15 + 0 = 15$

3. $2 + 8 = 8 + 2$

4. $(6 + 1) + 4 = 6 + (1 + 4)$

5. $(3 + 2) + 5 = 3 + (2 + 5)$

6. $9 + 4 = 4 + 9$

7. $2 + (3 + 0) = (2 + 3) + 0$

8. $(5 + 3) + 0 = 5 + 3$

9. $(4 + 3) + 6 = 6 + (4 + 3)$

10. $2 + (7 + 3) = 2 + (3 + 7)$

1 Add using the principles learned in this section. You may want to check your answers using a calculator.

11. $\begin{array}{r}2\\8\\5\\+5\\\hline\end{array}$ **12.** $\begin{array}{r}3\\7\\4\\+6\\\hline\end{array}$ **13.** $\begin{array}{r}1\\8\\3\\9\\+2\\\hline\end{array}$ **14.** $\begin{array}{r}7\\9\\5\\5\\+3\\\hline\end{array}$ **15.** $\begin{array}{r}7\\5\\3\\5\\6\\+4\\\hline\end{array}$

16. $\begin{array}{r}3\\4\\5\\7\\4\\+1\\\hline\end{array}$ **17.** $\begin{array}{r}6\\5\\8\\1\\4\\+7\\\hline\end{array}$ **18.** $\begin{array}{r}5\\4\\5\\2\\8\\+1\\\hline\end{array}$ **19.** $\begin{array}{r}54\\+32\\\hline\end{array}$ **20.** $\begin{array}{r}75\\+24\\\hline\end{array}$

21. $\begin{array}{r}49\\+21\\\hline\end{array}$ **22.** $\begin{array}{r}32\\+48\\\hline\end{array}$ **23.** $\begin{array}{r}43\\+72\\\hline\end{array}$ **24.** $\begin{array}{r}86\\+31\\\hline\end{array}$ **25.** $\begin{array}{r}83\\+29\\\hline\end{array}$

26. $\begin{array}{r}65\\+35\\\hline\end{array}$ **27.** $\begin{array}{r}23\\45\\+53\\\hline\end{array}$ **28.** $\begin{array}{r}42\\38\\+44\\\hline\end{array}$ **29.** $\begin{array}{r}55\\48\\+25\\\hline\end{array}$ **30.** $\begin{array}{r}57\\39\\+86\\\hline\end{array}$

31. $\begin{array}{r}536\\+243\\\hline\end{array}$ **32.** $\begin{array}{r}614\\+325\\\hline\end{array}$ **33.** $\begin{array}{r}842\\+139\\\hline\end{array}$ **34.** $\begin{array}{r}546\\+317\\\hline\end{array}$ **35.** $\begin{array}{r}285\\+646\\\hline\end{array}$

36. $\begin{array}{r}327\\+585\\\hline\end{array}$ **37.** $\begin{array}{r}783\\+269\\\hline\end{array}$ **38.** $\begin{array}{r}646\\+798\\\hline\end{array}$ **39.** $\begin{array}{r}63\\456\\+547\\\hline\end{array}$ **40.** $\begin{array}{r}38\\419\\+762\\\hline\end{array}$

41. $\begin{array}{r}75\\796\\+489\\\hline\end{array}$ **42.** $\begin{array}{r}37\\677\\+586\\\hline\end{array}$ **43.** $\begin{array}{r}219\\564\\+835\\\hline\end{array}$ **44.** $\begin{array}{r}614\\328\\+597\\\hline\end{array}$ **45.** $\begin{array}{r}829\\507\\+695\\\hline\end{array}$

46. $\begin{array}{r}185\\647\\+576\\\hline\end{array}$ **47.** $\begin{array}{r}2{,}511\\5{,}699\\+3{,}068\\\hline\end{array}$ **48.** $\begin{array}{r}1{,}554\\2{,}795\\+4{,}106\\\hline\end{array}$ **49.** $\begin{array}{r}2{,}436\\986\\4{,}205\\+\quad 87\\\hline\end{array}$ **50.** $\begin{array}{r}24{,}682\\19{,}428\\564\\+83{,}046\\\hline\end{array}$

51.

$$\begin{array}{r} 98 \\ 12{,}564 \\ 759 \\ +\ 5{,}685 \\ \hline \end{array}$$

52.

$$\begin{array}{r} 32{,}048 \\ 673 \\ 8{,}567 \\ +14{,}412 \\ \hline \end{array}$$

Write in column form and add.

53. Find the sum of 46; 198; 2,305; and 1,795.

54. Find the sum of 463; 3,488; 76; and 5,654.

55. Find the sum of 39; 7,255; 664; and 2,325.

56. Find the sum of 719; 1,682; 908; and 4,265.

57. Find the sum of 3,069; 624; 1,253; and 376.

58. Find the sum of 58; 472; 3,804; and 2,196.

59. Find the sum of 515; 724; 36; 1,485; and 99,611.

60. Find the sum of 36; 843; 6,207; 3,124; and 618.

61. Helen bought a book for $15, a pen for $4, and a calculator for $21. What was the total cost of the three items?

62. Jim earned $164 the first week, $159 the second week, and $208 the third week. What were his total earnings for the three weeks?

63. A man mixed 35 pounds of one type of grass seed with 48 pounds of another type of grass seed. How many pounds of mixture does he have?

64. A man weighing 195 pounds and a woman weighing 120 pounds ride in a car that weighs 3,248 pounds without any passengers. What is the total weight of the car and passengers?

65. A merchant pays $65 for an item and wishes to make a profit of $23 when selling it. How much should the item sell for?

66. A man borrows $2,690 for one year and agrees to pay $325 interest. How much does he need to pay at the end of the year?

67. A woman drove from Mobile to Atlanta, a distance of 353 miles. She then drove 676 miles to Baltimore. What was the total distance that she drove?

68. During one month a person's electric bill was $176, the telephone bill was $38, and the water bill was $17. What was the total of the three bills?

69. In the presidential election of 1960 John Kennedy received 303 electoral votes, Richard Nixon received 219, and Harry Byrd received 15. What was the total number of electoral votes cast in the election?

70. In California the estimated 1986 population of Los Angeles was 3,215,506. That of San Diego was 1,013,400 and of San Jose was 713,385. Find the total population of the three cities.

1–5 Subtracting Whole Numbers

OBJECTIVES

Upon completing this section you should be able to:

1. Understand the relationship between subtraction and addition.
2. Subtract using "borrowing."
3. Subtract when zeros occur in the minuend.
4. Use subtraction in solving applications.

In section 1–4 the basic facts of addition were emphasized. If you know those facts well, you already know the basic facts necessary for subtraction. We use the symbol "$-$" to indicate subtraction of whole numbers.

1 Relationship Between Subtraction and Addition

Example 1 $9 - 6 = 3$

This number sentence states that "nine minus six is three" or "the difference of nine and six is three."

But how do we know that nine minus six is three? We are actually using the *addition* fact that $6 + 3 = 9$. In other words, the question "What is the difference of nine and six?" is the same as "What number added to six will give a result of nine?" A formal statement of this fact is given in the following rule.

> If a, b, and c represent any whole numbers and if $a - b = c$, then it is also true that $b + c = a$.

In other words, we only need the facts from addition to perform the operation of subtraction.

Example 2 Find 784 − 432.

Solution As in addition we will place the numbers in columns.

IMPORTANT! Place the number you are subtracting on the bottom.

$$\begin{array}{r} 784 \\ -432 \\ \hline \end{array}$$

Remember that only like quantities can be added. This is also true for subtraction. We subtract ones from ones, tens from tens, and so on. Thus we obtain

$$\begin{array}{r} 784 \\ -432 \\ \hline 352. \end{array}$$

We can check our answer by adding 432 and 352 to obtain 784.

Note that the check is an application of the above rule.

$$\textit{Check:}\quad \begin{array}{r} 432 \\ +352 \\ \hline 784 \end{array}$$

It is a good practice to always check your answer.

⦿ ***SKILL CHECK 1*** Find 674 − 542.

Example 3 Find the difference of 425 and 213.

Solution The number being subtracted follows the word "and," so the subtraction should be written 425 − 213.

$$\begin{array}{r} 425 \\ -213 \\ \hline 212 \end{array}$$

Checking will verify the correctness of your answer.

$$\textit{Check:}\quad \begin{array}{r} 213 \\ +212 \\ \hline 425 \end{array}$$

⦿ ***Skill Check Answers***

1. 132
2. 222

⦿ ***SKILL CHECK 2*** Find the difference of 536 and 314.

Exercise 1–5–1

1 Subtract and check.

1. 8 − 5 **2.** 9 − 4 **3.** 7 − 4 **4.** 6 − 5 **5.** 18 − 7

6. 19 − 6 **7.** 15 − 5 **8.** 27 − 7 **9.** 57 − 35 **10.** 49 − 25

11. $\begin{array}{r} 74 \\ -53 \\ \hline \end{array}$ **12.** $\begin{array}{r} 46 \\ -33 \\ \hline \end{array}$ **13.** $\begin{array}{r} 38 \\ -27 \\ \hline \end{array}$ **14.** $\begin{array}{r} 59 \\ -48 \\ \hline \end{array}$ **15.** $\begin{array}{r} 584 \\ -263 \\ \hline \end{array}$

16. $\begin{array}{r} 453 \\ -241 \\ \hline \end{array}$

17. $\begin{array}{r} 593 \\ -361 \\ \hline \end{array}$

18. $\begin{array}{r} 879 \\ -565 \\ \hline \end{array}$

19. $\begin{array}{r} 936 \\ -416 \\ \hline \end{array}$

20. $\begin{array}{r} 745 \\ -525 \\ \hline \end{array}$

21. Find the difference of 9 and 2.

22. Find the difference of 25 and 13.

23. Find the difference of 86 and 54.

24. Find the difference of 263 and 142.

25. Find the difference of 574 and 473.

26. Find the difference of 958 and 557.

27. Find the difference of 396 and 153.

28. Find the difference of 477 and 256.

29. Find the difference of 958 and 658.

30. Find the difference of 894 and 594.

• Concept Enrichment

31. Is subtraction commutative? That is, does $a - b = b - a$?

32. Is subtraction associative? That is, does $(a - b) - c = a - (b - c)$?

2 Borrowing

In the previous exercise set all digits in the top number (**minuend**) were larger than those in the bottom number (**subtrahend**). When this is not true, we must use a technique known as **borrowing.** For example, suppose we wish to find the difference of 54 and 9. We first place the two numbers in columns.

$$\begin{array}{r} 54 \\ -\ 9 \\ \hline \end{array}$$

$$\begin{array}{rl} 54 & \text{(minuend)} \\ -\ 9 & \text{(subtrahend)} \\ \hline \end{array}$$

We now have a problem. We know that 54 is larger than 9 but we cannot subtract in the ones column. That is, we cannot subtract 9 from 4. Solving this problem requires *borrowing* and we proceed as follows.

From previous work we know that 54 = 5 tens + 4 ones
= 4 tens + 10 ones + 4 ones
= 4 tens + 14 ones.

Remember, this is expanded form.

Then

$$\begin{array}{r} 54 = 4 \text{ tens} + 14 \text{ ones} \\ -9 = \phantom{4 \text{ tens}} - \ 9 \text{ ones.} \\ \hline 4 \text{ tens} + \ 5 \text{ ones.} \end{array}$$

The result is
So 54 − 9 = 45.

This entire process is simplified by writing the problem in this manner.

$$\begin{array}{r} {\scriptstyle 4\,14} \\ \not{5}4 \\ -\ 9 \\ \hline 45 \end{array}$$

We borrow 1 from the 5 in the tens column (we cross out the 5 and replace it with a 4 to indicate we have borrowed 1). This becomes 10 in the ones column. Together with the 4 already in the ones column we have a total of 14 ones. We then use our addition facts to subtract.

Don't forget to check.

Check:
$$\begin{array}{r} 9 \\ +45 \\ \hline 54 \end{array}$$

Example 4 Subtract:
$$\begin{array}{r} 238 \\ -\ 59 \\ \hline \end{array}$$

Solution First, since we cannot subtract 9 from 8, we *borrow* one 10 from the tens column. We thus have 18 ones. Now subtracting 9 from 18 gives us 9.

Again, we cross out the 3 and replace it with a 2 to indicate we borrowed from it.

$$\begin{array}{r} {\scriptstyle 2\,18} \\ 2\not{3}8 \\ -\ 59 \\ \hline 9 \end{array}$$

Working with the tens digits, we cannot subtract 5 tens from 2 tens, so we must borrow one 100 from the hundreds column to obtain 12 tens. We then complete the subtraction to obtain

Cross out the 2 and replace it with 1.

$$\begin{array}{r} {\scriptstyle 12} \\ {\scriptstyle 1\,2\,18} \\ \not{2}\not{3}8 \\ -\ 59 \\ \hline 179 \end{array}$$

Check:
$$\begin{array}{r} 59 \\ +179 \\ \hline 238 \end{array}$$

It is more efficient if we can do the borrowing mentally. However, you should write whatever is necessary to assist you in obtaining the correct answer.

This will come with practice.

Example 5 Subtract:
$$\begin{array}{r} 834 \\ -265 \\ \hline \end{array}$$

Solution

You may wish to write only this much.

$$\begin{array}{r} {\scriptstyle 7\,2} \\ \not{8}\not{3}4 \\ -265 \\ \hline 569 \end{array}$$

We borrowed from the 3 tens to obtain 14 ones but did not write the 14. We also borrowed from the 8 hundreds to obtain 12 tens but did not write the 12. However, it is always a good idea to strike through as we did the 3 and 8 so that we will not forget that we have borrowed.

We changed the 3 to 2, and the 8 to 7.

Check: $\begin{array}{r} 265 \\ +569 \\ \hline 834 \end{array}$

⊙ ***SKILL CHECK 3*** Subtract and check: $\begin{array}{r} 956 \\ -387 \\ \hline \end{array}$

⊙ ***Skill Check Answer***

3. 569

Exercise 1–5–2

2 Subtract and check.

1. 23 − 8

2. 13 − 6

3. 34 − 16

4. 44 − 27

5. $\begin{array}{r} 31 \\ -13 \\ \hline \end{array}$

6. $\begin{array}{r} 53 \\ -35 \\ \hline \end{array}$

7. $\begin{array}{r} 52 \\ -35 \\ \hline \end{array}$

8. $\begin{array}{r} 61 \\ -38 \\ \hline \end{array}$

9. $\begin{array}{r} 524 \\ -\ 56 \\ \hline \end{array}$

10. $\begin{array}{r} 322 \\ -\ 35 \\ \hline \end{array}$

11. $\begin{array}{r} 283 \\ -\ 95 \\ \hline \end{array}$

12. $\begin{array}{r} 234 \\ -\ 55 \\ \hline \end{array}$

13. $\begin{array}{r} 333 \\ -157 \\ \hline \end{array}$

14. $\begin{array}{r} 555 \\ -268 \\ \hline \end{array}$

15. $\begin{array}{r} 812 \\ -648 \\ \hline \end{array}$

16. $\begin{array}{r} 713 \\ -569 \\ \hline \end{array}$

17. $\begin{array}{r} 352 \\ -264 \\ \hline \end{array}$

18. $\begin{array}{r} 258 \\ -179 \\ \hline \end{array}$

19. $\begin{array}{r} 316 \\ -198 \\ \hline \end{array}$

20. $\begin{array}{r} 413 \\ -284 \\ \hline \end{array}$

21. Find the difference of 35 and 7.

22. Find the difference of 43 and 8.

23. Find the difference of 56 and 28.

24. Find the difference of 43 and 26.

25. Find the difference of 234 and 68.

26. Find the difference of 432 and 79.

27. Find the difference of 324 and 165.

28. Find the difference of 535 and 367.

29. Find the difference of 611 and 573.

30. Find the difference of 711 and 695.

3 Zeros in the Minuend

When zeros occur in the minuend, the problem is a little more complex.

Example 6 Subtract: $\begin{array}{r} 703 \\ -264 \\ \hline \end{array}$

703 (minuend)
−264 (subtrahend)
zero tens

Solution Since we cannot subtract 4 ones from 3 ones, we must borrow a 10 from the tens column. But there aren't any tens in that column. What we must do is borrow 1 hundred from the hundreds column and enter it into the tens column as 10 tens.

The 7 is replaced with 6.

$$\begin{array}{r} {\scriptstyle 6\,10} \\ \not{7}03 \\ -264 \\ \hline \end{array}$$

We can now borrow 1 ten from the tens column and obtain 13 ones in the ones column. This enables us to subtract.

Again, it is a good idea to strike out the columns borrowed from. Here 10 is changed to 9.

$$\begin{array}{r} {\scriptstyle 9} \\ {\scriptstyle 6\,\not{10}13} \\ \not{7}0\,3 \\ -264 \\ \hline 439 \end{array}$$

Check: $\begin{array}{r} 264 \\ +439 \\ \hline 703 \end{array}$

⊙ ***SKILL CHECK 4*** Subtract and check: $\begin{array}{r} 804 \\ -375 \\ \hline \end{array}$

Example 7 Subtract: $\begin{array}{r} 2{,}003 \\ -\ \ 514 \\ \hline \end{array}$

Move from right to left until you find a number to borrow from.

Solution This time we must first borrow one from the thousands place to obtain 10 hundreds, then one from the hundreds place to get 10 tens, and then one from the tens place to have enough ones to subtract.

$$\begin{array}{r} {\scriptstyle 9\ 9} \\ {\scriptstyle 1\,\not{10}\not{10}13} \\ \not{2}{,}0\,0\,3 \\ -\ \ 514 \\ \hline 1{,}489 \end{array}$$

⊙ ***Skill Check Answer***

4. 429

Check:
$$\begin{array}{r} 514 \\ +1{,}489 \\ \hline 2{,}003 \end{array}$$

◉ ***SKILL CHECK 5*** Subtract and check:
$$\begin{array}{r} 3{,}005 \\ -\ \ 726 \\ \hline \end{array}$$

Example 8 Subtract:
$$\begin{array}{r} 729 \\ -254 \\ \hline \end{array}$$

Solution
$$\begin{array}{r} {}^{6\,12} \\ \not{7}29 \\ -254 \\ \hline 475 \end{array}$$

Check:
$$\begin{array}{r} 254 \\ +475 \\ \hline 729 \end{array}$$

Notice that in this example we did not need to borrow to subtract in the ones column, but we had to borrow a hundred so we could subtract in the tens column.

We could subtract 4 from 9 without borrowing.

◉ ***SKILL CHECK 6*** Subtract and check:
$$\begin{array}{r} 839 \\ -342 \\ \hline \end{array}$$

It is only necessary to borrow when the digit in any column of the minuend is smaller than the digit in the corresponding column in the subtrahend.

4 Applications

Example 9 A television set regularly priced at $565 is reduced by $78. What is the new price?

Solution In this problem the key words "reduced by" tell us that we must subtract.

To find the new price we must subtract 78 from 565.

$$\begin{array}{r} 565 \\ -\ \ 78 \\ \hline 487 \end{array}$$

Thus the new price is $487.

Check this answer.

◉ ***SKILL CHECK 7*** A CD player regularly priced at $345 is reduced by $67. What is the new price?

Example 10 Bill's grandfather sent him $100 as a birthday gift. His grandfather's instructions were "Buy yourself a pair of tennis shoes and shorts and then place the remainder in your savings account." Bill bought shoes for $32 and shorts for $16. How much did he have left to place in his savings account?

Solution This word problem is more involved than those you have solved in previous sections. As in all word problems, read and read again if necessary until you clearly understand the principles involved.

First we see that we must find how much money Bill spent. To do this we add the amounts spent on shoes and shorts.

amount for shoes	$32
amount for shorts	+$16
total amount spent	$48

If the problem had asked "How much did Bill spend?" we would have our answer. But the question is "How much is left of the $100

◉ ***Skill Check Answers***

5. 2,279
6. 497
7. $278

WARNING

When solving a word problem, always look back to see that you have answered the question asked.

for savings?" To find this amount we must subtract the amount spent from the original amount.

original amount	$100
amount spent	−$ 48
amount left for savings	$ 52

⊙ ***SKILL CHECK 8*** Sarah received $100 as a birthday gift. She bought a book for $33 and a calculator for $28. How much was left from the $100?

⊙ ***Skill Check Answer***

8. $39

Exercise 1–5–3

Subtract and check.

3

1. 30 − 8 **2.** 40 − 7 **3.** 203 − 47 **4.** 504 − 39 **5.** 506 − 19

6. 708 − 59 **7.** 801 − 423 **8.** 501 − 232 **9.** 2,005 − 389 **10.** 2,004 − 226

11. 4,203 − 147 **12.** 3,405 − 386 **13.** 468 − 93 **14.** 524 − 62 **15.** 345 − 126

16. 857 − 318 **17.** 408 − 299 **18.** 707 − 399 **19.** 4,060 − 792 **20.** 5,040 − 839

21. 316 − 118 **22.** 417 − 219 **23.** 10,000 − 9,768 **24.** 10,000 − 9,492 **25.** 4,307 − 298

26. 2,804 − 796 **27.** 55,007 − 43,198 **28.** 68,003 − 16,917 **29.** 3,015 − 1,807 **30.** 6,014 − 4,608

31. Find the difference of 306 and 58.

32. Find the difference of 701 and 43.

33. Find the difference of 1,036 and 938.

34. Find the difference of 2,042 and 744.

35. Find the difference of 1,000 and 564.

36. Find the difference of 1,000 and 405.

37. Find the difference of 6,040 and 5,234.

38. Find the difference of 8,050 and 7,953.

4

39. There are 31 days in March. After March 8 how many days remain in the month?

40. A basketball team scored 110 points in a game. If they scored 48 points during the first half, how many points did they score in the second half?

41. A student read 128 pages of a book containing 415 pages. How many pages remain to be read?

42. It was announced five days before a football game that there were 4,168 unsold tickets for a stadium that has a seating capacity of 83,104. How many tickets had been sold?

43. The temperature rose from 38 degrees in the morning to 64 degrees in the afternoon. How many degrees did the temperature rise?

44. The regular price of an item is $123. A store is having a sale and offers a discount of $35. What is the sale price of the item?

45. The Holland Tunnel is 8,557 feet long. The Hampton Roads Tunnel, near Norfolk, has a length of 7,479 feet. How much longer is the Holland Tunnel?

46. A car with a regular price of $15,098 is offered on sale for $13,999. How much is the discount?

47. The Empire State Building in New York is 1,250 feet tall. The Transamerica Pyramid in San Francisco is 853 feet tall. How much taller is the Empire State Building?

48. The height of Mt. Everest is 29,028 feet and that of Mt. Whitney is 14,494 feet. How much taller is Mt. Everest?

49. A student purchased a book for $23 and a calculator for $19. How much change does the student receive from $50?

50. A trucker drives 427 miles the first day and 515 miles the second day. If the total mileage of the trip is 1,297 miles, how many miles must the trucker drive the third day to complete the trip?

1–6 Algebraic Expressions

OBJECTIVES

Upon completing this section you should be able to:

1. Distinguish between terms and factors of an algebraic expression.
2. Understand the words base and coefficient.

A **numerical expression** consists of numbers rather than letters. In the previous two sections we have dealt with expressions containing only numbers. When letters are used to represent numbers, they and the symbols for operations are referred to as **algebraic symbols.** These letters are of two main types, *variables* and *constants*. A **variable** is a letter that can take on various values while a **constant** is a quantity whose value does not change. An expression containing numbers and algebraic symbols is called an **algebraic expression.**

Example 1 $5 + 3$ is a numerical expression.

Example 2 $x + 3$ is an algebraic expression.

1 Terms and Factors

We now wish to mention some important terminology that is used when we speak of algebraic expressions.

Recall that we first introduced the word *term* in section 1–1.

> When an algebraic expression is composed of parts connected by addition or subtraction signs, these parts are called the **terms** of the expression.

In $a + b$ the terms are a and b.

Example 3 $a + b$ has two terms.

⊙ ***SKILL CHECK 1*** How many terms does the expression $x + 3$ have?

Example 4 $2x + 5y + 3$ has three terms: $2x$, $5y$, and 3.

⊙ ***SKILL CHECK 2*** $3a + 4b + 2$ has how many terms?

> When an algebraic expression is composed of parts to be multiplied, these parts are called the **factors** of the expression.

⊙ ***Skill Check Answers***

1. 2
2. 3

Example 5 ab has two factors, a and b.

It is very important to be able to distinguish between terms and factors. Rules that apply to terms will *not,* in general, apply to factors. When naming terms or factors it is necessary to regard the entire expression.

From now on through all of algebra you will be using the words *term* and *factor.* Make sure you understand the definitions.

Example 6 The expression $3x + 2y - 7$ has three terms. Note that a term may contain factors (for example, the first term of this expression, $3x$, contains two factors)—but the *entire* expression is made up of terms.

Example 7 $5xyz$ is one term made up of factors.

It should be noted that there are more than just four factors in the expression $5xyz$. The obvious factors are 5, x, y, and z. But $5x$ is also a factor. Other factors are $5xy$, xy, $5yz$, xz, $5z$, and so on.

2 Base and Coefficient

In section 1–3 we learned that $x^3 = (x)(x)(x)$, and that the 3 is called an exponent. We now wish to introduce two new words—*base* and *coefficient.*

In an expression such as $5x^4$

5 is the **coefficient**
x is the **base**
4 is the **exponent**

(5)(x)(4) — exponent, base, coefficient

$5x^4$ means $5(x)(x)(x)(x)$

Note that the base is the only factor affected by the exponent.

Example 8 In the expression ax^3

a is the coefficient
x is the base
3 is the exponent
ax^3 means $(a)(x)(x)(x)$

Example 9 $5xy^2$ means $5(x)(y)(y)$

The base of the exponent 2 is y.

◉ ***SKILL CHECK 3*** $4ab^2$ means ?

When we write a letter such as x, it will be understood that the coefficient is 1 and the exponent is 1. This can be very important in many operations.

$$x \text{ means } 1x^1$$

Example 10 $3x$ means $3x^1$.

The exponent 1 is usually not written.

◉ ***SKILL CHECK 4*** $5a$ means $5a^?$

Example 11 x^4 means $1x^4$.

The coefficient, 1, is understood and usually we do not bother to write it in the expression.

◉ ***SKILL CHECK 5*** x^3 means $?x^3$

◉ ***Skill Check Answers***

3. $4(a)(b)(b)$
4. $5a^1$
5. $1x^3$

Exercise 1–6–1

1 State the number of terms in each expression.

1. $x + y$

2. $2x + y$

3. $2x$

4. $8xy$

5. $x + 3$

6. $25 - a$

7. $3x^2$

8. $x^2 + 1$

9. $5a - b + 3c$

10. $x - y - 3$

11. $3x^2 - 2x + 4$

12. $4a^2$

13. $a + 2b - c + 1$

14. $4a^2 + ab - 2b$

15. $6x^3 - 4x^2 + x - 8$

16. $3a^2 - 3a$

17. $3a^4b^2$

18. $10x^2yx^3$

19. $x + 5x^5 - 3x^2$

20. $2a - 3b^2 + 4ab^2$

2 Name the coefficient, the base, and the exponent.

21. $3x^2$

22. $5a^3$

23. $12y$

24. b^7

• Concept Enrichment

25. Explain the distinction between terms and factors.

26. Explain the distinction between a numerical expression and an algebraic expression.

1–7 Adding and Subtracting Like Terms

OBJECTIVES

Upon completing this section you should be able to:

1. Identify like terms.
2. Add and subtract like terms.

In the previous section we learned how to distinguish between terms and factors in an algebraic expression. In this section we will use some special facts concerning terms. First we will define *like terms*.

1 Like Terms

Like terms are terms that have exactly the same literal factors.

Example 1 $5x$ and $3x$ are like terms since they have the same literal factors. (x is the literal factor in each term.)

◉ ***SKILL CHECK 1*** Are $2x$ and $5x$ like terms?

Example 2 $5x$ and $3y$ are *not* like terms since the literal factors are not the same. (x and y are the literal factors.)

◉ ***SKILL CHECK 2*** Are $3a$ and $5b$ like terms?

Example 3 $3x^2y$ and x^2y are like terms since they have the same literal factors (x^2 and y).

◉ ***SKILL CHECK 3*** Are $2ab^2$ and $4ab^2$ like terms?

Example 4 $3x^2y$ and $2xy$ are *not* like terms. They have different literal factors. (Note that x^2 and x are *not* the same.)

Note that having the same letters is not enough, they must have the same exponents as well.

◉ ***SKILL CHECK 4*** Are $2ab$ and $6a^2b$ like terms?

◉ ***Skill Check Answers***

1. yes 2. no
3. yes 4. no

Exercise 1-7-1

1

1. Are $4x$ and $7x$ like terms?
2. Are $3a$ and $5a$ like terms?
3. Are a and $2a$ like terms?
4. Are $4b$ and b like terms?
5. Are $3x$ and $3y$ like terms?
6. Are $8a$ and $8b$ like terms?
7. Are $3x$ and $5x^2$ like terms?
8. Are a^2 and $2a$ like terms?
9. Are $4x^2y$ and $3x^2y$ like terms?
10. Are $5a^2$ and $2a^2b$ like terms?

When the coefficient of a term is a number, it is referred to as the **numerical coefficient.** The numerical coefficient of $5ab$ is 5. Remember that the numerical coefficient of terms such as x, x^2, and a^2b is understood to be 1.

Recall that $x = 1x$.

2 Adding and Subtracting Like Terms

To add or subtract like terms we must make use of a previously mentioned property of numbers. Suppose we want to add $2(3) + 5(3)$. The result would be

$$\begin{aligned} 2(3) + 5(3) &= 6 + 15 \\ &= 21. \end{aligned}$$

2 threes + 5 threes = 7 threes

If we also recognize that 2(3) means 2 threes and 5(3) means 5 threes then we have a total of 7 threes. We can write

$$\begin{aligned} 2(3) + 5(3) &= (2 + 5)(3) \\ &= 7(3) \\ &= 21. \end{aligned}$$

We have obtained the same result. This illustrates the important principle known as the **distributive property,** which will be discussed in the next chapter. We will use it here however to add and subtract like terms. For this purpose the principle can be stated as follows.

> If a, b, and c represent any three numbers, then
> $$ac + bc = (a + b)c \text{ or}$$
> $$ac - bc = (a - b)c.$$

Here $a = 2$, $b = 4$, and $c = 5$.

Example 5 $2(5) + 4(5) = (2 + 4)(5)$
$= 6(5)$
$= 30$

Here $a = 2$, $b = 3$, and $c = x$.

Example 6 $2x + 3x = (2 + 3)x$
$= 5x$

⦿ ***SKILL CHECK 5*** $3x + 5x = ?$

Here $c = x^2$.

Example 7 $3x^2 + x^2 = (3 + 1)x^2$
$= 4x^2$

⦿ ***SKILL CHECK 6*** $2a^2 + 3a^2 = ?$

Here $c = a^2b$.

Example 8 $8a^2b - 5a^2b = (8 - 5)a^2b$
$= 3a^2b$

⦿ ***SKILL CHECK 7*** $5x^2y - 2x^2y = ?$

Here $c = a^2b^2c$.

Example 9 $7a^2b^2c - 3a^2b^2c = (7 - 3)a^2b^2c$
$= 4a^2b^2c$

⦿ ***SKILL CHECK 8*** $9x^2y^2z - 5x^2y^2z = ?$

Only like terms can be added or subtracted.

> To add or subtract like terms add or subtract the numerical coefficients to obtain the coefficient of the like common literal factors.

Example 10 $7A + 2A = 9A$

$7 + 2 = 9$

We add the numerical coefficients 7 and 2 to obtain 9, the coefficient of the common literal factor A.

⦿ ***SKILL CHECK 9*** $6B + 3B = ?$

⦿ ***Skill Check Answers***

5. $(3 + 5)x = 8x$
6. $(2 + 3)a^2 = 5a^2$
7. $(5 - 2)x^2y = 3x^2y$
8. $(9 - 5)x^2y^2z = 4x^2y^2z$
9. $9B$

Example 11 $7A + 3a - 2A + 5a = 5A + 8a$

WARNING

Many students make the mistake of interchanging capital and lowercase letters in an expression. In algebra a capital "*A*" and a lowercase "*a*" are considered just as different as *x* and *y*. Therefore terms such as 7*A* and 3*a* would *not* be considered like terms. Be consistent with the letters you use.

⦿ ***SKILL CHECK 10*** $8B + 4b - 3B + 2b = ?$

Example 12 $3x^2y + 5xy - 2xy + 3x^2y = 6x^2y + 3xy$

⦿ ***SKILL CHECK 11*** $5a^2b + 7ab - 3a^2b + 2ab = ?$

Example 13 $7x + 2y + x + 5y - 3y = 8x + 4y$

Notice that the numerical coefficient of x is 1 and in this example must be added to the 7 of $7x$.

⦿ ***SKILL CHECK 12*** $6a + 3b + a + 4b - 2b = ?$

⦿ ***Skill Check Answers***

10. $5B + 6b$
11. $2a^2b + 9ab$
12. $7a + 5b$

Exercise 1-7-2

2 Simplify by adding and subtracting like terms.

1. $4x + 7x$

2. $5a + 8a$

3. $9x - 5x$

4. $11a - 6a$

5. $3a + a$

6. $x + 5x$

7. $5x^2 + 3x^2$

8. $3x^2 + x^2$

9. $7a^3 - a^3$

10. $10a^3 - 5a^3$

11. $4xy + 12xy$

12. $12ab - 5ab$

13. $8xy - 7xy$

14. $3a^2b + 8a^2b$

15. $9a + 6b - 5a$

16. $5x - 4y + 6x$

17. $6ab^2 + 9a^2b + 5ab^2$

18. $13xy^2 - 7x^2y - 5xy^2$

19. $14abc + 3ab + 8abc - 2ab$

20. $4xy + 9x^2y + 5xy - 3x^2y$

21. $5ab + 11ac - 4ab - 10ac$

22. $6x^2y + 21xy^2 + 4x^2y - 17xy^2$

23. $13a^2 + 14a - 11a^2 - 6a + a^2$

24. $8xyz + 3xz - 5xyz - 2xz - 3xyz$

25. $14x^3y + 3x^3y^2 - 9x^3y + 5x^3y^2 - 5x^3y$

26. $5a^2b + 10ab - 4a^2b - 4ab + 2ab^2$

27. $4x^2y - 5xy^2 + 15x^2y^2 + 3x^2y - 6x^2y^2$

28. $15ab^2 + 9a^2b - 5ab + 4a^2b - 6ab^2$

29. $12M + 2m + 6m - 11M$

30. $8T + 9t - 7T - 8t$

• Concept Enrichment

31. What property is illustrated by the statement $3x + 8x = (3 + 8)x$?

32. What is wrong with the statement $2a^2b^3 + 5ab^2 = 7a^3b^5$?

1–8 Rounding Whole Numbers

OBJECTIVE

Upon completing this section you should be able to:

1. Round whole numbers to any place value.

In many situations it may be desirable to use a close estimate of a number rather than the exact number. However, there must be exact rules for obtaining this estimate. The process is known as *rounding*.

1 Rounding

Suppose an automobile is priced at $12,859. Would you say it cost "about $13,000"? Why? This illustrates the basic idea of **rounding.** We want to go to the *nearest* number. Sometimes we want the *nearest ten, nearest hundred,* or *nearest thousand,* and so on. So we need a rule that will work for all cases.

IMPORTANT! The digit in the desired place *never* decreases.

> To round a number to any place value if the digit in the next place to the right is 5 or greater, the digit in the desired place is increased by one. If it is less than 5, the digit in the desired place remains the same. All digits to the right of the desired place are changed to zero.

3 6 8 = 3 7 0

This digit is 5 or more so 6 is increased to 7.

Example 1 Round 368 to the nearest ten.

Solution Since 8 is in the next place to the right of the tens place and it is greater than 5, 368 rounds as 370 to the nearest ten.

◉ SKILL CHECK 1 Round 438 to the nearest ten.

◉ Skill Check Answer

1. 440

Example 2 Round 6,439 to the nearest hundred.

Solution Since 3 is in the next place to the right of the hundreds place and is less than 5, 6,439 rounds as 6,400 to the nearest hundred.

6 4 3 9 = 6 4 0 0

This digit is less than 5 so 4 is not changed.

◉ ***SKILL CHECK 2*** Round 2,628 to the nearest hundred.

Example 3 Round 25,738 to the nearest thousand.

Solution Since 7 is in the next place to the right of the thousands place, 25,738 rounds as 26,000 to the nearest thousand.

2 5 7 3 8

This digit will be increased by 1.

◉ ***SKILL CHECK 3*** Round 46,832 to the nearest thousand.

Example 4 Round 25,738 to the nearest ten thousand.

Solution Since 5 is in the next place to the right of the ten thousands place, 25,738 rounds as 30,000 to the nearest ten thousand.

2 5 7 3 8

This digit will be increased by 1.

◉ ***SKILL CHECK 4*** Round 46,832 to the nearest ten thousand.

Example 5 Round 5,538,278 to the nearest million.

Solution Since 5 is in the next place to the right of the millions place, 5,538,278 rounds as 6,000,000 to the nearest million.

5 5 3 8 2 7 8

Digit to the right of the millions place.

◉ ***SKILL CHECK 5*** Round 7,531,468 to the nearest million.

Example 6 Round 999 to the nearest hundred.

Solution Since 9 is in the next place to the right of the hundreds place, we add one to the hundreds place giving ten hundreds. We therefore place a one in the thousands place. Thus 999 rounds as 1,000 to the nearest hundred.

9 9 9

This digit will be increased by 1.

◉ ***SKILL CHECK 6*** Round 9,999 to the nearest thousand.

◉ ***Skill Check Answers***

2. 2,600
3. 47,000
4. 50,000
5. 8,000,000
6. 10,000

Exercise 1–8–1

1 Round to the nearest ten.

1. 63 **2.** 86 **3.** 28 **4.** 44 **5.** 517

6. 388 **7.** 293 **8.** 794 **9.** 4,206 **10.** 1,387

Round to the nearest hundred.

11. 384 **12.** 596 **13.** 249 **14.** 156 **15.** 2,352

16. 5,233 **17.** 1,063 **18.** 3,049 **19.** 6,954 **20.** 4,980

Round to the nearest thousand.

21. 58,473 **22.** 23,258 **23.** 25,581 **24.** 88,741 **25.** 34,699

26. 36,282 **27.** 40,455 **28.** 70,501 **29.** 19,458 **30.** 59,628

Round to the nearest ten thousand.

31. 135,000 **32.** 315,000 **33.** 455,133 **34.** 629,514 **35.** 263,594

36. 583,685 **37.** 337,999 **38.** 142,815 **39.** 595,416 **40.** 294,956

Round 158,463 to the indicated place value.

41. Nearest ten **42.** Nearest thousand **43.** Nearest hundred

44. Nearest ten thousand **45.** Nearest hundred thousand

Round 295,997 to the indicated place value.

46. Nearest ten **47.** Nearest ten thousand **48.** Nearest hundred

49. Nearest thousand **50.** Nearest hundred thousand

Round 5,555,555 to the indicated place value.

51. Nearest ten **52.** Nearest hundred thousand **53.** Nearest hundred

54. Nearest thousand **55.** Nearest ten thousand **56.** Nearest million

Round 9,999 to the indicated place value.

57. Nearest ten

58. Nearest hundred

59. Nearest thousand

60. Nearest ten thousand

CHAPTER 1 SUMMARY

Key Words

Section 1–1

- A **sum** is obtained when adding.
- A **difference** is obtained when subtracting.
- A **product** is obtained when multiplying.
- A **quotient** is obtained when dividing.
- A **variable** is a letter that can take on various numerical values.

Section 1–2

- The **digits** 0, 1, 2, 3, 4, 5, 6, 7, 8, 9 are used to write all numbers.
- Each digit has a **digital value** and a **place value.**
- A **number line** is useful for visualizing the relative positions of numbers.
- **Inequality symbols** are used to indicate the relative position of one number to another on the number line.

Section 1–3

- An **exponent** is used to indicate how many times a number is used as a factor.
- The **zero power** of any number (except zero) is 1.
- Our number system is a **base ten system** with each place value ten times as large as the place to its right.

Section 1–4

- An **algorithm** is a step-by-step procedure for obtaining an answer.
- The **additive identity** is zero. For any number a, $a + 0 = a$.
- The **commutative property of addition** is $a + b = b + a$ for any two numbers a and b.
- Addition is a **binary** operation.
- The **associative property of addition** is $(a + b) + c = a + (b + c)$ for any numbers a, b, and c.

Section 1–5

- **Subtraction** is based on the addition facts.
- **Borrowing** in subtraction is founded on the base ten system.

Section 1–6

- **Algebraic symbols** are letters used to represent numbers, and the symbols for operations.
- A **variable** is a letter that can take on various values.
- A **constant** is a quantity whose value does not change.
- An **algebraic expression** is an expression that contains numbers and algebraic symbols.
- **Terms** of an algebraic expression are connected by addition or subtraction signs.
- **Factors** of an algebraic expression are parts to be multiplied.
- In the expression ax^3
  - a is the **coefficient**
  - x is the **base**
  - 3 is the **exponent**

Section 1–7

- **Like terms** are terms that have exactly the same literal factors.
- A **numerical coefficient** of a term is a number that is a factor of the term.
- The **distributive property** allows us to add and subtract like terms.

Section 1–8

- **Rounding** is a process used to obtain an estimate of a number.

Procedures

Section 1–3

- To write a number in expanded form the digital value is multiplied by the place value.

Section 1–4

- Only like quantities can be added.

Section 1–5

- Only like quantities can be subtracted.

Section 1–7

- To add or subtract like terms add or subtract the numerical coefficients to obtain the coefficient of the like common literal factors.

Section 1–8

- In rounding to any place value, if the digit in the next place to the right is 5 or greater, the digit in the desired place is increased by one. If it is less than 5, the digit in the desired place remains the same. All digits to the right of the desired place are changed to zero.

CHAPTER 1 REVIEW

Name the operation or operations indicated in each expression.

1. $x \div 3$

2. $8x$

3. $5a - 2$

4. $6a + 7$

5. $3 + \frac{x}{y}$

6. $\frac{a}{b} - 1$

7. What is the place value of the digit 4 in the number 10,546?

8. What is the place value of the digit 2 in the number 12,385?

9. What is the place value of the digit 2 in the number 618,324,705?

10. What is the place value of the digit 7 in the number 704,251,638?

11. Write the number 10,325 in words.

12. Write the number 21,364 in words.

13. Write the number 724,803,241 in words.

14. Write the number 105,063,004 in words.

15. Write "five hundred twenty-three thousand, two hundred one" as a number.

16. Write "four million, eighty-six thousand, twenty-one" as a number.

Replace the question mark with the inequality symbol positioned to indicate a true statement.

17. 8 ? 0

18. 12 ? 13

19. 3,099 ? 3,103

20. 504 ? 509

21. Write the number 827 in expanded form.

22. Write the number 468 in expanded form.

23. Write the number 40,396 in expanded form.

24. Write the number 21,508 in expanded form.

25. Add: $\begin{array}{r} 8 \\ 9 \\ 1 \\ 7 \\ +3 \\ \hline \end{array}$

26. Add: $\begin{array}{r} 3 \\ 8 \\ 7 \\ 4 \\ +2 \\ \hline \end{array}$

27. Add: $\begin{array}{r} 34 \\ +63 \\ \hline \end{array}$

28. Add: $\begin{array}{r} 28 \\ +51 \\ \hline \end{array}$

Add.

29. $\begin{array}{r} 68 \\ +55 \\ \hline \end{array}$

30. $\begin{array}{r} 39 \\ +61 \\ \hline \end{array}$

31. $\begin{array}{r} 28 \\ 43 \\ +39 \\ \hline \end{array}$

32. $\begin{array}{r} 53 \\ 27 \\ +49 \\ \hline \end{array}$

33. $\begin{array}{r} 216 \\ +389 \\ \hline \end{array}$

34. $\begin{array}{r} 546 \\ +375 \\ \hline \end{array}$

35. $\begin{array}{r} 329 \\ +671 \\ \hline \end{array}$

36. $\begin{array}{r} 485 \\ +515 \\ \hline \end{array}$

37. $\begin{array}{r} 304 \\ 685 \\ +293 \\ \hline \end{array}$

38. $\begin{array}{r} 416 \\ 398 \\ +502 \\ \hline \end{array}$

39. $\begin{array}{r} 2{,}428 \\ 1{,}373 \\ 454 \\ +4{,}162 \\ \hline \end{array}$

40. $\begin{array}{r} 3{,}129 \\ 4{,}361 \\ 283 \\ +1{,}256 \\ \hline \end{array}$

41. Find the sum of 2,064; 348; 38; and 8,269.

42. Find the sum of 1,906; 86; 9,271; and 568.

Subtract.

43. $\begin{array}{r} 587 \\ -354 \\ \hline \end{array}$

44. $\begin{array}{r} 674 \\ -541 \\ \hline \end{array}$

45. $\begin{array}{r} 324 \\ -\ 65 \\ \hline \end{array}$

46. $\begin{array}{r} 236 \\ -\ 57 \\ \hline \end{array}$

47. $\begin{array}{r} 653 \\ -456 \\ \hline \end{array}$

48. $\begin{array}{r} 575 \\ -276 \\ \hline \end{array}$

49. $\begin{array}{r} 4{,}604 \\ -1{,}426 \\ \hline \end{array}$

50. $\begin{array}{r} 7{,}507 \\ -4{,}319 \\ \hline \end{array}$

51. Find the difference of 35,027 and 12,848.

52. Find the difference of 15,304 and 7,219.

State the number of terms in the expression.

53. $3x + 5$

54. $5a - 4$

55. $6x^2$

56. $x^2 - 1$

57. $2x^2 - 3x + 5$

58. $a^3 + 5a^2 + 3bx$

Add or subtract like terms.

59. $4a + 5a - 6b$

60. $3x + 5y - x$

61. $5x^2 + 3x - 4y + 6x$

62. $2xy + 3x^2y + 5xy - x^2y$

63. $3a^2b + 9a^2 - 2a^2b - 5a^2$

64. $20xy^2 + 4xy - 15xy^2 - 3xy - 5xy^2$

65. Round 346 to the nearest ten.

66. Round 524 to the nearest ten.

67. Round 1,735 to the nearest ten.

68. Round 3,685 to the nearest ten.

69. Round 21,346 to the nearest hundred.

70. Round 55,663 to the nearest hundred.

71. Round 45,568 to the nearest thousand.

72. Round 38,519 to the nearest thousand.

73. Round 461,528 to the nearest ten thousand.

74. Round 256,104 to the nearest ten thousand.

Write the number in the sentence in words.

75. The Amazon River is 3,915 miles long.

76. One kilogram of seawater contains 10,561 milligrams of sodium.

Write the number in the sentence using digits.

77. The cruise ship Queen Elizabeth II can carry one thousand, eight hundred fifteen passengers.

78. In 1985 the population of Alaska was estimated at five hundred twenty-one thousand people.

Solve.

79. You receive three checks for \$218, \$46, and \$109 and wish to deposit them in your bank account. What is the total deposit?

80. A person borrows \$875 and agrees to pay back \$1,003 at the end of one year. How much interest does the person pay?

CHAPTER 1 TEST

1. Name the operation indicated in the expression $\frac{a}{b}$.

2. Name the operations indicated in the expression $4x - 3$.

3. What is the place value of the digit 5 in the number 25,604?

4. Write the number 52,619 in words.

5. Replace the question mark in 18 ? 23 with the inequality symbol positioned to indicate a true statement.

6. Write the number 3,005 in expanded form.

7. Add: $\begin{array}{r} 475 \\ +528 \\ \hline \end{array}$

8. Add: $\begin{array}{r} 268 \\ 723 \\ +356 \\ \hline \end{array}$

9. Find the sum of 3,196; 38; 506; and 1,624.

10. Subtract: $\begin{array}{r} 843 \\ -548 \\ \hline \end{array}$

11. Subtract: $\begin{array}{r} 608 \\ -199 \\ \hline \end{array}$

12. Find the difference of 10,135 and 6,128.

13. State the number of terms in the expression $5a^3 - 2x + 4$.

14. Name the exponent in the expression $4x$.

15. Simplify: $9a - 5a$

16. Simplify: $7x + x$

17. Simplify: $4x^2y + 7xy + 5x^2y - 2x^2y - xy$

18. Round the number 375 to the nearest ten.

19. Round the number 12,496 to the nearest thousand.

20. During one month there were three hundred eighty-seven thousand, nine hundred six college students on spring break in Daytona Beach, Florida. Write this number using digits.

21. After climbing a total of 13,800 feet, mountain climbers still needed to travel another 6,506 feet to reach the summit of Mt. McKinley in Alaska. How high is Mt. McKinley?

22. A consumer purchased a new car costing $12,092 but received a discount of $3,254 on a trade-in for an old car. What was the balance due on the new car?

CHAPTER

2 Pretest

Answer as many of the following problems as you can before starting this chapter. When you finish the chapter, take the test at the end and compare your scores to see how much you have learned.

1. Multiply: 18(24)

2. Multiply: $\begin{array}{r} 2{,}369 \\ \times\ 305 \\ \hline \end{array}$

3. Multiply: $257(10^3)$

4. Divide: $50 \div 6$

5. Divide: $8\overline{)512}$

6. Divide: $23\overline{)1{,}097}$

7. Evaluate: $10 - (3 + 4)$

8. Evaluate: $10 - 3 + 4$

9. Evaluate: $4 + 38 \div 2 - 3^2 \times 2 - 2$

10. Evaluate: $5[2(7 - 4) - 4]$

11. Evaluate $3x^2$ if $x = 2$.

12. Evaluate $(3x)^2$ if $x = 2$.

13. Evaluate $3a^2b + 4ac - 6bc$ if $a = 5$, $b = 3$, $c = 1$.

14. Simplify: x^2x^5

15. Multiply: $(4a^2b)(2ab^3)$

16. Find the prime factorization of 180.

17. Find the greatest common factor of 189 and 210.

18. Find the least common multiple of $14x^2$ and $21xy$.

19. The perimeter of a rectangle is given by the formula $P = 2\ell + 2w$. Find P when $\ell = 16$ and $w = 9$.

20. $3,072 is to be shared equally among 12 people. What is each person's share?

21. If there are 24 hours in one day, how many hours are there in 31 days?

22. A person has a choice of purchasing a video camera for $795 cash or making 24 payments of $38 each. How much is saved by paying cash?

C H A P T E R

2 Algebraic Expressions Involving Whole Numbers

Mathematics affects our daily lives in almost every conceivable way. Items in a supermarket are identified by a universal product code that involves multiplication and addition of whole numbers.

STUDY HELP

As you study and work the examples, skill checks, and exercises, make note of anything that is not totally clear. In this way you can ask your instructor for specific help rather than just saying "I don't understand."

In chapter 1 we studied the operations of addition and subtraction of whole numbers and used the facts we learned in our study of some algebraic topics. In this chapter we will learn to evaluate some algebraic expressions but to do this our first task is to make sure we understand the operations of multiplication and division of whole numbers.

2–1 Multiplying Whole Numbers

OBJECTIVES

Upon completing this section you should be able to:

1. Recognize some properties of multiplication.
2. Multiply any two whole numbers using the long multiplication algorithm.
3. Use multiplication in solving applications.
4. Use a short method for multiplying by a power of 10.

The operation of multiplication can be thought of as repeated addition. For instance, if we wish to multiply 7×5 we can think of this as seven 5s, or $5 + 5 + 5 + 5 + 5 + 5 + 5$. This sum is 35, so we say $7 \times 5 = 35$. We could also interpret 7×5 as five 7s, or $7 + 7 + 7 + 7 + 7$, which also has a sum of 35.

1 Properties of Multiplication

The fact that seven 5s (7×5) gives the same result as five 7s (5×7) illustrates an important property of multiplication called the commutative property. A formal statement of this property follows.

> **Commutative Property of Multiplication**
> If a and b represent any two numbers, then $ab = ba$.

SKILL CHECK 1 What property is illustrated by $(2)(3) = (3)(2)$?

Recall that addition is also commutative, that is $a + b = b + a$ for any numbers a and b.

For example, $5 + 0 = 5$.

We should also recall that zero is the **additive identity** and that for any number a it is true that $a + 0 = a$. In multiplication we also have an identity. It is the number 1.

The product of any number and 1 is the number itself. For example, $5(1) = 5$.

> **Multiplicative Identity**
> If a represents any number, then $a(1) = a$.

SKILL CHECK 2 What property is illustrated by $5(1) = 5$?

Multiplication is also associative. For instance

$$(3 \times 2) \times 7 = 3 \times (2 \times 7).$$

If we evaluate the left side, we have

$$\begin{aligned}(3 \times 2) \times 7 &= 6 \times 7 \\ &= 42.\end{aligned}$$

The right side gives us

$$\begin{aligned}3 \times (2 \times 7) &= 3 \times 14 \\ &= 42.\end{aligned}$$

Find the value of each side: $(4 \times 5) \times 6 = 4 \times (5 \times 6)$

> **Associative Property of Multiplication**
> If a, b, and c represent any three numbers, then $(ab)c = a(bc)$.

SKILL CHECK 3 What property is illustrated by $(2 \times 3) \times 5 = 2 \times (3 \times 5)$?

Skill Check Answers

1. commutative property of multiplication
2. multiplicative identity
3. associative property of multiplication

When preparing for the operation of addition, we pointed out that there are a limited number of basic facts necessary for being competent in addition. The same is true for multiplication. Since all whole numbers are written by the use of ten digits, we need only to know the results of multiplying any two of these digits. Once again we must emphasize the importance of knowing the basic facts. Satisfactory progress cannot be made in the operation of multiplying whole numbers unless you are proficient in these facts.

Recall that the result of multiplying two numbers is called the *product*.

2 Long Multiplication Algorithm

A process (or algorithm) of multiplying any two whole numbers is called **long multiplication.** Suppose we want to find the product of 7(32). If we look at this as repeated addition, we could say we want the sum of thirty-two 7s. Now, if we remember that $32 = 30 + 2$, we could say that $7(32) = 30$ sevens $+ 2$ sevens. In other words, $7(32) = 7(30 + 2)$, which is also $7(30) + 7(2)$. This illustrates a very important property, called the distributive property of multiplication over addition.

Placing thirty-two 7s in a column and adding them would take some time. We could place seven 32s in a column and add them.

Distributive Property of Multiplication over Addition
If a, b, and c are any three numbers, then $a(b + c) = ab + ac$.

Notice that this property allows us to write a product as a sum and vice versa.

Sometimes this property or law is referred to simply as the distributive property. Now let's apply this property in the following example.

Example 1 Multiply 7(32).

Solution

$$7(32) = 7(30 + 2) = 7(30) + 7(2)$$

Now $7(30) = 7(3 \text{ tens}) = 21 \text{ tens} = 210$.
Also, $7(2) = 14$ from our basic multiplication facts. So we have

$$7(32) = 7(30) + 7(2) = 210 + 14 = 224.$$

Now let's examine the same problem using the multiplication algorithm. We first write the two numbers to be multiplied in columns.

$$\begin{array}{r} 32 \\ \underline{\times\ 7} \end{array}$$

We now multiply 7(2) and obtain 14.

$$\begin{array}{r} 32 \\ \underline{\times\ 7} \\ 14 \end{array}$$

We also multiply 7(3) and get 21. But we must note that the 3 in 32 is in the tens column, so the 21 is 21 tens, making it necessary to place the 1 of 21 in the tens column. We then add the 14 ones and 21 tens to obtain the answer of 224.

7(3 tens) = 21 tens.

$$\begin{array}{r} 32 \\ \underline{\times\ 7} \\ 14 \\ \underline{21} \\ 224 \end{array}$$

We could further refine the work involved by a process called **carrying.** We first say that $7(2) = 14$. We then place the 4 in the ones column and carry the 1 to the tens column.

The digit 1 in 14 is in the tens place.

$$\begin{array}{r} 1 \\ 32 \\ \underline{\times\ 7} \\ 4 \end{array}$$

We now multiply 7(3) to get 21 and add the 1 that was carried to get 22.

$$\begin{array}{r} 1 \\ 32 \\ \times\ 7 \\ \hline 224 \end{array}$$

It will be helpful if you learn to carry mentally as soon as possible.

Example 2 Find the product of 27(32).

Solution We first place the numbers in columns.

Make sure you have tens under tens, and so on.

$$\begin{array}{r} 32 \\ \times 27 \\ \hline \end{array}$$

We now multiply 7(32) to obtain 224 as in the previous example.

$$\begin{array}{r} 32 \\ \times 27 \\ \hline 224 \end{array}$$

We next multiply by the 2 of 27. Since 2 is in the tens column, we are actually multiplying by 2 tens or 20. This requires that our first number $2(2) = 4$ be placed in the tens column.

This is actually $20 \times 2 = 40$.

$$\begin{array}{r} 32 \\ \times 27 \\ \hline 224 \\ 4 \end{array}$$

We must now multiply the 3 tens by the 2 tens or $2(10) \times 3(10)$. This gives us 6(100) or 600. So we say that $2(3) = 6$, but we place the 6 in the hundreds column. Adding now gives 864 as the product of 27 and 32.

$$2(10) \times 3(10) = 2(3)(10)(10)$$

by the commutative property.

$$\begin{array}{r} 32 \\ \times 27 \\ \hline 224 \\ 64 \\ \hline 864 \end{array}$$

Example 3 Find the product of 283 and 42.

Solution

$$\begin{array}{r} 283 \\ \times\ 42 \\ \hline 566 \end{array}$$

- 2(3)
- 2(8) = 16; write the 6 and carry 1.
- 2(2) = 4, plus the 1 carried to give 5.

$$\begin{array}{r} 283 \\ \times\ 42 \\ \hline 566 \\ 1132 \end{array}$$

- 4(3) = 12; write the 2 and carry 1.
- 4(8) = 32, plus 1 carried equals 33. Write the 3 and carry 3.
- 4(2) = 8, plus 3 carried equals 11.

Remember that when we say 4(3) here we are actually multiplying by 4 tens.

Adding now gives 11,886 as the product of 283 and 42.

$$\begin{array}{r} 283 \\ \times\ 42 \\ \hline 566 \\ 11\,32 \\ \hline 11{,}886 \end{array}$$

◉ ***SKILL CHECK 4*** Find the product of 352 and 43.

In multiplication you will find that neatness is a great help. In fact many errors are made by not keeping columns straight or by placing numbers in the wrong column. The best rule to remember is: place the rightmost digit of the product in the same column as the digit by which you are multiplying.

WARNING

Most students who make errors may have multiplied correctly but have placed digits in the wrong column.

Example 4 Find the product of 347 and 432.

Solution

```
    347
   ×432
    694 ←— 4 is placed in the same column as 2.
  10 41 ←— 1 is placed in the same column as 3.
  138 8 ←— 8 is placed in the same column as 4.
  149,904
```

4 is the rightmost digit of the product when multiplied by 2.

◉ ***SKILL CHECK 5*** Find the product of 352 and 234.

Example 5 Find the product of 385 and 203.

Solution

```
    385
   ×203
   1 155
  00 0
  77 0
  78,155
```

Note that we need not multiply by the 0 of 203 since only zeros will result. Notice again that $2(5) = 10$, so the 0 (rightmost digit of the product) is placed in the same column as the 2.

```
    385
   ×203
   1 155
  77 0
  78,155
```

Be very careful about placing the digits in the correct column.

◉ ***SKILL CHECK 6*** Find the product of 364 and 304.

A technique known as **estimation** is often helpful in determining if our answer is of the right magnitude. That is, it gives us a "ballpark" figure for the size of our answer.

To estimate we round the numbers to be operated on. In example 5 we would round the numbers to the nearest hundred. We round 385 to 400 and 203 to 200. Multiplying these numbers we obtain

Recall rounding from section 1-8.

```
    400
   ×200
    000
   0 00
  80 0
  80,000.
```

The correct answer should therefore be in the neighborhood of 80,000. We saw that the actual answer was 78,155. Thus we have a way of checking to see if our answer is reasonable.

Example 6 Find the product of 2,165 and 3,002.

◉ ***Skill Check Answers***

4. 15,136
5. 82,368
6. 110,656

Solution First we will use estimation to determine the approximate size of the product. Rounding 2,165 to 2,000 and 3,002 to 3,000, we find the product of these two numbers.

```
   2,000
  ×3,000
   0 000
  00 00
 000 0
6 000
6,000,000
```

Thus the product should be in the neighborhood of 6,000,000.

Now using the multiplication algorithm on the actual numbers we have

```
   2,165
  ×3,002
   4 330
6 495
6,499,330.
```

If it helps to keep columns straight, you can place zeros in the columns with zero multipliers.

```
   2,165
  ×3,002
   4 330
6 495 00
6,499,330
```

We see our answer agrees with the magnitude of our estimation.

⊙ ***SKILL CHECK 7*** Find the product of 2,214 and 4,003.

Example 7 Find the product of 241 and 382. Round the answer to the nearest ten.

Use estimation to approximate the answer.

Solution

```
   241
  ×382
   482
 19 28
 72 3
 92,062
```

Rounding the product to the nearest ten gives 92,060.

⊙ ***SKILL CHECK 8*** Find the product of 231 and 362. Round the answer to the nearest ten.

Example 8 Find the sum of 252(43) and 1,405(302).

Solution Read carefully! When you see the word *sum,* you know you must add. But don't make the mistake of reading *sum* and then adding all numbers you see in the problem. This problem involves the operation of addition as well as of multiplication. We are asked for the sum of two products.

Many problems require more than one operation.

We first find the indicated products.

```
   252        1,405
 ×  43      ×   302
   756        2 810
 10 08      421 5
 10,836     424,310
```

Recall that the word "sum" means "add."

The final step is to find the sum of these two products.

```
  10,836
+424,310
 435,146
```

⊙ ***SKILL CHECK 9*** Find the sum of 231(32) and 1,324(203).

⊙ ***Skill Check Answers***

7. 8,862,642
8. 83,620
9. 276,164

3 Applications

Example 9 If a car travels at a constant rate of 55 miles per hour for 4 hours, how far has it traveled?

Solution If a car travels 55 miles in 1 hour, then in 4 hours it will travel 55(4) miles. Thus

$$\begin{array}{r} 55 \\ \times\ 4 \\ \hline 220 \end{array}$$

The car has traveled 220 miles.

⊙ ***SKILL CHECK 10*** If a car travels at a constant rate of 45 miles per hour for 3 hours, how far has it traveled?

Example 10 A cyclist sets out on a trip of 83 miles and travels at an average rate of 18 miles per hour for the first 3 hours. How many miles remain to be traveled to complete the trip?

Solution We should reason in this manner. If the average rate is 18 miles per hour, then in 3 hours the cyclist will travel 18(3) miles.

$$\begin{array}{r} 18 \\ \times\ 3 \\ \hline 54 \end{array} \longleftarrow \text{Distance traveled so far.}$$

The cyclist has traveled 54 miles of the needed 83 miles. To find the number of miles remaining we must subtract 54 from 83.

$$\begin{array}{r} 83 \\ -54 \\ \hline 29 \end{array} \longleftarrow \text{Distance still to go.}$$

The cyclist must travel 29 miles to complete the trip.

⊙ ***SKILL CHECK 11*** A cyclist starts on a trip of 78 miles and travels at an average rate of 15 miles per hour for the first 4 hours. How many miles remain to be traveled to complete the trip?

⊙ ***Skill Check Answers***

10. 135 miles
11. 18 miles

Exercise 2–1–1

1 Name the property illustrated by the statement.

1. $3 \times 1 = 3$ **2.** $5 \times 6 = 6 \times 5$ **3.** $3(4) = 4(3)$ **4.** $10(1) = 10$

5. $3 \times (4 \times 5) = (3 \times 4) \times 5$ **6.** $(4 \times 2) \times 6 = 4 \times (2 \times 6)$ **7.** $(4 \times 1) \times 5 = 5 \times (4 \times 1)$

8. $2 \times (8 \times 1) = 2 \times 8$ **9.** $ab(1) = ab$ **10.** $2 \times (9 \times 4) = (2 \times 9) \times 4$

2 Find each product. Use estimation to determine the reasonableness of the size of your answer. You may want to check your answers using a calculator.

11. $\begin{array}{r} 35 \\ \times\ 5 \\ \hline \end{array}$	**12.** $\begin{array}{r} 26 \\ \times\ 3 \\ \hline \end{array}$	**13.** $\begin{array}{r} 73 \\ \times\ 4 \\ \hline \end{array}$	**14.** $\begin{array}{r} 36 \\ \times\ 5 \\ \hline \end{array}$
15. $\begin{array}{r} 17 \\ \times\ 9 \\ \hline \end{array}$	**16.** $\begin{array}{r} 25 \\ \times\ 8 \\ \hline \end{array}$	**17.** $\begin{array}{r} 123 \\ \times\ 4 \\ \hline \end{array}$	**18.** $\begin{array}{r} 215 \\ \times\ 6 \\ \hline \end{array}$
19. $\begin{array}{r} 425 \\ \times\ 5 \\ \hline \end{array}$	**20.** $\begin{array}{r} 518 \\ \times\ 6 \\ \hline \end{array}$	**21.** $\begin{array}{r} 21 \\ \times 34 \\ \hline \end{array}$	**22.** $\begin{array}{r} 16 \\ \times 32 \\ \hline \end{array}$
23. $\begin{array}{r} 28 \\ \times 19 \\ \hline \end{array}$	**24.** $\begin{array}{r} 37 \\ \times 18 \\ \hline \end{array}$	**25.** $\begin{array}{r} 72 \\ \times 24 \\ \hline \end{array}$	**26.** $\begin{array}{r} 63 \\ \times 24 \\ \hline \end{array}$
27. $\begin{array}{r} 234 \\ \times\ 16 \\ \hline \end{array}$	**28.** $\begin{array}{r} 521 \\ \times\ 24 \\ \hline \end{array}$	**29.** $\begin{array}{r} 374 \\ \times\ 36 \\ \hline \end{array}$	**30.** $\begin{array}{r} 267 \\ \times\ 37 \\ \hline \end{array}$
31. $\begin{array}{r} 478 \\ \times\ 69 \\ \hline \end{array}$	**32.** $\begin{array}{r} 687 \\ \times\ 76 \\ \hline \end{array}$	**33.** $\begin{array}{r} 39 \\ \times 10 \\ \hline \end{array}$	**34.** $\begin{array}{r} 41 \\ \times 10 \\ \hline \end{array}$
35. $\begin{array}{r} 154 \\ \times\ 10 \\ \hline \end{array}$	**36.** $\begin{array}{r} 523 \\ \times\ 10 \\ \hline \end{array}$	**37.** $\begin{array}{r} 124 \\ \times 232 \\ \hline \end{array}$	**38.** $\begin{array}{r} 314 \\ \times 163 \\ \hline \end{array}$
39. $\begin{array}{r} 328 \\ \times 439 \\ \hline \end{array}$	**40.** $\begin{array}{r} 765 \\ \times 278 \\ \hline \end{array}$	**41.** $\begin{array}{r} 325 \\ \times 104 \\ \hline \end{array}$	**42.** $\begin{array}{r} 634 \\ \times 205 \\ \hline \end{array}$

43. 416×100

44. 328×100

45. 724×100

46. 813×100

47. $2,325 \times 164$

48. $1,419 \times 236$

49. $3,014 \times 213$

50. $7,304 \times 219$

51. $4,625 \times 304$

52. $2,419 \times 507$

53. $2,041 \times 703$

54. $5,052 \times 209$

55. $3,001 \times 425$

56. $2,008 \times 196$

57. $2,153 \times 1,409$

58. $6,244 \times 1,306$

59. $2,512 \times 1,000$

60. $3,416 \times 1,000$

61. $3,205 \times 10$

62. $5,214 \times 10$

Round the product to the nearest ten.

63. 38×46

64. 207×56

Round the product to the nearest hundred.

65. 27×63

66. 512×38

Round the product to the nearest thousand.

67. $\begin{array}{r} 78 \\ \times 34 \\ \hline \end{array}$

68. $\begin{array}{r} 216 \\ \times\ 95 \\ \hline \end{array}$

Round the product to the nearest ten thousand.

69. $\begin{array}{r} 413 \\ \times\ 67 \\ \hline \end{array}$

70. $\begin{array}{r} 512 \\ \times\ 86 \\ \hline \end{array}$

71. Find the sum of 324(116) and 1,204(639).

72. Find the sum of the following products.

$\begin{array}{r} 63 \\ \times 81 \\ \hline \end{array}$ $\begin{array}{r} 560 \\ \times\ 34 \\ \hline \end{array}$ $\begin{array}{r} 384 \\ \times\ 76 \\ \hline \end{array}$ $\begin{array}{r} 401 \\ \times 306 \\ \hline \end{array}$

73. Find the difference of 165(73) and 403(25).

74. Find the difference of 2,419(201) and 1,264(306).

3

75. A man buys six bags of grass seed at $15 per bag. What is the total cost?

76. If a person earns $275 per week, how much is earned in 26 weeks?

77. If you drove a constant rate of 45 miles per hour, how far would you travel in 14 hours?

78. If a certain car gets 32 miles per gallon, how far can it go on 15 gallons?

79. If fencing costs $4 per foot, what is the cost of 193 feet?

80. Each home in a homeowners association is assessed $192 per year. If there are 47 homes in the association, how much money should be collected per year?

81. A car is priced at $8,995 cash or 36 payments of $328 each. What is the total of the payments? How much is saved by paying cash?

82. A woman owes $3,200. She pays $112 per month for 24 months. How much does she still owe?

4 Powers of Ten

In the preceding exercise set some of the problems involved multiplying by a power of ten (that is, 10, 100, 1,000, and so on). A short method of multiplying by a power of ten is discussed here.

Notice that when we write powers of ten using exponents the exponent gives the number of zeros. For instance,

$$10^1 = 10 \longleftarrow \text{One zero}$$
$$10^2 = (10)(10) = 100 \longleftarrow \text{Two zeros}$$
$$10^3 = (10)(10)(10) = 1{,}000. \longleftarrow \text{Three zeros}$$

If we wish to multiply (39)(10) or $(39)(10^1)$, we note that the product 390 is 39 followed by *one* zero. In multiplying (325)(100) or $(325)(10^2)$ we note that the product 32,500 is 325 followed by *two* zeros. This leads us to the shortcut method of multiplying by a power of ten.

> The product of a whole number and a power of ten is the whole number followed by the number of zeros in the power of ten.

Instead of placing the numbers to be multiplied in columns, we write them horizontally as shown in the following examples.

Example 11 Find the product (24)(10).

Solution The product is 24 followed by one zero.

$$(24)(10) = 24(10^1) = 240$$

10^1 indicates one zero.

Example 12 Find the product (65)(100).

Solution The product is 65 followed by two zeros.

$$(65)(100) = (65)(10^2) = 6{,}500$$

10^2 indicates two zeros.

Example 13 Find the product (214)(10,000).

Solution The product is 214 followed by four zeros.

$$(214)(10{,}000) = (214)(10^4) = 2{,}140{,}000$$

10^4 indicates four zeros.

Exercise 2–1–2

4 Find each product.

1. (12)(10) **2.** (81)(10) **3.** (211)(10) **4.** (601)(10)

5. (32)(100) **6.** (316)(100) **7.** $205(10^2)$ **8.** $1{,}216(10^2)$

9. $132(1{,}000)$ **10.** $204(10^3)$ **11.** $16(10^4)$ **12.** $345(10{,}000)$

13. $503(100{,}000)$ **14.** $62(10^5)$ **15.** $2{,}091(1{,}000)$ **16.** $5{,}117(1{,}000)$

17. $20(10^2)$ **18.** $80(10^3)$ **19.** $610(10^4)$ **20.** $100(10^4)$

2–2 Dividing Whole Numbers

OBJECTIVES

Upon completing this section you should be able to:

1. Understand the relationship between division and multiplication.
2. Divide by a one-digit number using the long division algorithm.
3. Divide by a number consisting of more than one digit.

The last of the four basic operations on whole numbers that we will study is division. Just as we found subtraction defined in terms of addition, we will find that division can be defined in terms of multiplication.

1 Relationship Between Division and Multiplication

Example 1 Suppose that a man sent \$15 to be divided equally among his five grandchildren. How much would each grandchild receive?

Solution The problem is simply, "How much is 15 divided by 5, or $15 \div 5$?" The answer is 3.

But the real question in this simple problem is, "How do we know that $15 \div 5 = 3$?" The answer to this question actually gives us the definition of division, for we know that $15 \div 5 = 3$ because $5(3) = 15$. In other words, we use our multiplication facts to divide.

In example 1, $a = 15$, $b = 5$, and $c = 3$.

> If a, b, and c represent whole numbers and if $a \div b = c$, then it must be true that $a = bc$.

For any number c $(0)(c) = 0$, *not* 15.

$0 = (0)(c)$ is true for *any* number c.

It is important to notice that division by zero is never possible. For instance, suppose there is some number c such that $15 \div 0 = c$. Then by the definition of division it must be true that $15 = (0)(c)$. But there isn't any number whose product with 0 is 15 since any number multiplied by 0 yields a result of 0. We could ask, "What is the result of $0 \div 0$?" If there is some number c such that $0 \div 0 = c$, then by the definition of division we have $0 = (0)(c)$. But we can see that c could be any number since the product of 0 and any number is 0. For these reasons we agree not to allow division by zero.

Since you already know the basic facts of multiplication, it is not necessary to learn another set of basic facts for division.

The question is:

7 times what number is 42?

Example 2 Solve: $42 \div 7 = ?$

Solution We know that $42 \div 7 = 6$ because $7(6) = 42$.

◉ ***SKILL CHECK 1*** Solve: $20 \div 5$

9 times what number is 72?

Example 3 Solve $72 \div 9 = ?$

Solution $72 \div 9 = 8$ because $9(8) = 72$.

◉ ***Skill Check Answer***

1. 4

Other ways of indicating $72 \div 9$ are $\frac{72}{9}$ and $9\overline{)72}$.

In the example $72 \div 9 = 8$, 72 is called the **dividend,** 9 is called the **divisor,** and 8 is called the **quotient.**

Dividend is the number being divided.
Divisor is the number doing the dividing.
Quotient is the answer.

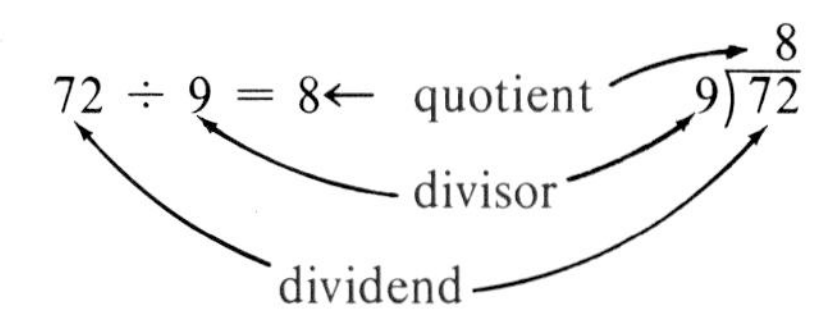

Of course, not all division problems will come out even. For example, if we wish to find $45 \div 7$, we try to find a number that will multiply by 7 to give a product of 45. There is no such whole number, so we find the largest whole number that can be multiplied by 7 to give a product less than 45.

What is the largest whole number that can be multiplied by 7 without the product being larger than 45?

Whatever is left over is indicated as a **remainder,** usually denoted by the letter R.

Example 4 $45 \div 7 = 6$ with a remainder of 3, or $45 \div 7 = 6$ R 3.
To check we find the product of 6 and 7 and then add the remainder 3. That is, $6(7) + 3 = 42 + 3 = 45$.

Example 5 $58 \div 6 = 9$ R 4 because $6(7) + 4 = 54 + 4 = 58$.

⊙ ***SKILL CHECK 2*** Solve: $38 \div 7$

⊙ ***Skill Check Answer***
2. 5 R 3

Exercise 2–2–1

1 The following division problems will require only the basic facts of multiplication.

1. $8 \div 2$ **2.** $6 \div 3$ **3.** $15 \div 3$ **4.** $18 \div 6$ **5.** $21 \div 3$

6. $35 \div 5$ **7.** $49 \div 7$ **8.** $56 \div 7$ **9.** $63 \div 9$ **10.** $28 \div 4$

11. $84 \div 9$ **12.** $42 \div 5$ **13.** $17 \div 2$ **14.** $19 \div 2$ **15.** $28 \div 3$

16. $49 \div 8$ **17.** $73 \div 9$ **18.** $77 \div 8$ **19.** $47 \div 5$ **20.** $33 \div 4$

2 Dividing by a One-Digit Number

Suppose that we did not know any of the multiplication facts. Would it still be possible to find the solution of $15 \div 5$? Just as multiplication could be thought of as repeated addition, it is also possible to think of division as repeated subtraction. In other words, $15 \div 5$ can be thought of as, "How many times can 5 be subtracted from 15?"

Find $12 \div 3$ using repeated subtraction.

$$\begin{array}{rl} 15 & \\ -\ 5 & \text{first time} \\ \hline 10 & \\ -\ 5 & \text{second time} \\ \hline 5 & \\ -\ 5 & \text{third time} \\ \hline 0 & \end{array}$$

We therefore see that 5 can be subtracted from 15 three times.
Thus $15 \div 5 = 3$.

Example 6 Solve: $45 \div 7$ by repeated subtraction.

Find $1{,}000 \div 10$ using repeated subtraction. (Long, isn't it?)

Solution

$$\begin{array}{rl} 45 & \\ -\ \ 7 & \text{first time} \\ \hline 38 & \\ -\ \ 7 & \text{second time} \\ \hline 31 & \\ -\ \ 7 & \text{third time} \\ \hline 24 & \\ -\ \ 7 & \text{fourth time} \\ \hline 17 & \\ -\ \ 7 & \text{fifth time} \\ \hline 10 & \\ -\ \ 7 & \text{sixth time} \\ \hline 3 & \end{array}$$

Since we can no longer subtract 7, we have 3 as a remainder. Therefore, $45 \div 7 = 6 \text{ R } 3$.

Remember, an algorithm is a shortcut method.

The **long-division algorithm** is based on the idea of repeated subtraction. We will show its use by an example.

Example 7 Solve: $45 \div 7$ using the long-division algorithm.

Solution We first write the problem using the long-division symbol.

$$7\overline{)45}$$

We now estimate the largest number of times 7 will divide 45. We call this estimate the **trial quotient.** Example 6 has already given us a clue as to the best estimate, so we will use 6 as our trial quotient.

$$\begin{array}{r} 6 \\ 7\overline{)45} \end{array}$$

We then multiply 6(7) to get 42, which is placed under the 45. We then subtract 42 from 45 and obtain 3, which is the remainder.

$$\begin{array}{r} 6 \\ 7\overline{)45} \\ -42 \\ \hline 3 \end{array}$$

Check by multiplying 6(7) and adding 3.

Thus our quotient is 6 R 3.

Notice that $6(7) = 42$ is subtracted from 45. We have actually subtracted six 7s at once. We have a remainder of 3. If our remainder had been larger than 7, it would have indicated that the trial quotient was too small.

IMPORTANT! The remainder cannot be larger than or equal to the divisor.

$$\begin{array}{r} 5 \\ 7\overline{)45} \\ -35 \\ \hline 10 \end{array}$$

Suppose that we had guessed a trial quotient of 5. Then the remainder would still contain a 7. Our trial quotient is therefore too small.

If we guess too high for a trial quotient, it will not be possible to subtract.

$$\begin{array}{r} 7 \\ 7\overline{)45} \\ -49 \\ \hline \end{array}$$

If we had guessed a trial quotient of 7, we could not subtract 49 from 45. Our trial quotient is too large.

Example 8 Solve: $96 \div 4$

Solution

$$\begin{array}{r} 2 \\ 4\overline{)96} \\ \underline{8} \\ 1 \end{array}$$

First we divide 4 into 9 and multiply the trial quotient 2 by 4 and place the 8 under the 9. Subtracting, we obtain 1.

Since we are dividing 4 into 9, make sure the 2 is placed in the same column as the 9.

$$\begin{array}{r} 24 \\ 4\overline{)96} \\ \underline{8} \\ 16 \\ \underline{16} \end{array}$$

We now bring the 6 down with the 1 and divide 4 into 16. Multiplying the trial quotient 4 by 4 we obtain 16. Subtracting, we have no remainder. Thus $96 \div 4 = 24$.

Check by multiplying 4(24).

WARNING

Many students obtain wrong answers by not placing digits correctly.

A very important part of the long-division algorithm is the proper placement of the trial quotient. Notice in example 8 that the 2 was placed over the 9 when we divided 4 into 9.

Example 9 Solve: $112 \div 8$

Solution

$$\begin{array}{r} 1 \\ 8\overline{)112} \\ \underline{-8} \\ 3 \end{array}$$

Since 8 cannot be divided into 1, we divide 8 into 11. We now multiply 1(8) and subtract.

Notice that the trial quotient 1 is placed over the last digit of 11.

$$\begin{array}{r} 14 \\ 8\overline{)112} \\ \underline{-8} \\ 32 \\ \underline{32} \end{array}$$

Bringing down the 2 gives our next division as $32 \div 8$. Since $4(8) = 32$ we have no remainder. So $112 \div 8 = 14$.

Check: $8(14) =$

◉ ***SKILL CHECK 3*** Solve: $112 \div 7$

Example 10 Find the quotient of $7\overline{)205}$.

Solution

$$\begin{array}{r} 2 \\ 7\overline{)205} \\ \underline{14} \\ 6 \end{array}$$

7 will not divide into 2 so we must divide 7 into 20, giving 2 with a remainder of 6.

$$\begin{array}{r} 29 \\ 7\overline{)205} \\ \underline{14} \\ 65 \\ \underline{63} \\ 2 \end{array}$$

We now bring down the 5 and divide 7 into 65. Our nearest trial quotient is 9. Subtracting, we have a remainder of 2. So the quotient is 29 R 2.

Check!

◉ ***SKILL CHECK 4*** Find the quotient of $4\overline{)101}$.

Example 11 Find the quotient of $6\overline{)2,341}$.

Solution

$$\begin{array}{r} 390 \text{ R } 1 \\ 6\overline{)2,341} \\ \underline{18} \\ 54 \\ \underline{54} \\ 1 \end{array}$$

Note that the first trial quotient 3 is placed over the 3 in 23.

It is very important to note that each digit in the dividend to the right of the placement of the first trial quotient must have a digit over it in the answer. Thus, in example 11 when we cannot divide 6 into 1 we must place a zero over the 1.

Many students would incorrectly have given the quotient as 39 R 1.

◉ ***SKILL CHECK 5*** Find the quotient of $7\overline{)1,753}$.

◉ ***Skill Check Answers***

3. 16
4. 25 R 1
5. 250 R 3

Example 12 Find the quotient of $5\overline{)21,035}$.

Solution

$$\begin{array}{r} 4\,207 \\ 5\overline{)21,035} \\ \underline{20} \\ 10 \\ \underline{10} \\ 3 \\ \underline{0} \\ 35 \\ \underline{35} \end{array}$$

Do you see why the 0 showed up in this answer? Never bring down more than one digit at a time.

◉ ***SKILL CHECK 6*** Find the quotient of $6\overline{)25,812}$.

Example 13 At snack time a kindergarten teacher opened a box of 35 cookies. She said, "I'm going to give each of you the same number of cookies and have the remainder for myself." If she had 8 students, how many cookies did each one receive and how many were left for the teacher?

Solution

$$\begin{array}{r} 4 \text{ R } 3 \\ 8\overline{)35}\phantom{\text{ R } 3} \\ \underline{32}\phantom{\text{ R } 3} \\ 3\phantom{\text{ R } 3} \end{array}$$

Each student had 4 cookies and the teacher had 3.

Remember that we must always answer the question asked in a word problem. Note that this problem asks two questions and both must be answered.

◉ ***SKILL CHECK 7*** If a box of 50 cookies is shared equally among 8 people, how many does each receive and how many are left over?

◉ ***Skill Check Answers***

6. 4,302
7. 6, 2 left over

Exercise 2–2–2

2 Find each quotient.

1. $3\overline{)36}$ **2.** $2\overline{)28}$ **3.** $2\overline{)46}$ **4.** $3\overline{)69}$

5. $3\overline{)45}$ **6.** $4\overline{)96}$ **7.** $7\overline{)91}$ **8.** $6\overline{)84}$

9. $5\overline{)79}$ **10.** $4\overline{)61}$ **11.** $6\overline{)215}$ **12.** $5\overline{)321}$

13. $2\overline{)509}$ **14.** $3\overline{)701}$ **15.** $7\overline{)450}$ **16.** $6\overline{)840}$

17. $4\overline{)1,682}$ **18.** $2\overline{)1,261}$ **19.** $6\overline{)1,816}$ **20.** $4\overline{)2,805}$

21. $5\overline{)15,535}$ **22.** $6\overline{)25,230}$ **23.** $4\overline{)20,206}$ **24.** $7\overline{)35,276}$

• Concept Enrichment

25. If $152 is to be divided equally among four people, how much does each person receive?

26. An eight-ounce container of dairy cream costs 48 cents. What is the price per ounce?

27. In a livestock yard 464 head of cattle are to be equally divided into eight pens. How many cattle should be placed in each pen?

28. If one pie serves 8 people, how many pies are needed to serve 256 people?

29. If you owe $336 and agree to pay it in six equal monthly payments, how much do you pay each month?

30. If an RV van gets 9 miles to a gallon of fuel, how many gallons are needed to drive 207 miles?

31. A box of 32 chocolates is equally divided among five people. How many chocolates does each person receive and how many are left over?

32. How many 8-ounce glasses of milk can be obtained from 410 ounces? How many ounces are left over?

33. You drove 182 miles and used seven gallons of gasoline. How many miles per gallon did you get?

34. Four neighbors decide to split the cost equally for a satellite TV system. If the total cost of the system is $6,048, how much is each neighbor's share?

35. Forty hamburgers are divided equally among nine people. How many hamburgers does each person receive and how many are left over?

36. Three fishermen caught 74 fish. They divide them equally. How many fish are left over?

3 Dividing by a Number Having More Than One Digit

When dividing by a number consisting of more than one digit, the long-division process does not change. There is an increase in difficulty, however, in finding the proper trial quotient.

Example 14 Find the quotient of $22\overline{)653}$.

Solution Since 22 will not divide into 6 we must first divide 65 by 22.

This means our trial quotient is too large.

$$\begin{array}{r} 3 \\ 22\overline{)653} \\ \underline{66} \end{array}$$

If we first guess at 3 for the trial quotient, we find that we cannot subtract.

We then try 2 as our trial quotient.

Don't hesitate to try trial quotients. Only practice will enable you to do well at it.

$$\begin{array}{r} 2 \\ 22\overline{)653} \\ \underline{44} \\ 21 \end{array}$$

In this case we can subtract and the remainder (21) is smaller than the divisor (22).

$$\begin{array}{r} 2 \\ 22\overline{)653} \\ \underline{44} \\ 213 \end{array}$$

We bring down the 3 and must now make a guess as to 213 divided by 22.

37 is larger than 22. The remainder must be smaller than the divisor.

$$\begin{array}{r} 28 \\ 22\overline{)653} \\ \underline{44} \\ 213 \\ \underline{176} \\ 37 \end{array}$$

If we guess 8 as the trial quotient, we find our remainder is too large.

This gives us the correct result.

$$\begin{array}{r} 29 \text{ R } 15 \\ 22\overline{)653}\phantom{\text{ R } 15} \\ \underline{44}\phantom{\text{ R } 15} \\ 213\phantom{\text{ R } 15} \\ \underline{198}\phantom{\text{ R } 15} \\ 15\phantom{\text{ R } 15} \end{array}$$

If we try 9, we find the remainder to be 15, which is smaller than the divisor.

⊙ ***SKILL CHECK 8*** Find the quotient of $22\overline{)756}$.

We can also use estimation in division to determine if our answer is of the right magnitude.

Example 15 Find the quotient of $514\overline{)29{,}298}$.

Solution We must first find, from the left, a number into which 514 can divide. The first such number is 2,929 and our trial quotient is 5.

Note that the 5 is placed over the 9.

$$\begin{array}{r} 5 \\ 514\overline{)29{,}298} \\ \underline{25\ 70} \\ 3\ 59 \end{array}$$

The remainder is less than 514 so we bring down the next digit (8).

⊙ ***Skill Check Answer***

8. 34 R 8

```
      57
514)29,298
    25 70
     3 598
     3 598
```

Our next trial quotient is 7 and we find no remainder, so our quotient is 57.

To estimate the answer we would round 514 to 500 and 29,298 to 30,000. Then dividing, we have

```
        60
500)30,000
    30 00
       00
       00
        0.
```

We thus expect the answer to be somewhere near 60.

◉ ***SKILL CHECK 9*** Find the quotient of $316\overline{)12,324}$.

Example 16 Find the quotient of $44\overline{)88,132}$.

Solution

```
     2   ←——— Our first trial quotient is 2.
44)88,132
   88
     1   ←——— Bring down the next digit 1.
```

```
     2 0   ←——— 44 will not divide 1 so we place a zero in
44)88,132       the answer.
   88
     1
     0
     13   ←——— Bring down the next digit 3.
```

```
     2 00   ←——— 44 will not divide 13 so we place a zero in
44)88,132        the answer.
   88
     1
     0
     13
      0
     132   ←——— Bring down the next digit 2.
```

```
     2 003   ←——— Now 44 will divide 132 and using a trial
44)88,132         quotient of 3 gives us no remainder.
   88
     1
     0
     13
      0
     132
     132
       0
```

Be careful to note again that each digit in the dividend to the right of the first trial divisor must have a digit above it in the answer.

◉ ***SKILL CHECK 10*** Find the quotient of $43\overline{)86,301}$.

◉ ***Skill Check Answers***

9. 39
10. 2,007

Example 17 A vat contains 7,500 ounces of a soft drink. How many 12 ounce cans can be filled from this vat?

Use estimation to determine the approximate size of the answer.

Solution We should think this way. Each can will use 12 ounces of the soft drink so we need to determine how many times 12 will divide into 7,500.

$$\begin{array}{r} 625 \\ 12\overline{)7{,}500} \\ \underline{72} \\ 30 \\ \underline{24} \\ 60 \\ \underline{60} \\ 0 \end{array}$$

So we can obtain 625 cans of soft drink from the 7,500 ounces.

◉ SKILL CHECK 11 How many 12 ounce cans of soft drink can be filled from a vat containing 9,636 ounces?

◉ Skill Check Answer

11. 803 cans

Exercise 2–2–3

Find each quotient. Use estimation to determine the reasonableness of the size of your answer.

1. $13\overline{)468}$ **2.** $12\overline{)648}$ **3.** $11\overline{)693}$ **4.** $17\overline{)799}$

5. $32\overline{)1{,}358}$ **6.** $41\overline{)960}$ **7.** $47\overline{)14{,}310}$ **8.** $39\overline{)23{,}477}$

9. $17\overline{)5{,}780}$ **10.** $21\overline{)14{,}490}$ **11.** $10\overline{)340}$ **12.** $10\overline{)690}$

13. $20\overline{)11{,}260}$ **14.** $30\overline{)21{,}660}$ **15.** $28\overline{)3{,}226}$ **16.** $34\overline{)7{,}525}$

17. $53\overline{)21{,}369}$ **18.** $61\overline{)12{,}525}$ **19.** $71\overline{)156{,}413}$ **20.** $68\overline{)137{,}700}$

21. $511\overline{)27{,}083}$

22. $314\overline{)20{,}096}$

23. $256\overline{)21{,}339}$

24. $443\overline{)33{,}769}$

25. $100\overline{)28{,}500}$

26. $100\overline{)58{,}200}$

27. $43\overline{)129{,}172}$

28. $52\overline{)104{,}364}$

29. $33\overline{)66{,}185}$

30. $47\overline{)94{,}218}$

31. Martha earns $36,144 in one year. How much does she earn each month?

32. Twenty-two homeowners will pay equal shares of a tax bill for $4,510. How much does each person owe?

33. A case of soda contains 24 cans. How many cases can be filled from 1,350 cans? How many cans are left over?

34. How many 32-ounce bottles can be filled from 7,720 ounces? How many ounces are left over?

35. Total of the car payments for three years is $6,228. How much is each monthly payment?

36. A machine shop can make 25 auto parts per day. How many days will it take to make 1,300 parts?

37. How many gallons of gasoline are needed to travel 1,716 miles if the car gets 22 miles per gallon?

38. If the tax on a home is $1,260 per year, how much is it per month?

39. A tennis club purchases tennis balls by the gross (144 tennis balls in a gross). If the club needs 3,000 tennis balls, how many gross should they buy? (*Hint:* It is all right if there are a few tennis balls left over.)

40. If one package of weiners makes twelve hot dogs, how many packages are needed for 260 hot dogs?

41. If a car traveled 768 miles and used 32 gallons of gasoline, how many miles to the gallon did the car get?

42. If a person earns $475 per week, how many weeks will it take that person to earn $16,000?

• Concept Enrichment

43. Divide 782 by 63 using the long division algorithm. Check by using repeated subtraction.

44. Why can't we divide by 0?

2–3 Grouping Symbols and Order of Operations

1 Grouping Symbols

We will often find parentheses ( ), brackets [ ], and braces { } used in number expressions. All three of these are used as grouping symbols. An expression enclosed in grouping symbols is treated as if it were a single number.

Example 1 $12 - (3 + 4)$ indicates that the sum $(3 + 4)$ is subtracted from 12.

$$\begin{aligned} 12 - (3 + 4) &= 12 - 7 \\ &= 5 \end{aligned}$$

◉ ***SKILL CHECK 1*** What does $8 - (2 + 3)$ indicate?

Example 2 $7(5 + 3)$ indicates that the sum $(5 + 3)$ is multiplied by 7.

$$\begin{aligned} 7(5 + 3) &= 7(8) \\ &= 56 \end{aligned}$$

◉ ***SKILL CHECK 2*** What does $3(2 + 5)$ indicate?

In general, operations within the grouping symbols are performed first.

Brackets and braces can be used as well as parentheses. $12 - [3 + 4]$ and $12 - \{3 + 4\}$ mean the same as $12 - (3 + 4)$. Parentheses are most commonly used when no other grouping symbols are involved.

OBJECTIVES

Upon completing this section you should be able to:

1. Perform operations in the correct order indicated by grouping symbols.
2. Perform operations in a specific order when grouping symbols are not used.
3. Evaluate expressions involving more than one set of grouping symbols.

◉ ***Skill Check Answers***

1. the sum $(2 + 3)$ subtracted from 8
2. the sum $(2 + 5)$ multiplied by 3

Exercise 2–3–1

1 Find the value of each expression.

1. $8 - (5 + 2)$ **2.** $14 - (8 + 3)$ **3.** $9 - (3 - 1)$

4. $20 - (13 - 9)$ **5.** $(8 - 2) + 4$ **6.** $(18 - 12) + 6$

7. $(10 - 7) - 1$ **8.** $(15 - 8) - 5$ **9.** $18 + (6 - 3)$

10. $21 - (11 + 4)$ **11.** $(15 + 4) - 16$ **12.** $34 - (16 + 15)$

13. $(15 - 6) + 29$ **14.** $58 - (24 + 16)$ **15.** $63 - (91 - 52)$

16. $49 - (38 - 19)$ **17.** $(12 - 5) + (14 - 6)$ **18.** $(16 - 7) - (14 - 6)$

19. $3(6 + 8)$

20. $5(4 + 7)$

21. $4(11 - 5)$

22. $7(24 - 19)$

23. $(13 - 6)(4 + 10)$

24. $(5 - 2)(5 + 2)$

25. $(18 - 2) - (10 + 5)$

26. $(10 + 14) - (18 + 6)$

27. $(23 - 15) - (14 - 9)$

28. $(34 - 28) + (72 - 66)$

29. $(14 + 3) + (43 - 18)$

30. $(31 - 8) - (7 + 4)$

2 Order of Operations

We stated earlier in this section that operations within grouping symbols must be performed first. We will now establish rules that apply when no grouping symbols occur in an expression.

An expression may have a different value if grouping symbols are not used. We saw that

$$12 - (3 + 4) = 5,$$

but if parentheses are not used, the value of the expression becomes

$$12 - 3 + 4 = 13.$$

This is due to the following rule.

Evaluate the following:

$18 - (4 + 6) =$

$18 - 4 + 6 =$

> If an expression without grouping symbols contains only additions and subtractions, these operations are performed in order from left to right.

Example 3 $10 + 5 - 3 + 8 = 15 - 3 + 8$
$= 12 + 8$
$= 20$

⦿ ***SKILL CHECK 3*** Evaluate: $11 + 3 - 2 + 4$

Example 4 $6 - 4 + 11 - 4 = 2 + 11 - 4$
$= 13 - 4$
$= 9$

⦿ ***SKILL CHECK 4*** Evaluate: $8 - 3 + 10 - 4$

Example 5 $5 + 9 - 7 - 3 + 4 - 8 = 14 - 7 - 3 + 4 - 8$
$= 7 - 3 + 4 - 8$
$= 4 + 4 - 8$
$= 8 - 8$
$= 0$

⦿ ***SKILL CHECK 5*** Evaluate: $7 + 9 - 6 - 2 + 3 - 11$

⦿ ***Skill Check Answers***

3. 16
4. 11
5. 0

If an expression contains operations other than just addition and subtraction, we use the following rule.

> *If no grouping symbols occur in an expression to be evaluated,* the operations follow this order: First exponents; next multiplication and division as they occur left to right; then addition and subtraction as they occur left to right.

WARNING

Until this process is more familiar, it is best to do one operation per step.

Example 6 Evaluate: $2 \times 5^2 + 3$

Solution We must first evaluate 5^2, then multiply by 2, and then add 3. Recall from chapter 1 that an exponent of a number indicates how many times that number is used as a factor. In this case $5^2 = (5)(5) = 25$.

Thus
$$\begin{aligned} 2 \times 5^2 + 3 &= 2 \times 25 + 3 \\ &= 50 + 3 \\ &= 53. \end{aligned}$$

First exponents; then multiplication; then addition.

⊙ *SKILL CHECK 6* Evaluate: $3 \times 2^2 + 1$

Example 7 Evaluate: $6^2 \div 2 \times 3 + 6$

Solution
$$\begin{aligned} 6^2 \div 2 \times 3 + 6 &= 36 \div 2 \times 3 + 6 \\ &= 18 \times 3 + 6 \\ &= 54 + 6 \\ &= 60 \end{aligned}$$

Notice that division is performed before multiplication because it is to the left of multiplication.

⊙ *SKILL CHECK 7* Evaluate: $6^2 \div 3 \times 4 + 2$

Example 8 Evaluate: $5 + 3 \times 10 \div 2 + 1$

Solution
$$\begin{aligned} 5 + 3 \times 10 \div 2 + 1 &= 5 + 30 \div 2 + 1 \\ &= 5 + 15 + 1 \\ &= 21 \end{aligned}$$

⊙ *SKILL CHECK 8* Evaluate: $4 + 3 \times 8 \div 4 + 2$

⊙ *Skill Check Answers*

6. 13
7. 50
8. 12

Exercise 2–3–2

2 Evaluate each expression.

1. $5 + 3 + 7$

2. $8 - 4 + 6$

3. $5 + 9 - 14$

4. $3 + 7 - 8$

5. $12 - 8 + 2 - 4$

6. $11 + 4 - 6 - 3$

7. $4 + 19 - 8 + 1$

8. $13 - 10 - 1 + 5$

9. $15 + 37 - 15 + 3$

10. $7 - 5 + 3 - 5$

11. $4 \times 3 + 6$

12. $8 \times 3 + 5$

13. $12 + 6 \times 2^2$

14. $15 - 2^3 \times 4$

15. $7 + 3 \times 5 - 2^3$

16. $3 + 4 \times 3 + 2$

17. $15 + 10 \div 5 - 3$

18. $6 + 21 \div 3 - 5$

19. $4 \div 2 + 8 \times 3^2$

20. $40 \div 2^2 - 5 \times 4$

21. $3 \times 5 - 9 \div 3 + 5$

22. $12 \div 2 - 2 \times 3 + 1$

23. $6 + 4 \times 5 - 3$

24. $6 + 4 \times (5 - 3)$

25. $(6 + 4) \times 5 - 3$

26. $(6 + 4) \times (5 - 3)$

27. $3 + 2^3 \times 8 \div 4 - 5$

28. $4 + 27 \div 3^2 \times 2 - 6$

29. $6 \times 5 \div 3 + 12$

30. $24 \div 6 \times 2 - 8$

• Concept Enrichment

31. Find the error in this solution.

$$\begin{aligned} 3 \times 6 - 4 \div 2 - 1 &= 3 \times 2 \div 2 - 1 \\ &= 6 \div 2 - 1 \\ &= 3 - 1 \\ &= 2 \end{aligned}$$

32. Find the error in this solution.

$$\begin{aligned} 2 \times 8 + 12 \div 6 - 2 &= 16 + 12 \div 6 - 2 \\ &= 16 + 12 \div 4 \\ &= 16 + 3 \\ &= 19 \end{aligned}$$

3 More Than One Set of Grouping Symbols

Remember that the different symbols are all used for the same purpose—to group numbers.

Sometimes more than one set of grouping symbols is needed in an expression. When this occurs we use brackets or braces along with parentheses for clarification. For instance, $5 + [7 - (2 + 1)]$ could be written using only parentheses, but $5 + (7 - (2 + 1))$ is not as clear at first glance. Therefore, we alternate the symbols to avoid confusion. To evaluate such an expression we use the following rule.

Again, make sure you do only *one* operation at a time. This is very important in order to avoid errors.

> When simplifying an expression containing grouping symbols within grouping symbols, remove the *innermost* set of symbols first.

Example 9 To evaluate $5 + [7 - (2 + 1)]$ we simplify the innermost set of symbols, namely $(2 + 1)$. Writing $(2 + 1)$ as 3 we now obtain

$$\begin{aligned} 5 + [7 - (2 + 1)] &= 5 + [7 - 3] \\ &= 5 + 4 \\ &= 9. \end{aligned}$$

⦿ ***SKILL CHECK 9*** Evaluate: $4 + [8 - (3 + 2)]$

Example 10 $$\begin{aligned} [11 - (9 - 5)] - 4 &= [11 - 4] - 4 \\ &= 7 - 4 \\ &= 3 \end{aligned}$$

⦿ ***SKILL CHECK 10*** Evaluate: $[10 - (6 - 4)] - 2$

Example 11 $$\begin{aligned} 18 - \{5 + [4(5 - 2) - 7]\} &= 18 - \{5 + [4(3) - 7]\} \\ &= 18 - \{5 + [12 - 7]\} \\ &= 18 - \{5 + 5\} \\ &= 18 - 10 \\ &= 8 \end{aligned}$$

Notice that we start with $(5 - 2)$ since it is the innermost set of symbols.

⦿ ***SKILL CHECK 11*** Evaluate: $20 - \{6 + [3(7 - 3) - 2]\}$

⦿ ***Skill Check Answers***

9. 7
10. 6
11. 4

Exercise 2–3–3

3 Evaluate each expression.

1. $4 + [9 - (3 + 4)]$

2. $5 + [10 - (4 + 3)]$

3. $14 - [10 - (5 - 1)]$

4. $10 - [14 - (6 - 1)]$

5. $[(8 + 5) - 6] - 3$

6. $[(10 + 4) - 8] - 6$

7. $4[8 + (11 - 3)]$

8. $3[6 + (9 - 5)]$

9. $2 + [9 - (3 - 1)]$

10. $16 + [13 - (6 - 2)]$

11. $10 - [(6 - 4) - 1]$

12. $23 - [(12 - 3) - 5]$

13. $5 + 3[16 - (11 + 3)]$

14. $10 + 2[24 + (8 - 2)]$

15. $7 + [15 + 3(4 - 1)]$

16. $18 - [10 + 2(8 - 5)]$

17. $[2(6 - 3) + 4] - 5$

18. $[3(7 - 3) + 2] - 9$

19. $6 + 2[8 + 3(6 - 3)]$

20. $19 - 3[18 - 4(7 - 4)]$

21. $[8 - (5 + 2)] + 3(15 - 6)$

22. $2(19 + 4) - [4(8 + 2) + 3]$

23. $3[4 + (3 - 1)] + 5[12 - (8 + 2)]$

24. $6[4(10 - 7) - 3] - 4[16 - 2(3 + 1)]$

25. $2 + [10 - (3 + 6) + 5]$

26. $8 - [13 - (2 + 4) - 6]$

27. $2\{5 + 3[14 - (6 + 2)]\}$

28. $4\{[2(6 - 3) + 4] - 10\}$

29. $16 - \{15 - [2(5 + 2) - 4] + 6\}$

30. $2[4(7 - 3) + 1] - \{12 + 3[14 - 2(8 - 5)] - 5\}$

2–4 Evaluating Algebraic Expressions

OBJECTIVES

Upon completing this section you should be able to:

1. Evaluate algebraic expressions by substituting numbers for letters.
2. Apply this process to evaluating formulas.

1 Substituting Numbers for Letters

The **principle of substitution** states that any quantity may be substituted for its equal in any process. This principle is used extensively in algebra and we will use it here to evaluate algebraic expressions. In this process we substitute numbers for letters and find a numerical value.

Example 1 Evaluate $x + 3$ if $x = 5$.

Solution We substitute 5 for x in the expression obtaining

$$5 + 3 = 8.$$

⊙ SKILL CHECK 1 Evaluate $x + 3$ if $x = 2$.

Example 2 Evaluate $4a - 1$ if $a = 3$.

Solution We substitute 3 for a in the expression obtaining

$$\begin{aligned} 4(3) - 1 &= 12 - 1 \\ &= 11. \end{aligned}$$

⊙ SKILL CHECK 2 Evaluate $4a - 1$ if $a = 5$.

When substituting a number for a letter, it is a good practice to enclose the number in parentheses so that the proper operation will be performed.

Example 3 Evaluate $x^2 + 2x$ if $x = 3$.

Solution We substitute 3 for x in the expression obtaining

$$(3)^2 + 2(3).$$

Evaluating this, we obtain

$$\begin{aligned} (3)^2 + 2(3) &= 9 + 6 \\ &= 15. \end{aligned}$$

Remember that in an algebraic expression the letters are merely holding a place for various numbers that may be assigned to them. For that reason, these letters are sometimes called *place holders* or *variables*.

⊙ SKILL CHECK 3 Evaluate $x^2 + 2x$ if $x = 4$.

In section 1–6 we introduced the words base, coefficient, and exponent and stated that only the base is affected by the exponent. Notice how grouping symbols discussed in the previous section can determine the base in an expression.

Example 4 Evaluate $3x^2$ when $x = 2$.

Solution

$$\begin{aligned} 3x^2 &= 3(x)(x) \\ &= 3(2)(2) \\ &= 3(4) \\ &= 12 \end{aligned}$$

Notice that only the literal factor x is affected by the exponent 2.

⊙ SKILL CHECK 4 Evaluate $3x^2$ when $x = 5$.

Example 5 Evaluate $(3x)^2$ when $x = 2$.

Solution

$$\begin{aligned} (3x)^2 &= (3x)(3x) \\ &= [3(2)][3(2)] \\ &= (6)(6) \\ &= 36 \end{aligned}$$

Here $3x$ is the base, thus $3x$ is to be multiplied by itself.

⊙ SKILL CHECK 5 Evaluate $(3x)^2$ when $x = 5$.

Example 6 Evaluate $3a^3 - 2a^2 + a$ if $a = 2$.

Solution We substitute 2 for a in the expression obtaining

$$\begin{aligned} 3(2)^3 - 2(2)^2 + 2 &= 3(8) - 2(4) + 2 \\ &= 24 - 8 + 2 \\ &= 18. \end{aligned}$$

⊙ SKILL CHECK 6 Evaluate $3a^3 - 2a^2 + a$ if $a = 3$.

Example 7 Evaluate $3x + 7y$ if $x = 4$ and $y = 7$.

Solution

$$\begin{aligned} 3x + 7y &= 3(4) + 7(7) \\ &= 12 + 49 \\ &= 61 \end{aligned}$$

Here we must make two different substitutions.

⊙ SKILL CHECK 7 Evaluate $3x + 7y$ if $x = 2$ and $y = 3$.

⊙ Skill Check Answers

1. 5
2. 19
3. 24
4. 75
5. 225
6. 66
7. 27

Example 8 Evaluate $3x^2 - 5y$ if $x = 5$ and $y = 2$.

Solution

$$\begin{aligned} 3x^2 - 5y &= 3(5)^2 - 5(2) \\ &= 3(25) - 5(2) \\ &= 75 - 10 \\ &= 65 \end{aligned}$$

⊙ ***SKILL CHECK 8*** Evaluate $3x^2 - 5y$ if $x = 3$ and $y = 2$.

Example 9 Evaluate $a^2 + b^2$ if $a = 3$ and $b = 2$.

Solution

$$\begin{aligned} a^2 + b^2 &= (3)^2 + (2)^2 \\ &= 9 + 4 \\ &= 13 \end{aligned}$$

⊙ ***SKILL CHECK 9*** Evaluate $a^2 + b^2$ if $a = 4$ and $b = 3$.

Example 10 Evaluate $(a + b)^2$ if $a = 3$ and $b = 2$.

Solution

$$\begin{aligned} (a + b)^2 &= (3 + 2)^2 \\ &= (5)^2 \\ &= 25 \end{aligned}$$

⊙ ***SKILL CHECK 10*** Evaluate $(a + b)^2$ if $a = 4$ and $b = 3$.

If the same letter appears in more than one term of an expression it must have the same value each time it occurs. For instance, in the expression $2x + 3xy$ the x could not have one value in the first term and another in the second term.

Example 11 Evaluate $5x^2y - 3y - z^2$ if $x = 3$, $y = 2$, and $z = 5$.

Notice that the number 2 was substituted for y in both places it occurred in the expression.

Solution

$$\begin{aligned} 5x^2y - 3y - z^2 &= 5(3)^2(2) - 3(2) - (5)^2 \\ &= 5(3)(3)(2) - 3(2) - (5)(5) \\ &= 90 - 6 - 25 \\ &= 59 \end{aligned}$$

⊙ ***SKILL CHECK 11*** Evaluate $5x^2y - 3y - z^2$ if $x = 2$, $y = 3$, and $z = 4$.

Note: "Evaluate" will always mean that you are to obtain a *numerical value.*

2 Formulas

One of the most common uses of evaluating algebraic expressions is in working with formulas.

Example 12 The perimeter P (distance around) of a rectangle is found by using the formula $P = 2\ell + 2w$, where ℓ represents the length and w represents the width.

To find the perimeter of a rectangle we need to add up all four sides, but since the opposite sides are equal, we only need to double the length and double the width and then find their sum.

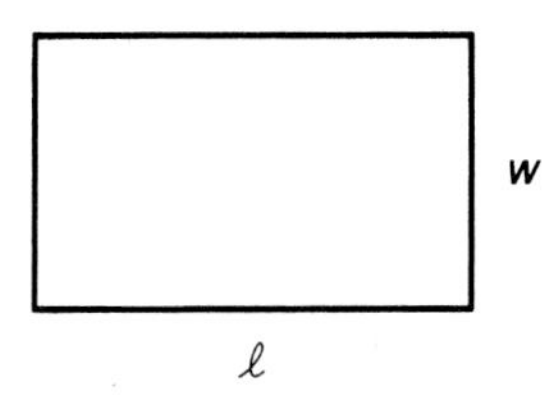

If the length of a rectangle is 10 and the width is 6, we may find the perimeter by substituting 10 for ℓ and 6 for w.

Solution

$$P = 2(10) + 2(6)$$

(10: value for length; 6: value for width)

⊙ ***Skill Check Answers***

8. 17
9. 25
10. 49
11. 35

Then evaluating, we obtain

$$\begin{aligned} P &= 2(10) + 2(6) \\ &= 20 + 12 \\ &= 32. \end{aligned}$$

◉ SKILL CHECK 12 Find the perimeter of a rectangle if the length is 12 and the width is 5.

◉ *Skill Check Answer*

12. 34

Exercise 2–4–1

1 Evaluate.

1. x^4 if $x = 2$
2. a^5 if $a = 3$
3. $3a^2$ if $a = 5$
4. $8x^3$ if $x = 4$
5. $x^2 + x$ if $x = 7$
6. $a^4 - 3a$ if $a = 2$
7. $2a^3 - 3a$ if $a = 6$
8. $4x^5 + 3x$ if $x = 3$
9. $3x^4 + x^3 - x^2$ if $x = 3$
10. $5x^5 - 2x^3 + x^2$ if $x = 2$
11. $x + y$ if $x = 3$ and $y = 5$
12. $2a - b$ if $a = 9$ and $b = 10$
13. $5x + 3y$ if $x = 6$ and $y = 4$
14. $5x^2y$ if $x = 3$ and $y = 2$
15. $4a^3b^2$ if $a = 2$ and $b = 3$
16. $a^3 - a^2$ if $a = 5$
17. $3x^2 - x$ if $x = 4$
18. $x^3 - y^3$ if $x = 5$ and $y = 4$
19. $3a^4 + b^2$ if $a = 2$ and $b = 5$
20. $2x^2yz^3$ if $x = 3$, $y = 7$, and $z = 2$
21. $3a^5b^2 - 4ab^3$ if $a = 2$ and $b = 3$
22. $x^2 + 3x - 8$ if $x = 5$
23. $4a^2 - 2a + 5$ if $a = 3$
24. $4a^2bc - 5a + 4bc$ if $a = 5$, $b = 3$, and $c = 6$
25. $6x^3y + 2z - 5xy^2$ if $x = 5$, $y = 2$, and $z = 1$
26. $10a^4b^8c^5d^9$ if $a = 4$, $b = 2$, $c = 0$, and $d = 1$

27. $5x^3y + xy^2z - 10z$ if $x = 4$, $y = 5$, and $z = 10$

28. $6a^2b^3 - 10ab + b^2$ if $a = 5$ and $b = 3$

29. $4x^4y^5z + 11xz - 4y$ if $x = 3$, $y = 0$, and $z = 2$

30. $5x^3y + 7x^2z - 3yz$ if $x = 2$, $y = 3$, and $z = 4$

31. The perimeter of a rectangle is given by $P = 2\ell + 2w$. Find P when $\ell = 12$ and $w = 7$.

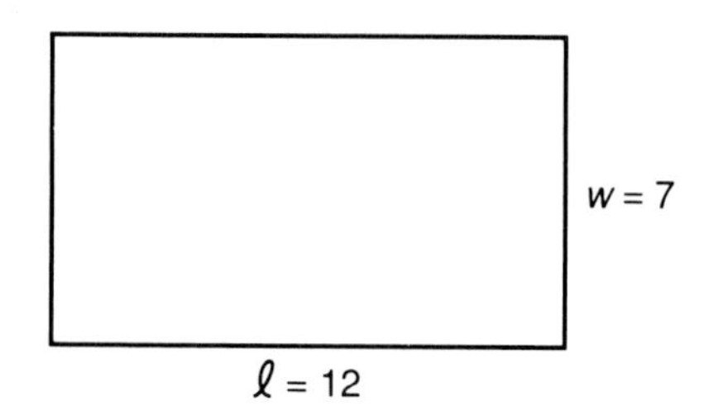

32. The area A of a rectangle is given by $A = bh$, where b represents the length of the base and h represents the height. Find A when $b = 9$ and $h = 3$.

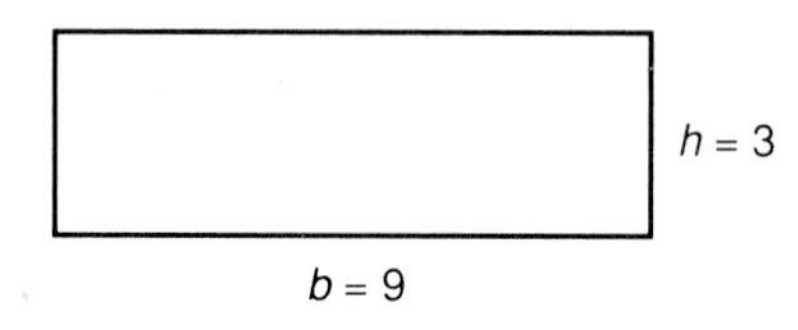

33. The perimeter P of a square is given by $P = 4s$, where s represents the length of one side. Find P when $s = 8$.

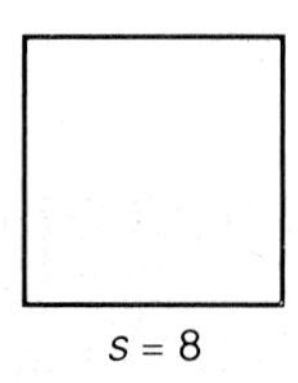

34. The area A of a square is given by $A = s^2$. Find A when $s = 7$.

35. The perimeter P of a triangle is given by the formula $P = a + b + c$, where a, b, and c represent the lengths of the three sides. Find P when $a = 14$, $b = 10$, and $c = 17$.

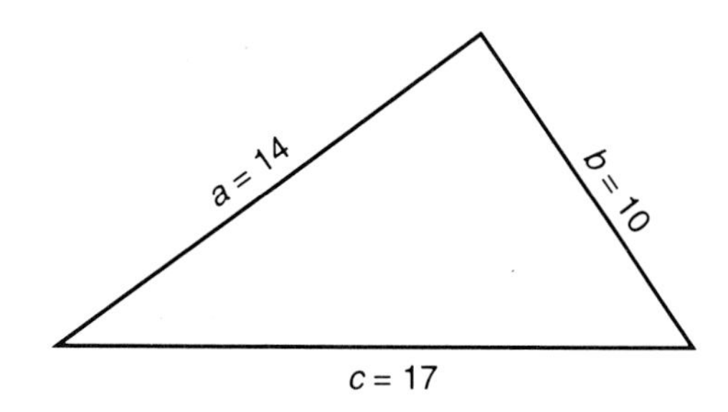

36. A distance formula from physics is $d = rt$. Find d when $r = 55$ and $t = 4$.

37. A force formula from physics is $F = ma$. Find F when $m = 120$ and $a = 32$.

38. The volume V of a rectangular solid is given by $V = xyz$. Find V when $x = 3$, $y = 2$, and $z = 4$.

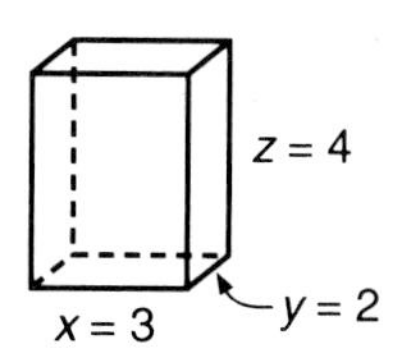

39. The volume V of a cube is given by $V = s^3$. Find V when $s = 5$.

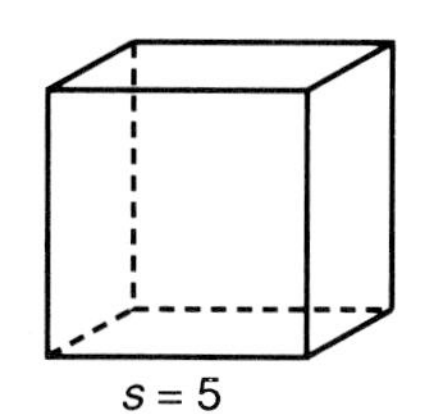

40. A distance formula for a free-falling object is given by $s = gt^2 \div 2$. Find s when $g = 32$ and $t = 5$.

• Concept Enrichment

41. The **Universal Product Code** (**UPC**) appears on many items purchased in supermarkets and other stores. The code consists of twelve digits that are written and also bar coded on the product. The first eleven digits identify the country, the manufacturer, and the product. The last digit is called the **check digit.** Suppose the twelve digits are represented by the first twelve letters of the alphabet $a - l$. The twelfth digit l (check digit) must be selected so that the value of the expression

$$3a + b + 3c + d + 3e + f + 3g + h + 3i + j + 3k + l$$

is divisible by 10 with no remainder.
Verify that this is true by substituting the UPC for the following **Hormel** product in the above expression.

In this case $a = 0$, $b = 3$, $c = 7$, $d = 6$, $e = 0$, $f = 0$, $g = 0$, $h = 7$, $i = 0$, $j = 6$, $k = 0$, $l = 7$.

42. For many products the check digit is not written although it is bar coded. Determine what the check digit is for the following product.

2–5 The Multiplication Law of Exponents

OBJECTIVE

Upon completing this section you should be able to:

1. Correctly apply the multiplication law of exponents.

In section 1–6 we introduced the concept of exponent and defined the words base and coefficient. In the previous section we substituted number values in expressions containing exponents. We now wish to establish a very important law of exponents. This law is derived directly from the definition of a whole number exponent.

1 Multiplication Law of Exponents

Example 1 Multiply: $(x^3)(x^5)$

Solution From the definition we have

$$x^3 = (x)(x)(x) \text{ and } x^5 = (x)(x)(x)(x)(x)$$
$$\text{So } (x^3)(x^5) = [(x)(x)(x)][(x)(x)(x)(x)(x)]$$
$$= (x)(x)(x)(x)(x)(x)(x)(x)$$
$$= x^8.$$

There are eight x's being multiplied together. In other words, x is a factor eight times.

If we generalize example 1 to the product of $x^a x^b$, we have

$$x^a = \underbrace{(x)(x)(x)\ldots(x)}_{a \text{ factors}}$$
$$x^b = \underbrace{(x)(x)(x)\ldots(x)}_{b \text{ factors}}$$

and

$$x^a x^b = [\underbrace{(x)(x)(x)\ldots(x)}_{a \text{ factors}}][\underbrace{(x)(x)(x)\ldots(x)}_{b \text{ factors}}].$$

So

$$x^a x^b = x^{a+b}.$$

To multiply factors with the same base add the exponents. The base remains unchanged.

Multiplication Law of Exponents

$$x^a x^b = x^{a+b}$$

For any rule, law, or formula we must always be very careful to meet the conditions required before attempting to apply it. Note in the above law that the base is the same in both factors. This law applies only when this condition is met.

Example 2 $x^4x^2 = x^6$

Example 3 $y^3y^4 = y^7$

Example 4 $a^2a^5 = a^7$

◉ ***SKILL CHECK 1*** Multiply: a^3a^6

These factors do not have the same base.

Example 5 $x^3y^4 = ?$ (Rule does NOT apply.)

An exponent of 1 is not usually written. When we write x the exponent is assumed: $x = x^1$. This fact is necessary to apply the laws of exponents.

$(x^2)(x) = (x^2)(x^1) = x^3$

Example 6 $(x^2)(x) = x^3$

◉ ***SKILL CHECK 2*** Multiply: $(x^4)(x)$

Example 7 $(x^4)(x)(x^2) = x^7$

Example 8 $(y^3)(y^2)(y) = y^6$

◉ ***SKILL CHECK 3*** Multiply: $(y^4)(y^3)(y)$

◉ ***Skill Check Answers***

1. a^9
2. x^5
3. y^8

If an expression contains the product of different bases, we apply the law to those bases that are alike.

Example 9 $(x^4)(y^2)(x^3)(y^4) = (x^4)(x^3)(y^2)(y^4)$
$= x^7y^6$

These factors have the same base.
$(x^4)(y^2)(x^3)(y^4)$
These factors have the same base.

Example 10 $(x^3)(y)(x^3) = (x^3)(x^3)(y)$
$= x^6y$

Example 11 $(x^2)(y^2)(z^2)(x^4)(y) = (x^2)(x^4)(y^2)(y)(z^2)$
$= x^6y^3z^2$

◉ ***SKILL CHECK 4*** Multiply: $(x^2)(y^4)(z^3)(x^3)(y)$

◉ ***Skill Check Answer***

4. $x^5y^5z^3$

Exercise 2–5–1

1 Apply the multiplication law of exponents to simplify each expression.

1. $(x^2)(x^5)$ **2.** $(a^3)(a^4)$ **3.** $(a^4)(a^2)$ **4.** $(x^2)(x^3)$

5. $(y^3)(y^8)$ **6.** $(a^2)(a^6)$ **7.** $(x^5)(x^9)$ **8.** $(y^7)(y^2)$

9. $(w^3)(w^9)$ **10.** $(b^4)(b^5)$ **11.** $(a)(a^4)$ **12.** $(c)(c^8)$

13. $(x^3)(x)(x^4)$ **14.** $(y^5)(y^4)(y)$ **15.** $(a^3)(b^4)$ **16.** $(a)(a^3)(a^6)$

17. $(x^2)(x^4)(y)$ **18.** $(w)(z^4)(w^2)$ **19.** $(a^4)(b^3)(a)$ **20.** $(x^4)(y^5)$

21. $(x^3)(y^4)(x)$ **22.** $(a)(b^9)(a^3)$ **23.** $(a^3)(b)(a^2)$ **24.** $(x^4)(y)(x)(y)$

25. $(x^4)(x)(y^3)(x^8)$ **26.** $(x^3)(y^5)(x^2)(y^4)$ **27.** $(a^3)(b^2)(a^5)(b^3)$

28. $(a)(b^4)(a^5)(b)$ **29.** $(a^2)(b)(c^4)(a)(c^2)(a^3)$ **30.** $(x^5)(y)(z^2)(x)(y^4)(z)$

• Concept Enrichment

31. What is wrong with the statement $x^2x^3 = x^6$?

32. Verify that $(xy)^3 = x^3y^3$.

2–6 Multiplying Monomials

OBJECTIVES

Upon completing this section you should be able to:

1. Recognize a monomial.
2. Find the product of two or more monomials.

Now that we have established the law for multiplying expressions when exponents are involved, we wish to use it to multiply algebraic expressions called monomials.

1 Monomials

A **monomial** is an algebraic expression consisting of a single term composed of a product of a numerical coefficient and literal factors (variables). These variables must have whole number exponents.

Example 1 $3x^2$ is a monomial. It is a single term composed of a numerical coefficient 3 and a variable x having a whole number exponent 2.

Example 2 $3x^2 + y$ is *not* a monomial. This expression contains two terms.

Example 3 $5x^2y^3$ is a monomial.

◉ SKILL CHECK 1 Is $2x^3y$ a monomial?

The definition states the expression must be a product.

Example 4 $\frac{x^2y}{z}$ is *not* a monomial. This expression involves the operation of division by the variable z.

◉ SKILL CHECK 2 Is $\frac{x}{y}$ a monomial?

◉ Skill Check Answers

1. yes
2. no

Exercise 2–6–1

1 Determine if the expression is a monomial.

1. $2a$
2. $5x$
3. x^2
4. $3a^2$
5. $x + 5$
6. $a - 3$
7. $4a^2b^3c$
8. $2x^3yz^2$
9. $\frac{3a}{b}$
10. $\frac{5x^2}{y}$

2 Multiplying Monomials

Do you remember the multiplication law of exponents?

To find the product of two monomials multiply the numerical coefficients and apply the multiplication law of exponents to the literal factors (variables).

Multiply 5 times 3 and add the exponents of x.

Example 5
$$(5x^3)(3x^5) = (5)(3)(x^3)(x^5) = 15x^8$$

◉ SKILL CHECK 3 Multiply: $(2x^3)(3x^4)$

Remember, if an exponent is not written, an exponent of 1 is understood.

Example 6
$$(2x^2y)(6xy^3) = (2)(6)(x^2)(x)(y)(y^3) = 12x^3y^4$$

◉ SKILL CHECK 4 Multiply: $(3x^2y)(5xy^4)$

◉ Skill Check Answers

3. $6x^7$
4. $15x^3y^5$

Example 7 $(3x^2y^3z)(4x^4y^6)(x^3z) = 12x^9y^9z^2$

◉ ***SKILL CHECK 5*** Multiply: $(2x^3y^2z)(3x^2y^5)(x^2z^2)$

◉ ***Skill Check Answer***

5. $6x^7y^7z^3$

Exercise 2–6–2

Find the product of each expression.

1. $(2x^2)(5x^3)$
2. $(3a^2)(2a^4)$
3. $(4x^3)(3x^5)$
4. $(2b^3)(5b^2)$
5. $(3x^2)(7x^3)$
6. $(4a^3)(3a^5)$
7. $(11a^3)(6a^7)$
8. $(9x^2)(5x)$
9. $(8a^2)(4b^3)$
10. $(3x^4)(x^3)$
11. $(7xy)(5xy)$
12. $(4ab)(6ab)$
13. $(2x^2y)(3xy^3)$
14. $(3a^3b)(7ab^5)$
15. $(6a^2b)(5a^3c)$
16. $(8xy^4)(6y^3z^3)$
17. $(2x^2)(5x^5)(3x^4)$
18. $(3a^4)(4a^3)(a^5)$
19. $(7a^2b^2)(2ab)(3ab)$
20. $(5x^3y)(3xy^2)(2xy)$
21. $(6xy)(2x^2)(4y^5)$
22. $(2a^2b)(a^3b)(3b^4)$
23. $(11ab)(3b^2c)(5a^2c^3)$
24. $(6xy^2)(2y^3z)(4x^3z^2)$
25. $(3x^2y^3)(8xy^5z^2)(5yz)$
26. $(2a^3b)(4ab^2c^2)(3c^4)$
27. $(6x^2)(5y^2)(2z)$
28. $(3a^3)(2b^2)(5c^2)$
29. $(2a^2b)(5b^2c^2)(6c^3d^2)$
30. $(5x^4z)(x^5y^4z^2)(9yz^4)$

• Concept Enrichment

31. Why isn't $\frac{1}{x}$ a monomial?

32. Is $\frac{2x^2}{3}$ a monomial?

2–7 Factors and Prime Factorization

OBJECTIVES

Upon completing this section you should be able to:

1. Find all factors of a given number.
2. Determine whether or not a given number is prime.
3. Find the prime factorization of a given number.

In chapter 1 and the preceding sections of this chapter we have learned the four basic operations on whole numbers. In the next two sections we wish to introduce some very important concepts that will soon be used as we study a set of numbers called fractions.

1 Factors

If a and b are whole numbers, a is said to be a **factor** of b if a divides b with no remainder.

Example 1 The factors of 12 are 1, 2, 3, 4, 6, and 12.

Notice that each of these whole numbers divides 12 with no remainder.

Example 2 Find all factors of 20.

Solution When we are asked to find *all* factors of a number we must have a system that will not leave out any numbers. The best way to arrive at all factors is to start with the number 1 and try each successive digit. In this way we will find the factors two at a time.

That is, try 1, then 2, then 3, and so on.

We first divide 20 by 1. It divides exactly 20 times. Therefore, 1 and 20 are factors.

$$1\overline{)20}\quad 20$$

Any nonzero number always has 1 and itself as factors.

Next we divide 20 by 2, obtaining a quotient of 10. So 2 and 10 are factors.

$$2\overline{)20}\quad 10$$

Notice that if 2 divides 20 ten times, then 10 divides 20 two times. Thus we find two factors at once.

$$10\overline{)20}\quad 2$$

If we try to divide 20 by 3 we have a remainder, so 3 is not a factor.

$$3\overline{)20}\quad 6,\quad 20-18=2$$

We next divide 20 by 4 obtaining 5. Thus 4 and 5 are factors.

All the factors of 20 are 1, 2, 4, 5, 10, 20.

Why don't we try the next number larger than 4?

◉ ***SKILL CHECK 1*** Find all factors of 24.

Example 3 Find all factors of 48.

Solution They are 1 and 48, 2 and 24, 3 and 16, 4 and 12, 6 and 8.

When we are obtaining the factors two at a time, we can stop when the numbers meet. For instance, in this case if we tried digits above 6 we would next get 8 and 6, which we already have.

WARNING

A common error when listing all factors is to forget 1 and the number itself.

◉ ***SKILL CHECK 2*** Find all factors of 50.

Example 4 Find all factors of 100.

Solution They are 1 and 100, 2 and 50, 4 and 25, 5 and 20, and 10.

When we try 3, 6, 7, 8, 9 none of them will divide 100 without a remainder.

Do you see why we can stop at 10? (See example 3.)

Example 5 Find all factors of 7.

Solution 1 and 7 are the only factors.

◉ ***SKILL CHECK 3*** Find all factors of 13.

◉ ***Skill Check Answers***

1. 1,2,3,4,6,8,12,24
2. 1,2,5,10,25,50
3. 1,13

Example 6 Find all factors of 23.

Solution 1 and 23 are the only factors.

The last two examples indicate that some numbers have exactly two different or distinct factors. This is a special class of numbers called *primes*.

2 Prime Numbers

> A **prime number** is a whole number that has exactly two distinct factors.

Example 7 5 is a prime number since it has exactly two distinct factors. They are 1 and 5.

◉ ***SKILL CHECK 4*** Is 11 a prime number? Why?

Example 8 15 is not a prime number since it has more than two distinct factors. They are 1, 3, 5, and 15.

◉ ***SKILL CHECK 5*** Is 6 a prime number? Why?

The first nine prime numbers are 2, 3, 5, 7, 11, 13, 17, 19, and 23. Euclid, a famous mathematician (ca. 300 BC), proved that there is no limit to the number of primes. But as of this date, no one has ever found a formula that will give all the primes.

Note that 2 (an even number) is prime. Are there any other even numbers that are prime? Why?

There is a method called the *Sieve of Eratosthenes* that will give all primes up to any given number we choose. This method is a simple device that removes all the numbers having more than two factors (up to whatever number we choose) and leaves any number that can be divided only by 1 and itself.

Here is a Sieve of Eratosthenes giving us the primes less than 50.

~~1~~	(2)	(3)	~~4~~	5	~~6~~	7	~~8~~	~~9~~	~~10~~
11	~~12~~	13	~~14~~	~~15~~	~~16~~	17	~~18~~	19	~~20~~
~~21~~	~~22~~	23	~~24~~	~~25~~	~~26~~	~~27~~	~~28~~	29	~~30~~
31	~~32~~	~~33~~	~~34~~	~~35~~	~~36~~	37	~~38~~	~~39~~	~~40~~
41	~~42~~	43	~~44~~	~~45~~	~~46~~	47	~~48~~	~~49~~	~~50~~

We have followed these steps.

Step 1 Cross out the number 1 since it has only one divisor.

Remember, a prime has two *different* divisors.

Step 2 Circle 2 and then cross out all numbers which have a factor of 2. Such a number would be divisible by 1, 2, and itself. That would be too many divisors to be prime.

Step 3 Now circle 3 and cross out all numbers which have a factor of 3.

Again, these numbers have too many divisors. Note that some of the numbers that have 3 as a factor were already eliminated because they have a factor of 2.

Step 4 Circle the next number not crossed out and cross out all numbers divisible by that number without a remainder.

Step 5 Continue this process until you reach a number that multiplied by itself will give 50 or more.

In this case we don't need to go beyond 7.

When this process is completed all numbers not crossed out will be prime. They are 2, 3, 5, 7, 11, 13, 17, 19, 23, 29, 31, 37, 41, 43, 47.

You will find a list of all prime numbers less than 200 on the inside front cover of this book. These are provided for reference when needed for future problems.

◉ ***Skill Check Answers***

4. Yes. It has only two distinct factors, 1 and 11.
5. No. It has more than two distinct factors.

Exercise 2–7–1

1 Find all factors of each number.

1. 4 **2.** 9 **3.** 13 **4.** 17

5. 10 **6.** 22 **7.** 27 **8.** 51

9. 42 **10.** 88

2 Determine whether or not each number is prime.

11. 3 **12.** 2 **13.** 27 **14.** 33

15. 23 **16.** 37 **17.** 41 **18.** 49

19. 87 **20.** 43

• Concept Enrichment

21. Is 1 a prime number? Explain.

22. Use the Sieve of Eratosthenes to find all primes less than 300.

We must now make a distinction between listing the factors of a number and factoring a number.

Factoring a number means to show the number as a product of two or more numbers.

Example 9 Factor 36.

Solution Our answer could be given several different ways.

Note that the statement "Factor 36" is not the same as the statement "Find factors of 36."

$$36 = 1 \times 36$$
$$36 = 6 \times 6$$
$$36 = 3 \times 12$$
$$36 = 2 \times 3 \times 6$$
$$36 = 3 \times 3 \times 4$$
and so on

3 Prime Factorization

The **prime factorization** of a number means expressing the number as a product of primes.

Remember, product means multiply.

Example 10 Find the prime factorization of 36.

Solution

$$36 = 2 \times 2 \times 3 \times 3$$

Notice that each factor is a prime and the product is 36.

A very important statement that we will accept but not attempt to prove is the fundamental theorem of arithmetic. When you read this theorem you will see why we said "the" prime factorization of 36 rather than "a" prime factorization of 36.

A *theorem* is a statement that is based on mathematical proof.

Fundamental Theorem of Arithmetic
Any whole number that is not a prime can be factored into primes in exactly one way except for the order of the primes.

Example 11 $36 = 2 \times 2 \times 3 \times 3$ and no other product of primes can give us 36. We could say $36 = 2 \times 3 \times 2 \times 3$ but this is the same group of primes in a different order.

Remember the commutative property of multiplication?

We will discuss two methods of prime factorization. The first is commonly called the **tree method.**

Example 12 Factor 48 into primes.

Solution First find any two numbers whose product is 48. For instance, we could write 48 as 4×12. Now factor each of these factors if they are not already prime, drawing down any that are already prime.

There are several ways you could start this tree. Try 6×8.

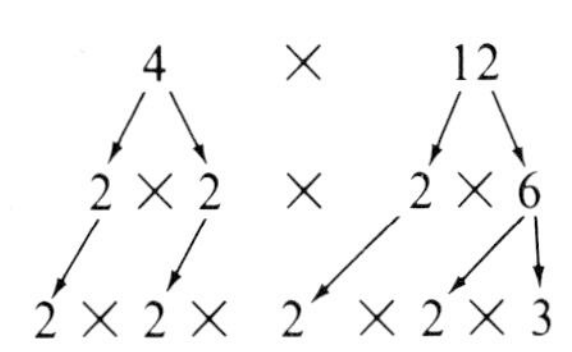

When all numbers are primes, we have the prime factorization.

Notice that no matter how we start, we will still arrive at the same prime factorization.

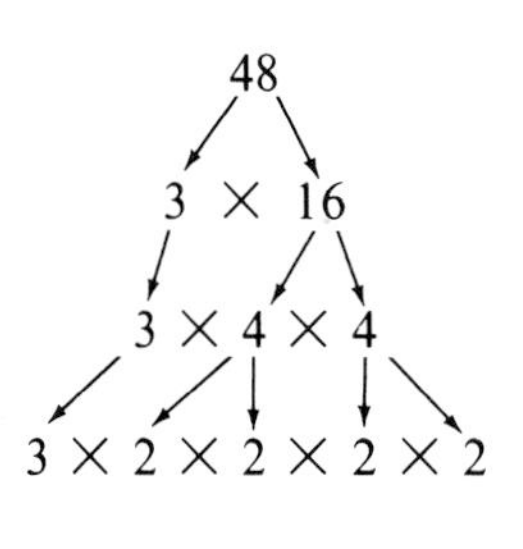

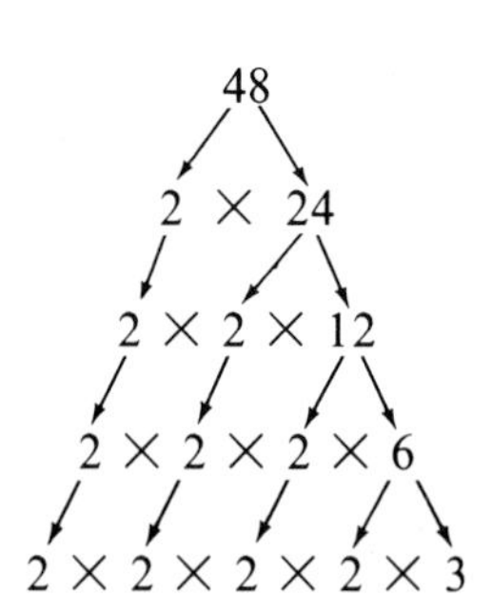

Another method of prime factorization that you may wish to use is repeated division by primes.

Example 13 Find the prime factorization of 48.

It will generally be more straightforward if you start with the smallest prime.

Solution

2|48
24

We first divide by 2 placing our answer *below* the 48.

prime divisors →
2|48
2|24
2|12
2|6
3|3
1

We divide by 2 again and place the answer below the 24. Then by 2 again and again until we can no longer divide by 2. Then we divide by the next highest prime possible, and so on, until our final answer is 1. When this is done, the list of prime divisors is the prime factorization of 48. Thus $48 = (2)(2)(2)(2)(3) = (2^4)(3)$.

How does the result compare to that in the tree method?

⊙ ***SKILL CHECK 6*** Find the prime factorization of 24.

Example 14 Find the prime factorization of 108.

Solution

2|108
2|54
3|27
3|9
3|3
1

Thus $108 = (2)(2)(3)(3)(3) = (2^2)3^3$.

Multiply to check.

Always start with the list of primes: 2, 3, 5, 7, and so on, and divide each as many times as possible before going on to the next.

⊙ ***SKILL CHECK 7*** Find the prime factorization of 72.

Example 15 Find the prime factorization of 882.

Solution

2|882
3|441
3|147
7|49
7|7
1

$882 = (2)(3)(3)(7)(7) = (2)(3^2)(7^2)$

Again, check by multiplying the prime factors.

Notice we used 2, then 3, and since 5 would give a remainder, we next went to 7. If 7 had not been a factor, we would have tried 11, 13, and so on.

⊙ ***Skill Check Answers***

6. $(2^3)(3)$
7. $(2^3)(3^2)$

◉ ***SKILL CHECK 8*** Find the prime factorization of 84.

Example 16 Find the prime factorization of 2,244.

Solution

$$
\begin{array}{r|l}
2 & 2244 \\
2 & 1122 \\
3 & 561 \\
11 & 187 \\
17 & 17 \\
 & 1
\end{array}
$$

$2{,}244 = (2)(2)(3)(11)(17)$

$= (2^2)(3)(11)(17)$

If you are not sure when dividing mentally, use scratch paper and long division.

◉ ***SKILL CHECK 9*** Find the prime factorization of 1,092.

◉ ***Skill Check Answers***

8. $(2^2)(3)(7)$
9. $(2^2)(3)(7)(13)$

Exercise 2–7–2

3 Find the prime factorization of each number.

1. 21 **2.** 15 **3.** 12 **4.** 18

5. 16 **6.** 81 **7.** 60 **8.** 90

9. 56 **10.** 40 **11.** 45 **12.** 75

13. 70 **14.** 110 **15.** 77 **16.** 91

17. 196 **18.** 126 **19.** 390 **20.** 330

21. 1,155 **22.** 2,695 **23.** 187 **24.** 209

25. 374

26. 627

27. 1,615

28. 1,309

29. 2,728

30. 3,330

31. 5,166

32. 9,975

33. 2,772

34. 1,980

35. 128

36. 432

37. 6,435

38. 6,622

39. 13,158

40. 32,085

41. 1,575

42. 1,456

43. 1,960

44. 1,375

45. 6,825

46. 945

47. 847

48. 612

49. 38,709

50. 153,615

• Concept Enrichment

51. What is meant by prime factorization of a number?

52. Distinguish between factors of a number and factoring a number.

2-8 Greatest Common Factor and Least Common Multiple

1 Greatest Common Factor

OBJECTIVES

Upon completing this section you should be able to:

1. Find the greatest common factor of a set of numbers or monomials.
2. Find the least common multiple of a set of numbers or monomials.

Another concept that will be needed in future work is the *greatest common factor* (abbreviated GCF). In the previous section we listed the factors of 36 and also the factors of 48. Look again at these factors.

36: 1, 2, 3, 4, 6, 9, 12, 18, 36
48: 1, 2, 3, 4, 6, 8, 12, 16, 24, 48

Now from these two lists pick out all numbers common to both. They are 1, 2, 3, 4, 6, and 12. The largest of these "common" factors is 12. So the greatest common factor or GCF of 36 and 48 is 12. We can, of course, find the GCF of any group of numbers.

Example 1 Find the GCF of 36, 42, and 90.

Solution

36: 1, 2, 3, 4, (6), 9, 12, 18, 36
42: 1, 2, 3, (6), 7, 14, 21, 42
90: 1, 2, 3, 5, (6), 9, 10, 15, 30, 45, 90

Note that 1, 2, and 3 are also common factors but 6 is the largest.

Looking at the three lists we find that the GCF of 36, 42, and 90 is 6.

There is a better and shorter method of obtaining the GCF.

> The **greatest common factor** of two or more numbers is the product of all prime factors common to the numbers.

Example 2 Find the GCF of 36, 42, and 90.

Solution We first factor each number into primes.

$$36 = 2 \times 2 \times 3 \times 3$$
$$42 = 2 \times 3 \times 7$$
$$90 = 2 \times 3 \times 3 \times 5$$

It may be helpful to circle the common primes.

The primes common to all three numbers are 2 and 3 so the GCF = (2)(3) = 6.

◉ ***SKILL CHECK 1*** Find the GCF of 12, 18, and 30.

Example 3 Find the GCF of 36 and 48.

Solution

$$36 = 2 \times 2 \times 3 \times 3$$
$$48 = 2 \times 2 \times 2 \times 2 \times 3$$

Note that 2 is a factor twice in each number.

The common primes are 2, 2, and 3 so the GCF = (2)(2)(3) = 12.

◉ ***SKILL CHECK 2*** Find the GCF of 18 and 45.

> If two whole numbers have no common prime factors, their GCF is 1 and the numbers are said to be **relatively prime.**

◉ ***Skill Check Answers***

1. 6
2. 9

Example 4 Find the GCF of 21 and 55.

Solution

$$21 = 3 \times 7$$
$$55 = 5 \times 11$$

Remember, 1 is always a factor of every number.

We see there are no common primes so in this case the GCF = 1. Therefore 21 and 55 are relatively prime.

⊙ ***SKILL CHECK 3*** Find the GCF of 14 and 39.

We now wish to find the greatest common factor of a set of monomials containing literal factors or variables. As we did with numbers, we want the product of the greatest number of factors common to the monomials.

Example 5 Find the GCF of x^2 and x^5.

$x^2 = x \cdot x$
$x^5 = x \cdot x \cdot x \cdot x \cdot x$

Solution From the definition of exponent we know that $x^2 = (x)(x)$ and $x^5 = (x)(x)(x)(x)(x)$. We see that the factors of x^2 are the greatest number of factors contained in both x^2 and x^5. So the GCF = x^2.

⊙ ***SKILL CHECK 4*** Find the GCF of x^3 and x^7.

Example 6 Find the GCF of a^3b^2 and a^2b^6.

$a^3b^2 = a \cdot a \cdot a \cdot b \cdot b$
$a^2b^6 = a \cdot a \cdot b \cdot b \cdot b \cdot b \cdot b \cdot b$

Solution The factors of a^2 are contained in a^3 and the factors of b^2 are contained in b^6. So the GCF = a^2b^2.

⊙ ***SKILL CHECK 5*** Find the GCF of a^2b^5 and a^3b^4.

We now find the GCF of monomials that contain variables having coefficients other than the number 1.

Example 7 Find the GCF of $4x^2y$ and $6x^3$.

y is not a common factor.

Solution We find that the GCF of 4 and 6 is 2. Also, the factors of x^2 are contained in x^3. y is *not* a factor of both terms. So the GCF = $2x^2$.

⊙ ***SKILL CHECK 6*** Find the GCF of $12x^3y$ and $30x^2$.

⊙ ***Skill Check Answers***

3. 1
4. x^3
5. a^2b^4
6. $6x^2$

From the previous examples we see that the greatest common factor of a set of monomials is the product of the GCF of the numerical coefficients and the GCF of the variables.

Exercise 2–8–1

1 Find the GCF of each set of numbers or monomials.

1. 12 and 20 **2.** 15 and 18 **3.** 28 and 30

4. 36 and 42 **5.** 8 and 15 **6.** 8 and 16

7. 42 and 84 **8.** 18 and 54 **9.** 54 and 72

10. 90 and 105

11. 35 and 88

12. 66 and 154

13. 84 and 105

14. 91 and 99

15. 108 and 126

16. 168 and 252

17. 117 and 315

18. 165 and 630

19. x^2 and x^7

20. a^3 and a^4

21. a^3b and a^2b^5

22. x^4y and x^3y^5

23. $84x^4$ and $105x^6$

24. $91a^3$ and $99a^2$

25. $3xy^3$ and $6x^2y^2$

26. $5x^2y^3$ and $10xy^5$

27. $117x^3y$ and $315x^4z$

28. $165a^4b^2$ and $630b^4c$

29. 36, 40, and 60

30. 24, 40, and 63

31. 72, 84, and 108

32. 30, 40, 63

33. $84a^2$, $126a$, and $252a^3$

34. $99x^4$, $126x^8$, and $189x^5$

35. 42, 45, and 56

36. 24, 60, and 84

37. $72x^3y^2z$, $180x^4y^3z^2$, and $216x^5yz^4$

38. $180a^4bc^2$, and $210a^5b^2c^4$, and $630a^7bc^2$

39. 252, 441, and 630

40. 270, 720, and 1,260

2 Least Common Multiple

The concept of *least common multiple* (abbreviated LCM) of a set of whole numbers will also be of use in future work with fractions.

Multiples of a number *a* are the numbers obtained by multiplying the number *a* by the whole numbers 1, 2, 3, 4, 5, ...

We say the list of multiples is infinite.

From the definition we can see that the list of multiples of a number would never end since the list of whole numbers never ends.

Example 8 List several multiples of 3.

$(3)(1) = 3$
$(3)(2) = 6$
$(3)(3) = 9$
and so on

Solution Such numbers would be 3, 6, 9, 12, 15, 18, 21, 24, . . . , where the dots indicate we could go on and on.

Now look at the following lists of some multiples of 3 and 5.

3: 3, 6, 9, 12, (15) 18, 21, 24, 27, (30) 33, 36, 39, 42, (45) . . .

5: 5, 10, (15) 20, 25, (30) 35, 40, (45) 50, 55, 60, 65, 70, . . .

Observing these lists we see that 3 and 5 have some multiples in common such as 15, 30, and 45. There would be more if our lists continued. Our interest, however, is really in the *least* of the multiples in both lists—in this case 15. In other words, the LCM of 3 and 5 is 15.

15 is the smallest number that is a multiple of 3 and 5.

The **least common multiple** of a set of numbers is the smallest number that is a multiple of each number in the set.

Let's now look at an example that will explain a method of finding the LCM of any set of numbers.

Example 9 Find the LCM of 12, 15, and 18.

Solution First note that by definition this unknown number must contain 12 as a factor, 15 as a factor, and 18 as a factor. We now factor 12, 15, and 18 into primes.

Some students simply multiply (12)(15)(18). This would give a multiple of the three numbers but it would not be the *least* common multiple.

$$12 = (2)(2)(3)$$
$$15 = (3)(5)$$
$$18 = (2)(3)(3)$$

Now since we know that all of these numbers must be factors of the LCM and that no other prime factorization is possible, we can see that (2)(2)(3) must be part of the LCM. This is also true for (3)(5) and (2)(3)(3). So we "build" the LCM in the following manner.

Write the first number as a product of primes $12 = (2)(2)(3)$. These factors must be part of the LCM. Now we look at the prime factorization of the second number $15 = (3)(5)$. We see that the factor 3 is already a factor of the first number but the factor 5 is not. Therefore we need to include 5 as a factor so we write $(2)(2)(3)(5)$ as a necessary part of the LCM.

We now look at the third number in prime factored form $18 = (2)(3)(3)$. The factor 3 appears twice in this number so we must have 3 as a factor twice in our LCM. This gives us $(2)(2)(3)(3)(5) = 180$ as the LCM of 12, 15, and 18.

Always check your final answer to make sure all the prime factors are in it.

Example 10 Find the LCM of 21 and 30.

Solution

$$21 = (3)(7)$$
$$30 = (2)(3)(5)$$

We start with the first number $(3)(7)$ and multiply it by $(2)(5)$, which are the other primes necessary to ensure that the second number is part of the LCM. Thus the LCM $= (2)(3)(5)(7) = 210$.

⊙ ***SKILL CHECK 7*** Find the LCM of 15 and 42.

Example 11 Find the LCM of 4 and 14.

Solution

$$4 = (2)(2)$$
$$14 = (2)(7)$$
$$\text{LCM} = (2)(2)(7) = 28$$

⊙ ***SKILL CHECK 8*** Find the LCM of 9 and 21.

We now consider the least common multiple of a set of monomials containing variables.

Example 12 Find the LCM of x^2 and x^5.

Solution The LCM must contain all factors of each monomial. In other words, the LCM of x^2 and x^5 must contain $(x)(x)$ and $(x)(x)(x)(x)(x)$. The factors of the lower power (x^2) are factors of the higher power (x^5), so the LCM of x^2 and x^5 is x^5.

$x^2 = (x)(x)$

$x^5 = (x)(x)(x)(x)(x)$

⊙ ***SKILL CHECK 9*** Find the LCM of x^3 and x^7.

Example 13 Find the LCM of ax^2 and a^2y^3.

Solution

$$ax^2 = (a)(x)(x) \text{ and}$$
$$a^2y^3 = (a)(a)(y)(y)(y)$$
$$\text{LCM} = (a)(a)(x)(x)(y)(y)(y)$$
$$= a^2x^2y^3$$

⊙ ***SKILL CHECK 10*** Find the LCM of ax^3 and a^2x^2.

Note in the two previous examples that the LCM is the product of the highest powers of the variables that occur in the set.

Example 14 Find the LCM of $4x^2$ and $6xy$.

Solution The LCM of 4 and 6 is 12, so the LCM of $4x^2$ and $6xy$ is $12x^2y$.

x^2 and y are the highest power of each variable involved.

⊙ ***SKILL CHECK 11*** Find the LCM of $6x^3$ and $15x^2y$.

From these examples we see that the least common multiple of a set of monomials is the product of the LCM of the numerical coefficients and the LCM of the variables.

⊙ ***Skill Check Answers***

7. 210
8. 63
9. x^7
10. a^2x^3
11. $30x^3y$

Exercise 2–8–2

2 Find the LCM of each set of numbers or monomials.

1. 2 and 3

2. 3 and 5

3. 4 and 6

4. 6 and 9

5. 6 and 12

6. 6 and 15

7. 15 and 18

8. 18 and 24

9. $9x^3$ and $24x^2$

10. $12a^2$ and $16a^4$

11. $12a^3b^2$ and $18a^2b^3$

12. $10xy^5$ and $14x^2y$

13. 24 and 36

14. 25 and 40

15. $24x^2$ and $30y$

16. $42x^2$ and $70y^3$

17. 3, 4, and 6

18. 2, 8, and 12

19. 3, 6, and 8

20. 12, 18, and 20

21. 12, 20, and 24

22. 15, 24, and 36

23. $9a^3b^2$, $12ab^2$, and $14ab$

24. $8x^2y$, $24xy^3$, and $60xy^2$

25. $27x^5y$, $45x^2y^2$, and $54x^2y^7$

26. $18a^4b$, $20ab^4$, and $24a^3b^3$

27. 18, 32, and 54

28. 9, 36, and 63

29. 14, 34, and 60

30. 16, 40, and 64

31. $12a^3b$, $16a^2bc$, $18abc$

32. $6xyz$, $18x^2y$, $34z^2$

33. 3, 6, 9, and 15

34. 6, 8, 10, and 12

35. $8a$, $9b^2$, $12c$, and $18a^2c$

36. $4x$, $8x^2y$, $12xz$, and $16y^2z$

37. 12, 15, 18, and 24

38. 12, 18, 20, and 36

39. $9x^2$, 18, $27xy^2$, and 36

40. 8, $12a^3b$, 36, and $60a^2c^2$

• Concept Enrichment

41. Distinguish between a multiple of a number and the least common multiple of a set of numbers.

42. Distinguish between the greatest common factor and the least common multiple of a set of numbers.

CHAPTER 2 SUMMARY

Key Words

Section 2–1

- The **commutative property of multiplication** is $ab = ba$ for any two numbers a and b.
- The **multiplicative identity** is 1. For any number a, $a(1) = a$.
- The **associative property of multiplication** is $(ab)c = a(bc)$ for any numbers a, b, and c.
- The **distributive property** of multiplication over addition is $a(b + c) = ab + ac$ for any numbers a, b, and c.
- The technique of **estimation** enables us to determine if our answer is reasonable.

Section 2–2

- The **dividend** is the number being divided.
- The **divisor** is the number doing the dividing.
- The **quotient** is the result of division.
- The **remainder** is the number (less than the divisor) left over when dividing.
- The **long division algorithm** is based on repeated subtraction.

Section 2–3

- **Parentheses, brackets,** and **braces** are used as grouping symbols.

Section 2–4

- The **principle of substitution** states that any quantity may be substituted for its equal in any process.

Section 2–5

- The **multiplication law of exponents** is

$$x^a x^b = x^{a+b}.$$

Section 2–6

- A **monomial** is an algebraic expression consisting of a single term composed of a product of a numerical coefficient and literal factors (variables). These variables must have whole number exponents.

Section 2–7

- A **prime number** is a whole number that has exactly two distinct factors.
- **Factoring** a number is expressing it as a product of two or more numbers.
- The **prime factorization** of a number is expressing it as a product of primes.
- The **fundamental theorem of arithmetic** states that there is only one prime factorization of a number.

Section 2–8

- The **greatest common factor** of two or more numbers is the product of all prime factors common to the numbers.
- **Relatively prime numbers** have no common prime factors.
- **Multiples** of a number are obtained by multiplying that number by the whole numbers 1, 2, 3, ...
- The **least common multiple** of a set of numbers is the smallest number that is a multiple of each number in the set.

Procedures

Section 2–2

- Multiplication and division are related by the fact that if $a \div b = c$, then $a = bc$.

Section 2–3

- Operations within grouping symbols are performed first.
- If no grouping symbols occur in an expression, multiplication and division are performed from left to right, and then addition and subtraction from left to right.
- When simplifying an expression containing grouping symbols within grouping symbols, remove the innermost set of symbols first.

Section 2–6

- To find the product of two monomials multiply the numerical coefficients and apply the multiplication law of exponents to the literal factors.

Section 2–7

- To find the prime factorization of a number, use the *tree method* or the *repeated division by primes method.*

Section 2–8

- To find the greatest common factor of a set of monomials use the product of the GCF of the numerical coefficients and the GCF of the variables.
- To find the least common multiple of a set of monomials use the product of the LCM of the numerical coefficients and the LCM of the variables.

CHAPTER 2 REVIEW

Multiply.

1. $\begin{array}{r} 83 \\ \times\ 6 \\ \hline \end{array}$ **2.** $\begin{array}{r} 75 \\ \times\ 8 \\ \hline \end{array}$ **3.** $\begin{array}{r} 34 \\ \times 23 \\ \hline \end{array}$ **4.** $\begin{array}{r} 27 \\ \times 34 \\ \hline \end{array}$

5. $\begin{array}{r} 325 \\ \times\ 34 \\ \hline \end{array}$ **6.** $\begin{array}{r} 519 \\ \times\ 27 \\ \hline \end{array}$ **7.** $\begin{array}{r} 228 \\ \times\ 10 \\ \hline \end{array}$ **8.** $\begin{array}{r} 349 \\ \times\ 10 \\ \hline \end{array}$

9. $\begin{array}{r} 426 \\ \times 153 \\ \hline \end{array}$ **10.** $\begin{array}{r} 365 \\ \times 218 \\ \hline \end{array}$ **11.** $\begin{array}{r} 548 \\ \times 100 \\ \hline \end{array}$ **12.** $\begin{array}{r} 634 \\ \times 100 \\ \hline \end{array}$

13. $\begin{array}{r} 1{,}345 \\ \times\ 268 \\ \hline \end{array}$ **14.** $\begin{array}{r} 5{,}418 \\ \times\ 325 \\ \hline \end{array}$ **15.** $\begin{array}{r} 2{,}041 \\ \times\ 304 \\ \hline \end{array}$ **16.** $\begin{array}{r} 1{,}605 \\ \times\ 704 \\ \hline \end{array}$

Divide.

17. $6\overline{)162}$ **18.** $3\overline{)138}$ **19.** $4\overline{)1{,}564}$ **20.** $7\overline{)3{,}311}$

21. $5\overline{)213}$ **22.** $8\overline{)284}$ **23.** $14\overline{)784}$ **24.** $12\overline{)756}$

25. $21\overline{)531}$

26. $24\overline{)785}$

27. $33\overline{)66,185}$

28. $47\overline{)94,218}$

29. $43\overline{)129,215}$

30. $53\overline{)212,318}$

31. $100\overline{)64,000}$

32. $100\overline{)85,300}$

Evaluate each expression.

33. $15 - (6 + 5)$

34. $31 - (15 - 9)$

35. $(16 + 4) - (18 - 13)$

36. $(8 + 3)(14 - 6)$

37. $21 - 8 + 5 - 7$

38. $12 + 9 - 11 + 2$

39. $8 + 5 \times 3^2 - 4$

40. $27 - 14 \div 2$

41. $5 + 3[4 + (8 - 5)]$

42. $12 - 2[17 - (6 + 5)]$

43. $2\{6[(5 - 2) - 2] + 4\}$

44. $25 - \{15 - [3(4 - 2) - 6]\}$

45. Evaluate $5a^3$ if $a = 2$.

46. Evaluate $(5a)^3$ if $a = 2$.

47. Evaluate $3a^2b$ if $a = 3$ and $b = 2$.

48. Evaluate $4a^2b$ if $a = 3$ and $b = 2$.

49. Evaluate $2x^3 + 4y^2$ if $x = 2$ and $y = 5$.

50. Evaluate $a^2 + 3ab - b^2$ if $a = 6$ and $b = 4$.

51. Evaluate $5x^3y + 3yz - 10xy^4$ if $x = 0$, $y = 4$, and $z = 5$.

52. Evaluate $2a^2bc + 4bc^2 - 5ac$ if $a = 1$, $b = 3$, and $c = 10$.

53. Simplify: a^4a^9

54. Simplify: x^3x^8

55. Simplify: $(x^2)(y)(x^3)(y^3)$

56. Simplify: $(a^3)(b^2)(a)(b^4)$

57. Simplify: $(3x^2)(5x^4)$

58. Simplify: $(4a^4)(5a^3)$

59. Simplify: $(2ab^2)(8ab)$

60. Simplify: $(3xy^2)(4x^2y^5)$

61. Find the prime factorization of 132.

62. Find the prime factorization of 84.

63. Find the prime factorization of 1,260.

64. Find the prime factorization of 4,158.

65. Find the greatest common factor of 22 and 26.

66. Find the greatest common factor of 12 and 44.

67. Find the greatest common factor of $56x^2y$ and $60xy^2$.

68. Find the greatest common factor of $210ab^3$ and $462a^2b^2$.

69. Find the greatest common factor of 60, 132, and 210.

70. Find the greatest common factor of 84, 120, and 396.

71. Find the least common multiple of 4 and 5.

72. Find the least common multiple of 3 and 8.

73. Find the least common multiple of $6a^2b^3$ and $8ab^2$.

74. Find the least common multiple of $12x^3y^2$ and $15x^2y^3$.

75. Find the least common multiple of 9, 15, and 20.

76. Find the least common multiple of 24, 27, and 36.

77. Find the least common multiple of $6xy$, $8x^2$, 9, and $12y^2z$.

78. Find the least common multiple of $8a^3b$, $12ab^2$, 16, and $24ac$.

79. If your monthly salary was $1,245, how much would you make in a year?

80. Real estate taxes on a building amount to $168 per month. How much is that per year?

81. If a Boeing 727 aircraft can carry 189 passengers, how many planes would it take to carry 2,400 passengers?

82. A man purchases a car for $10,295. He pays $4,000 down. The finance charge on the balance is $2,921. How much are the monthly payments if it is paid off in 36 months?

CHAPTER 2 TEST

1. Multiply: 32(24)

2. Multiply: $\begin{array}{r} 1{,}235 \\ \times\ \ 406 \\ \hline \end{array}$

3. Multiply: $321(10^5)$

4. Divide: $64 \div 7$

5. Divide: $7\overline{)406}$

6. Divide: $31\overline{)1{,}413}$

7. Evaluate: $16 - (9 + 2)$

8. Evaluate: $16 - 9 + 2$

9. Evaluate: $8 + 4 \times 3 - 12 \div 2^2 - 1$

10. Evaluate: $2[3(4 - 1) - 5]$

11. Evaluate $2a^3$ if $a = 3$.

12. Evaluate $(2a)^3$ if $a = 3$.

13. Evaluate $3x^2y + 5xz - 4yz^2$ if $x = 4$, $y = 2$, and $z = 3$.

14. Simplify: $(a^3)(b)(a^4)(b)$

15. Multiply: $(3a^3b^2)(4ab^3)$

16. Find the prime factorization of 408.

17. Find the greatest common factor of $126x^2y^2$ and $140xy^3$.

18. Find the least common multiple of 12 and 15.

19. The area of a rectangle is given by the formula $A = bh$. Find A when $b = 12$ and $h = 7$.

20. Total payments for a compact disc player for two years was $696. If each monthly payment was for the same amount, how much was each payment?

21. If there are 16 slices of cheese in a package, how many slices are in 12 packages?

22. Susan wants to save a total of $2,000. If she saves $65 a week for 18 weeks, how much more does she need to save to meet her goal?

CHAPTERS 1–2

CUMULATIVE TEST

1. Name the operation indicated in the expression (3)(4).

2. What is the place value of the digit 8 in the number 483,600?

3. Multiply: $\begin{array}{r} 521 \\ \times\ 63 \\ \hline \end{array}$

4. Divide: $32\overline{)6{,}528}$

5. Write the number 30,016 in words.

6. Evaluate: $3 + 2[6 - (4 - 1)]$

7. Evaluate $2x^3$ if $x = 2$.

8. Replace the question mark in 30 ? 28 with the inequality symbol positioned to indicate a true statement.

9. Add: $\begin{array}{r} 36 \\ 294 \\ +109 \\ \hline \end{array}$

10. How many terms are there in the expression $a^2 - 3a + 6ab - 2$?

11. Evaluate $ab^2 + 4ac - 2a^2$ if $a = 1$, $b = 3$, and $c = 5$.

12. Find the prime factorization of 168.

13. Subtract: $\begin{array}{r} 3{,}002 \\ -\ \ 435 \\ \hline \end{array}$

14. Simplify: $3x + 4y - x + 5y$

15. Simplify: $(2x^2y)(7xy^5)$

16. Round 506,491 to the nearest ten thousand.

17. Find the greatest common factor of $18a^2b^3$ and $30ab^4$.

18. Find the least common multiple of 6, 10, and 15.

19. During one semester there were twenty eight thousand, four hundred three students enrolled in a particular college. Write this number using digits.

20. The total balance to be paid for an audio system is \$1,152. How many months will it take to pay off the balance if \$24 of it is paid each month?

CHAPTER

3 Pretest

Answer as many of the following problems as you can before starting this chapter. When you finish the chapter, take the test at the end and compare your scores to see how much you have learned.

1. Reduce: $\frac{24a^3b}{84a^2b^4}$

2. Change $9\frac{11}{12}$ to an improper fraction.

3. Find the missing numerator: $\frac{3}{16} = \frac{?}{64}$

4. Find the LCD of $\frac{1}{5}, \frac{2}{3}, \frac{7}{9}$.

Perform the indicated operation. Give all answers in reduced form.

5. $\left(\frac{1}{3}\right)\left(\frac{2}{5}\right)$

6. $\frac{3b}{7a} \div \frac{5b^2}{6a^2}$

7. $\frac{2}{11} + \frac{7}{11}$

8. $\frac{3}{8} - \frac{1}{8}$

9. $3\frac{3}{4} \times 2\frac{2}{5}$

10. $\frac{4}{7xy^2} + \frac{3}{8x^2y}$

11. $\frac{9a^2}{11} \div 6a^2$

12. $\frac{3}{5a} - \frac{1}{4b}$

13. $\frac{7}{12} + \frac{5}{9} + \frac{3}{4}$

14. $\left(\frac{3}{8}\right)\left(\frac{4}{11}\right)$

15. $5\frac{3}{4} - 2\frac{1}{3}$

16. $\frac{4}{9} \div \frac{14}{15}$

17. $\left(\frac{6c}{7ab}\right)\left(\frac{14a^2b}{27c^2}\right)$

18. $3\frac{2}{3} + 1\frac{7}{12} + 6\frac{8}{9}$

19. $3\frac{3}{4} \div 6\frac{2}{3}$

20. Replace the question mark in $\frac{7}{9}\ ?\ \frac{8}{11}$ with either $=$, $<$, or $>$ to indicate a true statement.

21. If $\frac{1}{3}$ of a person's income is spent for rent, how much rent is paid by a person who earns $1,425 per month?

22. If a truck can carry $\frac{3}{4}$ ton of gravel, how many loads will it take to carry 24 tons?

23. If it takes $5\frac{2}{3}$ hours to drive a certain distance, how much time remains after traveling $3\frac{3}{5}$ hours?

C H A P T E R

3 Fractions

Pythagoras (572–501 BC) discovered that the underlying mathematics of the musical scale is based on common fractions.

STUDY HELP

New concepts in mathematics are always based on previous ideas. Failure to master a particular concept will prevent you from understanding some subsequent idea. Therefore, be sure that you accomplish each objective before proceeding.

In the previous two chapters we have studied the four basic operations on whole numbers and used them to work with certain algebraic expressions. In this chapter we will be concerned with another set of numbers called fractions. Whole numbers and fractions made up of whole numbers are together called the **numbers of arithmetic.**

3–1 Simplifying Fractions

OBJECTIVE

Upon completing this section you should be able to:

1. Use the fundamental principle of fractions to reduce common and algebraic fractions.

Since the fractions we will study in this chapter are composed of whole numbers and variables representing whole numbers, the operations learned in the previous two chapters form the foundation for operations on this new set of numbers.

> In the set of the numbers of arithmetic, a **common fraction** is the indicated quotient of two whole numbers.

Examples of common fractions are $\frac{3}{5}, \frac{1}{2}, \frac{5}{8}, \frac{7}{3}$, and $\frac{2}{2}$. These fractions could be written as $3 \div 5$, $1 \div 2$, $5 \div 8$, and so on, but the usual method is to write one number over the other with a division bar between them. Some special names make it easier to talk about fractions.

Division or fractional bar → $\frac{a}{b}$ ← Top number is called the **numerator**
← Bottom number is called the **denominator**

Fractions can be classified as *proper* or *improper.*

$\frac{1}{3}, \frac{3}{5}, \frac{7}{8}$ are proper fractions.

$\frac{5}{3}, \frac{7}{4}, \frac{5}{5}$ are improper fractions.

If the numerator is smaller than the denominator, we call the indicated division a **proper fraction.**

If the numerator is equal to or greater than the denominator, we call the indicated division an **improper fraction.**

A fraction is read as a division.

Example 1 $\frac{3}{4}$ is read "three divided by four" or "three-fourths."

Example 2 $\frac{1}{2}$ is read "one divided by two" or "one-half."

Example 3 $\frac{2}{3}$ is read "two divided by three" or "two-thirds."

Which one of these examples is an improper fraction?

Example 4 $\frac{12}{7}$ is read "twelve divided by seven" or "twelve-sevenths."

In chapter 1 we used the number line to order the whole numbers. Fractions can also be positioned on the number line.

$\frac{1}{4}$ $\frac{3}{2}$ $\frac{8}{3}$ $\frac{19}{4}$

0 1 2 3 4 5

We now wish to define *algebraic fractions.*

> An **algebraic fraction** is a fraction in which the numerator or denominator (or both) contains one or more variables.

Examples of algebraic fractions are $\frac{3}{a}$, $\frac{2x}{y}$, $\frac{3a}{2b}$, and $\frac{x^2}{a^2b}$.

Fractions that are equal in value may be written in many ways. For instance, if a dime is $\frac{1}{10}$ of a dollar and you have 5 dimes, then you have $\frac{5}{10}$ of a dollar. If a quarter is $\frac{1}{4}$ of a dollar and you have 2 quarters, then you have $\frac{2}{4}$ of a dollar. If a nickel is $\frac{1}{20}$ of a dollar and you have 10 nickels, then you have $\frac{10}{20}$ of a dollar. In each case you have one-half of a dollar. In other words, $\frac{5}{10}$, $\frac{2}{4}$, $\frac{10}{20}$, and $\frac{1}{2}$ all represent the same value. This illustrates the following very important principle.

Fundamental Principle of Fractions

If $\frac{a}{b}$ is a fraction and c is any number except zero, then $\frac{a}{b} = \frac{ac}{bc}$.

If two quantities are equal, we can write them in either order (reflexive property).

So we can also write

$$\frac{ac}{bc} = \frac{a}{b}.$$

Notice that this principle descends directly from the property for multiplicative identity. That is, $\frac{a}{b} \cdot 1 = \frac{a}{b}$ and since $\frac{c}{c} = 1$, $c \neq 0$, we have

$$\frac{a}{b} = \frac{a}{b} \cdot 1 = \frac{a}{b} \cdot \frac{c}{c} = \frac{ac}{bc}.$$

In other words, the principle states that we can multiply both the numerator and denominator of a fraction by the same nonzero number and the value of the fraction remains the same.

Example 5 $\frac{5}{8} = \frac{(5)(2)}{(8)(2)} = \frac{10}{16}$

Multiply both numerator and denominator by 2.

Example 6 $\frac{5}{8} = \frac{(5)(3)}{(8)(3)} = \frac{15}{24}$

Multiply both numerator and denominator by 3.

Example 7 $\frac{2}{3} = \frac{(2)(5)}{(3)(5)} = \frac{10}{15}$

Multiply both numerator and denominator by 5.

Example 8 $\frac{x}{y} = \frac{(x)(a)}{(y)(a)} = \frac{ax}{ay}$

Multiply numerator and denominator by a. We agree to write the factors in alphabetical order.

Example 9 $\frac{3x^2}{4y^2} = \frac{(3x^2)(2x)}{(4y^2)(2x)} = \frac{6x^3}{8xy^2}$

Multiply numerator and denominator by $2x$.

1 Reducing Fractions

We wish now to use the fundamental principle of fractions to *simplify* or *reduce* fractions.

A fraction $\frac{a}{b}$ is in **simplified** or **reduced form** if a and b have no factor in common except the number 1.

a and b represent any whole numbers, except b cannot be zero.

Example 10 $\frac{2}{3}$, $\frac{4}{7}$, $\frac{x}{y}$, and $\frac{3a^2}{8b^2}$ are in simplified form because 1 is the only factor common to both the numerator and denominator.

$\frac{4}{6}=\frac{(2)(2)}{(3)(2)}$

2 is a factor common to the numerator and denominator.

Example 11 $\frac{4}{6}, \frac{9}{12}, \frac{3x^2}{2x}$, and $\frac{5a^2}{10b^2}$ are *not* in simplified form. (Can you find a factor of the numerator and denominator in each fraction that is not just the number 1?)

> To reduce or simplify a fraction use the fundamental principle of fractions to divide all factors common to the numerator and denominator.

Example 12 Simplify $\frac{12}{18}$.

Solution Using mental arithmetic you may notice immediately that 6 is a factor of 12 and 18.

This is the fundamental principle of fractions.

$$\frac{12}{18}=\frac{(\not{6})(2)}{(\not{6})(3)}=\frac{2}{3}$$

If you do not immediately see a common factor, use prime factorization from chapter 2 and factor the numerator and denominator into primes. Then divide the like primes.

$$\frac{12}{18}=\frac{(\not{2})(2)(\not{3})}{(\not{2})(3)(\not{3})}=\frac{2}{3}$$

So $\frac{12}{18}=\frac{2}{3}$ in reduced form. Notice that when we divide 2 into 2 and 3 into 3 we are left with a factor of 1. This is not necessary to write unless 1 is the only number left in the numerator.

◉ ***SKILL CHECK 1*** Simplify: $\frac{10}{15}$

Example 13 Reduce: $\frac{7}{14}$

Notice the importance of writing the 1 in this case.

Solution

$$\frac{7}{14}=\frac{\not{7}}{(\not{7})(2)}=\frac{1}{2}$$

◉ ***SKILL CHECK 2*** Reduce: $\frac{2}{6}$

The method of simplifying fractions is the same if variables are involved.

Example 14 Simplify: $\frac{x^5}{x^3}$

This uses the definition of exponent.

Solution $\frac{x^5}{x^3}=\frac{(\not{x})(\not{x})(\not{x})(x)(x)}{(\not{x})(\not{x})\not{x})}=\frac{(x)(x)}{1}=\frac{x^2}{1}=x^2$

◉ ***SKILL CHECK 3*** Simplify: $\frac{x^6}{x^2}$

◉ ***Skill Check Answers***

1. $\frac{2}{3}$
2. $\frac{1}{3}$
3. x^4

Example 15 Simplify: $\frac{12xy}{28x^3y^2}$

Solution

$$\frac{12xy}{28x^3y^2} = \frac{(\not{4})(3)(\not{x})(\not{y})}{(\not{4})(7)(\not{x})(x)(x)(\not{y})(y)}$$

$$= \frac{3}{(7)(x)(x)(y)}$$

$$= \frac{3}{7x^2y}$$

We divide like factors—numbers and/or letters.

⊙ ***SKILL CHECK 4*** Simplify: $\frac{6xy}{15x^2y^3}$

Example 16 Reduce: $\frac{15a^2}{22b^2}$

Solution

$$\frac{15a^2}{22b^2} = \frac{(3)(5)(a)(a)}{(2)(11)(b)(b)} = \frac{15a^2}{22b^2}$$

We factor the numbers into primes to make sure there are no common factors.

Since there are no common factors, the fraction is already in reduced form.

⊙ ***SKILL CHECK 5*** Reduce: $\frac{10a^3}{21b}$

⊙ ***Skill Check Answers***

4. $\frac{2}{5xy^2}$
5. $\frac{10a^3}{21b}$

Simplifying or reducing fractions is important because fractional answers should be given in simplest form.

Exercise 3–1–1

1 Reduce.

1. $\frac{4}{6}$ **2.** $\frac{6}{9}$ **3.** $\frac{9}{12}$ **4.** $\frac{6}{8}$

5. $\frac{4}{10}$ **6.** $\frac{6}{15}$ **7.** $\frac{6}{10}$ **8.** $\frac{8}{10}$

9. $\frac{x^3}{x}$ **10.** $\frac{a^4}{a^2}$ **11.** $\frac{a^3}{a^5}$ **12.** $\frac{x}{x^3}$

13. $\frac{15}{35}$ **14.** $\frac{12}{15}$ **15.** $\frac{8}{12}$ **16.** $\frac{12}{18}$

17. $\frac{3x}{6x^2}$

18. $\frac{3x}{9x^3}$

19. $\frac{5a^2}{20a}$

20. $\frac{5a^4}{10a}$

21. $\frac{7}{21}$

22. $\frac{5}{15}$

23. $\frac{2}{18}$

24. $\frac{9}{18}$

25. $\frac{4xy^3}{16xy^3}$

26. $\frac{9x^2y^2}{36x^4y^2}$

27. $\frac{6}{12}$

28. $\frac{8}{24}$

29. $\frac{12ab}{16a^2b}$

30. $\frac{8a^2b}{24ab^2}$

31. $\frac{12xy^2}{18x^2y}$

32. $\frac{24x^2y^3}{28x^3y^2}$

33. $\frac{36}{42}$

34. $\frac{16}{72}$

35. $\frac{24}{64}$

36. $\frac{15}{32}$

37. $\frac{10x^2}{27y^2}$

38. $\frac{45ab}{54bc}$

39. $\frac{32}{88}$

40. $\frac{12}{78}$

41. $\frac{28a^2}{39b}$

42. $\frac{20x^3}{50y}$

43. $\frac{20x^3y}{65x^2}$

44. $\frac{24a^4b^2}{64a^5b}$

45. $\frac{42}{72}$

46. $\frac{96}{132}$

47. $\frac{24a^5b}{54a^3b^3}$

48. $\frac{66x^2yz^3}{154xyz}$

49. $\frac{36}{54}$

50. $\frac{36}{91}$

51. $\frac{24}{112}$

52. $\frac{72}{90}$

53. $\frac{63x^4y}{68xy^5}$

54. $\frac{36a^2xy}{96by}$

55. $\frac{108x^2y^6}{144x^3y}$

56. $\frac{64xy^5z^6}{144x^4y^2z}$

57. $\frac{108}{198}$ **58.** $\frac{168}{273}$ **59.** $\frac{196}{336}$ **60.** $\frac{176}{288}$

- **Concept Enrichment**

61. Explain the distinction between a proper and improper fraction.

62. Explain how the multiplicative identity property leads to the fundamental principle of fractions.

3–2 Multiplying Fractions

OBJECTIVES

Upon completing this section you should be able to:

1. Multiply fractions and give the product in reduced form.
2. Use multiplication of fractions to solve applications.

Now that we have learned to reduce or simplify fractions we will proceed to the basic operations on them. The first of the four basic operations we will study is multiplication.

1 Multiplication of Fractions

Multiplication of one fraction by another occurs in many situations. As a simple illustration, suppose you have $\frac{1}{4}$ of a pizza that you wish to share equally with a friend. Each of you would get $\frac{1}{2}$ of the $\frac{1}{4}$ which would be $\frac{1}{8}$ of the pizza. In other words, $\left(\frac{1}{2}\right)\left(\frac{1}{4}\right) = \frac{1}{8}$. This illustration relates to the following definition.

> If $\frac{a}{b}$ and $\frac{c}{d}$ are fractions, then their product is $\frac{ac}{bd}$.

This definition stated simply is *to multiply two fractions multiply numerators to get the numerator of the product, and multiply denominators to get the denominator of the product.*

Example 1 $\left(\frac{2}{3}\right)\left(\frac{5}{7}\right) = \frac{(2)(5)}{(3)(7)} = \frac{10}{21}$

Example 2 $\left(\frac{5}{2}\right)\left(\frac{17}{21}\right) = \frac{(5)(17)}{(2)(21)} = \frac{85}{42}$

◉ ***SKILL CHECK 1*** Multiply: $\left(\frac{2}{5}\right)\left(\frac{3}{7}\right)$

Always giving the product in reduced form sometimes requires more than just multiplying as in the preceding examples. Consider the following.

Answers should *always* be in reduced form.

◉ ***Skill Check Answer***

1. $\frac{6}{35}$

Example 3 $\left(\frac{2}{5}\right)\left(\frac{3}{4}\right) = \frac{(2)(3)}{(5)(4)} = \frac{6}{20}$

We note that the fraction $\frac{6}{20}$ can be reduced. So we factor and obtain

This answer is reduced.

$$\frac{6}{20} = \frac{(\not{2})(3)}{(\not{2})(2)(5)} = \frac{3}{10}.$$

In example 3 we multiplied (2)(3) to get 6, and then factored 6 to get (2)(3). This, of course, is wasted motion or unnecessary work. To avoid this we actually reduce the answer before we find it. We use the following rule.

This is a direct use of the fundamental principle of fractions.

> In an indicated multiplication of two or more fractions any factor of any numerator can be divided by a like factor of any denominator.

Example 4 $\left(\frac{2}{3}\right)\left(\frac{6}{7}\right) = \left(\frac{2}{\not{3}}\right)\left[\frac{(2)(\not{3})}{7}\right] = \frac{4}{7}$

A more efficient way of writing this example is

Instead of factoring the 6, we merely divide the 3 into it.

$$\left(\frac{2}{\not{3}}\right)\left(\frac{\overset{2}{\not{6}}}{7}\right) = \frac{4}{7}.$$

Notice that we simply divided the factor 3 into 6 and wrote the quotient above it.

$21 \div 7 = 3$
$12 \div 4 = 3$

Example 5 $\left(\frac{3}{7}\right)\left(\frac{21}{12}\right) = \left(\frac{\not{3}}{\not{7}}\right)\left(\frac{\overset{3}{\not{21}}}{\underset{4}{\not{12}}}\right) = \frac{3}{4}$

⊙ ***SKILL CHECK 2*** Multiply: $\left(\frac{3}{4}\right)\left(\frac{8}{15}\right)$

We use the same technique for multiplying algebraic fractions.

We again make use of the multiplication law of exponents.

Example 6 $\left(\frac{2x}{3y}\right)\left(\frac{4x^2}{7y^3}\right) = \frac{(2)(4)(x)(x)(x)}{(3)(7)(y)(y)(y)(y)} = \frac{8x^3}{21y^4}$

⊙ ***SKILL CHECK 3*** Multiply: $\left(\frac{2x}{5y}\right)\left(\frac{3x}{y}\right)$

$21 \div 7 = 3$
$12 \div 4 = 3$

Example 7 $\left(\frac{4a^3}{21b^2}\right)\left(\frac{7b}{12a^2}\right) = \frac{(\not{4})(\not{7})(\not{a})(\not{a})(a)(\not{b})}{(\underset{3}{\not{21}})(\underset{3}{\not{12}})(\not{b})(b)(\not{a})(\not{a})}$

$$= \frac{a}{9b}$$

$4 \div 2 = 2$
$9 \div 3 = 3$

Example 8 $\left(\frac{3a^2b}{4}\right)\left(\frac{2b^2}{9a^3}\right) = \frac{(\not{3})(\not{2})(\not{a})(\not{a})(b)(b)(b)}{(\underset{2}{\not{4}})(\underset{3}{\not{9}})(\not{a})(\not{a})(a)}$

$$= \frac{b^3}{6a}$$

⊙ ***SKILL CHECK 4*** Multiply: $\left(\frac{2ab^2}{15}\right)\left(\frac{3b}{8a^2}\right)$

⊙ ***Skill Check Answers***

2. $\frac{2}{5}$
3. $\frac{6x^2}{5y^2}$
4. $\frac{b^3}{20a}$

The number of fractions being multiplied does not change the rule or method.

Example 9 $\left(\frac{2}{3}\right)\left(\frac{9}{11}\right)\left(\frac{5}{4}\right) = \left(\frac{\cancel{2}}{\cancel{3}}\right)\left(\frac{\cancel{9}^{3}}{11}\right)\left(\frac{5}{\cancel{4}_{2}}\right) = \frac{15}{22}$

$9 \div 3 = 3$
$4 \div 2 = 2$

⊙ ***SKILL CHECK 5*** Multiply: $\left(\frac{2}{3}\right)\left(\frac{9}{13}\right)\left(\frac{5}{6}\right)$

Example 10 $\left(\frac{x^2}{a}\right)\left(\frac{2y}{ab}\right)\left(\frac{a}{x^4}\right) = \frac{(\cancel{x})(\cancel{x})(2)(y)(\cancel{a})}{(\cancel{a})(a)(b)(\cancel{x})(\cancel{x})(x)(x)}$

$= \frac{2y}{abx^2}$

⊙ ***SKILL CHECK 6*** Multiply: $\left(\frac{x^2}{a}\right)\left(\frac{3y}{x}\right)\left(\frac{a^2}{x^2}\right)$

It is important to recognize that the set of whole numbers can be written as fractions. For instance, the whole number 5 can be written as $\frac{5}{1}, \frac{10}{2}, \frac{15}{3}$, and so on. So multiplication of a fraction by a whole number follows the same rule.

How many different ways could you write 5 as a fraction?

Example 11 Find $\frac{2}{3}$ of 6.

In word problems, *of* usually means multiply.

Solution $\left(\frac{2}{3}\right)(6) = \left(\frac{2}{\cancel{3}}\right)\left(\frac{\cancel{6}^{2}}{1}\right) = \frac{4}{1} = 4$

Notice that the whole number is written with a denominator of 1.

⊙ ***SKILL CHECK 7*** Find $\frac{3}{4}$ of 12.

Example 12 $\left(\frac{5}{8}\right)(4)\left(\frac{3}{5}\right) = \left(\frac{\cancel{5}}{\cancel{8}_{2}}\right)\left(\frac{\cancel{4}}{1}\right)\left(\frac{3}{\cancel{5}}\right) = \frac{3}{2}$

⊙ ***SKILL CHECK 8*** Multiply: $\left(\frac{3}{8}\right)(10)\left(\frac{2}{3}\right)$

Example 13 $\frac{3x^2}{2ab}(4a^3) = \left(\frac{3x^2}{2ab}\right)\left(\frac{4a^3}{1}\right) = \frac{(3)(\cancel{4}^{2})(x)(x)(\cancel{a})(a)(a)}{(\cancel{2})(\cancel{a})(b)}$

$= \frac{6a^2x^2}{b}$

⊙ ***SKILL CHECK 9*** Multiply: $\left(\frac{2x^2}{3ab}\right)(6a^2)$

2 Applications

Fractions are often used in solving word problems. To solve a problem stated in words first carefully read the problem. Determine what is being asked for. That is, what do we need to find? Then consider how the given information can be used to find the answer.

⊙ ***Skill Check Answers***

5. $\frac{5}{13}$
6. $\frac{3ay}{x}$
7. 9
8. $\frac{5}{2}$
9. $\frac{4ax^2}{b}$

Example 14 Bob and his friend earned $24 mowing a lawn and each took one-half of the total earned. Bob gave one-third of his half to his sister for her birthday. How much did his sister get?

Analyze the problem carefully.

Solution Reading this problem carefully we see we are asked to find the amount of money that Bob's sister receives. Before we can find how much she receives we must first determine how much Bob earned. Bob and his friend each earned one-half the total earnings. Thus Bob earned one-half of $24.

$$\left(\frac{1}{\not{2}}\right)\left(\frac{\overset{12}{\not{24}}}{1}\right) = \$12$$

You could also write this problem in one step as

$$\left(\frac{1}{3}\right)\left(\frac{1}{2}\right)(24).$$

Bob gave his sister $\frac{1}{3}$ of the amount he earned so the amount he gave her is

$$\left(\frac{1}{\not{3}}\right)\left(\frac{\overset{4}{\not{12}}}{1}\right) = \$4.$$

◉ *SKILL CHECK 10* Solve the previous example by multiplying: $\left(\frac{1}{3}\right)\left(\frac{1}{2}\right)(24)$

Example 15 A man decided to give one-tenth of his income to charity. He also designated that one-third of the amount for charities will go to the Red Cross. What part of his income goes to the Red Cross?

Solution We are asked to find the fractional part of the man's income that is given to the Red Cross. The information tells us that one-tenth of the man's income is given to all charities.

$$\frac{1}{10} \text{ to charities}$$

To find the part that goes to the Red Cross we must find $\frac{1}{3}$ of $\frac{1}{10}$.

$$\left(\frac{1}{3}\right)\left(\frac{1}{10}\right) = \frac{1}{30} \text{ to Red Cross}$$

◉ *SKILL CHECK 11* If Carol gives one-tenth of her income to charity and two-thirds of that amount goes to the United Way, what part of her income goes to the United Way?

◉ *Skill Check Answers*

10. $4

11. $\frac{1}{15}$

Exercise 3–2–1

1 Multiply and simplify.

1. $\left(\frac{1}{2}\right)\left(\frac{3}{5}\right)$
2. $\left(\frac{1}{3}\right)\left(\frac{2}{5}\right)$
3. $\left(\frac{1}{4}\right)\left(\frac{1}{2}\right)$

4. $\left(\frac{1}{3}\right)\left(\frac{1}{6}\right)$

5. $\left(\frac{2}{3}\right)\left(\frac{5}{7}\right)$

6. $\left(\frac{3}{5}\right)\left(\frac{2}{11}\right)$

7. $\left(\frac{3}{4}\right)\left(\frac{3}{5}\right)$

8. $\left(\frac{2}{3}\right)\left(\frac{2}{5}\right)$

9. $\left(\frac{a^2}{b^3}\right)\left(\frac{a}{b}\right)$

10. $\left(\frac{a}{b^2}\right)\left(\frac{a}{b}\right)$

11. $\left(\frac{x}{y}\right)\left(\frac{x^3}{y}\right)$

12. $\left(\frac{x^2}{y^3}\right)\left(\frac{x}{y^2}\right)$

13. $\left(\frac{2}{5}\right)\left(\frac{5}{7}\right)$

14. $\left(\frac{3}{8}\right)\left(\frac{1}{3}\right)$

15. $\left(\frac{5}{6}\right)\left(\frac{6}{11}\right)$

16. $\left(\frac{3}{4}\right)\left(\frac{4}{7}\right)$

17. $\left(\frac{x^2}{y}\right)\left(\frac{y^3}{x}\right)$

18. $\left(\frac{x}{y}\right)\left(\frac{y^4}{x^2}\right)$

19. $\left(\frac{a^2}{b^3}\right)\left(\frac{b^4}{a^3}\right)$

20. $\left(\frac{a}{b}\right)\left(\frac{b^2}{a^2}\right)$

21. $\left(\frac{1}{2}\right)\left(\frac{4}{5}\right)$

22. $\left(\frac{2}{3}\right)\left(\frac{6}{7}\right)$

23. $\left(\frac{14}{3}\right)\left(\frac{5}{7}\right)$

24. $\left(\frac{12}{13}\right)\left(\frac{1}{4}\right)$

25. $\left(\frac{x}{5}\right)\left(\frac{10}{x^2}\right)$

26. $\left(\frac{x^3}{4}\right)\left(\frac{8}{x^5}\right)$

27. $\left(\frac{a^2}{5}\right)\left(\frac{3}{a}\right)$

28. $\left(\frac{a^3}{8}\right)\left(\frac{5}{a^2}\right)$

29. $\left(\frac{1}{2}\right)\left(\frac{1}{4}\right)\left(\frac{3}{5}\right)$

30. $\left(\frac{1}{3}\right)\left(\frac{2}{5}\right)\left(\frac{1}{7}\right)$

31. $\left(\frac{1}{3a^2}\right)\left(\frac{9a^3}{5}\right)\left(\frac{1}{2a}\right)$

32. $\left(\frac{2x^4}{3}\right)\left(\frac{1}{10x}\right)\left(\frac{1}{5x^2}\right)$

33. $\left(\frac{3a}{4b}\right)\left(\frac{2b^3}{3a^2}\right)$

34. $\left(\frac{2b^2}{5a^2}\right)\left(\frac{5a}{8b^5}\right)$

35. $\left(\frac{4}{10}\right)\left(\frac{2}{5}\right)$

36. $\left(\frac{3}{4}\right)\left(\frac{12}{15}\right)$

37. $\left(\frac{2a}{3y}\right)\left(\frac{3y^2}{4}\right)\left(\frac{1}{9a^3}\right)$

38. $\left(\frac{1}{4x}\right)\left(\frac{2x^3}{3y}\right)\left(\frac{6y^2}{8x}\right)$

39. $\left(\frac{4}{7}\right)\left(\frac{21}{22}\right)$

40. $\left(\frac{7}{8}\right)\left(\frac{18}{21}\right)$

41. $\left(\frac{3a}{16b^2}\right)\left(\frac{6a^2b^2}{9ab}\right)$

42. $\left(\frac{4x^2}{18y}\right)\left(\frac{6xy^3}{20x^3y}\right)$

43. $\left(\frac{21}{32}\right)\left(\frac{18}{28}\right)\left(\frac{16}{40}\right)$

44. $\left(\frac{14}{36}\right)\left(\frac{9}{24}\right)\left(\frac{6}{21}\right)$

45. $\left(\frac{4}{9}\right)\left(\frac{30}{50}\right)$

46. $\left(\frac{16}{21}\right)\left(\frac{28}{38}\right)$

47. $\left(\frac{15a}{28b^2}\right)\left(\frac{14b^3}{25a^2b}\right)$

48. $\left(\frac{15x}{17y}\right)\left(\frac{51xy}{54x^2}\right)$

49. $\left(\frac{21}{14}\right)\left(\frac{10}{15}\right)$

50. $\left(\frac{30}{18}\right)\left(\frac{9}{15}\right)$

51. $\left(\frac{2}{3}\right)(6)$

52. $\left(\frac{1}{2}\right)(4)$

53. $\left(\frac{a^2b}{4ab}\right)(12b^2)$

54. $\left(\frac{3a}{4xy}\right)(8x^2y)$

55. $\left(\frac{3a}{5a^2b}\right)(15a^3b^3)$

56. $\left(\frac{2x^4}{3x^2y}\right)(18x^2y^2)$

57. $\left(\frac{1}{2}\right)(6)\left(\frac{5}{3}\right)$

58. $\left(\frac{3}{5}\right)(10)\left(\frac{3}{8}\right)$

59. $\left(\frac{3x}{4y^2}\right)(12y)\left(\frac{xy}{9x^2}\right)$

60. $\left(\frac{5a^2}{12b^3}\right)(16b^4)\left(\frac{3b}{20ab}\right)$

61. Find $\frac{3}{4}$ of 80.

62. Find $\frac{2}{5}$ of 35.

63. Find $\frac{3}{7}$ of 21.

64. Find $\frac{1}{8}$ of 96.

65. Find $\frac{3}{5}$ of $\frac{15}{21}$.

66. Find $\frac{2}{3}$ of $\frac{9}{14}$.

67. Find $\frac{12}{13}$ of $\frac{39}{36}$.

68. Find $\frac{4}{11}$ of $\frac{121}{122}$.

69. Find $\frac{1}{12}$ of 156.

70. Find $\frac{3}{7}$ of $\frac{28}{36}$.

2

71. A bottle of a certain perfume contains $\frac{1}{2}$ ounce. How many ounces are there in 36 bottles?

72. A can of vegetables weighs $\frac{2}{3}$ pound. What is the weight of 48 cans?

73. A 2-liter bottle of soft drink is $\frac{1}{4}$ full. How much is in the bottle?

74. A man owned $\frac{1}{4}$ interest in a business. If he sold $\frac{1}{2}$ of his share, how much does he now own?

75. A certain recipe requires $\frac{2}{3}$ cup of sugar, $\frac{1}{4}$ teaspoon of vanilla, and $\frac{1}{2}$ cup of flour. How much of each ingredient is needed for one-half the recipe?

76. A rectangular field measures $\frac{2}{3}$ mile by $\frac{3}{8}$ mile. What is the area of the field?

77. Janet earns $225 per week. She spends $\frac{1}{3}$ of her earnings for rent. How much does she spend for rent each week?

78. The scale on a map reads that 1 inch represents 400 miles. How many miles are represented by $\frac{3}{8}$ inch?

- **Concept Enrichment**

79. Mike's pickup truck can carry $\frac{3}{4}$ ton of gravel. If he carries six loads and is paid \$20 per ton, how much does he earn?

80. A man gave $\frac{1}{10}$ of his income to charity. He gave $\frac{1}{3}$ the amount for charity to the United Fund. If his income was \$36,000, how much did he give to the United Fund?

3–3 Dividing Fractions

OBJECTIVES

Upon completing this section you should be able to:

1. Simplify a complex fraction.
2. Divide fractions.
3. Use division of fractions to solve applications.

The second basic operation on fractions we wish to study is the operation of division. In chapter 2 we saw that division of whole numbers was related to multiplication. The same is true for fractions and to establish a rule for dividing them we must first look at the *reciprocal* of a number. Observe the following multiplications.

$$\left(\frac{2}{3}\right)\left(\frac{3}{2}\right) = \left(\frac{\overset{1}{\cancel{2}}}{\underset{1}{\cancel{3}}}\right)\left(\frac{\overset{1}{\cancel{3}}}{\underset{1}{\cancel{2}}}\right) = 1$$

$$(5)\left(\frac{1}{5}\right) = \left(\frac{\cancel{5}}{1}\right)\left(\frac{1}{\cancel{5}}\right) = 1$$

$$\left(\frac{x}{y}\right)\left(\frac{y}{x}\right) = \left(\frac{\overset{1}{\cancel{x}}}{\underset{1}{\cancel{y}}}\right)\left(\frac{\overset{1}{\cancel{y}}}{\underset{1}{\cancel{x}}}\right) = 1$$

The above multiplications lead us to the following definition.

> If a and b are numbers and if $ab = 1$, then a and b are **reciprocals** of each other. Every number, except zero, has a reciprocal.

Do you see why zero could not have a reciprocal? See section 2–2.

Notice that if a and b are not zero, the reciprocal of a fraction

$$\frac{a}{b} \text{ is } \frac{b}{a} \text{ since } \frac{a}{b} \times \frac{b}{a} = 1.$$

What is the reciprocal of the number 1?

We sometimes call the reciprocal the **multiplicative inverse.**

Example 1 $\frac{5}{8}$ and $\frac{8}{5}$ are reciprocals because $\frac{5}{8} \times \frac{8}{5} = 1$.

Example 2 $\frac{x^2}{a}$ and $\frac{a}{x^2}$ are reciprocals because $\frac{x^2}{a} \times \frac{a}{x^2} = 1$.

Example 3 b^3 and $\frac{1}{b^3}$ are reciprocals because $b^3 \times \frac{1}{b^3} = 1$.

1 Complex Fractions

Now consider this division problem.

Example 4 Find the quotient: $\frac{2}{3} \div \frac{5}{8}$

Solution Since the fraction bar always means division we can write $\frac{2}{3} \div \frac{5}{8}$ as

$a \div b$ and $\frac{a}{b}$ mean the same thing.

$$\frac{\frac{2}{3}}{\frac{5}{8}}.$$

We now have what is called a **complex fraction,** that is, a fraction containing fractions. To simplify the complex fraction we will use the fundamental principle of fractions and multiply both the numerator and denominator of the complex fraction by a number that will yield 1 in the denominator. In other words, we multiply by the reciprocal of the denominator of the complex fraction. In this case we multiply by the reciprocal of $\frac{5}{8}$, which is $\frac{8}{5}$.

Remember, $\frac{a}{b} = \frac{ac}{bc}$.

We multiply the numerator and denominator by $\frac{8}{5}$ and simplify.

$$\frac{\frac{2}{3}}{\frac{5}{8}} = \frac{\left(\frac{2}{3}\right)\left(\frac{8}{5}\right)}{\left(\frac{5}{8}\right)\left(\frac{8}{5}\right)} = \frac{\left(\frac{2}{3}\right)\left(\frac{8}{5}\right)}{1} = \left(\frac{2}{3}\right)\left(\frac{8}{5}\right)$$

So $\frac{2}{3} \div \frac{5}{8} = \left(\frac{2}{3}\right)\left(\frac{8}{5}\right) = \frac{16}{15}$.

SKILL CHECK 1 Find the quotient: $\frac{\frac{1}{2}}{\frac{3}{5}}$

2 Dividing Fractions

Example 4 is an illustration of the following rule.

> If $\frac{a}{b}$ and $\frac{c}{d}$ are fractions, then
>
> $$\frac{a}{b} \div \frac{c}{d} = \left(\frac{a}{b}\right)\left(\frac{d}{c}\right).$$

WARNING

Be sure to invert the *divisor,* which is the number to the right of the $\div$ sign.

In simpler words, if we wish to divide two fractions we change the problem to multiplication by inverting the divisor. We even shorten the statement to "invert and multiply."

Example 5 Divide: $\frac{5}{8} \div \frac{1}{2}$

$\frac{1}{2}$ is the divisor.

Solution

$$\frac{5}{8} \div \frac{1}{2} = \left(\frac{5}{8}\right)\left(\frac{2}{1}\right) = \left(\frac{5}{\cancel{8}_4}\right)\left(\frac{\cancel{2}^1}{1}\right) = \frac{5}{4}$$

Skill Check Answer

1. $\frac{5}{6}$

⊙ ***SKILL CHECK 2*** Divide: $\frac{3}{4} \div \frac{1}{2}$

Example 6 Find the quotient: $\frac{2}{3} \div \frac{5}{6}$

Which is the divisor?

Solution

$$\frac{2}{3} \div \frac{5}{6} = \left(\frac{2}{\cancel{3}_1}\right)\left(\frac{\cancel{6}^2}{5}\right) = \frac{4}{5}$$

Example 7 Divide: $7 \div \frac{3}{4}$

Solution

$$7 \div \frac{3}{4} = \left(\frac{7}{1}\right)\left(\frac{4}{3}\right) = \frac{28}{3}$$

Remember, $7 = \frac{7}{1}$.

⊙ ***SKILL CHECK 3*** Divide: $5 \div \frac{2}{3}$

Example 8 Find the quotient: $\frac{4}{5} \div 6$

Solution

$$\frac{4}{5} \div 6 = \left(\frac{\cancel{4}^2}{5}\right)\left(\frac{1}{\cancel{6}_3}\right) = \frac{2}{15}$$

⊙ ***SKILL CHECK 4*** Find the quotient: $\frac{3}{5} \div 6$

Example 9 Divide: $\frac{2ax^2}{5y} \div \frac{10x^3}{y^2}$

Solution

$$\frac{2ax^2}{5y} \div \frac{10x^3}{y^2} = \left(\frac{2ax^2}{5y}\right)\left(\frac{y^2}{10x^3}\right)$$

$$= \frac{(\cancel{2})(a)(\cancel{x})(\cancel{x})(\cancel{y})(y)}{(5)(\cancel{10}_5)(\cancel{y})(\cancel{x})(\cancel{x})(x)}$$

$$= \frac{ay}{25x}$$

⊙ ***SKILL CHECK 5*** Divide: $\frac{2ax}{3y} \div \frac{4x^2}{y^3}$

Example 10 Divide: $3x^2 \div \frac{12x^2}{7}$

Solution

$$3x^2 \div \frac{12x^2}{7} = \left(\frac{3x^2}{1}\right)\left(\frac{7}{12x^2}\right)$$

$$= \frac{(\cancel{3})(7)(\cancel{x})(\cancel{x})}{(\cancel{12}_4)(\cancel{x})(\cancel{x})}$$

$$= \frac{7}{4}$$

⊙ ***SKILL CHECK 6*** Divide: $5x^2 \div \frac{10x^2}{3}$

⊙ ***Skill Check Answers***

2. $\frac{3}{2}$
3. $\frac{15}{2}$
4. $\frac{1}{10}$
5. $\frac{ay^2}{6x}$
6. $\frac{3}{2}$

Example 11 Divide: $\frac{14a^2b^2}{5} \div 7a^3$

Solution

$$\frac{14a^2b^2}{5} \div 7a^3 = \left(\frac{14a^2b^2}{5}\right)\left(\frac{1}{7a^3}\right)$$

$$= \frac{(\overset{2}{\cancel{14}})(\cancel{a})(\cancel{a})(b)(b)}{(5)(7)(\cancel{a})(\cancel{a})(a)}$$

$$= \frac{2b^2}{5a}$$

⦿ ***SKILL CHECK 7*** Divide: $\frac{12a^3b^2}{7} \div 6ab^3$

3 Applications

Example 12 How many $\frac{1}{4}$-pound candy bars are there in a 3-pound box?

Solution If the box contains 3 pounds of candy and each bar weighs $\frac{1}{4}$ pound, then the question is how many times will $\frac{1}{4}$ divide into 3?

Be careful when deciding which number is the divisor.

$$3 \div \frac{1}{4} = \left(\frac{3}{1}\right)\left(\frac{4}{1}\right) = 12$$

There are therefore 12 bars in the box.

⦿ ***SKILL CHECK 8*** How many $\frac{2}{5}$-pound candy bars are there in a 4-pound box?

⦿ ***Skill Check Answers***

7. $\frac{2a^2}{7b}$
8. 10 bars

Exercise 3–3–1

1 Simplify.

1. $\frac{\frac{2}{3}}{\frac{3}{4}}$ **2.** $\frac{\frac{2}{3}}{\frac{5}{7}}$ **3.** $\frac{\frac{1}{2}}{\frac{2}{3}}$ **4.** $\frac{\frac{1}{3}}{\frac{3}{4}}$

5. $\frac{5}{\frac{1}{4}}$ **6.** $\frac{3}{\frac{2}{3}}$ **7.** $\frac{\frac{1}{4}}{5}$ **8.** $\frac{\frac{2}{3}}{3}$

9. $\dfrac{\frac{3}{5}}{\frac{9}{10}}$ **10.** $\dfrac{\frac{2}{7}}{\frac{6}{35}}$ **11.** $\dfrac{\frac{5}{6}}{\frac{15}{18}}$ **12.** $\dfrac{\frac{3}{7}}{\frac{15}{28}}$

2 Find each quotient and simplify.

13. $\frac{1}{2} \div \frac{3}{4}$ **14.** $\frac{1}{3} \div \frac{5}{6}$ **15.** $\frac{3}{4} \div \frac{1}{2}$

16. $\frac{5}{6} \div \frac{1}{3}$ **17.** $\frac{x^2}{y} \div \frac{x}{y}$ **18.** $\frac{a}{b^2} \div \frac{a^2}{b}$

19. $\frac{4a}{9b} \div \frac{2a^2}{3b^2}$ **20.** $\frac{9a^2}{16b} \div \frac{3a}{4b^2}$ **21.** $\frac{5}{9} \div \frac{2}{5}$

22. $\frac{1}{3} \div \frac{3}{5}$ **23.** $\frac{4}{3} \div \frac{8}{6}$ **24.** $\frac{4}{5} \div \frac{3}{10}$

25. $\frac{5x^3}{6a} \div \frac{6ax^2}{7a^2}$ **26.** $\frac{3a}{8b} \div \frac{8a^2x}{5by}$ **27.** $\frac{6a^2}{11by} \div \frac{4a}{11b^2y}$

28. $\frac{4x^4}{9y^2} \div \frac{2ax^2}{3y^3}$ **29.** $\frac{5}{8} \div \frac{7}{8}$ **30.** $\frac{4}{7} \div \frac{3}{7}$

31. $8 \div \frac{4}{5}$

32. $6 \div \frac{2}{3}$

33. $\frac{3}{4} \div \frac{7}{8}$

34. $\frac{5}{12} \div \frac{2}{9}$

35. $\frac{6ax^2}{7y} \div \frac{7x^3}{8y^2}$

36. $\frac{2a^2b}{9x} \div \frac{9a}{10x^2}$

37. $5a^2 \div \frac{2a}{5}$

38. $9x^3 \div \frac{2x}{3}$

39. $\frac{2}{5} \div 5$

40. $\frac{2}{3} \div 9$

41. $\frac{3}{4} \div \frac{3}{8}$

42. $\frac{5}{9} \div \frac{5}{3}$

43. $\frac{13x^2y^3}{28} \div 52xy^2$

44. $\frac{11x^2y^2}{20} \div 44y^3$

45. $16 \div \frac{8}{9}$

46. $30 \div \frac{6}{5}$

47. $\frac{3}{8} \div \frac{5}{6}$

48. $3 \div \frac{6}{7}$

49. $\frac{7a^2}{24b} \div \frac{21ax}{18b^2y}$

50. $\frac{6x^3}{25a} \div \frac{3bx}{10a^2}$

51. $\frac{4}{5} \div 8$

52. $\frac{12}{13} \div 4$

53. $\frac{7}{18} \div \frac{7}{12}$

54. $\frac{8}{15} \div \frac{6}{35}$

55. $\frac{5ax^3}{28by} \div \frac{25a^2x}{42b^2y^2}$

56. $\frac{7a^2c^2}{11b^2} \div \frac{14a^3c}{33bd}$

57. $\frac{11a^2c}{60b^4d} \div \frac{44ac^3}{50b^3d^2}$

58. $\frac{12x^2y}{21a^2b^2} \div \frac{16xy^3}{35ab^2}$

59. $\frac{17}{35} \div \frac{51}{56}$

60. $\frac{11}{38} \div \frac{22}{19}$

61. $\frac{33}{7} \div \frac{11}{21}$

62. $\frac{24}{27} \div \frac{8}{9}$

63. $\frac{18}{7} \div \frac{9}{21}$

64. $\frac{6}{11} \div \frac{2}{3}$

65. $\frac{15}{21} \div \frac{3}{14}$

66. $\frac{4}{15} \div \frac{9}{20}$

67. $\frac{4a^2}{25b} \div \frac{12b^2}{35a}$

68. $\frac{32x}{15y} \div \frac{16ay}{25x}$

69. $\frac{5}{21} \div \frac{15}{49}$

70. $\frac{14}{15} \div \frac{35}{36}$

3

71. Each student is served $\frac{1}{2}$ pint of milk. How many students can be served from 50 pints of milk?

72. How many "quarter-pounder" hamburgers can be made from 12 pounds of hamburger?

73. Jean has a 10-yard spool of ribbon from which she wants to make some bows. If it takes $\frac{2}{3}$ yard of ribbon to make each bow, how many bows can she make?

74. Bob wants to buy twelve liters of wine for a party but the store only has $\frac{3}{4}$-liter bottles. How many bottles will it take to get twelve liters?

75. A furlong is $\frac{1}{8}$ mile. How many furlongs are there in $\frac{3}{4}$ mile?

76. A certain tablet contains $\frac{5}{8}$ grain of medicine. How many tablets can be made from 40 grains?

77. A person with a $\frac{3}{4}$-gallon pail must fill a 24-gallon tank. How many full pails will it take?

78. A $\frac{3}{4}$-acre tract of land is to be divided into six equal sections. How much of an acre will each section be?

79. A truck can carry $\frac{3}{4}$ ton of gravel. How many loads will it take to carry 12 tons?

80. A car travels $\frac{7}{10}$ of a mile per minute. How long does it take to travel 14 miles?

• Concept Enrichment

81. The product of any number and its multiplicative inverse will always result in what value?

82. Why doesn't zero have a reciprocal?

3–4 Adding and Subtracting Fractions

In the two previous sections we studied the operations of multiplication and division of fractions. We now wish to present the remaining two basic operations, addition and subtraction.

OBJECTIVES

Upon completing this section you should be able to:

1. Add and subtract like fractions.
2. Find the LCD of two or more fractions.
3. Change a fraction to an equivalent fraction.
4. Compare the relative size of two fractions.
5. Add and subtract any two fractions.

1 Add and Subtract Like Fractions

In chapter 1 we stated that in mathematics only like quantities can be added or subtracted. This indicates that we need the following definition of like fractions.

> Two fractions are said to be **like fractions** if and only if they have the same denominator.

We say the fractions have a *common* denominator.

$\frac{2}{3}$ and $\frac{5}{3}$ are like fractions. — 3 is the common denominator.

$\frac{7}{x}$ and $\frac{5}{x}$ are like fractions. — x is the common denominator.

$\frac{2}{3}$ and $\frac{2}{5}$ are *not* like fractions. — No common denominator.

$\frac{a}{x}$ and $\frac{a}{y}$ are *not* like fractions. — No common denominator.

In section 1–7 we added like terms such as $3x + 7x = 10x$ where x represents any number. If we think of the fraction $\frac{3}{8}$ as $3\left(\frac{1}{8}\right)$ and the fraction $\frac{7}{8}$ as $7\left(\frac{1}{8}\right)$, then $\frac{3}{8} + \frac{7}{8} = 3\left(\frac{1}{8}\right) + 7\left(\frac{1}{8}\right) = 10\left(\frac{1}{8}\right) = \frac{10}{8}$.
This illustrates the following rules for adding and subtracting like fractions.

> $$\frac{a}{x} + \frac{b}{x} = \frac{a+b}{x}$$
>
> To add like fractions place the sum of the numerators over the common denominator.

$$\frac{a}{x} - \frac{b}{x} = \frac{a-b}{x}$$

To subtract like fractions place the difference of the numerators over the common denominator.

Example 1 $\frac{2}{3} + \frac{5}{3} = \frac{2+5}{3} = \frac{7}{3}$

⦿ ***SKILL CHECK 1*** Add: $\frac{1}{5} + \frac{2}{5}$

Example 2 $\frac{7}{x} - \frac{5}{x} = \frac{7-5}{x} = \frac{2}{x}$

⦿ ***SKILL CHECK 2*** Subtract: $\frac{4}{a} - \frac{1}{a}$

Note that $\frac{6}{2}$ reduces to 3.

Example 3 $\frac{1}{2} + \frac{5}{2} = \frac{6}{2} = 3$

⦿ ***SKILL CHECK 3*** Add: $\frac{1}{3} + \frac{5}{3}$

Example 4 $\frac{2x}{a} + \frac{x}{a} = \frac{2x+x}{a} = \frac{3x}{a}$

⦿ ***SKILL CHECK 4*** Add: $\frac{2a}{3} + \frac{a}{3}$

Note that if the numerators of algebraic fractions are not like terms, they cannot be added or subtracted.

Example 5 $\frac{a^2}{x} + \frac{2b^2}{x} = \frac{a^2+2b^2}{x}$

⦿ ***SKILL CHECK 5*** Add: $\frac{a}{x} + \frac{3}{x}$

Example 6 $\frac{5}{xy} - \frac{b}{xy} = \frac{5-b}{xy}$

⦿ ***SKILL CHECK 6*** Subtract: $\frac{a}{xy} - \frac{b}{xy}$

Reduce the answer if possible.

Example 7 $\frac{5}{8} + \frac{7}{8} = \frac{12}{8} = \frac{3}{2}$

Notice that the sum or difference of two fractions may be reduced even though neither of the fractions can be reduced. All fractional answers should be in reduced form, if possible.

⦿ ***SKILL CHECK 7*** Subtract: $\frac{7}{8} - \frac{5}{8}$

⦿ ***Skill Check Answers***

1. $\frac{3}{5}$
2. $\frac{3}{a}$
3. 2
4. a
5. $\frac{a+3}{x}$
6. $\frac{a-b}{xy}$
7. $\frac{1}{4}$

Exercise 3-4-1

1 Perform the indicated operation and simplify.

1. $\frac{1}{5} + \frac{2}{5}$

2. $\frac{2}{7} + \frac{3}{7}$

3. $\frac{3}{5} - \frac{1}{5}$

4. $\frac{5}{9} - \frac{4}{9}$

5. $\frac{5}{x} + \frac{1}{x}$

6. $\frac{1}{a} + \frac{5}{a}$

7. $\frac{a}{7} + \frac{b}{7}$

8. $\frac{x}{5} + \frac{2y}{5}$

9. $\frac{a}{9} - \frac{2b}{9}$

10. $\frac{x}{8} - \frac{x^2}{8}$

11. $\frac{11}{12} - \frac{5}{12}$

12. $\frac{13}{15} - \frac{3}{15}$

13. $\frac{4x}{5} + \frac{6x}{5}$

14. $\frac{5x}{8} + \frac{11x}{8}$

15. $\frac{22}{a} - \frac{9}{a}$

16. $\frac{25}{x} - \frac{9}{x}$

17. $\frac{2x}{15} + \frac{3x}{15}$

18. $\frac{a}{45} + \frac{2a}{45}$

19. $\frac{2a}{xy} - \frac{b}{xy}$

20. $\frac{3x}{ab} - \frac{2y}{ab}$

21. $\frac{11}{2x} + \frac{7}{2x}$

22. $\frac{13}{7x} + \frac{1}{7x}$

23. $\frac{31}{36} - \frac{13}{36}$

24. $\frac{43}{42} - \frac{29}{42}$

25. $\frac{31}{18a} - \frac{15}{18a}$

26. $\frac{39}{33x} - \frac{17}{33x}$

27. $\frac{27}{15} + \frac{18}{15}$

28. $\frac{23}{14} + \frac{33}{14}$

29. $\frac{11}{40ab} + \frac{15}{40ab}$

30. $\frac{13}{27xy} + \frac{5}{27xy}$

31. $\frac{31}{19x} - \frac{12}{19x}$

32. $\frac{41}{24a} - \frac{17}{24a}$

33. $\frac{5}{8a} - \frac{3}{8a}$

34. $\frac{7}{9ab} - \frac{4}{9ab}$

35. $\frac{39a}{41b} + \frac{43a}{41b}$

36. $\frac{31x}{26y} + \frac{21x}{26y}$

37. $\frac{31}{38} + \frac{1}{38}$

38. $\frac{17}{28} + \frac{1}{28}$

39. $\frac{15a}{32bc} - \frac{3a}{32bc}$

40. $\frac{29x}{46yz} + \frac{11x}{46yz}$

2 Finding the LCD

The preceding examples and exercises all involved the sum or difference of like fractions. But what about the sum of two fractions such as $\frac{1}{3}$ and $\frac{1}{2}$? Obviously these two fractions cannot be added in their present form because they are not alike.

Remember, only like fractions can be added or subtracted.

Recalling the fundamental principle of fractions, we know that $\frac{1}{3}$ can be written in many equivalent forms.

$$\frac{1}{3} = \frac{2}{6} = \frac{3}{9} = \frac{4}{12} = \frac{5}{15}, \text{ and so on}$$

There are infinitely many ways to write a fraction.

Also

$$\frac{1}{2} = \frac{2}{4} = \frac{3}{6} = \frac{4}{8} = \frac{5}{10} = \frac{6}{12}, \text{ and so on.}$$

In the lists of ways to write $\frac{1}{3}$ and $\frac{1}{2}$ we find some ways in which the fractions are expressed as like fractions. Notice that $\frac{1}{3}$ is $\frac{2}{6}$ and $\frac{1}{2}$ is $\frac{3}{6}$. Also that $\frac{1}{3}$ is $\frac{4}{12}$ and $\frac{1}{2}$ is $\frac{6}{12}$. In these forms they can be added (or subtracted) and this is our clue to adding and subtracting unlike fractions.

$\frac{2}{6}$ and $\frac{3}{6}$ are like fractions.

$$\frac{1}{3} + \frac{1}{2} = \frac{2}{6} + \frac{3}{6} = \frac{5}{6}$$

or

$$\frac{1}{3} + \frac{1}{2} = \frac{4}{12} + \frac{6}{12} = \frac{10}{12} = \frac{5}{6}$$

Note that here we have to reduce.

Notice that if we do not find the least denominator that makes $\frac{1}{3}$ and $\frac{1}{2}$ like fractions, it will be necessary to reduce the sum. This will always be true so it is to our advantage to find the least number that is a common denominator. In other words, we want the *least common denominator.*

Even if we find the least common denominator it may sometimes be necessary to reduce the sum.

> The **least common denominator** (LCD) of two or more fractions is the least common multiple of their denominators.

Refer to LCM in chapter 2.

Example 8 Find the LCD of $\frac{5}{12}$ and $\frac{7}{18}$.

By definition, the LCD of the two fractions will be the LCM of the denominators 12 and 18.

Solution Remember to find the LCM of 12 and 18 we factor each number into primes.

$$12 = 2 \times 2 \times 3$$
$$18 = 2 \times 3 \times 3$$
$$\text{LCM} = 2 \times 2 \times 3 \times 3 = 36$$

So the LCD of $\frac{5}{12}$ and $\frac{7}{18}$ is 36.

⊙ ***SKILL CHECK 8*** Find the LCD of $\frac{5}{6}$ and $\frac{3}{10}$.

⊙ ***Skill Check Answer***

8. 30

Example 9 Find the LCD of $\frac{5x}{a^2}$ and $\frac{5}{a^3b}$.

Solution The least common denominator is the least common multiple of the denominators. From chapter 2 we know that we must use the highest power of each letter involved. So the LCD of $\frac{5x}{a^2}$ and $\frac{5}{a^3b}$ is a a^3b.

⊙ ***SKILL CHECK 9*** Find the LCD of $\frac{3x}{a^3}$ and $\frac{4}{a^2b}$.

Example 10 Find the LCD of $\frac{3}{5x^2}$ and $\frac{5}{3xy}$.

Solution The LCM of 3 and 5 is 15. The LCM of x^2 and xy is x^2y. So the LCD of $\frac{3}{5x^2}$ and $\frac{5}{3xy}$ is $15x^2y$.

⊙ ***SKILL CHECK 10*** Find the LCD of $\frac{2}{3x^3}$ and $\frac{3}{4x^2y}$.

⊙ ***Skill Check Answers***

9. a^3b
10. $12x^3y$

Exercise 3–4–2

2 Find the LCD of each set of fractions.

1. $\frac{1}{2}$ and $\frac{1}{3}$

2. $\frac{1}{3}$ and $\frac{1}{5}$

3. $\frac{1}{2}$ and $\frac{5}{6}$

4. $\frac{2}{3}$ and $\frac{1}{6}$

5. $\frac{1}{a}$ and $\frac{4}{b}$

6. $\frac{4}{x}$ and $\frac{5}{y}$

7. $\frac{5}{x^2}$ and $\frac{1}{xy}$

8. $\frac{3}{a}$ and $\frac{7}{a^2b}$

9. $\frac{3x}{16}$ and $\frac{11y}{24}$

10. $\frac{7a}{20}$ and $\frac{11a^2}{36}$

11. $\frac{x}{12}$ and $\frac{13}{42}$

12. $\frac{3x}{14}$ and $\frac{16}{21}$

13. $\frac{5}{2x}$ and $\frac{7}{5x3y}$

14. $\frac{5}{3a^2}$ and $\frac{3}{4ab^2}$

15. $\frac{1}{2}$, $\frac{5}{6}$, and $\frac{1}{9}$

16. $\frac{1}{2}$, $\frac{3}{8}$, and $\frac{5}{6}$

17. $\frac{1}{6}$, $\frac{3}{8}$, and $\frac{4}{9}$

18. $\frac{1}{3}$, $\frac{5}{6}$, and $\frac{1}{10}$

19. $\frac{1}{x}$, $\frac{a}{y}$, and $\frac{1}{z}$

20. $\frac{1}{a}$, $\frac{2}{b}$, and $\frac{x}{c}$

21. $\frac{1}{2x}$, $\frac{1}{5y}$, and $\frac{b}{y^2}$

22. $\frac{1}{3x^2}$, $\frac{1}{6xy}$, and $\frac{z}{2y^2}$

23. $\frac{2}{a^2}$, $\frac{x}{ab}$, and $\frac{3}{4}$

24. $\frac{x}{abc}$, $\frac{2}{ab^2}$, and $\frac{1}{5a^2}$

25. $\frac{a}{12}$, $\frac{3}{5a^2b}$, and $\frac{1}{2ab^3}$

26. $\frac{2}{xy^3}$, $\frac{a}{6xz}$, and $\frac{4}{15x^2}$

27. $\frac{1}{2x}$, $\frac{5}{4x^2y}$, and $\frac{1}{3xy}$

28. $\frac{3}{4a^2}$, $\frac{1}{7a^2b}$, and $\frac{2}{8ab^2}$

29. $\frac{5}{12x^2}$, $\frac{a}{16xy}$, and $\frac{7}{18xy^2}$

30. $\frac{5a}{12x^3y}$, $\frac{4}{15x^2}$, and $\frac{6}{36xy^2z}$

3 Equivalent Fractions

The fundamental principle of fractions can be used to change a fraction to an equivalent fraction with a different denominator.

$\frac{a}{b} = \frac{ac}{bc}$

Example 11 Change $\frac{3}{4}$ to a fraction with a denominator of 12.

Solution

$$\frac{3}{4} = \frac{?}{12}$$

We know that we can multiply the numerator and denominator of $\frac{3}{4}$ by the same nonzero number. Our task then is to determine what number multiplied by 4 (the original denominator) will give 12 (the new denominator). We divide 4 into 12 to find the number 3. Multiplying the numerator and denominator by 3, we have

$\frac{3}{4} = \frac{3 \times ?}{4 \times ?} = \frac{?}{12}$

$$\frac{3}{4} = \frac{(3)(3)}{(4)(3)} = \frac{9}{12}.$$

Reduce $\frac{9}{12}$ and see if you get $\frac{3}{4}$.

Example 12 Change $\frac{3}{4}$ to a fraction with a denominator of 20.

Solution

$$\frac{3}{4} = \frac{?}{20}$$

We divide 4 into 20 and see that 4 must be multiplied by 5 to obtain the new denominator. So the numerator 3 must also be multiplied by 5.

$$\frac{3}{4} = \frac{(3)(5)}{(4)(5)} = \frac{15}{20}$$

Again, reduce the answer as a check.

Example 13 $\frac{7}{8} = \frac{?}{24}$

Multiply the numerator and denominator by 3.

Solution

$$\frac{7}{8} = \frac{(7)(3)}{(8)(3)} = \frac{21}{24}$$

⊙ ***SKILL CHECK 11*** $\frac{3}{5} = \frac{?}{10}$

Example 14 $\frac{5}{x^2} = \frac{?}{x^5}$

$x^5 \div x^2 = x^3$

Solution We must multiply x^2 by some quantity to obtain x^5. We may obtain this quantity by dividing the new denominator x^5 by the old denominator x^2. We thus obtain x^3 as the quantity. So if we multiply numerator and denominator of the original fraction by x^3 we will obtain

$$\frac{5}{x^2} = \frac{(5)(x^3)}{(x^2)(x^3)} = \frac{5x^3}{x^5}.$$

⊙ ***SKILL CHECK 12*** $\frac{3}{x} = \frac{?}{x^3}$

Example 15 $\frac{x}{a^2} = \frac{?}{a^3b}$

$\frac{a^3b}{a^2} = ab$

Solution We first divide a^3b by a^2 to obtain ab. Thus if we multiply numerator and denominator of the original fraction by ab we obtain

$$\frac{x}{a^2} = \frac{(x)(ab)}{(a^2)(ab)} = \frac{abx}{a^3b}.$$

⊙ ***SKILL CHECK 13*** $\frac{x}{a^3} = \frac{?}{a^4b}$

Example 16 $\frac{7a}{4xy} = \frac{?}{12x^3y^2}$

$\frac{12x^3y^2}{4xy} = 3x^2y$

Solution Dividing $12x^3y^2$ by $4xy$, we obtain $3x^2y$. Thus

$$\frac{7a}{4xy} = \frac{(7a)(3x^2y)}{(4xy)(3x^2y)} = \frac{21ax^2y}{12x^3y^2}.$$

⊙ ***SKILL CHECK 14*** $\frac{2a}{3x^2y} = \frac{?}{15x^2y^2}$

⊙ ***Skill Check Answers***

11. 6
12. $3x^2$
13. abx
14. $10ay$

Exercise 3–4–3

3 Find the missing numerator.

1. $\frac{1}{2} = \frac{?}{6}$ **2.** $\frac{1}{2} = \frac{?}{8}$ **3.** $\frac{1}{3} = \frac{?}{12}$ **4.** $\frac{1}{5} = \frac{?}{10}$

5. $\frac{1}{x} = \frac{?}{x^2}$ **6.** $\frac{1}{x} = \frac{?}{x^3}$ **7.** $\frac{2}{3} = \frac{?}{24}$ **8.** $\frac{5}{6} = \frac{?}{24}$

9. $\frac{3}{ab^2} = \frac{?}{a^2b^4}$

10. $\frac{2}{a^2b} = \frac{?}{a^5b^3}$

11. $\frac{4}{3a} = \frac{?}{9a^2b}$

12. $\frac{3}{4x} = \frac{?}{20x^3y}$

13. $\frac{2}{7} = \frac{?}{63}$

14. $\frac{7}{9} = \frac{?}{72}$

15. $\frac{3}{11} = \frac{?}{44}$

16. $\frac{3}{5} = \frac{?}{30}$

17. $\frac{13}{6a} = \frac{?}{30a^2}$

18. $\frac{4}{7a} = \frac{?}{91a^4b}$

19. $\frac{4a}{9xy^2} = \frac{?}{108x^2y^5}$

20. $\frac{7x}{6a^2b} = \frac{?}{114a^4b^3}$

4 Comparing Fractions

In chapter 1 we saw that for two different whole numbers one was greater or larger than the other. The same is true for fractions. We can make the following general statement.

See section 1-2.

> For any two numbers of arithmetic a and b, either $a = b$, $a < b$, or $a > b$.

Recall the inequality symbol introduced in section 1-2.

This is sometimes referred to as the **trichotomy principle.**

When two fractions have a common denominator, it is fairly easy to tell which fraction is larger than the other. The larger fraction has the larger numerator. For instance, $\frac{5}{8} > \frac{3}{8}$ because the numerator 5 is larger than the numerator 3.

When two fractions do not have a common denominator it is a little more difficult, at least at a glance, to tell which is larger or perhaps that they are equal. To compare such fractions we must first change each fraction to an equivalent form having a common denominator.

Example 17 Replace the question mark in $\frac{4}{5} ? \frac{3}{4}$ with one of the symbols $=$, $<$, or $>$ to indicate a true statement.

Solution The LCD of 5 and 4 is 20.

$$\frac{4}{5} = \frac{16}{20} \text{ and } \frac{3}{4} = \frac{15}{20}$$

Which is larger, $\frac{16}{20}$ or $\frac{15}{20}$?

We see that the numerator 16 is larger than 15. Therefore $\frac{4}{5} > \frac{3}{4}$.

⊙ ***SKILL CHECK 15*** Replace the question mark in $\frac{3}{4} ? \frac{2}{3}$ with one of the symbols $=$, $<$, or $>$ to indicate a true statement.

⊙ ***Skill Check Answer***

15. $\frac{3}{4} > \frac{2}{3}$

Example 18 Replace the question mark in $\frac{8}{12}$? $\frac{10}{15}$ with one of the symbols $=$, $<$, or $>$ to indicate a true statement.

Solution The LCD of 12 and 15 is 60.

We could also reduce $\frac{8}{12}$ and $\frac{10}{15}$ obtaining $\frac{8}{12} = \frac{2}{3}$ and $\frac{10}{15} = \frac{2}{3}$.

$$\frac{8}{12} = \frac{40}{60} \text{ and } \frac{10}{15} = \frac{40}{60}$$

We see that the numerators are equal. Therefore $\frac{8}{12} = \frac{10}{15}$.

◉ ***Skill Check Answer***

16. $\frac{6}{9} = \frac{8}{12}$

◉ ***SKILL CHECK 16*** Replace the question mark in $\frac{6}{9}$? $\frac{8}{12}$ with one of the symbols $=$, $<$, or $>$ to indicate a true statement.

Exercise 3–4–4

4 Replace the question mark with one of the symbols $=$, $<$, or $>$ to indicate a true statement.

1. $\frac{1}{3}$? $\frac{2}{5}$

2. $\frac{2}{3}$? $\frac{4}{5}$

3. $\frac{3}{4}$? $\frac{5}{8}$

4. $\frac{3}{5}$? $\frac{4}{7}$

5. $\frac{2}{4}$? $\frac{6}{8}$

6. $\frac{3}{6}$? $\frac{3}{9}$

7. $\frac{3}{9}$? $\frac{5}{15}$

8. $\frac{3}{8}$? $\frac{2}{5}$

9. $\frac{5}{6}$? $\frac{6}{7}$

10. $\frac{5}{20}$? $\frac{3}{12}$

11. $\frac{5}{16}$? $\frac{3}{8}$

12. $\frac{11}{16}$? $\frac{23}{32}$

13. $\frac{8}{20}$? $\frac{6}{15}$

14. $\frac{7}{9}$? $\frac{16}{21}$

15. $\frac{3}{8}$? $\frac{4}{11}$

16. $\frac{15}{35}$? $\frac{6}{14}$

5 Adding and Subtracting Any Two Fractions

Since we now know how to find the LCD of any two or more fractions and also how to change a fraction to one with a new denominator, we can state a rule for adding any two or more fractions.

> To add any two or more fractions find the LCD and change each fraction to an equivalent fraction with that denominator. The fractions are now like fractions and can be added by using the rule for adding like fractions.

The LCD is the least common denominator.

Example 19 Add: $\frac{3}{5} + \frac{2}{3}$

Solution The LCD of 5 and 3 is 15.

$$\frac{3}{5} + \frac{2}{3} = \frac{9}{15} + \frac{10}{15} = \frac{19}{15}$$

$$\frac{3}{5} = \frac{(3)(3)}{(5)(3)} = \frac{9}{15}$$

and

$$\frac{2}{3} = \frac{(2)(5)}{(3)(5)} = \frac{10}{15}.$$

◉ ***SKILL CHECK 17*** Add: $\frac{1}{3} + \frac{1}{2}$

Example 20 Add: $\frac{5}{12} + \frac{1}{4}$

Solution The LCD of 12 and 4 is 12.

$$\frac{5}{12} + \frac{1}{4} = \frac{5}{12} + \frac{3}{12} = \frac{8}{12} = \frac{2}{3}$$

Remember to reduce when possible.

◉ ***SKILL CHECK 18*** Add: $\frac{3}{16} + \frac{1}{4}$

Example 21 Add: $\frac{1}{2} + \frac{2}{3} + \frac{5}{8}$

Solution The LCD of 2, 3, and 8 is 24.

$$\frac{1}{2} + \frac{2}{3} + \frac{5}{8} = \frac{12}{24} + \frac{16}{24} + \frac{15}{24} = \frac{43}{24}$$

◉ ***SKILL CHECK 19*** Add: $\frac{1}{3} + \frac{2}{5} + \frac{3}{10}$

Example 22 Add: $\frac{5}{x^2} + \frac{3}{x^5}$

Solution The LCD of x^2 and x^5 is x^5.

$$\frac{5}{x^2} + \frac{3}{x^5} = \frac{5x^3}{x^5} + \frac{3}{x^5} = \frac{5x^3 + 3}{x^5}$$

We need to multiply the numerator and denominator of the first fraction by x^3.

◉ ***SKILL CHECK 20*** Add: $\frac{1}{x^3} + \frac{2}{x^4}$

◉ ***Skill Check Answers***

17. $\frac{5}{6}$ 18. $\frac{7}{16}$
19. $\frac{31}{30}$ 20. $\frac{x + 2}{x^4}$

Example 23 Add: $\frac{a}{x^2y} + \frac{b}{xy^3}$

Solution The LCD is x^2y^3.

The numerator and denominator of the first fraction must be multiplied by y^2 and those of the second fraction by x.

$$\frac{a}{x^2y} + \frac{b}{xy^3} = \frac{(a)(y^2)}{(x^2y)(y^2)} + \frac{(b)(x)}{(xy^3)(x)}$$
$$= \frac{ay^2}{x^2y^3} + \frac{bx}{x^2y^3}$$
$$= \frac{ay^2 + bx}{x^2y^3}$$

⊙ ***SKILL CHECK 21*** Add: $\frac{a}{xy^2} + \frac{b}{x^2y}$

Example 24 Add: $\frac{2}{a} + \frac{2}{b} + \frac{2}{c}$

Here we are adding three fractions. We must find the LCD for all three.

Solution The LCD is abc.

$$\frac{2}{a} + \frac{2}{b} + \frac{2}{c} = \frac{2bc}{abc} + \frac{2ac}{abc} + \frac{2ab}{abc}$$
$$= \frac{2bc + 2ac + 2ab}{abc}$$

⊙ ***SKILL CHECK 22*** Add: $\frac{1}{x} + \frac{1}{y} + \frac{1}{z}$

This rule is similar to the rule for addition.

> To subtract one fraction from another find the LCD and change each fraction to an equivalent fraction with that denominator. The fractions are now like fractions and can be subtracted using the rule for subtracting like fractions.

Example 25 Subtract: $\frac{5}{8} - \frac{1}{2}$

Solution The LCD of 8 and 2 is 8.

$\frac{1}{2} = \frac{(1)(4)}{(2)(4)} = \frac{4}{8}$

$$\frac{5}{8} - \frac{1}{2} = \frac{5}{8} - \frac{4}{8} = \frac{1}{8}$$

⊙ ***SKILL CHECK 23*** Subtract: $\frac{3}{8} - \frac{1}{4}$

Example 26 Subtract: $\frac{5}{6} - \frac{11}{15}$

Solution The LCD of 6 and 15 is 30.

$$\frac{5}{6} - \frac{11}{15} = \frac{25}{30} - \frac{22}{30} = \frac{3}{30} = \frac{1}{10}$$

⊙ ***SKILL CHECK 24*** Subtract: $\frac{7}{8} - \frac{5}{6}$

⊙ ***Skill Check Answers***

21. $\frac{ax+by}{x^2y^2}$
22. $\frac{yz+xz+xy}{xyz}$
23. $\frac{1}{8}$
24. $\frac{1}{24}$

Example 27 Subtract: $\dfrac{x}{4a^2b} - \dfrac{y}{6ab^2}$

Solution The LCD is $12a^2b^2$.

$$\frac{x}{4a^2b} - \frac{y}{6ab^2} = \frac{3bx}{12a^2b^2} - \frac{2ay}{12a^2b^2}$$
$$= \frac{3bx - 2ay}{12a^2b^2}$$

We must multiply the numerator and denominator of the first fraction by $3b$ and those of the second fraction by $2a$.

⊙ ***SKILL CHECK 25*** Subtract: $\dfrac{x}{6ab^2} - \dfrac{2}{9a^2b}$

⊙ ***Skill Check Answer***

25. $\dfrac{3ax-4b}{18a^2b^2}$

Exercise 3–4–5

5 Perform the indicated operation and simplify.

1. $\dfrac{1}{2} + \dfrac{1}{5}$

2. $\dfrac{1}{3} + \dfrac{1}{4}$

3. $\dfrac{1}{3} + \dfrac{1}{5}$

4. $\dfrac{1}{2} + \dfrac{1}{7}$

5. $\dfrac{2}{x} + \dfrac{3}{y}$

6. $\dfrac{3}{a} + \dfrac{2}{b}$

7. $\dfrac{5}{8} - \dfrac{1}{3}$

8. $\dfrac{4}{7} - \dfrac{2}{5}$

9. $\dfrac{7}{9} - \dfrac{3}{4}$

10. $\dfrac{5}{3} - \dfrac{3}{8}$

11. $\dfrac{3}{5} + \dfrac{3}{10}$

12. $\dfrac{2}{3} + \dfrac{5}{6}$

13. $\dfrac{2}{a} - \dfrac{1}{a^2}$

14. $\dfrac{5}{x} - \dfrac{1}{x^2}$

15. $\dfrac{5}{9} - \dfrac{5}{12}$

16. $\dfrac{3}{5} - \dfrac{3}{8}$

17. $\dfrac{a}{x} + \dfrac{b}{xy}$

18. $\dfrac{b}{ax} + \dfrac{c}{ax^2}$

19. $\frac{7}{9} - \frac{3}{4}$

20. $\frac{11}{12} - \frac{5}{18}$

21. $\frac{1}{2} + \frac{7}{10}$

22. $\frac{3}{8} + \frac{5}{16}$

23. $\frac{5}{ax} - \frac{3}{x^2}$

24. $\frac{9}{x^2} - \frac{7}{xy}$

25. $\frac{3}{8} + \frac{7}{20}$

26. $\frac{5}{6} + \frac{2}{15}$

27. $\frac{1}{12} + \frac{7}{16}$

28. $\frac{5}{16} + \frac{11}{24}$

29. $\frac{a}{6x} + \frac{4}{15x^2}$

30. $\frac{2}{9x} + \frac{5b}{12x^2}$

31. $\frac{5}{6a} - \frac{5}{12ab}$

32. $\frac{3}{7b} - \frac{3}{14b^2}$

33. $\frac{5}{8} - \frac{3}{12}$

34. $\frac{5}{6} - \frac{4}{9}$

35. $\frac{4}{x^2y} - \frac{5}{xy^4}$

36. $\frac{5}{ab^2} - \frac{1}{a^3b}$

37. $\frac{1}{3y^2} + \frac{11}{6x^2y}$

38. $\frac{2}{3a^2} + \frac{5}{2ab}$

39. $\frac{3}{16} + \frac{11}{24}$

40. $\frac{7}{20} + \frac{11}{36}$

41. $\frac{4a}{9x} - \frac{5}{12x^2}$

42. $\frac{5}{6a} - \frac{7b}{9a^3}$

43. $\frac{5}{14} - \frac{4}{21}$

44. $\frac{5}{12} - \frac{3}{10}$

45. $\frac{1}{2} + \frac{1}{4} + \frac{1}{8}$

46. $\frac{1}{3} + \frac{1}{6} + \frac{1}{2}$

47. $\frac{3}{4} + \frac{2}{5} + \frac{3}{10}$

48. $\frac{3}{4} + \frac{5}{6} + \frac{3}{8}$

49. $\frac{1}{a} + \frac{3}{b} + \frac{4}{c}$

50. $\frac{1}{x} + \frac{5}{y} + \frac{1}{z}$

51. $\frac{1}{18} + \frac{4}{27}$

52. $\frac{4}{15} + \frac{7}{25}$

53. $\frac{7}{16a^2b} + \frac{5}{18ab^2}$

54. $\frac{5}{12xy^2} + \frac{9}{28x^3y}$

55. $\frac{7a}{8x^2y} - \frac{a}{2xy^5}$

56. $\frac{7x}{12ab^3} - \frac{x}{3a^2b}$

57. $\frac{3}{4} - \frac{5}{12}$

58. $\frac{8}{15} - \frac{1}{3}$

59. $\frac{2}{3} + \frac{5}{6} + \frac{4}{9}$

60. $\frac{1}{12} + \frac{5}{16} + \frac{1}{18}$

61. $\frac{5}{12a} + \frac{3}{16ab} + \frac{7}{32a^2b}$

62. $\frac{2}{9x^2y} + \frac{5}{12x^3} + \frac{3}{8xy^2}$

63. $\frac{11}{18} - \frac{4}{15}$

64. $\frac{11}{14} - \frac{13}{42}$

65. $\frac{34}{35xy^2z} - \frac{13a}{14x^2yz^2}$

66. $\frac{23x}{24a^2c} - \frac{17}{18ab^2c^2}$

67. $\frac{3}{10} + \frac{2}{15} + \frac{1}{21}$

68. $\frac{5}{12} + \frac{7}{18} + \frac{3}{32}$

69. $\frac{2}{3} + \frac{5}{12} - \frac{7}{8}$

70. $\frac{3}{4} + \frac{5}{6} - \frac{4}{9}$

• Concept Enrichment

71. One-third of a cup of water was added to a container that already had $\frac{3}{8}$ cup of water in it. What is the total amount of water in the container?

72. If $\frac{1}{4}$ ounce of a liquid is added to $\frac{5}{8}$ ounce of the same liquid, what is the total?

73. A hamburger that weighed $\frac{1}{4}$ pound before cooking was found to weigh $\frac{3}{16}$ pound after cooking. How much weight was lost?

74. A $\frac{1}{8}$-pound slice is removed from a $\frac{3}{4}$-pound block of butter. How much butter is left?

75. Al starts to walk to Julia's house, which is $\frac{7}{10}$ mile away. After he has walked $\frac{2}{5}$ of a mile, what distance must he still walk?

76. Margaret owns $\frac{3}{4}$ acre on which she wants to place an office building and parking lot. The zoning board declares that she cannot use $\frac{1}{6}$ acre of her property for that purpose. How much can she use?

77. A man own $\frac{7}{8}$ acre of land. He purchases an adjacent $\frac{2}{3}$-acre parcel. How much land does he now own?

78. If a person spends $\frac{1}{3}$ of the day working and $\frac{1}{4}$ of the day sleeping, how much of the day is left for other activities?

79. A company that sells gravel receives orders for $\frac{3}{4}$ ton, $\frac{5}{8}$ ton, and $\frac{1}{2}$ ton. What is the total tonnage of the three orders?

80. A woman gives $\frac{1}{10}$ of her salary to charity, $\frac{1}{4}$ is spent for rent, and $\frac{1}{6}$ for food and clothing. What total portion of her salary is used for these three purposes?

3–5 Mixed Numbers

OBJECTIVES

Upon completing this section you should be able to:

1. Change a mixed number to an improper fraction and vice versa.
2. Multiply and divide mixed numbers.
3. Add mixed numbers.
4. Subtract mixed numbers.

We have learned the four basic operations on whole numbers and fractions. We now turn our attention to a combination of the two called *mixed numbers*.

A **mixed number** is the indicated sum of a whole number and a proper fraction.

Example 1 $4\frac{2}{3}$, read "four and two-thirds," is a way of writing $4 + \frac{2}{3}$.

Example 2 $5\frac{1}{2}$, read "five and one-half," is a way of writing $5 + \frac{1}{2}$.

1 Changing Mixed Numbers to Improper Fractions and Vice Versa

These are examples of mixed numbers. Every improper fraction can be changed to a mixed number and every mixed number can be changed to an improper fraction.

Recall that in an improper fraction the numerator is larger than the denominator.

Example 3 Change $4\frac{2}{3}$ to an improper fraction.

Solution No new rule is needed here since $4\frac{2}{3}$ means

$4 + \frac{2}{3} = \frac{4}{1} + \frac{2}{3}$.

The LCD is 3. So

$$4\frac{2}{3} = 4 + \frac{2}{3} = \frac{4}{1} + \frac{2}{3} = \frac{12}{3} + \frac{2}{3} = \frac{14}{3}.$$

$4 = \frac{4}{1} = \frac{(4)(3)}{(1)(3)} = \frac{12}{3}$

Example 4 Change $5\frac{1}{2}$ to an improper fraction.

Solution $$5\frac{1}{2} = 5 + \frac{1}{2} = \frac{10}{2} + \frac{1}{2} = \frac{11}{2}$$

Since the denominator of the fractional part of the mixed number will always be the LCD, we can use a shortcut.

Example 5 Change $5\frac{1}{2}$ to an improper fraction.

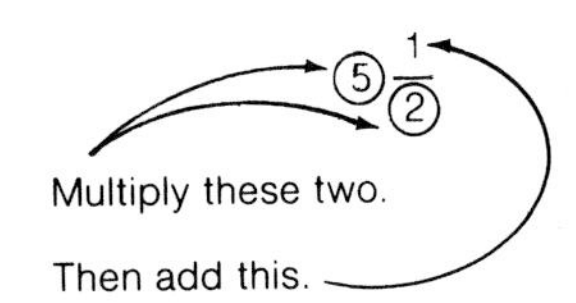

Solution The shortcut method involves multiplying the denominator (2) by the whole number (5) and then adding the product to the numerator (1). This gives a result of 11 in this case. Place this number over the original denominator (2) to obtain $5\frac{1}{2} = \frac{11}{2}$.

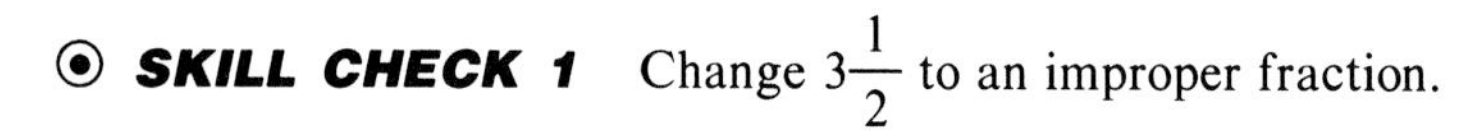

◉ ***SKILL CHECK 1*** Change $3\frac{1}{2}$ to an improper fraction.

Example 6 Change $7\frac{5}{8}$ to an improper fraction.

Multiply the whole number (7) and the denominator (8). Then add the numerator (5).

Solution $$7\frac{5}{8} = \frac{(8)(7) + 5}{8} = \frac{61}{8}.$$

Recall the order of operations—multiply and then add.

Example 7 $2\frac{3}{4} = \frac{(4)(2) + 3}{4} = \frac{11}{4}$

Note that the denominator remains the same.

Example 8 $9\frac{1}{6} = \frac{(6)(9) + 1}{6} = \frac{55}{6}$

◉ ***SKILL CHECK 2*** Change $5\frac{3}{8}$ to an improper fraction.

Recall division of whole numbers.

Changing an improper fraction to a mixed number uses division to obtain the whole number part and the remainder to obtain the fractional part.

Example 9 Change $\frac{39}{4}$ to a mixed number.

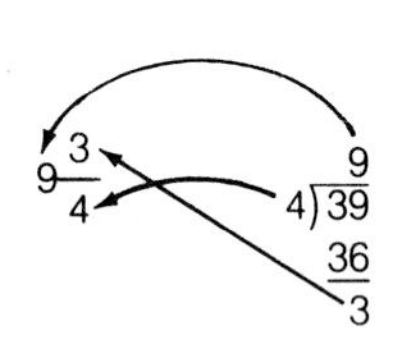

Solution We divide 4 into 39. This gives 9 with a remainder of 3. In other words.

$$\frac{39}{4} = \frac{36}{4} + \frac{3}{4} = 9 + \frac{3}{4} = 9\frac{3}{4}.$$

Example 10 Change $\frac{7}{3}$ to a mixed number.

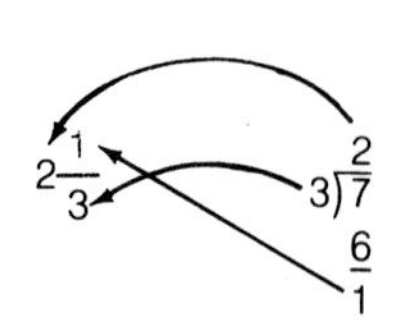

Solution $$7 \div 3 = 2\text{ R}1;\text{ so } \frac{7}{3} = 2\frac{1}{3}.$$

◉ ***SKILL CHECK 3*** Change $\frac{17}{5}$ to a mixed number.

Example 11 Change $\frac{38}{3}$ to a mixed number.

Solution $$38 \div 3 = 12\text{ R}2;\text{ so } \frac{38}{3} = 12\frac{2}{3}.$$

◉ ***Skill Check Answers***

1. $\frac{7}{2}$
2. $\frac{43}{8}$
3. $3\frac{2}{5}$

◉ ***SKILL CHECK 4*** Change $\frac{53}{4}$ to a mixed number.

◉ ***Skill Check Answer***

4. $13\frac{1}{4}$

Exercise 3–5–1

1 Change to an improper fraction.

1. $4\frac{1}{2}$ **2.** $2\frac{1}{8}$ **3.** $3\frac{2}{3}$

4. $6\frac{3}{4}$ **5.** $6\frac{5}{8}$ **6.** $9\frac{2}{3}$

7. $7\frac{5}{6}$ **8.** $8\frac{4}{9}$ **9.** $12\frac{1}{4}$

10. $13\frac{8}{9}$

Change to a mixed number.

11. $\frac{5}{2}$ **12.** $\frac{8}{3}$ **13.** $\frac{9}{2}$

14. $\frac{12}{5}$ **15.** $\frac{18}{7}$ **16.** $\frac{25}{6}$

17. $\frac{34}{5}$ **18.** $\frac{22}{3}$ **19.** $\frac{53}{4}$

20. $\frac{68}{7}$

2 Multiplying and Dividing Mixed Numbers

To multiply or divide mixed numbers change the mixed numbers to improper fractions and proceed by the rule for multiplying or dividing fractions.

Example 12 Multiply: $4\frac{1}{2} \times 3\frac{1}{3}$

First change the mixed numbers to improper fractions.

Solution $4\frac{1}{2} \times 3\frac{1}{3} = \left(\frac{\overset{3}{\cancel{9}}}{\cancel{2}}\right)\left(\frac{\overset{5}{\cancel{10}}}{\cancel{3}}\right) = 15$

⦿ ***SKILL CHECK 5*** Multiply: $4\frac{1}{5} \times 3\frac{1}{3}$

Example 13 Multiply: $7\frac{1}{8} \times 4\frac{1}{5}$

Nothing reduces here.

Solution $7\frac{1}{8} \times 4\frac{1}{5} = \left(\frac{57}{8}\right)\left(\frac{21}{5}\right) = \frac{1{,}197}{40} = 29\frac{37}{40}$

Example 14 Multiply: $5\frac{1}{3} \times 3\frac{5}{8}$

Notice again to reduce when possible.

Solution $5\frac{1}{3} \times 3\frac{5}{8} = \left(\frac{\overset{2}{\cancel{16}}}{3}\right)\left(\frac{29}{\cancel{8}}\right) = \frac{58}{3} = 19\frac{1}{3}$

⦿ ***SKILL CHECK 6*** Multiply: $3\frac{1}{8} \times 4\frac{3}{5}$

Example 15 Divide: $3\frac{1}{2} \div 2\frac{3}{4}$

Remember to "invert and multiply."

Solution $3\frac{1}{2} \div 2\frac{3}{4} = \frac{7}{2} \div \frac{11}{4} = \left(\frac{7}{\cancel{2}}\right)\left(\frac{\overset{2}{\cancel{4}}}{11}\right) = \frac{14}{11} = 1\frac{3}{11}$

⦿ ***SKILL CHECK 7*** Divide: $5\frac{1}{2} \div 2\frac{3}{4}$

Example 16 Divide: $19\frac{1}{2} \div 2$

Recall, $2 = \frac{2}{1}$.

Solution $19\frac{1}{2} \div 2 = \frac{39}{2} \div \frac{2}{1} = \left(\frac{39}{2}\right)\left(\frac{1}{2}\right) = \frac{39}{4} = 9\frac{3}{4}$

⦿ ***SKILL CHECK 8*** Divide: $25\frac{2}{3} \div 2$

Example 17 A motor home averages $7\frac{1}{2}$ miles per gallon of fuel. If the capacity of the fuel tank is $14\frac{3}{8}$ gallons, how far will the motor home travel on one full tank of fuel?

⦿ ***Skill Check Answers***

5. 14
6. $14\frac{3}{8}$
7. 2
8. $12\frac{5}{6}$

Solution In this problem we see that one gallon of fuel will be used for each unit of $7\frac{1}{2}$ miles. Since there are $14\frac{3}{8}$ gallons of fuel available we must multiply $7\frac{1}{2}$ by $14\frac{3}{8}$.

$$\left(7\frac{1}{2}\right)\left(14\frac{3}{8}\right) = \left(\frac{15}{2}\right)\left(\frac{115}{8}\right) = \frac{1725}{16} = 107\frac{13}{16} \text{ miles}$$

⦿ ***SKILL CHECK 9*** If a motor home containing $21\frac{3}{5}$ gallons of fuel averages $9\frac{1}{2}$ miles per gallon, how far will it travel on that amount of fuel?

⦿ ***Skill Check Answer***

9. $205\frac{1}{5}$ miles

Exercise 3–5–2

Perform the indicated operation. Give answers in mixed number form.

1. $2\frac{1}{2} \times 4\frac{1}{3}$

2. $3\frac{1}{2} \times 5\frac{1}{3}$

3. $3\frac{3}{8} \times 2\frac{1}{3}$

4. $4\frac{1}{2} \div 3\frac{1}{3}$

5. $5\frac{1}{8} \div 2\frac{3}{4}$

6. $7\frac{1}{2} \div 3\frac{5}{8}$

7. $8\frac{1}{3} \times 5\frac{2}{5}$

8. $4\frac{2}{5} \div 1\frac{4}{7}$

9. $10\frac{2}{3} \div 4$

10. $3\frac{4}{7} \times 4\frac{1}{5}$

11. $8 \times 3\frac{3}{4}$

12. $7 \div 4\frac{2}{3}$

13. $3\frac{1}{5} \times 4\frac{2}{7}$

14. $4\frac{5}{6} \times 1\frac{3}{5}$

15. $3\frac{4}{7} \times 1\frac{5}{8}$

16. $5 \div 1\frac{1}{4}$

17. $13\frac{1}{3} \div 3\frac{3}{4}$

18. $3\frac{3}{4} \times 2\frac{2}{5}$

19. $3\frac{3}{5} \times 4\frac{2}{3}$

20. $4\frac{3}{8} \div 5$

21. $3\frac{3}{5} \div 2\frac{4}{7}$

22. $5\frac{3}{8} \div 2\frac{4}{5}$

23. $2\frac{1}{3} \div 1\frac{3}{4} \times 3\frac{4}{5}$

24. $4\frac{2}{3} \times 3\frac{1}{4} \div 2\frac{5}{8}$

25. A man works $6\frac{1}{2}$ hours a day for five days. What is the total number of hours worked?

26. If a car averages $28\frac{2}{5}$ miles per gallon of gasoline, how far will it travel on $7\frac{1}{2}$ gallons?

27. A car traveled $76\frac{4}{5}$ miles on $2\frac{1}{4}$ gallons of gasoline. How many miles per gallon did it get?

28. What is the area of a room that is $16\frac{1}{2}$ feet long and $12\frac{2}{3}$ feet wide?

29. If $5\frac{1}{3}$ pounds of hamburger are divided into four equal portions, what is the weight of each portion?

30. Five overnight hikers decide to divide $23\frac{3}{4}$ pounds of equipment so that each of them carries the same weight. How much is each person's share?

3 Adding Mixed Numbers

Addition of mixed numbers again requires us to come back to the basic fact that only like quantities can be added. So we have the following rule.

> To add mixed numbers add the fractional parts to the fractional parts and the whole number parts to the whole number parts.

Make sure that the fractional parts have a common denominator.

Example 18 Add: $5\frac{3}{8} + 2\frac{1}{2}$

Solution We will add $\frac{3}{8}$ to $\frac{1}{2}$, and 5 to 2.

$$\begin{aligned}5\frac{3}{8} + 2\frac{1}{2} &= \left(5 + \frac{3}{8}\right) + \left(2 + \frac{1}{2}\right)\\ &= (5 + 2) + \left(\frac{3}{8} + \frac{1}{2}\right)\\ &= 7 + \frac{7}{8}\\ &= 7\frac{7}{8}\end{aligned}$$

The LCD is 8.

SKILL CHECK 10 Add: $3\frac{1}{3} + 4\frac{1}{2}$

Example 19 Add: $2\frac{1}{2} + 3\frac{5}{8} + 7\frac{1}{3}$

The LCD is 24.

Solution

$$2\frac{1}{2} + 3\frac{5}{8} + 7\frac{1}{3} = 2\frac{12}{24} + 3\frac{15}{24} + 7\frac{8}{24} = 12\frac{35}{24}$$

Note that $\frac{35}{24}$ is improper.

This answer contains an improper fraction. This is not an acceptable way to leave the answer, so we will simplify it.

Always give the fractional part of a mixed number as a *proper* fraction.

$$12\frac{35}{24} = 12 + \frac{35}{24} = 12 + 1 + \frac{11}{24} = 13\frac{11}{24}$$

Example 20 Add: $1\frac{7}{8} + 9\frac{3}{4} + 6\frac{4}{5}$

The LCD is 40.

Solution

$$1\frac{7}{8} + 9\frac{3}{4} + 6\frac{4}{5} = 1\frac{35}{40} + 9\frac{30}{40} + 6\frac{32}{40} = 16\frac{97}{40}$$
$$= 16 + 2 + \frac{17}{40} = 18\frac{17}{40}$$

SKILL CHECK 11 Add: $1\frac{3}{5} + 5\frac{2}{3} + 4\frac{5}{6}$

Example 21 A painter needs $7\frac{1}{2}$ gallons of paint for his house and $2\frac{3}{4}$ gallons for the garage. What is the total number of gallons needed for both?

Solution The word *total* indicates we must add. So we find the sum of $7\frac{1}{2}$ and $2\frac{3}{4}$.

The LCD is 4.

$$7\frac{1}{2} + 2\frac{3}{4} = 7\frac{2}{4} + 2\frac{3}{4}$$
$$= 9\frac{5}{4} = 9 + 1 + \frac{1}{4} = 10\frac{1}{4} \text{ gallons}$$

Skill Check Answers

10. $7\frac{5}{6}$
11. $12\frac{1}{10}$
12. $8\frac{5}{6}$ gallons

SKILL CHECK 12 A painter uses $3\frac{1}{3}$ gallons of paint for one project and $5\frac{1}{2}$ gallons for another. What is the total number of gallons used?

Exercise 3–5–3

3 Add and simplify.

1. $5\frac{1}{3} + 2\frac{1}{3}$ **2.** $3\frac{5}{8} + 9\frac{1}{8}$ **3.** $2\frac{1}{4} + 3\frac{1}{2}$

4. $3\frac{1}{5} + 4\frac{2}{3}$

5. $1\frac{2}{5} + 5\frac{1}{3}$

6. $7\frac{2}{3} + 5\frac{3}{4}$

7. $5\frac{3}{8} + 6\frac{3}{4}$

8. $12\frac{1}{5} + 2\frac{7}{8}$

9. $9\frac{3}{4} + 7\frac{5}{6}$

10. $14\frac{3}{5} + 11\frac{7}{10}$

11. $2\frac{1}{2} + 1\frac{1}{3} + 5\frac{1}{8}$

12. $3\frac{1}{4} + 2\frac{1}{6} + 1\frac{1}{8}$

13. $4\frac{2}{3} + 3\frac{1}{7} + 2\frac{2}{9}$

14. $6\frac{3}{4} + 4\frac{1}{12} + 2\frac{4}{9}$

15. $3\frac{3}{4} + 5\frac{9}{10} + 7\frac{5}{6}$

16. $4\frac{7}{8} + 10\frac{11}{12} + 8\frac{1}{6}$

17. If $6\frac{2}{3}$ cups of water are poured into a container that has $8\frac{3}{4}$ cups of water in it, what is the total number of cups of water in the container?

18. A woman decides to carpet her living room and hallway. She needs $14\frac{3}{4}$ square yards for the living room and $9\frac{1}{2}$ square yards for the hallway. What is the total number of square yards needed for both?

19. Bill drove for $5\frac{3}{4}$ hours on Monday, $6\frac{1}{3}$ hours on Tuesday, and $8\frac{1}{2}$ hours on Wednesday. What was the total number of hours that he drove for the three days?

20. A man worked $6\frac{1}{2}$ hours on Monday, $9\frac{1}{3}$ hours on Tuesday, $5\frac{2}{3}$ hours on Wednesday, $13\frac{1}{4}$ hours on Thursday, and $4\frac{3}{4}$ hours on Friday. What was the total number of hours he worked for the five days?

4 Subtracting Mixed Numbers

Subtraction of mixed numbers can involve the idea of borrowing that we used in dealing with whole numbers.

Example 22 Subtract: $5\frac{1}{4} - 1\frac{3}{4}$

Solution Notice that if we attempt to subtract the fractional parts we run into the fact that we cannot take $\frac{3}{4}$ from $\frac{1}{4}$.

We can write $5\frac{1}{4}$ as $5 + \frac{1}{4} = (4 + 1) + \frac{1}{4}$ $= 4 + \left(\frac{4}{4} + \frac{1}{4}\right) = 4\frac{5}{4}$. Then

$$5\frac{1}{4} - 1\frac{3}{4} = 4\frac{5}{4} - 1\frac{3}{4} = 3\frac{2}{4} = 3\frac{1}{2}.$$

⦿ ***SKILL CHECK 13*** Subtract: $4\frac{1}{3} - 1\frac{2}{3}$

⦿ ***Skill Check Answer***

13. $2\frac{2}{3}$

Example 23 Subtract: $25\frac{3}{8} - 5\frac{1}{4}$

Solution The LCD of 8 and 4 is 8.

$$25\frac{3}{8} - 5\frac{1}{4} = 25\frac{3}{8} - 5\frac{2}{8} = 20\frac{1}{8}$$

Notice no borrowing is necessary here.

⦿ ***SKILL CHECK 14*** Subtract: $9\frac{5}{6} - 4\frac{2}{3}$

Example 24 Subtract: $15\frac{2}{3} - 4\frac{7}{8}$

Solution

$$15\frac{2}{3} - 4\frac{7}{8} = 15\frac{16}{24} - 4\frac{21}{24} = 14\frac{40}{24} - 4\frac{21}{24} = 10\frac{19}{24}$$

We cannot subtract 21 from 16 so we write

$$15\frac{16}{24} = 14 + \frac{24}{24} + \frac{16}{24}$$
$$= 14\frac{40}{24}.$$

⦿ ***SKILL CHECK 15*** Subtract: $12\frac{3}{5} - 3\frac{2}{3}$

Example 25 A cook started to use a recipe that called for $2\frac{2}{3}$ cups of flour. It was discovered that only $1\frac{1}{2}$ cups of flour were available. How much flour is still needed for the recipe?

Solution We must subtract the amount available from the amount the recipe calls for.

$$2\frac{2}{3} - 1\frac{1}{2} = 2\frac{4}{6} - 1\frac{3}{6} = 1\frac{1}{6} \text{ cups}$$

The LCD is 6.

⦿ ***SKILL CHECK 16*** How much more flour is needed if a recipe calls for $3\frac{1}{2}$ cups and there are only $1\frac{2}{3}$ cups available?

⦿ ***Skill Check Answers***

14. $5\frac{1}{6}$ 15. $8\frac{14}{15}$

16. $1\frac{5}{6}$ cups

Exercise 3–5–4

4 Subtract and simplify.

1. $3\frac{3}{4} - 1\frac{1}{4}$

2. $4\frac{5}{8} - 3\frac{3}{8}$

3. $4\frac{1}{3} - 1\frac{2}{3}$

4. $5\frac{1}{5} - 2\frac{4}{5}$

5. $8\frac{1}{4} - 3\frac{3}{4}$

6. $12\frac{3}{8} - 7\frac{5}{8}$

7. $5\frac{1}{4} - 2\frac{1}{8}$

8. $7\frac{1}{2} - 5\frac{1}{4}$

9. $6\frac{3}{4} - 5\frac{2}{3}$

10. $4\frac{5}{8} - 3\frac{4}{5}$

11. $16\frac{3}{7} - 5\frac{4}{5}$

12. $14\frac{5}{8} - 6\frac{7}{9}$

13. $5 - 3\frac{2}{3}$

14. $9 - 6\frac{3}{4}$

15. $18\frac{5}{12} - 9\frac{5}{6}$

16. $24\frac{3}{8} - 17\frac{7}{12}$

17. A container has $30\frac{1}{4}$ ounces of water in it. If $7\frac{2}{3}$ ounces are poured out, how many ounces remain?

18. If a certain trip takes $8\frac{1}{3}$ hours, how much time remains after you have traveled $6\frac{3}{4}$ hours?

19. How much yard goods remain after $3\frac{3}{4}$ yards are cut from a bolt containing $11\frac{1}{2}$ yards?

20. A man owned $82\frac{1}{2}$ acres of land. If he sold $24\frac{2}{3}$ acres, how many acres does he have left?

• Concept Enrichment

21. A woman worked $8\frac{1}{2}$ hours on Monday, $6\frac{3}{4}$ hours on Tuesday, and $10\frac{1}{3}$ hours on Wednesday. How many more hours must she work to total 40 hours for the week?

22. A spool of ribbon contains 100 yards. $23\frac{1}{2}$ yards are used by one person. Another person takes enough to make five bows that require $2\frac{1}{3}$ yards for each bow. How many yards of ribbon remain on the spool?

CHAPTER 3 SUMMARY

Key Words

Section 3–1

- A **common fraction** is the indicated quotient of two whole numbers.
- Common fractions can be classified as **proper** or **improper.**
- An **algebraic fraction** is a fraction in which the numerator or denominator (or both) contains one or more variables.
- The **fundamental principle of fractions** is

$$\frac{a}{b} = \frac{ac}{bc} \text{ if } c \neq 0.$$

- A fraction is in **simplified** or **reduced form** if the numerator and denominator have no factor in common except the number 1.

Section 3–2

- The **product of two fractions** $\frac{a}{b}$ and $\frac{c}{d}$ is $\frac{ac}{bd}$.

Section 3–3

- Two numbers a and b are **reciprocals** if $ab = 1$.
- The reciprocal is sometimes called the **multiplicative inverse.**
- A **complex fraction** is a fraction that contains fractions in its numerator and/or denominator.

Section 3–4

- **Like fractions** are fractions having the same denominator.
- The **least common denominator** of two or more fractions is the least common multiple of their denominators.

- The **trichotomy principle** states that for any two numbers a and b, either $a = b$, $a < b$, or $a > b$.

Section 3–5

- A **mixed number** is a way of writing the sum of a whole number and a proper fraction.

Procedures

Section 3–1

- To reduce or simplify a fraction use the fundamental principle of fractions to divide all factors common to numerator and denominator.

Section 3–2

- In an indicated multiplication of two or more fractions any factor of any numerator can be divided by a like factor of any denominator.
- To solve a problem that is stated in words
  1. Read the problem carefully.
  2. Determine what is being asked for.
  3. Use the given information to answer the question.

Section 3–3

- To divide two fractions multiply by the reciprocal of the divisor.

Section 3–4

- To add like fractions place the sum of the numerators over the common denominator.
- To subtract like fractions place the difference of the numerators over the common denominator.
- To add or subtract fractions they must first be changed to like fractions.

Section 3–5

- To multiply or divide mixed numbers change the mixed numbers to improper fractions and perform the indicated operation.
- To add mixed numbers add whole numbers to whole numbers and fractions to fractions.
- Subtracting mixed numbers sometimes requires the process of borrowing.

CHAPTER 3 REVIEW

Reduce the fractions.

1. $\frac{12}{28}$

2. $\frac{8}{20}$

3. $\frac{18a}{48a^3}$

4. $\frac{45x^4}{105x}$

5. $\frac{54}{63}$

6. $\frac{60}{700}$

7. $\frac{36xy^6}{126x^2y^4}$

8. $\frac{189a^3b}{198ab^5}$

Find the missing numerator.

9. $\frac{3}{4} = \frac{?}{20}$

10. $\frac{4}{5} = \frac{?}{60}$

11. $\frac{11}{12a} = \frac{?}{72a^2b}$

12. $\frac{7a}{18x} = \frac{?}{126x^2y}$

13. Change $9\frac{6}{7}$ to an improper fraction.

14. Change $11\frac{3}{8}$ to an improper fraction.

15. Change $\frac{61}{8}$ to a mixed number.

16. Change $\frac{45}{4}$ to a mixed number.

Perform the indicated operation. Give all answers in reduced form. If answers are improper fractions, give them as mixed numbers.

17. $\left(\frac{3}{4}\right)\left(\frac{10}{27}\right)$

18. $\left(\frac{4}{5}\right)\left(\frac{15}{16}\right)$

19. $\frac{3}{5} \div \frac{9}{10}$

20. $\frac{4}{9} \div \frac{16}{27}$

21. $\frac{3}{4} + \frac{1}{6}$

22. $\frac{1}{9} + \frac{5}{6}$

23. $\frac{5}{8} - \frac{3}{8}$

24. $\frac{7}{10} - \frac{3}{10}$

25. $\frac{5}{13a} + \frac{7}{13a}$

26. $\frac{4}{11x^2} + \frac{6}{11x^2}$

27. $\left(\frac{5a}{9b}\right)\left(\frac{18b^4}{35a^2}\right)$

28. $\left(\frac{2x^2}{3y}\right)\left(\frac{9y^2}{10x^3}\right)$

29. $\frac{3x^2}{8} \div 9x^3$

30. $\frac{4a^3}{7} \div 10a$

31. $\left(\frac{4}{5}\right)\left(\frac{13}{16}\right)$

32. $\left(\frac{5}{6}\right)\left(\frac{18}{23}\right)$

33. $\frac{7}{10} + \frac{9}{10}$

34. $\frac{7}{8} + \frac{5}{8}$

35. $\frac{5}{7a} - \frac{2}{3b}$

36. $\frac{5}{6x} - \frac{3}{5y^2}$

37. $\frac{2x}{3a^2b} + \frac{5y}{6b}$

38. $\frac{5a}{7xy^2} + \frac{9}{14x^2y}$

39. $\left(\frac{5}{6a^2b}\right)(24ab^3)$

40. $\left(\frac{3}{8xy^4}\right)(32x^2y^2)$

41. $\frac{7}{8} \div \frac{3}{16}$

42. $\frac{3}{5} \div \frac{6}{7}$

43. $4\frac{5}{12} - 2\frac{3}{16}$

44. $5\frac{1}{6} - 3\frac{7}{15}$

45. $\frac{5a}{12x^2y} + \frac{7}{18xy^3}$

46. $\frac{4x}{9ab^2} + \frac{9}{12a^3b}$

47. $(7a)\left(\frac{13}{14a^2b}\right)$

48. $(5xy)\left(\frac{9}{10x^2y^3}\right)$

49. $4 \div 2\frac{2}{3}$

50. $7 \div 4\frac{2}{3}$

51. $3\frac{5}{6} + 2\frac{2}{3}$

52. $5\frac{1}{3} + 3\frac{3}{4}$

53. $\frac{7}{9xy^2} - \frac{5x}{12yz}$

54. $\frac{7a}{10b^2c^2} - \frac{8}{15ac}$

55. $2\frac{1}{2} \times 4$

56. $3\frac{1}{8} \times 5$

57. $6 \div \frac{2}{3}$

58. $12 \div \frac{3}{4}$

59. $\frac{5}{16} + \frac{7}{10}$

60. $\frac{8}{9} + \frac{11}{12}$

61. $9\frac{1}{6} - 2\frac{5}{8}$

62. $4\frac{3}{8} - 1\frac{5}{12}$

63. $1\frac{4}{5} \times 2\frac{2}{3}$

64. $4\frac{1}{2} \times 3\frac{3}{5}$

65. $\frac{5}{9a} \div \frac{3a}{5}$

66. $\frac{3}{4x} \div \frac{8x^2}{15}$

67. $\frac{1}{4} + \frac{3}{5} + \frac{7}{10}$

68. $\frac{3}{8} + \frac{7}{12} + \frac{5}{16}$

69. $3\frac{5}{12} - 2\frac{5}{8}$

70. $7\frac{2}{9} - 6\frac{1}{12}$

71. $4\frac{3}{8} \div \frac{21}{26}$

72. $3\frac{2}{3} \div \frac{22}{27}$

73. $\left(\frac{7y}{9x}\right)\left(\frac{x^2}{2y^2}\right)\left(\frac{4}{21xy}\right)$

74. $\left(\frac{3a^2}{8b}\right)\left(\frac{4}{9a}\right)\left(\frac{3b^2}{5a^2}\right)$

75. $7 - \frac{5}{8}$

76. $9 - \frac{3}{5}$

77. $2\frac{1}{8} + 1\frac{5}{6} + 5\frac{4}{9}$

78. $3\frac{3}{4} + 2\frac{7}{18} + 1\frac{5}{24}$

79. $6\frac{1}{2} \div 3\frac{1}{4}$

80. $5\frac{2}{3} \div 4\frac{1}{4}$

81. $8\frac{5}{14} - 1\frac{8}{21}$

82. $4\frac{1}{16} - 1\frac{1}{18}$

83. $2\frac{1}{2} \times 4\frac{1}{6} \times 1\frac{1}{5}$

84. $5\frac{1}{3} \times 3\frac{1}{4} \times 2\frac{5}{8}$

85. $1\frac{6}{15} \div \frac{7}{18}$

86. $2\frac{11}{12} \div 5\frac{1}{4}$

Replace the question mark with either $=$, $<$, or $>$ to indicate a true statement.

87. $\frac{3}{4}\ ?\ \frac{8}{11}$

88. $\frac{3}{7}\ ?\ \frac{4}{9}$

89. $\frac{3}{13}\ ?\ \frac{1}{4}$

90. $\frac{3}{9}\ ?\ \frac{5}{15}$

91. A family spends $\frac{1}{4}$ of its income for food. If the weekly income is \$600, how much is spent for food per week?

92. A girl pours 24 pails of water into an empty tank. If the pail holds $\frac{2}{3}$ gallon, how many gallons are in the tank?

93. A pump can deliver $\frac{3}{8}$ liter of water per second. How long will it take to pump 48 liters?

94. How many cartons of milk, each containing $\frac{1}{8}$ gallon, can be filled from 20 gallons?

95. A man works $8\frac{2}{3}$ hours a day for five days. What is the total number of hours worked?

96. Jean drove $7\frac{2}{3}$ hours on Monday, $8\frac{1}{2}$ hours on Tuesday, and $4\frac{1}{4}$ hours on Wednesday. What was the total number of hours she drove for the three days?

97. A piece of ribbon $5\frac{1}{3}$ yards long is cut from a spool containing $20\frac{3}{4}$ yards. How many yards remain on the spool?

98. If $2\frac{3}{4}$ liters of water are drained out of a car radiator containing $6\frac{1}{2}$ liters, how many liters are left in the radiator?

99. Harness horse A paced the first quarter-mile in $28\frac{2}{5}$ seconds. Horse B paced it in $27\frac{4}{5}$ seconds. How much faster was horse B?

100. A certain stock was selling in the morning at $39\frac{1}{2}$. In the afternoon it was selling at $42\frac{3}{8}$. How much had it gained?

CHAPTER 3 TEST

1. Reduce $\dfrac{54x^2y^2}{90xy^4}$

2. Change $11\dfrac{3}{5}$ to an improper fraction.

3. Find the missing numerator: $\dfrac{5}{19} = \dfrac{?}{57}$

4. Find the LCD of $\dfrac{2}{3}, \dfrac{5}{8}, \dfrac{3}{4}$.

Perform the indicated operation. Give all answers in reduced form.

5. $\left(\dfrac{2}{3}\right)\left(\dfrac{1}{7}\right)$

6. $\dfrac{3x^2}{5y} \div \dfrac{10x^3}{11y^2}$

7. $\dfrac{3}{13} + \dfrac{8}{13}$

8. $\dfrac{5}{9} - \dfrac{2}{9}$

9. $5\dfrac{1}{3} \times 2\dfrac{1}{4}$

10. $\dfrac{2}{5a^2b} + \dfrac{4}{9ac}$

11. $\dfrac{15x^2}{16} \div 10x$

12. $\dfrac{5}{6x} - \dfrac{4}{7y}$

13. $\dfrac{11}{12} + \dfrac{7}{18} + \dfrac{2}{3}$

14. $\left(\dfrac{2}{9}\right)\left(\dfrac{3}{5}\right)$

15. $3\dfrac{2}{3} - 1\dfrac{1}{8}$

16. $\dfrac{6}{35} \div \dfrac{9}{14}$

17. $\left(\frac{5x}{8y^2z}\right)\left(\frac{12yz}{25x^2}\right)$

18. $4\frac{5}{6} + 5\frac{7}{8} + 2\frac{11}{12}$

19. $4\frac{2}{3} \div 5\frac{3}{5}$

20. Replace the question mark in $\frac{5}{9}\ ?\ \frac{7}{12}$ with either $=$, $<$, or $>$ to indicate a true statement.

21. If a truck can carry $\frac{3}{4}$ ton of gravel, how many tons are carried in 12 loads?

22. A dose of a certain medicine contains $\frac{2}{3}$ grain. How many doses can be obtained from 18 grains?

23. A container has $21\frac{1}{4}$ ounces of liquid in it. If $9\frac{2}{3}$ ounces are poured out, how many ounces remain?

CHAPTERS 1–3 CUMULATIVE TEST

1. What is the place value of the digit 3 in the number 61,035?

2. Simplify: $2x + 8y - x - 3y$

3. Reduce: $\frac{24}{42}$

4. Add:
$$\begin{array}{r} 106 \\ 32 \\ +1{,}976 \\ \hline \end{array}$$

5. Round 46,501 to the nearest ten thousand.

6. Divide: $\frac{6}{7} \div 9$

7. Divide: $63\overline{)12{,}497}$

8. Subtract:
$$\begin{array}{r} 2{,}073 \\ -\ 875 \\ \hline \end{array}$$

9. Add: $\frac{4}{9} + \frac{1}{6}$

10. Find the prime factorization of 2,200.

11. How many terms are there in the expression $3x^2 - 2x + 5xy - 2 + y$?

12. Change $8\frac{2}{3}$ to an improper fraction.

13. Replace the question mark in 201 ? 300 with the inequality symbol positioned to indicate a true statement.

14. Multiply: $\left(\frac{2a^2}{3b^2}\right)\left(\frac{9b^3}{4a}\right)$

15. Find the least common multiple of 6, 10, and 24.

16. Multiply: $\begin{array}{r} 6{,}023 \\ \times \quad 87 \\ \hline \end{array}$

17. Simplify: $(3ab^2)(5ab)$

18. Evaluate $2x^2 + 5x - 3$ if $x = 3$.

19. Subtract: $7\frac{2}{3} - 3\frac{3}{5}$

20. To buy a car Mary finances $9,500 for 36 months. The monthly payment is $311. How much is she paying in interest?

CHAPTER

4 Pretest

Answer as many of the following problems as you can before starting this chapter. When you finish the chapter, take the test at the end and compare your scores to see how much you have learned.

1. Write 38.752 in expanded form.

2. Write 406.98 in words.

3. Round 5.6937 to the nearest hundredth.

4. Change 0.075 to a common fraction in simplest form.

5. Change $6\frac{3}{8}$ to a decimal rounded to three decimal places.

Perform the indicated operation. In division problems round the quotient to three decimal places.

6. $\begin{array}{r} 7.48 \\ +16.27 \\ \hline \end{array}$

7. $\begin{array}{r} 23.142 \\ -\ 8.355 \\ \hline \end{array}$

8. $7\overline{)25.2}$

9. $0.5\overline{)5.67}$

10. $(3.16ab^2)(0.5a^2b)$

11. $6.321 + 15.4 + 7.589$

12. $32.24x - 17.955x$

13. $183.4 \div 15$

14. $0.6\overline{)6.876}$

15. $\begin{array}{r} 135.8 \\ \times\ 1.07 \\ \hline \end{array}$

16. $\begin{array}{r} 61.05 \\ 417.876 \\ +\ 23.144 \\ \hline \end{array}$

17. $\begin{array}{r} 40.003 \\ -16.454 \\ \hline \end{array}$

18. $24\overline{)121.44}$

19. $\begin{array}{r} 34.04 \\ \times 0.067 \\ \hline \end{array}$

20. $11.2\overline{)0.0784}$

21. The balance in a checking account is $512.63. If three checks are written for $24.95, $16.32, and $204.65, what is the new balance?

22. The price of a ticket to a certain concert is $25.95. What is the cost of six tickets?

CHAPTER

4 Decimals

Expressing fractions as decimals is possible because of our base 10 number system which in turn depends on the use of zero, developed by Hindu mathematicians around A.D. 500.

Death Valley, California, has an average rainfall of only 1.63 inches.

STUDY HELP

Your study habits will improve if you designate a specific period of time for study. The most appropriate time for you is when you will be mentally alert and will not be interrupted or distracted. Better results are obtained when you are not tired or rushed.

In chapter 1 we noted that our number system is a base 10 number system. In expanded form we saw that each place value was ten times the place value to its right. In that chapter we were dealing only with whole numbers. In this chapter we will extend the same ideas to include a set of numbers called *decimals*.

4–1 Expanded Form and Reading and Writing Decimals

OBJECTIVES

Upon completing this section you should be able to:

1. Write decimals in expanded form.
2. Write a decimal in words and vice versa.

If a **decimal point** precedes a digit or set of digits, a fraction is indicated. Thus a number containing a decimal point is often referred to as a **decimal fraction**. It is also called a **decimal number** and most often referred to simply as a **decimal.** Each place after the decimal point has a specific value.

Example 1 0.7 is read "seven tenths." (This could be written as the common fraction $\frac{7}{10}$.)

If no other digit precedes the decimal point, a zero is placed in front of it to emphasize the decimal point.

Example 2 0.65 is read "sixty-five hundredths." (This could be written as the common fraction $\frac{65}{100}$.)

Example 3 3.6 is a mixed number and is read "three and six tenths." (This could be written as $3\frac{6}{10}$.)

1 Expanded Form

The fact that each place value is ten times the place value to its right is still true in decimal notation. Observe the following names of some place values.

Note that as you move from right to left the place values increase.

375.261
hundreds, tens, ones, tenths, hundredths, thousandths

Hundreds	Tens	Ones		Tenths	Hundredths	Thousandths
(10)(10)	10	1	•	$\frac{1}{10}$	$\frac{1}{(10)(10)}$	$\frac{1}{(10)(10)(10)}$

This is true on either side of the decimal point.

Notice that each place value when multiplied by 10 gives the place value of the next place to the left. For instance, the place value $\frac{1}{(10)(10)}$ or $\frac{1}{100}$, if multiplied by 10, gives $(10)\left(\frac{1}{100}\right) = \frac{1}{10}$, which is the next place value to the left. The expanded form of a decimal number or mixed number in decimal form uses this fact.

Recall expanded form from chapter 1.

Using exponents, the previous table is written as:

Hundreds	Tens	Ones		Tenths	Hundredths	Thousandths
10^2	10^1	10^0	•	$\frac{1}{10^1}$	$\frac{1}{10^2}$	$\frac{1}{10^3}$

Example 4 Write 7.35 in expanded form.

Solution $$7.35 = 7(10^0) + 3\left(\frac{1}{10^1}\right) + 5\left(\frac{1}{10^2}\right)$$

◉ ***SKILL CHECK 1*** Write 6.24 in expanded form.

Example 5 Write 25.301 in expanded form.

Solution $$25.301 = 2(10^1) + 5(10^0) + 3\left(\frac{1}{10^1}\right) + 0\left(\frac{1}{10^2}\right) + 1\left(\frac{1}{10^3}\right)$$

Again, note how each place value is ten times that to the right.

◉ ***SKILL CHECK 2*** Write 42.605 in expanded form.

2 Reading and Writing Decimals

In reading or writing decimal notation we must remember that the word *and* is used only for the decimal point. Also, we must remember that the names of places to the right of the decimal point represent fractions and always must end in *ths*.

Spell the name the same; just add *ths*.

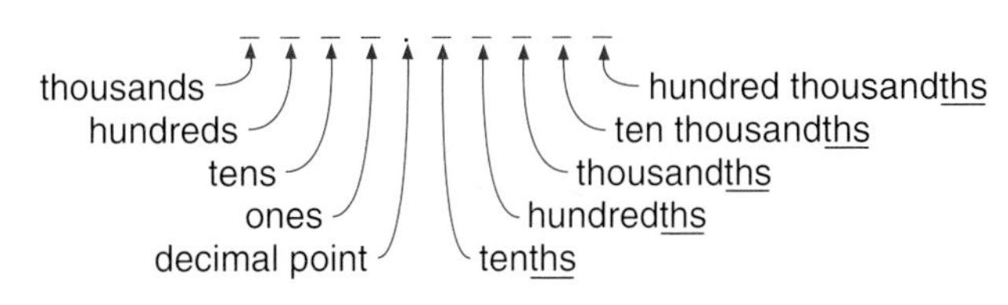

Example 6 27.36 is read "twenty-seven and thirty-six hundredths."

27.36 ← hundredths place

Example 7 0.003 is read "three thousandths."

0.003 ← thousandths place

◉ ***SKILL CHECK 3*** Write 41.056 in words.

Example 8 0.0125 is read "one hundred twenty-five ten thousandths."

0.0125 ← ten thousandths place

Example 9 200.017 is read "two hundred and seventeen thousandths."

Notice the word "and" for the decimal point in each of these examples.

◉ ***SKILL CHECK 4*** Write 304.06 in words.

Example 10 Express "twenty-seven and fifteen hundredths" in decimal notation.

Solution 27 is a whole number so it is placed before the decimal point. The 5 in 15 must be placed in the hundredths place. The decimal form is 27.15.

27.15 ← hundredths place

◉ ***SKILL CHECK 5*** Express "three and twenty-four hundredths" in decimal notation.

Example 11 Express "two hundred seven thousandths" in decimal notation.

Solution The 7 in 207 must be placed in the thousandths place. Therefore the decimal form is 0.207.

0.207 ← thousandths place

◉ ***Skill Check Answers***

1. $6(10^0) + 2\left(\frac{1}{10^1}\right) + 4\left(\frac{1}{10^2}\right)$
2. $4(10^1) + 2(10^0) + 6\left(\frac{1}{10^1}\right) + 0\left(\frac{1}{10^2}\right) + 5\left(\frac{1}{10^3}\right)$
3. forty-one and fifty-six thousandths
4. three hundred four and six hundredths
5. 3.24

0.017
thousandths place

Example 12 Express "seventeen thousandths" in decimal notation.

Solution The 7 in 17 must be placed in the thousandths place. Thus the decimal form is 0.017. Note that a zero is placed in the tenths place.

⦿ ***Skill Check Answer***

6. 0.204

⦿ ***SKILL CHECK 6*** Express "two hundred four thousandths" in decimal notation.

Exercise 4–1–1

1 Write the decimal numbers in expanded form.

1. 0.3 **2.** 0.24 **3.** 0.04

4. 0.135 **5.** 0.309 **6.** 0.004

7. 0.1053 **8.** 6.39 **9.** 38.124

10. 104.1005

2 Write the decimal numbers in words.

11. 0.6 **12.** 0.32 **13.** 0.129

14. 0.04 **15.** 0.802 **16.** 3.5

17. 27.03 **18.** 41.276 **19.** 16.0004

20. 100.0105

Express in decimal notation.

21. Four tenths

22. One hundred five thousandths

23. Three hundred sixty-one thousandths

24. Twenty-one and seven hundredths

25. One hundred twenty-five and two hundred four thousandths

26. Five hundred three and seventy-five hundredths

27. One hundred and seven thousandths

28. Three hundred and five ten thousandths

• Concept Enrichment

29. In writing a bank check the amount must be entered on the check in both numerals and words. If you write a check for $105.73, how would you write it in words?

30. If you write a check for $1,004.06, how would you write it in words?

4–2 Rounding Decimals

OBJECTIVE

Upon completing this section you should be able to:

1 Round a given number to any desired place.

In chapter 1 we established a rule for rounding whole numbers. That same rule applies to rounding decimals to any desired place. We will repeat that rule here for your convenience.

1 Rounding

> To round a number to any place value if the digit in the next place to the right is 5 or greater, the digit in the desired place value is increased by 1. If it is less than 5, the digit in the desired place remains the same. All digits to the right of the desired place are changed to zero.

See section 1–8 if you need a quick refresher.

Example 1 Round 2.3465 to the nearest hundredth.

Solution To round to the nearest hundredth we look at the digit in the next place to the right, which is the thousandths place. We see that it is 6, which is greater than 5. Thus the desired number is

2.3465 (6: thousandths place)

$$2.35$$

rounded to the nearest hundredth. It is understood that all digits to the right of 5 are zero and we do not write them.

◉ ***SKILL CHECK 1*** Round 4.2174 to the nearest hundredth.

◉ ***Skill Check Answer***

1. 4.22

Example 2 Round 2.3465 to the nearest tenth.

2.3465
hundredths place

Solution To round to the nearest tenth we look at the digit in the hundredths place and see that it is 4, which is less than 5. Thus the desired number is

2.3

rounded to the nearest tenth.

⊙ ***SKILL CHECK 2*** Round 4.2174 to the nearest tenth.

Example 3 Round 12.5764 to the nearest whole number.

12.5764
tenths place

Solution To round to the nearest whole number we look at the digit in the tenths place and see that it is 5, which by the rule means we increase the ones digit by 1 obtaining

13

rounded to the nearest whole number.

⊙ ***SKILL CHECK 3*** Round 23.5643 to the nearest whole number.

Example 4 Round 3.0149 to the nearest tenth.

3.0149
hundredths place

Solution To round to the nearest tenth we look at the digit in the hundredths place and see that it is 1, which is less than 5. Thus the desired number is

3.0

rounded to the nearest tenth. In this instance we need to write the zero in the tenths place to indicate that we have rounded to the nearest tenth.

⊙ ***SKILL CHECK 4*** Round 7.0468 to the nearest tenth.

Example 5 Round 2,674.821 to the nearest ten.

2,674.821
units place

Solution To round to the nearest ten we look at the digit in the units place and see that it is 4, which is less than 5. Thus the desired number is

2,670

rounded to the nearest ten.

⊙ ***SKILL CHECK 5*** Round 5,043.56 to the nearest ten.

⊙ ***Skill Check Answers***

2. 4.2
3. 24
4. 7.0
5. 5,040

Exercise 4–2–1

1

1. Round 0.4813 to the nearest tenth.
2. Round 0.4813 to the nearest hundredth.
3. Round 0.4813 to the nearest whole number.
4. Round 0.4813 to the nearest thousandth.
5. Round 3.4162 to the nearest thousandth.
6. Round 3.4162 to the nearest tenth.

7. Round 3.4162 to the nearest hundredth.

8. Round 3.4162 to the nearest whole number.

9. Round 12.3451 to the nearest hundredth.

10. Round 5.352 to the nearest tenth.

11. Round 10.54 to the nearest whole number.

12. Round 25.53 to the nearest whole number.

13. Round 1,306.152 to the nearest ten.

14. Round 0.58 to the nearest whole number.

15. Round 9.05365 to the nearest ten thousandth.

16. Round 3.027 to the nearest tenth.

17. Round 0.45 to the nearest whole number.

18. Round 2,305.384 to the nearest ten.

19. Round 2.005 to the nearest tenth.

20. Round 6.30074 to the nearest ten thousandth.

21. Round 0.100953 to the nearest ten thousandth.

22. Round 5,251.627 to the nearest hundred.

23. Round 0.16548 to the nearest whole number.

24. Round 5.002075 to the nearest ten thousandth.

25. Round 5,208.431 to the nearest hundred.

26. Round 0.4999 to the nearest whole number.

27. Round 0.2500147 to the nearest hundred thousandth.

28. Round 1.030555 to the nearest hundred thousandth.

29. Round 4.987 to the nearest tenth.

30. Round 12.0954 to the nearest hundredth.

31. Round 9.9999 to the nearest hundredth.

32. Round 9.9999 to the nearest thousandth.

• Concept Enrichment

33. In rounding the number 5.6138 to the nearest hundredth we write 5.60. Why can't we simply write 5.6 as the answer?

34. The Internal Revenue Service allows taxpayers to round amounts on their tax returns to the nearest dollar using the rounding method stated in this section. If a person's income for a tax year is $23,649.48, how should that be written on the tax return if it is rounded to the nearest dollar?

35. There are other methods of rounding that are often used. One method in business is to increase the desired rounding place to the next digit if it is followed by any nonzero digits regardless of their values. Thus $8.5312 rounded to the hundredths place is $8.54. Use this method to round $41.60024 to the hundredths place.

36. Computers often round a decimal number by deleting all digits to the right of the desired decimal place regardless of their values. Thus 15.0159659798 rounded to eight decimal places would be 15.01596597. Using this method, round 21.9999999999 to eight decimal places.

4–3 Adding and Subtracting Decimal Numbers

OBJECTIVES

Upon completing this section you should be able to:

1. Add and subtract decimal numbers.
2. Add and subtract like terms involving decimal numbers.
3. Use addition and subtraction of decimal numbers in applications.

1 Adding and Subtracting

We refer again to the basic fact that only like quantities can be added or subtracted. Since this is true we must add tenths to tenths, hundredths to hundredths, and so on. This can be accomplished by following this simple rule.

> When adding or subtracting decimal numbers in column form, the decimal points must be placed in the same column.

Example 1 Find the sum: 2.63 + 5.1 + 16.352

Solution Placing these numbers in column form we have

$$\begin{array}{r} 2.63 \\ 5.1 \\ +16.352 \\ \hline 24.082 \end{array}$$

Notice we are keeping tenths under tenths, hundredths under hundredths, and so on. Thus keeping the decimal points in the same column is like getting a common denominator.

Add in the usual manner. Be sure the decimal point in the answer is in the same column with the decimal points in the problem.

To keep columns of numbers straight it may be helpful to write zeros in the missing places.

$$\begin{array}{r} 2.630 \\ 5.100 \\ +16.352 \\ \hline 24.082 \end{array}$$

⦿ SKILL CHECK 1 Find the sum: 4.52 + 6.3 + 14.263

⦿ Skill Check Answer

1. 25.083

Example 2 Find the sum: $8 + 0.4 + 12.09 + 0.013$

Solution Placing these numbers in a column and writing in zeros we have

$$\begin{array}{r} 8.000 \\ 0.400 \\ 12.090 \\ +\ 0.013 \\ \hline 20.503. \end{array}$$

Note the whole number 8 is written as 8.000.

⦿ ***SKILL CHECK 2*** Find the sum: $9 + 0.5 + 11.03 + 0.074$

Decimal numbers may be added or subtracted without placing them in column form but it is not easy to make sure that only like place values are combined. So it is strongly suggested that the numbers be placed in columns.

Example 3 Subtract: $257.38 - 41.62$

Solution

$$\begin{array}{r} 257.38 \\ -\ 41.62 \\ \hline 215.76 \end{array}$$

Subtract the same way as with whole numbers.

⦿ ***SKILL CHECK 3*** Subtract: $348.29 - 35.76$

Example 4 Subtract: $15 - 0.002$

Solution

$$\begin{array}{r} 15.000 \\ -\ 0.002 \\ \hline 14.998 \end{array}$$

⦿ ***SKILL CHECK 4*** Subtract: $21 - 0.005$

2 Like Terms Involving Decimals

When adding or subtracting like terms, the numerical coefficients at times might be decimal numbers. Recall that to perform these operations we add or subtract the numerical coefficients to obtain the coefficient of the like common literal factors.

Example 5 $1.32x + 0.87x = 2.19x$

$$\begin{array}{r} 1.32 \\ +0.87 \\ \hline 2.19 \end{array}$$

⦿ ***SKILL CHECK 5*** Add: $3.45x + 0.76x$

Example 6 $5.03ab - 0.6ab = 4.43ab$

$$\begin{array}{r} 5.03 \\ -0.60 \\ \hline 4.43 \end{array}$$

⦿ ***SKILL CHECK 6*** Subtract: $9.01ab - 0.8ab$

3 Applications

The most common use of decimals is one we encounter almost every day: money. \$3.75 is read "three and seventy-five hundredths dollars" or "three dollars and seventy-five cents." \$0.37 is read "thirty-seven hundredths dollars" or "thirty-seven cents."

Cent comes from the Latin word *centum* meaning hundredths.

Example 7 Jim bought a pair of shoes for \$37.95, a shirt for \$14.25, and pants for \$32.50 (all taxes included). He gave the salesperson a one-hundred-dollar bill. How much change should he get?

Solution We first add the prices of the individual items to get the total cost of his purchase.

$$\begin{array}{r} \$37.95 \\ 14.25 \\ +\ 32.50 \\ \hline \$84.70 \end{array}$$

⦿ ***Skill Check Answers***

2. 20.604
3. 312.53
4. 20.995
5. $4.21x$
6. $8.21ab$

Then we subtract the total cost from \$100.00 to obtain the amount of change he should receive.

$$\begin{array}{r} \$100.00 \\ -\ 84.70 \\ \hline \$15.30 \end{array}$$

⊙ *Skill Check Answer*

7. \$18.10

⊙ *SKILL CHECK 7* Three items are purchased for \$23.95, \$16.25, and \$41.70. How much change should be received from \$100?

Exercise 4–3–1

1 Add.

1. $\begin{array}{r} 4.13 \\ +2.765 \\ \hline \end{array}$

2. $\begin{array}{r} 8.07 \\ +2.98 \\ \hline \end{array}$

3. $\begin{array}{r} 6.27 \\ +13.085 \\ \hline \end{array}$

4. $\begin{array}{r} 9.63 \\ +3.37 \\ \hline \end{array}$

5. $\begin{array}{r} 2.143 \\ 4.531 \\ +1.214 \\ \hline \end{array}$

6. $\begin{array}{r} 2.17 \\ 0.6343 \\ +4.516 \\ \hline \end{array}$

7. $\begin{array}{r} 2.053 \\ 11.67 \\ +\ 7.46 \\ \hline \end{array}$

8. $\begin{array}{r} 18.3 \\ 2.62 \\ +27.398 \\ \hline \end{array}$

9. $\begin{array}{r} 14.826 \\ 7.0154 \\ 13.27 \\ +35.8 \\ \hline \end{array}$

10. $\begin{array}{r} 4.0068 \\ 17.365 \\ 6.406 \\ +14.95 \\ \hline \end{array}$

Place in column form and add.

11. 9 + 0.9 + 0.17

12. 6 + 0.8 + 0.28

13. 1.1 + 5 + 13.94

14. 3.16 + 0.65 + 1

15. 21.67 + 3.75 + 2.61

16. 15.274 + 11.37 + 21.6

17. 6.38 + 24.2015 + 13.98

18. 28.076 + 18.3 + 7.65

19. $17 + 0.5 + 9.05 + 0.467$

20. $11 + 0.816 + 0.01 + 5.008$

21. $41.6725 + 16.087 + 33.47 + 12.09$

22. $25.8074 + 21.695 + 7.23 + 10.09$

Subtract.

23. $\begin{array}{r} 16.75 \\ -\ 4.62 \\ \hline \end{array}$

24. $\begin{array}{r} 12.32 \\ -\ 9.45 \\ \hline \end{array}$

25. $\begin{array}{r} 21.04 \\ -\ 6.21 \\ \hline \end{array}$

26. $\begin{array}{r} 18.53 \\ -12.47 \\ \hline \end{array}$

27. $\begin{array}{r} 16.427 \\ -11.348 \\ \hline \end{array}$

28. $\begin{array}{r} 13.214 \\ -\ 9.382 \\ \hline \end{array}$

29. $\begin{array}{r} 21.374 \\ -14.536 \\ \hline \end{array}$

30. $\begin{array}{r} 7.0034 \\ -4.2605 \\ \hline \end{array}$

31. $\begin{array}{r} 12.09 \\ -\ 8.352 \\ \hline \end{array}$

32. $\begin{array}{r} 17.34 \\ -11.005 \\ \hline \end{array}$

Place in column form and subtract.

33. $6 - 0.003$

34. $12 - 0.004$

35. $9 - 6.09$

36. $8 - 7.05$

37. $34.685 - 9.205$

38. $18.724 - 13.26$

39. $12.1 - 6.34$

40. $35.03 - 13.4761$

41. $12 - 9.704$

42. $7 - 5.632$

2 Perform the following operations.

43. $0.91a + 0.35a$

44. $1.08x + 0.96x$

45. $3.14y - 1.3y$

46. $1.6x - 0.34x$

47. $0.09a + 6.2b + 3.15a - 2.46b$

48. $4.68y + 2.04z - 2.4y - 1.6z$

3

49. If you have $46.04 in your bank account and deposit a check for $25.96, what is the new balance?

50. If you bought a book for $12.95 and a calculator for $29.75, what is the total cost for both?

51. Mary bought a book for $4.95. What is the change from a twenty-dollar bill?

52. Ellen's checkbook balance was $73.25. She wrote a check for $5.63. What is her new balance?

53. On a trip a person had to stop three times for fuel. The amounts purchased were 10.8 gallons, 9.6 gallons, and 12 gallons. What was the total amount of fuel purchased?

54. A certain medication was administered to a patient. The patient received 0.5 grams in the morning, 1.25 grams at noon, and 2 grams in the evening. What was the total amount administered?

55. At the start of a trip the car's odometer read 11,389.6 miles and at the end of the trip it read 12,305.1 miles. What was the distance of the trip?

56. A spool contains 50 meters of wire. If 18.05 meters are used how many meters of wire remain on the spool?

57. The regular price of a sport coat is $129.00. It is on sale at a discount of $25.80. What is the sale price?

58. Sally received four checks for $23.58, $6.45, $15.25, and $16.03. She deposits all four checks in her bank account. What is the total deposit?

• Concept Enrichment

59. Add: $12 + 0.003 + 1.79 + 8.657$. Round the sum to the nearest tenth.

60. What property of numbers is illustrated by $1.3x + 0.8x = (1.3 + 0.8)x$?

61. Jim went grocery shopping and bought a roast for \$17.63, milk for \$2.27, bread for \$1.39, and chicken for \$5.49. What change did he receive from \$40?

62. You are given a discount of \$4.95 on an item costing \$21.50. How much change will you receive from \$20?

63. A man's bank balance is \$283.42. He writes checks for \$27.95 and \$31.08. He also receives a check for \$38.27 that he deposits in his account. What is his new balance?

64. A woman's grocery bill is \$63.04. She has discount coupons for \$3.50, \$.65, and \$2.25. How much change will she receive from \$100.00?

4-4 Multiplying Decimal Numbers

OBJECTIVES

Upon completing this section you should be able to:

1. Multiply any two decimal numbers.
2. Multiply monomials involving decimal coefficients.
3. Use multiplication of decimals in applications.
4. Use a short method of multiplying a decimal by a power of 10.

The next operation with decimal numbers that we will examine is multiplication. Of course, the rules for this operation will differ from the rules for addition and subtraction. The first thing we notice is that in multiplication the decimal points do not have to be kept in the same column. We will look at some examples to establish the rule needed for multiplication.

1 Multiplying Decimals

Example 1 Multiply: $(3.75)(2.6)$

Solution We first write the problem in column form as if no decimal point existed.

$$\begin{array}{r} 3.75 \\ \times\ 2.6 \\ \hline 2250 \\ 750 \\ \hline 9750 \end{array}$$

Now the question is, "Where do we place the decimal point?" To answer this let's take a closer look at the problem. We have 3.75, which is three and seventy-five hundredths. In the last chapter this would have been written as $3\frac{75}{100}$. Also, we have 2.6, or two and six tenths, which we can write $2\frac{6}{10}$. Now if we look at $3\frac{75}{100} \times 2\frac{6}{10}$ as our problem, we can work it using the rules from the preceding chapter.

$$3\frac{75}{100} \times 2\frac{6}{10} = \left(\frac{375}{100}\right)\left(\frac{26}{10}\right)$$

$$= \frac{(375)(26)}{(100)(10)} = \frac{9,750}{1,000} = 9\frac{750}{1,000}$$

We could reduce these fractions but we want the answer over 1,000 so we can write it as a decimal.

In decimal form this would be 9.750.

If we look at the original numbers being multiplied, we see hundredths (3.75) being multiplied by tenths (2.6), so it is reasonable that the product should be thousandths.

If we multiply tenths by tenths, we get hundredths. If we multiply hundredths by hundredths, we get ten thousandths, and so on.

Example 2 Multiply: (0.6)(0.7)

Solution $$(0.6)(0.7) = \left(\frac{6}{10}\right)\left(\frac{7}{10}\right) = \frac{42}{100} = 0.42$$

Tenths times tenths equals hundredths.

⊙ ***SKILL CHECK 1*** Multiply: (0.4)(0.8)

Example 3 Multiply: (0.24)(0.12)

Solution $$(0.24)(0.12) = \left(\frac{24}{100}\right)\left(\frac{12}{100}\right) = \frac{288}{10,000} = 0.0288$$

Hundredths times hundredths equals ten thousandths.

⊙ ***Skill Check 2*** Multiply: (0.15)(0.23)

> When multiplying decimal numbers, the number of places to the right of the decimal point in the product must equal the total number of places to the right of the decimal point in both numbers to be multiplied.

Example 4 Multiply: (4.38)(7.1)

Solution 4.38 has two decimal places and 7.1 has one. Therefore the product must have three decimal places.

```
  4.38  ← 2 decimal places
 × 7.1  ← 1 decimal place
   438
 30 66
31.098  ← 3 decimal places in the product
```

After obtaining the product 31098, start at the right of 8 and count off three places to the left obtaining 31.098.

⊙ ***SKILL CHECK 3*** Multiply: (3.26)(4.1)

Example 5 Multiply: (0.761)(0.32)

Solution The product must have 3 + 2 = 5 decimal places.

```
   0.761
  ×0.32
   1522
  2283
0.24352
```

3 places plus 2 places.

5 places in the product.

⊙ ***Skill Check Answers***

1. 0.32 2. 0.0345 3. 13.366

⊙ ***SKILL CHECK 4*** Multiply: (0.471)(0.23)

Example 6 Multiply: (0.73)(0.045)

Solution Again the product must have five decimal places.

$$\begin{array}{r} 0.73 \\ \times 0.045 \\ \hline 365 \\ 292 \\ \hline 0.03285 \end{array}$$

Again, 2 + 3 = 5 places needed in the product.

Zero is needed here.

Notice here that to have five places to the right of the decimal point it was necessary to include a zero.

⊙ ***SKILL CHECK 5*** Multiply: (0.67)(0.035)

2 Multiplying Monomials Involving Decimal Coefficients

When multiplying monomials, the coefficients may sometimes be decimal numbers. Remember from chapter 2 that we multiply numerical coefficients and apply the multiplication law of exponents to the literal factors.

Example 7 $(0.7xy)(1.2x^2) = 0.84x^3y$

$(xy)(x^2) = x^3y$

⊙ ***SKILL CHECK 6*** Multiply: $(0.8xy)(3.4x^2y)$

3 Applications

Example 8 If chocolate bars cost $0.42 each, what is the cost of eight of them?

Solution We must multiply the cost of each bar by the number of bars.

$$\begin{array}{r} \$0.42 \\ \times \quad 8 \\ \hline \$3.36 \end{array}$$

2 + 0 = 2 places.

Notice in example 8 that a whole number times a decimal number still uses the same rule of the total number of places to the right of the decimal point. The decimal point can be considered to be to the right of the whole number. Thus 8 could be written 8.0.

⊙ ***SKILL CHECK 7*** What is the cost of twelve candy bars if they are priced at $0.48 each?

⊙ ***Skill Check Answers***

4. 0.10833
5. 0.02345
6. $2.72x^3y^2$
7. $5.76

Exercise 4-4-1

1 Multiply.

1. $\begin{array}{r} 3.2 \\ \times 0.4 \\ \hline \end{array}$ **2.** $\begin{array}{r} 8.5 \\ \times 0.2 \\ \hline \end{array}$ **3.** $\begin{array}{r} 0.8 \\ \times 0.4 \\ \hline \end{array}$ **4.** $\begin{array}{r} 0.24 \\ \times 0.03 \\ \hline \end{array}$ **5.** $\begin{array}{r} 1.32 \\ \times 0.7 \\ \hline \end{array}$

6. $\begin{array}{r} 4.34 \\ \times 2.6 \\ \hline \end{array}$ **7.** $\begin{array}{r} 7.05 \\ \times 0.24 \\ \hline \end{array}$ **8.** $\begin{array}{r} 12.62 \\ \times 0.07 \\ \hline \end{array}$ **9.** $\begin{array}{r} 31.6 \\ \times 0.412 \\ \hline \end{array}$ **10.** $\begin{array}{r} 3.172 \\ \times 0.43 \\ \hline \end{array}$

11. $\begin{array}{r} 5.27 \\ \times \quad 9 \\ \hline \end{array}$

12. $\begin{array}{r} 0.028 \\ \times\ 0.13 \\ \hline \end{array}$

13. $\begin{array}{r} 0.124 \\ \times\ 0.36 \\ \hline \end{array}$

14. $\begin{array}{r} 0.004 \\ \times\ 0.09 \\ \hline \end{array}$

15. $\begin{array}{r} 3.146 \\ \times \quad 10 \\ \hline \end{array}$

16. $\begin{array}{r} 7.139 \\ \times\ 100 \\ \hline \end{array}$

17. $\begin{array}{r} 13.217 \\ \times\ 1{,}000 \\ \hline \end{array}$

18. $\begin{array}{r} 14.62 \\ \times\ 0.1 \\ \hline \end{array}$

19. $\begin{array}{r} 68.13 \\ \times\ 0.01 \\ \hline \end{array}$

20. $\begin{array}{r} 45.62 \\ \times 0.001 \\ \hline \end{array}$

2

21. $(1.4x)(0.5x)$

22. $(3a)(1.5a^2)$

23. $(4a^2)(4.3a^3)$

24. $(3.12y^2)(0.5y^2)$

25. $(2.05xy)(0.4x^3)$

26. $(1.02a^2b)(3.5b^2)$

27. $(0.6a^2b)(1.5ab)$

28. $(1.4xy)(0.01xz)$

29. $(0.05x^2y)(1.4xy^3)$

30. $(1.2abc^2)(0.5ac)$

1

31. Find the product of 2.41 and 3.7.

32. Find the product of 3.29 and 2.6.

33. Find the product of 25.14 and 0.28.

34. Find the product of 12.89 and 0.35.

35. Find the product of 3.14 and 6.12.

36. Find the product of 4.27 and 3.18.

37. Find the product of 0.305 and 0.24.

38. Find the product of 0.704 and 0.45.

39. Find the product of 17.138 and 13.

40. Find the product of 15.231 and 25.

41. Find the product of 12.005 and 2.06.

42. Find the product of 21.015 and 4.02.

3

43. What is the cost of six pencils if they are priced at $0.55 each?

44. If the price of a candy bar is $0.45, find the cost of seven of them.

45. Find the cost of five cans of coffee if the price per can is $3.58.

46. What is the cost of twelve melons if they are priced at $1.29 each?

47. A person makes a monthly car payment of $219.58. How much is payed in 12 months?

48. The length of a dollar bill is 6.125 inches. What is the length of 38 dollar bills placed end to end?

49. If steak is priced at $5.98 per pound, what is the cost of a steak that weighs 1.74 pounds? (Round answer to the nearest cent.)

50. What is the cost of 3.28 pounds of hamburger if the price is $2.99 per pound? (Round answer to two decimal places.)

51. What is the area of a room that is 10.25 feet wide and 15.5 feet long? (Round answer to two decimal places.) Use the formula $A = \ell w$.

52. Find the area of a lot 100.8 feet by 75.6 feet. (Round answer to one decimal place.) Use the formula $A = \ell w$.

• Concept Enrichment

53. Consider the problem (6.15)(8.3). Without actually multiplying can you estimate the size of the answer?

54. A rectangular TV screen measures 0.41 meters high by 0.54 meters wide. Using the area formula $A = hw$, we obtain

$$\begin{aligned} A &= hw \\ &= (0.41\text{ m})(0.54\text{ m}) \\ &= 0.2214\text{ square meters.} \end{aligned}$$

Can this answer be accurate to four decimal places when the sides of the screen were measured accurate to only two decimal places? What should be done with this answer?

55. If ten gallons of gasoline were purchased at 95.9 cents per gallon, how much change was received from a twenty-dollar bill? (Be careful with the decimal point!)

56. A shopper bought five pounds of ground beef at $1.99 per pound and three cans of green beans at $0.75 per can. How much change was received from $20?

4 A Short Method of Multiplying by a Power of 10

A short method of multiplying a decimal number by a power of ten is given below.

Recall that in chapter 2 we multiplied whole numbers by powers of ten.

To multiply a decimal number by a power of ten move the decimal point to the *right* a number of places equal to the number of zeros in the power of ten. Remember that the exponent of ten indicates the number of zeros that are in a power of ten.

Example 9 Find the product of $3.142(10^2)$.

Solution We are multiplying by 100 so we move the decimal point *two* places to the right.

The exponent tells us how many places to the right the decimal point is moved.

$$3.142(10^2) = 314.2$$

⊙ ***SKILL CHECK 8*** Find the product: $5.368(10^2)$

Example 10 Find the product: $12.63(10^3)$

Solution We move the decimal point *three* places to the right.

$$12.63(10^3) = 12{,}630$$

⊙ ***SKILL CHECK 9*** Find the product: $24.3(10^3)$

⊙ ***Skill Check Answers***

8. 536.8
9. 24,300

Exercise 4–4–2

4 Find the products.

1. $3.5(10)$
2. $16.4(10)$
3. $16.13(10)$
4. $12.531(10)$
5. $2.417(10^2)$
6. $8.261(10^2)$
7. $301.2(10^2)$
8. $72.5(10^2)$
9. $4.21(10^3)$
10. $9.16(10^3)$
11. $8.259(10)$
12. $5.117(10)$
13. $0.319(10^2)$
14. $0.173(10^2)$
15. $0.041(10^2)$
16. $0.061(10^2)$
17. $21.351(10^4)$
18. $6.492(10^4)$
19. $0.014(10)$
20. $0.0003(10^2)$

4-5 Dividing Decimal Numbers by Whole Numbers

OBJECTIVES

Upon completing this section you should be able to:

1. Divide a decimal number by a whole number.
2. Round the quotient to any desired place.
3. Use a short method of dividing a decimal by a power of 10.
4. Use division of decimals in applications.

The operation of division will basically follow the long-division algorithm for whole numbers. We will first divide decimal numbers by whole numbers.

1 Dividing Decimals by Whole Numbers

Example 1 Divide 0.36 by 4.

Solution 0.36 (thirty-six hundredths) can be written as the common fraction $\frac{36}{100}$. If we divide this by 4, we have

$$\frac{36}{100} \div 4 = \left(\frac{\overset{9}{\cancel{36}}}{100}\right)\left(\frac{1}{\cancel{4}}\right) = \frac{9}{100}.$$

So $0.36 \div 4 = 0.09$.

Writing this in long-division form, we have

$$4\overline{)0.36.} \quad 0.09$$

0.09 ← quotient
0.36. ← dividend
4 ← divisor

This decimal is ten thousandths.

Example 2 Divide 0.7851 by 3.

Solution This problem written using common fractions is

$$\frac{7{,}851}{10{,}000} \div 3 = \left(\frac{\overset{2{,}617}{\cancel{7{,}851}}}{10{,}000}\right)\left(\frac{1}{\cancel{3}}\right) = \frac{2{,}617}{10{,}000}.$$

So $0.7851 \div 3 = 0.2617$.

If we write the division in long-division form, we have

$$\begin{array}{r} 0.2617 \\ 3\overline{)0.7851} \end{array}$$

Recall that the *dividend* is the number being divided.

> To divide a decimal number by a whole number use the long-division algorithm and place the decimal point in the quotient directly above the decimal point in the dividend.

Example 3 Divide 7.435 by 5 using long division.

First place the decimal point; then divide.

Solution

$$\begin{array}{r} 1.487 \\ 5\overline{)7.435} \\ \underline{5} \\ 2\,4 \\ \underline{2\,0} \\ 43 \\ \underline{40} \\ 35 \\ \underline{35} \\ 0 \end{array}$$

⊙ ***SKILL CHECK 1*** Divide 6.325 by 5.

⊙ ***Skill Check Answer***

1. 1.265

Example 4 Divide 0.00384 by 4.

Solution

```
    0.00096
4)0.00384
       36
        24
        24
         0
```

Note that zeros are needed in the quotient to put the digits in their proper place values.

⊙ ***SKILL CHECK 2*** Divide 0.00324 by 6.

Each digit to the right of the decimal point in the dividend must have a corresponding digit in the quotient.

Sometimes it is necessary to increase the number of decimal places to more than appear in the problem. One situation in which this occurs is when the quotient does not come out exactly. Adding more decimal places can sometimes give us an answer with no remainder. At other times we do it to round our answer.

We can always place additional zeros to the right of a decimal number without changing its value.

Example 5 Divide 22.4 by 5.

Solution

```
   4.4
5)22.4
  20
   2 4
   2 0
     4
```

We write a zero in the next place value in the dividend and continue dividing.

```
   4.48
5)22.40
  20
   2 4
   2 0
     40
     40
      0
```

Do you see that 22.4 and 22.40 have the same value?

⊙ ***SKILL CHECK 3*** Divide 17.9 by 5.

2 Rounding the Quotient

We now look at a combination of division and rounding that we studied in section 4–2.

Example 6 Find 5.325 divided by 8. Round the quotient to three decimal places.

Recall the rule for rounding.

Solution

```
  0.6656 = 0.666 (rounded to three places)
8)5.3250
  4 8
    52
    48
     45
     40
      50
      48
       2
```

Notice that the instructions stated to round to three decimal places. Why then divide to four places? The answer to the question is simply that, if we are going to round to three places, we must have one more place to the right of the third place to apply the rounding rule.

If we wanted to round to four places we would carry the quotient to five places, and so on.

⊙ ***Skill Check Answers***

2. 0.00054 3. 3.58

> If a division is to be rounded to a certain number of decimal places, the quotient must be carried to one more place to correctly round to the desired place.

We must carry the quotient to three decimal places.

We can place as many zeros after the decimal point as we wish.

Example 7 Find 7 divided by 6. Round the quotient to hundredths.

Solution We first note that the decimal point is always understood to be to the right of any whole number. Therefore we may write 7 as 7.000 to give us three decimal places so we can round to two places.

$$\begin{array}{r} 1.166 \\ 6\overline{)7.000} \\ \underline{6} \\ 1\,0 \\ \underline{6} \\ 40 \\ \underline{36} \\ 40 \\ \underline{36} \\ 4 \end{array} = 1.17 \text{ (rounded to hundredths)}$$

Skill Check Answer

4. 1.286

SKILL CHECK 4 Find 9 divided by 7. Round the quotient to thousandths.

Exercise 4–5–1

1 Divide so that there is no remainder.

1. $4\overline{)2.8}$ **2.** $5\overline{)8.5}$ **3.** $9\overline{)31.5}$ **4.** $6\overline{)8.04}$

5. $8\overline{)56.32}$ **6.** $12\overline{)1.68}$ **7.** $6\overline{)0.102}$ **8.** $7\overline{)0.0126}$

9. $8\overline{)128.32}$ **10.** $6\overline{)60.54}$ **11.** $21\overline{)84.273}$ **12.** $34\overline{)170.918}$

13. $10\overline{)14.83}$ **14.** $10\overline{)28.64}$ **15.** $100\overline{)31.265}$ **16.** $100\overline{)73.468}$

2 Round to two decimal places.

17. $6\overline{)29}$ **18.** $3\overline{)1.4}$ **19.** $8\overline{)5.3}$ **20.** $7\overline{)15.31}$

21. $9\overline{)44.6}$ **22.** $12\overline{)39.2}$ **23.** $16\overline{)54.7}$ **24.** $11\overline{)23.07}$

25. $48\overline{)0.304}$ **26.** $34\overline{)0.285}$

Round to thousandths.

27. $7\overline{)24.32}$ **28.** $8\overline{)20.15}$ **29.** $13\overline{)40}$ **30.** $17\overline{)50}$

31. $103\overline{)294}$ **32.** $108\overline{)531}$ **33.** $24\overline{)0.72}$ **34.** $23\overline{)0.56}$

• Concept Enrichment

35. When one number is divided by another, what is the name given to the answer?

36. How many decimal places must we obtain in a quotient if we are going to round it to six decimal places?

37. In the problem 2 ÷ 3, if we did not round the answer, how many decimal places would there be in the quotient?

38. In the problem 0.719 ÷ 12 would you expect the quotient to be greater than 1 or less than 1? Why?

3 A Short Method of Dividing by a Power of 10

A short method of dividing a decimal number by a power of ten is given below.

> To divide a decimal number by a power of ten move the decimal point to the *left* a number of places equal to the number of zeros in the power of ten.

Example 8 Divide: 231.5 ÷ 10

Solution We move the decimal point one place to the left.

$$231.5 \div 10 = 23.15$$

◉ ***SKILL CHECK 5*** Divide: 516.4 ÷ 10

Example 9 Divide: $1.32 \div 10^3$

Solution We move the decimal point three places to the left.

$$1.32 \div 1{,}000 = 0.00132$$

Note that zeros were included to properly position the decimal point.

◉ ***SKILL CHECK 6*** Divide: $3.06 \div 10^3$

The exponent indicates the number of places to the left to move the decimal point.

◉ ***Skill Check Answers***

5. 51.64
6. 0.00306

Exercise 4–5–2

3 Divide.

1. 18.3 ÷ 10 **2.** 143.2 ÷ 10 **3.** $412.6 \div 10^2$ **4.** $17.31 \div 10^2$

5. 1.36 ÷ 10 **6.** 1.05 ÷ 10 **7.** $12.94 \div 10^2$ **8.** $14.52 \div 10^2$

9. $8.31 \div 10^2$ **10.** 0.014 ÷ 10 **11.** $29.3 \div 10^3$ **12.** $513.7 \div 10^3$

13. 0.001 ÷ 10 **14.** $0.035 \div 10^2$ **15.** $1.003 \div 10^2$ **16.** 0.139 ÷ 10

17. $349.6 \div 10^4$ **18.** $81.63 \div 10^4$ **19.** $0.0204 \div 10$ **20.** $0.0194 \div 10^2$

4 Applications

One very important use of division by whole numbers is in finding an **arithmetic mean.** An arithmetic mean, sometimes referred to simply as a **mean,** is a type of **average** that is obtained by finding the sum of a set of scores or measurements and then dividing this sum by the number of scores in the set.

Almost everyone, at one time or another, wants to find an average.

Example 10 A math professor gave a test to 14 students with the following grade results: 56, 73, 62, 91, 87, 64, 72, 84, 97, 77, 72, 40, 83, and 90. What was the average grade rounded to one decimal place?

Solution We first find the sum of all the grades, obtaining a result of 1,048. Next we divide the sum by the number of grades.

$$\begin{array}{r} 74.85 \\ 14\overline{)1{,}048.00} \\ \underline{98} \\ 68 \\ \underline{56} \\ 12\,0 \\ \underline{11\,2} \\ 80 \\ \underline{70} \\ 10 \end{array} = 74.9 \text{ (rounded to one place)}$$

$$\begin{array}{r} 56 \\ 73 \\ 62 \\ 91 \\ 87 \\ 64 \\ 72 \\ 84 \\ 97 \\ 77 \\ 72 \\ 40 \\ 83 \\ \underline{+\ 90} \\ 1{,}048 \end{array}$$

⊙ ***SKILL CHECK 7*** A student took five tests and received scores of 78, 89, 95, 82, and 93. What was the average grade rounded to the nearest whole number?

Example 11 One September the rainfall in Miami, Florida was 23 inches for the month. What was the average daily rainfall to the nearest tenth of an inch?

How many days in September?

Solution To find the average daily rainfall we must divide the amount of rainfall for the entire month by the number of days in the month.

$$\begin{array}{r} 0.76 \\ 30\overline{)23.00} \\ \underline{21\,0} \\ 2\,00 \\ \underline{1\,80} \\ 20 \end{array} = 0.8 \text{ inch per day}$$

Tenth means one decimal place, so we must divide to two places to round.

⊙ ***SKILL CHECK 8*** If the total rainfall in September was 7 inches, what was the average daily rainfall rounded to the nearest tenth of an inch?

⊙ ***Skill Check Answers***

7. 87
8. 0.2 in.

Exercise 4–5–3

4

1. Sam took five tests in a math class during the semester. His scores were 73, 79, 85, 94, and 88. What was his average rounded to tenths?

2. The average annual rainfall for Atlanta is 48.66 inches. What is the average per month rounded to two decimal places?

3. During one season, Larry Bird of the Boston Celtics scored 1,745 points in 82 games. What was his average number of points per game rounded to one decimal place?

4. During one season, the New York Islanders hockey team played 80 games and scored 91 points. How many points per game did they average rounded to two decimal places?

5. Thirty-four people donated a total of $211.00. What was the average donation per person rounded to the nearest cent?

6. Twenty-three people invested a total of $2,506.00 in a stock. What was the average investment per person rounded to the nearest cent?

7. A certain company has three employees. Employee A earns $18,595.00 per year, employee B earns $20,400.00, and employee C earns $17,620.00. What is the average earnings for an employee of the company rounded to the nearest cent?

8. Barbara earned $135.00 the first week, $194.50 the second, and $203.45 the third. What was her average earnings per week rounded to the nearest cent?

9. The ages of seven people in a group are 18, 23, 18, 34, 55, 23, and 18 years. What is the average age rounded to one decimal place?

10. The weights of eight people are 120, 230, 145, 105, 114, 176, 168, and 125 pounds. What is the average weight per person rounded to two decimal places?

11. A man drove 452.7 miles on Monday, 341.2 miles on Tuesday, 103.8 miles on Wednesday, 316.4 miles on Thursday, and 265.0 miles on Friday. What was his average mileage per day for the five days rounded to the nearest tenth?

12. Helen worked 8.4 hours on Monday, 7.0 hours on Tuesday, 5.5 hours on Wednesday, 7.6 hours on Thursday, and 7.8 hours on Friday. What was her average number of hours per day rounded to the the nearest tenth?

4-6 Dividing by Decimal Numbers

1 Dividing by a Decimal

OBJECTIVE

Upon completing this section you should be able to:

1 Divide any whole number or decimal number by a decimal number.

We will now examine how to divide by a decimal. Consider the following problem.

Example 1 Divide: $7.834 \div 0.02$

Solution $7.834 \div 0.02$ can be written $\frac{7.834}{0.02}$ since the fractional bar always means division. The fundamental principle of fractions allows us to multiply both the numerator and denominator by the same nonzero number. Since we already know how to divide by a whole number, we will multiply both the numerator and denominator by a number that will make our problem one of dividing by a whole number.

Recall the fundamental principle of fractions:

$$\left(\frac{a}{b}\right)\left(\frac{c}{c}\right) = \frac{ac}{bc}.$$

$$\left(\frac{7.834}{0.02}\right)\left(\frac{100}{100}\right) = \frac{783.4}{2}$$

$0.02 \times 100 = 2$, which is a whole number.

Thus we can always change division by a decimal to division by a whole number. So in long-division form $0.02\overline{)7.834}$ is now $2\overline{)783.4}$.

$$\begin{array}{r} 391.7 \\ 2\overline{)783.4} \\ \underline{6} \\ 18 \\ \underline{18} \\ 3 \\ \underline{2} \\ 1\,4 \\ \underline{1\,4} \\ 0 \end{array}$$

This is now the same type of problem as in section 4-5.

So the fact is that we never actually divide by a decimal number because we always make our divisor a whole number.

Example 2 Divide: $17.28 \div 1.2$

Solution If we multiply both of these numbers by 10, we have

$$\left(\frac{17.28}{1.2}\right)\left(\frac{10}{10}\right) = \frac{172.8}{12}.$$

$(1.2)(10) = 12$, which again is a whole number.

Placing this in long-division form, we have

$$
\begin{array}{r}
14.4 \\
12\overline{)172.8} \\
\underline{12} \\
52 \\
\underline{48} \\
4\,8 \\
\underline{4\,8} \\
0
\end{array}
$$

The number used to multiply in each case will be 10; 100; 1,000; 10,000; and so on. Multiplying by these numbers simply moves the decimal point to the right a number of spaces equal to the number of zeros.

divisor $\overline{)\text{dividend}}$

> To divide by a decimal number move the decimal point in the divisor all the way to the right until the divisor is a whole number. Then move the decimal point in the dividend the same number of places to the right.

Example 3 Divide: $0.03\overline{)52.152}$

Solution We move the decimal point two places to the right in the divisor and in the dividend. We now have

It is not necessary to place the decimal point after the 3 since whole numbers are not usually written with the decimal point.

$$
\begin{array}{r}
1{,}738.4 \\
3\overline{)5{,}215.2} \\
\underline{3} \\
2\,2 \\
\underline{2\,1} \\
11 \\
\underline{9} \\
25 \\
\underline{24} \\
1\,2 \\
\underline{1\,2} \\
0
\end{array}
$$

⊙ ***SKILL CHECK 1*** Divide: $0.02\overline{)46.568}$

Example 4 Divide: $3.725\overline{)78}$ (Round the answer to one decimal place.)

Note that the decimal point in the dividend is understood to be after the whole number 78.

Solution We move the decimal point three places to the right in both the divisor and dividend. Zeros are included in the dividend to give enough places to move the decimal point. We then have

$$
\begin{array}{r}
20.93 \\
3{,}725\overline{)78{,}000.00} \\
\underline{74\,50} \\
3\,500\,0 \\
\underline{3\,352\,5} \\
147\,50 \\
\underline{111\,75} \\
35\,75
\end{array}
= 20.9 \text{ (rounded to one place).}
$$

⊙ ***Skill Check Answers***

1. 2,328.4
2. 22.5

⊙ ***SKILL CHECK 2*** Divide: $4.135\overline{)93}$ (Round the answer to one decimal place.)

Example 5 Divide: $0.3\overline{)5}$ (Round the answer to two decimal places.)

Solution Moving the decimal point one place to the right in both the divisor and dividend, we have

$$\begin{array}{r} 16.666 \\ 3\overline{)50.000} \\ \underline{3} \\ 20 \\ \underline{18} \\ 2\,0 \\ \underline{1\,8} \\ 20 \\ \underline{18} \\ 20 \\ \underline{18} \\ 2 \end{array} = 16.67 \text{ (rounded to two places).}$$

⊙ *SKILL CHECK 3* Divide: $0.3\overline{)7}$ (Round the answer to two decimal places.)

Example 6 If 6.5 gallons of gasoline cost $7.28, what is the cost per gallon?

Solution To find the cost per gallon we must divide the total cost by the number of gallons.

$$6.5\overline{)7.28}$$

Moving the decimal point one place to the right in both the divisor and dividend, we obtain

$$\begin{array}{r} 1.12 \\ 65\overline{)72.80} \\ \underline{65} \\ 7\,8 \\ \underline{6\,5} \\ 1\,30 \\ \underline{1\,30} \\ 0 \end{array}$$

The cost per gallon is $1.12.

⊙ *SKILL CHECK 4* If 11 gallons of super unleaded gasoline cost $14.96, what was the price per gallon?

⊙ *Skill Check Answers*

3. 23.33
4. $1.36

Exercise 4–6–1

1 Divide.

1. $0.2\overline{)0.7}$

2. $0.4\overline{)0.6}$

3. $0.4\overline{)1.24}$

4. $0.3\overline{)1.23}$

5. $0.02\overline{)0.502}$

6. $0.02\overline{)0.326}$

7. $0.07\overline{)0.742}$

8. $0.05\overline{)0.715}$

9. $0.21\overline{)0.4494}$

10. $0.17\overline{)0.6154}$

11. $0.32\overline{)1.184}$

12. $0.41\overline{)2.214}$

13. $3.6\overline{)29.592}$

14. $4.3\overline{)27.133}$

15. $1.28\overline{)5.2096}$

16. $3.62\overline{)7.3486}$

17. $0.015\overline{)0.00021}$

18. $0.013\overline{)0.000338}$

19. $0.218\overline{)0.00109}$

20. $0.173\overline{)0.001211}$

21. $0.1\overline{)3.15}$

22. $0.1\overline{)2.43}$

23. $0.01\overline{)16.2}$

24. $0.01\overline{)48.5}$

Round to one decimal place.

25. $0.7\overline{)23.4}$

26. $0.8\overline{)34.6}$

27. $0.03\overline{)8.612}$

28. $0.04\overline{)9.027}$

29. $1.6\overline{)17.24}$

30. $3.7\overline{)29.16}$

31. $3.8\overline{)88}$

32. $4.3\overline{)90}$

33. $0.32\overline{)68.1}$

34. $0.27\overline{)50.3}$

35. $1.61\overline{)4.3}$

36. $2.13\overline{)8.7}$

Round to hundredths.

37. $0.7\overline{)8}$

38. $0.9\overline{)11}$

39. $2.6\overline{)18.32}$

40. $4.2\overline{)22.41}$

41. $2.137\overline{)45}$

42. $1.037\overline{)14}$

43. $5.013\overline{)21.43}$

44. $3.028\overline{)17.52}$

45. If 5.5 pounds of apples cost $3.80, what is the cost per pound rounded to the nearest cent?

46. If a 6.34-pound roast cost $25.23, what is the cost per pound rounded to the nearest cent?

47. If a car travels 347.5 miles on 14.6 gallons of gasoline, how many miles did it get per gallon rounded to the nearest tenth?

48. After traveling 382.3 miles with your car you fill the tank and find it used 11.8 gallons of gasoline. How many miles per gallon did the car get rounded to the nearest tenth?

49. Mark Spitz swam 100 meters in 51.20 seconds. How many meters per second did he average? Round your answer to two decimal places.

50. Anne Henning skated 500 meters in 43.3 seconds. How many meters per second did she average? Round your answer to the nearest tenth.

• **Concept Enrichment**

51. What principle allows us to move the decimal point the same number of places in both the dividend and the divisor?

52. In the problem $10.362 \div 2.31$ how many places must the decimal point be moved in the dividend and the divisor?

4–7 Interchanging Common Fractions and Decimals

OBJECTIVES

Upon completing this section you should be able to:

1. Change a decimal to a common fraction.
2. Change a common fraction to a decimal.
3. Perform calculations involving decimals, fractions, and whole numbers.

In this chapter and in chapter 3 we learned the four basic operations on common fractions and decimal fractions. It is sometimes desirable to change common fractions to decimals or decimals to common fractions.

1 Changing Decimals to Common Fractions

First we will discuss changing decimals to common fractions. The process comes from the fact that the number of places to the right of the decimal point gives us the denominator of the common fraction.

1. One decimal place means tenths, so the denominator is 10.
2. Two decimal places means hundredths, so the denominator is 100.
3. Three decimal places means thousandths, so the denominator is 1,000. And so on.

Example 1 Change 0.375 to a common fraction.

Three decimal places means thousandths.

Solution 0.375 has three decimal places and is read "three hundred seventy-five thousandths." Writing this as a common fraction, we have

$$0.375 = \frac{375}{1{,}000}.$$

Now we must refer back to the fact that we always reduce a fractional answer to simplest form. So

Always reduce to simplest form.

$$0.375 = \frac{375}{1{,}000} = \frac{(3)(\not{5})(\not{5})(\not{5})}{(2)(2)(2)(\not{5})(\not{5})(\not{5})} = \frac{3}{8}.$$

⦿ ***SKILL CHECK 1*** Change 0.625 to a common fraction.

Example 2 Change 0.22 to a common fraction.

Two decimal places means hundredths.

Solution 0.22 or twenty-two hundredths is written

$$0.22 = \frac{22}{100} = \frac{11}{50}.$$

⦿ ***SKILL CHECK 2*** Change 0.44 to a common fraction.

If digits other than zero appear to the left of the decimal point, we will have a mixed number.

⦿ ***Skill Check Answers***

1. $\frac{5}{8}$
2. $\frac{11}{25}$

Example 3 Change 3.75 to a mixed number.

Solution Three and seventy-five hundredths is written

$$3.75 = 3\frac{75}{100} = 3\frac{3}{4}.$$

Remember, 3.75 is read as "three and seventy-five hundredths."

⦿ ***SKILL CHECK 3*** Change 5.25 to a mixed number.

⦿ ***Skill Check Answer***

3. $5\frac{1}{4}$

Exercise 4–7–1

1 Change to a simplified common fraction or mixed number.

1. 0.7 **2.** 0.11 **3.** 0.4 **4.** 0.16 **5.** 0.34

6. 0.55 **7.** 2.9 **8.** 4.8 **9.** 0.124 **10.** 0.025

11. 0.238 **12.** 0.144 **13.** 0.416 **14.** 0.732 **15.** 7.104

16. 3.496 **17.** 11.775 **18.** 19.016 **19.** 0.1152 **20.** 18.1104

2 Changing a Common Fraction to Decimal Form

The task of changing a common fraction to a decimal depends on the fact that the fractional bar indicates division. For instance, $\frac{3}{4}$ can be written $4\overline{)3}$.

Example 4 Change $\frac{3}{4}$ to a decimal.

Solution Writing this in long-division form, we have

$$\begin{array}{r} 0.75 \\ 4\overline{)3.00} \\ \underline{2\ 8} \\ 20 \\ \underline{20} \\ 0 \end{array}$$

We use the long-division algorithm.

The answer in example 4 did not have a remainder. If there had been a remainder, we could have continued dividing.

⊙ ***SKILL CHECK 4*** Change $\frac{3}{5}$ to a decimal.

When changing a common fraction to decimal form, if the division algorithm leads to a zero remainder, we have a **terminating decimal.** If the decimal does not terminate, a digit or group of digits will repeat and we have a **repeating decimal.** If the denominator of a common fraction contains any factor other than 2 or 5, the decimal form will be a repeating decimal.

Example 5 Change $\frac{2}{3}$ to a decimal.

The three dots indicate that the digit 6 will repeat over and over indefinitely.

Solution

$$\begin{array}{r} 0.666\ldots \\ 3\overline{)2.000} \\ \underline{1\,8} \\ 20 \\ \underline{18} \\ 20 \\ \underline{18} \\ 2 \end{array}$$

This is an example of a repeating decimal. In this case the 6 keeps repeating forever. We sometimes place a bar over the repeating digit or digits to indicate that they repeat indefinitely. We could thus write 0.666 . . . as $0.\overline{6}$ to indicate the repetition.

⊙ ***SKILL CHECK 5*** Change $\frac{1}{3}$ to a decimal.

Example 6 Change $\frac{3}{11}$ to a repeating decimal.

Solution

$$\begin{array}{r} 0.2727\ldots \\ 11\overline{)3.0000} \\ \underline{2\,2} \\ 80 \\ \underline{77} \\ 30 \\ \underline{22} \\ 80 \\ \underline{77} \\ 3 \end{array}$$

We see that there are two repeating digits. The decimal form can be written $0.\overline{27}$.

⊙ ***SKILL CHECK 6*** Change $\frac{5}{11}$ to a repeating decimal.

If we are going to round an answer in decimal form, we need to know how many places are needed. Remember that we must divide to one more place than the desired place.

⊙ ***Skill Check Answers***

4. 0.6
5. $0.\overline{3}$
6. $0.\overline{45}$

Example 7 Change $\frac{7}{9}$ to a decimal rounded to three places.

Solution

$$\begin{array}{r} 0.7777 \\ 9\overline{)7.0000} \\ \underline{6\,3} \\ 70 \\ \underline{63} \\ 70 \\ \underline{63} \\ 70 \\ \underline{63} \\ 7 \end{array} = 0.778 \text{ (rounded to three places)}$$

We must carry the quotient to four places in order to round to three places.

If we were not rounding, this answer could be written $0.\overline{7}$.

⊙ ***SKILL CHECK 7*** Change $\frac{5}{9}$ to a decimal rounded to three places.

An improper fraction is treated the same way.

Example 8 Change $\frac{12}{7}$ to a decimal rounded to two places.

Solution

$$\begin{array}{r} 1.714 \\ 7\overline{)12.000} \\ \underline{7} \\ 5\,0 \\ \underline{4\,9} \\ 10 \\ \underline{7} \\ 30 \\ \underline{28} \\ 2 \end{array} = 1.71 \text{ (rounded to two places)}$$

Always divide the denominator *into* the numerator.

⊙ ***SKILL CHECK 8*** Change $\frac{10}{7}$ to a decimal rounded to two places.

Example 9 Change $2\frac{5}{8}$ to decimal form rounded to three places.

Solution First change the mixed number to an improper fraction.

$$2\frac{5}{8} = \frac{21}{8}$$

Then

$$\begin{array}{r} 2.625 \\ 8\overline{)21.000} \\ \underline{16} \\ 5\,0 \\ \underline{4\,8} \\ 20 \\ \underline{16} \\ 40 \\ \underline{40} \\ 0 \end{array}$$

⊙ ***SKILL CHECK 9*** Change $2\frac{7}{8}$ to decimal form rounded to three places.

⊙ ***Skill Check Answers***

7. 0.556
8. 1.43
9. 2.875

Exercise 4–7–2

A Change to either a terminating or repeating decimal.

1. $\frac{1}{2}$ **2.** $\frac{1}{4}$ **3.** $\frac{4}{5}$ **4.** $\frac{1}{5}$ **5.** $\frac{1}{8}$

6. $\frac{3}{8}$ **7.** $\frac{1}{6}$ **8.** $\frac{5}{6}$ **9.** $\frac{2}{11}$ **10.** $\frac{7}{33}$

11. $\frac{3}{7}$ **12.** $\frac{1}{7}$ **13.** $3\frac{1}{3}$ **14.** $4\frac{2}{5}$ **15.** $\frac{11}{4}$

Change to a decimal number. If the decimal does not terminate within four places, round to thousandths.

16. $\frac{15}{8}$ **17.** $\frac{3}{14}$ **18.** $\frac{7}{16}$ **19.** $5\frac{17}{21}$ **20.** $7\frac{18}{25}$

21. $\frac{120}{11}$ **22.** $\frac{235}{18}$ **23.** $\frac{348}{103}$ **24.** $\frac{795}{114}$

• Concept Enrichment

25. Without actually dividing, explain why the decimal form of $\frac{1}{8}$ will terminate.

26. Without dividing, explain why the decimal form of $\frac{1}{30}$ will repeat.

27. Round the decimal $0.\overline{6}$ to three decimal places.

28. Round the decimal $1.\overline{352}$ to four decimal places.

3 Calculations Involving Decimals, Common Fractions, and Whole Numbers

Some calculations may involve common fractions, decimals, and whole numbers. We follow the same rules for order of operations when evaluating such expressions.

Example 10

$$\begin{aligned}(3.2)(0.5) + 24 \div 0.6 + 1.4 &= 1.6 + 24 \div 0.6 + 1.4 \\ &= 1.6 + 40 + 1.4 \\ &= 43\end{aligned}$$

Recall, multiplication and division must be performed before addition and subtraction.

Example 11

$$\begin{aligned}6.1 + \frac{1}{2}(0.5) - 1.35 \div \frac{3}{4} &= 6.1 + 0.25 - 1.35 \div \frac{3}{4} \\ &= 6.1 + 0.25 - 1.8 \\ &= 6.35 - 1.8 \\ &= 4.55\end{aligned}$$

⦿ ***SKILL CHECK 10*** Evaluate in decimal form:

$$9.4 + \frac{1}{2}(1.5) - 3.24 \div \frac{2}{3}$$

⦿ ***Skill Check Answer***

10. 5.29

Exercise 4-7-3

3 Evaluate and express the answer in decimal form.

1. $\frac{1}{4} + 0.08 \div 2$

2. $\frac{3}{4} - 0.36 \div 3$

3. $1.5 \div 2\frac{1}{2} + 5$

4. $3.2 \div 1\frac{1}{4} + 4$

5. $3(1.04) \div 0.2 + \frac{3}{4}$

6. $5(2.4) \div 0.12 - 6$

7. $8.12(0.5) - 6.9 \div 3$

8. $\frac{1}{2}(5.1) + 2.3 \div \frac{1}{4}$

9. $\frac{2}{3}(5.1) + 2.5(0.3)$

10. $\frac{1}{2}(0.5) + 1.6 \div \frac{1}{3}$

11. $3.4 + 0.18 \div 2 - 2.1$

12. $15.2 - 2.14 \div 0.2 + 2.18$

13. $4 + 0.012(5.1) \div 2$

14. $5 + 3.1(1.05) \div 3$

15. $3\frac{2}{5} + 2.64 \div 2.5 - 1.92$

16. $3.1 + 1.95 \div 1\frac{1}{2} - 2.8$

17. $4 \div 2\frac{1}{2} + 6.12 - \frac{2}{3}(5.7)$

18. $17 \div 3\frac{2}{5} - \frac{1}{3}(2.37) + 5.13$

19. $11.2 \div 1.4 - \frac{1}{2}(5.6) + \frac{1}{3}(41.7)$

20. $7 + 0.2(0.15) - \frac{3}{4}(6.12) + 0.153 \div 0.05$

CHAPTER 4 SUMMARY

Key Words

Section 4–1

- A **decimal point** is used in a number to indicate a fraction and is read "and."
- A number containing a decimal point is referred to as a **decimal fraction, decimal number,** or simply a **decimal.**

Section 4–5

- An **arithmetic mean,** sometimes referred to simply as a **mean,** is a type of **average.**

Section 4–7

- When the division algorithm leads to a zero remainder, we have a **terminating decimal.**
- If the decimal form of a common fraction does not terminate, a digit or group of digits will repeat and we have a **repeating decimal.**

Procedures

Section 4–2

- To round a decimal number to any place value if the digit in the next place to the right is 5 or greater, the digit in the desired place value is increased by 1. If it is less than 5, the digit in the desired place remains the same. All digits to the right of the desired place are changed to zero.
- To add or subtract decimal numbers place them in column form with the decimal points in the same column.

Section 4–4

- When multiplying decimal numbers, the number of places to the right of the decimal point in the product must equal the total number of places to the right of the decimal point in both numbers to be multiplied.
- To multiply a decimal number by a power of ten move the decimal point to the right a number of places equal to the power of ten.

Section 4–5

- To divide a decimal number by a whole number use the long-division algorithm and place the decimal point in the quotient directly above the decimal point in the dividend.
- If a division is to be rounded to a certain number of decimal places, the quotient must be carried to one more place to correctly round to the desired place.
- To divide a decimal number by a power of ten move the decimal point to the left a number of places equal to the power of ten.
- To find the arithmetic mean or average of a set of scores divide their sum by the number of scores in the set.

Section 4–6

- To divide by a decimal number move the decimal point in the divisor all the way to the right so that the divisor is a whole number. Then move the decimal point in the dividend the same number of places to the right.

Section 4–7

- A common fraction is changed to a decimal by dividing the numerator by the denominator.

CHAPTER 4 REVIEW

1. Write 0.24 in expanded form.

2. Write 0.07 in expanded form.

3. Write 7.3074 in expanded form.

4. Write 12.1308 in expanded form.

5. Write 0.49 in words.

6. Write 0.513 in words.

7. Write 15.075 in words.

8. Write 8.079 in words.

9. Express "four hundred three thousandths" in decimal notation.

10. Express "two hundred eight ten thousandths" in decimal notation.

11. Round 0.8145 to the nearest hundredth.

12. Round 3.619 to the nearest whole number.

13. Round 6.3965 to the nearest thousandth.

14. Round 4.082 to the nearest tenth.

15. Round 0.9999 to the nearest hundredth.

16. Round 0.4495 to the nearest hundredth.

17. Round 0.548 to the nearest tenth.

18. Round 0.1496 to the nearest thousandth.

19. Round 9.523 to the nearest whole number.

20. Round 1.5555 to the nearest thousandth.

21. Change 0.83 to a common fraction.

22. Change 0.07 to a common fraction.

23. Change 0.288 to a common fraction.

24. Change 0.304 to a common fraction.

25. Change 17.48 to a mixed number.

26. Change 24.125 to a mixed number.

27. Change $\frac{23}{24}$ to a decimal rounded to three places.

28. Change $\frac{11}{34}$ to a decimal rounded to three places.

29. Change $5\frac{7}{9}$ to a decimal rounded to three places.

30. Change $11\frac{3}{8}$ to a decimal rounded to three places.

Perform the indicated operation. In division problems round the answer to three decimal places.

31. $\begin{array}{r} 3.14 \\ \times 0.02 \\ \hline \end{array}$

32. $\begin{array}{r} 48.17 \\ +54.34 \\ \hline \end{array}$

33. $4\overline{)0.72}$

34. $0.3\overline{)29.1}$

35. $\begin{array}{r} 42.38 \\ +16.92 \\ \hline \end{array}$

36. $\begin{array}{r} 5.23 \\ \times 0.04 \\ \hline \end{array}$

37. $0.7\overline{)58.6}$

38. $6\overline{)1.44}$

39. $\begin{array}{r} 32.8 \\ -\ 5.14 \\ \hline \end{array}$

40. $20.6 - 8.34$

41. $7.34 + 8.5$

42. $51.041x - 36.28x$

43. $6.15x + 13.056x$

44. $0.04\overline{)0.005}$

45. $\begin{array}{r} 12.35 \\ \times\ 0.18 \\ \hline \end{array}$

46. $6.7ab + 9.31ab$

47. $(0.008)(0.05)$

48. $14\overline{)64.68}$

49. $0.09\overline{)0.306}$

50. $\begin{array}{r} 14.29 \\ \times\ 0.14 \\ \hline \end{array}$

51. $\begin{array}{r} 40.029 \\ -16.84\ \\ \hline \end{array}$

52. $(0.014xy)(0.07x^2y)$

53. $(1.5ab^2)(0.38ab)$

54. $\begin{array}{r} 138.65 \\ +\ 92.77 \\ \hline \end{array}$

55. $\begin{array}{r} 32.47 \\ \times\ 100 \\ \hline \end{array}$

56. $6.2\overline{)0.0186}$

57. $\begin{array}{r} 325.74 \\ +134.66 \\ \hline \end{array}$

58. $\begin{array}{r} 24.01 \\ -13.09 \\ \hline \end{array}$

59. $\begin{array}{r} 132.16 \\ -\ 94.39 \\ \hline \end{array}$

60. $\begin{array}{r} 64.17 \\ \times\ 100 \\ \hline \end{array}$

61. $4.7\overline{)0.0376}$

62. $\begin{array}{r} 69.043 \\ +58.958 \\ \hline \end{array}$

63. $17\overline{)39.6}$

64. $\begin{array}{r} 540.38 \\ -359.49 \\ \hline \end{array}$

65. $\begin{array}{r} 18.126 \\ \times\ 4.38 \\ \hline \end{array}$

66. $49.13 + 135.624 + 28.97$

67. $47.138 + 14.69$

68. $23\overline{)78.4}$

69. $\begin{array}{r} 35.003 \\ -13.999 \\ \hline \end{array}$

70. $(28.019)(1.27)$

71. $2.14\overline{)62.4}$

72. $5.23\overline{)18.9}$

73. $\begin{array}{r} 37.52 \\ 468.054 \\ +\ 7.967 \\ \hline \end{array}$

74. $\begin{array}{r} 68.004 \\ -48.995 \\ \hline \end{array}$

75. $\begin{array}{r} 125.08 \\ \times\ 0.156 \\ \hline \end{array}$

76. $\begin{array}{r} 319.26 \\ \times\ 0.347 \\ \hline \end{array}$

77. $15\overline{)397.12}$

78. $490.28ab + 6.847ab + 46.37ab$

79. $\begin{array}{r} 284.08 \\ -139.09 \\ \hline \end{array}$

80. $\begin{array}{r} 405.16 \\ -285.77 \\ \hline \end{array}$

81. $64.79xy + 153.8xy + 29.047xy$

82. $35\overline{)782.12}$

83. $\begin{array}{r} 47.16 \\ -29.362 \\ \hline \end{array}$

84. $36\overline{)584.7}$

85. $\begin{array}{r} 5.284 \\ \times\ 2.09 \\ \hline \end{array}$

86. $7.4\overline{)81.6664}$

87. $42\overline{)129.4}$

88. $34.35 - 16.054$

89. $8.7\overline{)26.7438}$

90. $\begin{array}{r} 4.635 \\ \times\ 7.04 \\ \hline \end{array}$

91. A woman bought some party goods for her daughter's birthday. She paid $3.98 for plates, $2.65 for napkins, and $5.89 for a game. What was the total cost?

92. A man bought a tire for his car for $69.88. He also bought a battery for $74.99 and some oil for $10.98. What was the total cost?

93. A bank balance is $543.24. If a check is written for $295.65, what is the new balance?

94. A bank balance is $216.44. If a check is written for $136.85, what is the new balance?

95. If you bought three items for $6.25, $2.38, and $4.28, how much change would you receive from $20?

96. The balance in a checking account is $304.28. If three checks are written for $64.12, $18.39, and $112.18, what is the new balance?

97. Chuck steak is on sale for $3.19 per pound. What is the cost of a 5.34-pound steak? Round answer to nearest cent.

98. What is the cost of a 21.15-pound turkey if the price is $0.79 per pound? Round answer to nearest cent.

99. What is the cost of four tires if they are priced at $69.98 each?

100. If the price of a certain material is $6.98 per yard, what is the cost of 5.5 yards?

101. The average annual rainfall for Boston is 41.55 inches. What is the average per month? Round the answer to two decimal places.

102. A student has test scores of 71, 87, 56, 74, 83, 92, 95, 81, 78, 80, and 86. What is the average rounded to two decimal places?

103. During a storm there were 5.41 inches of rain in four hours. What was the average rainfall per hour rounded to the nearest hundredth?

104. During six football games a running back rushed for 105, 86, 114, 72, 124, and 94 yards. What was the average per game rounded to the nearest tenth of a yard?

105. A package of ground beef weighing 3.16 pounds cost $6.60. What is the cost per pound rounded to the nearest cent?

106. Irena Szewinska of Poland ran the 400-meter dash in 49.29 seconds. What was her average number of meters per second rounded to two decimal places?

CHAPTER 4 TEST

1. Write 52.649 in expanded form.

2. Write 298.37 in words.

3. Round 8.6531 to the nearest tenth.

4. Change 0.112 to a common fraction in simplest form.

5. Change $13\frac{5}{7}$ to a decimal rounded to three decimal places.

Perform the indicated operation. In division problems round the quotient to three decimal places.

6. $\begin{array}{r} 32.79 \\ +\ 9.63 \\ \hline \end{array}$

7. $\begin{array}{r} 68.142 \\ -\ 9.374 \\ \hline \end{array}$

8. $9\overline{)66.6}$

9. $0.6\overline{)9.138}$

10. $(6.28xy)(0.8xy^2)$

11. $7.465 + 18.84 + 9.356$

12. $27.15a - 9.758a$

13. $460.3 \div 21$

14. $0.7\overline{)5.838}$

15. $\begin{array}{r} 240.15 \\ \times\ \ 5.6 \\ \hline \end{array}$

16. $\begin{array}{r} 28.64 \\ 321.523 \\ +\ 36.048 \\ \hline \end{array}$

17. $\begin{array}{r} 53.007 \\ -29.138 \\ \hline \end{array}$

18. $18\overline{)127.44}$

19. $\begin{array}{r} 28.03 \\ \underline{\times 0.049} \end{array}$

20. $7.3\overline{)0.0657}$

21. Three items are purchased for $8.19, $2.57, and $4.21. How much change is received from $20?

22. What is the cost of 4.65 pounds of ground beef that is priced at $2.28 per pound? Round answer to nearest cent.

CHAPTERS 1–4 CUMULATIVE TEST

1. Round 8,306 to the nearest ten.

2. Add: $\begin{array}{r} 1{,}906 \\ 242 \\ +3{,}054 \\ \hline \end{array}$

3. Multiply: $\begin{array}{r} 2{,}516 \\ \times\ 218 \\ \hline \end{array}$

4. Find the greatest common factor of $70x^2y^3$ and $84x^3y$.

5. Find the least common multiple of 16 and 24.

6. Reduce $\dfrac{84a^2bc}{90ab^3c^2}$.

7. Find the LCD of $\dfrac{2}{3}, \dfrac{1}{12}, \dfrac{3}{8}$.

8. Divide: $\dfrac{3x^2}{8y^3} \div \dfrac{9x}{10ay}$

9. Subtract: $\dfrac{3}{4ab^2} - \dfrac{5}{12a^2b}$

10. Multiply: $3\frac{1}{2} \times 4\frac{2}{7}$

11. Change 0.264 to a common fraction in simplest form.

12. Multiply: $\begin{array}{r} 25.3 \\ \times 0.42 \\ \hline \end{array}$

13. Divide: $21.4\overline{)93.518}$

14. Multiply: $(3x^2y)(5x^3y^5)$

15. Evaluate: $2[4 + (6 - 2)] + 3(5 - 1)$

16. If a person made 24 equal payments that totaled \$718.80, how much was each payment?

17. State the number of terms in the expression $3x^2 - 4x + 1$.

18. Evaluate $x^3 - 2x$ if $x = 5$.

19. Evaluate $3a^2 - 5b + 2ab$ if $a = 4$ and $b = 1$.

20. Combine like terms:
$25x^2y + 3y^2 - 5x^2y + y^2 - 11x^2y$

CHAPTER

5 Pretest

Answer as many of the following problems as you can before starting this chapter. When you finish the chapter, take the test at the end and compare your scores to see how much you have learned.

1. Find the measure of an angle whose complement measures $47°$.

2. The following diagram shows two parallel lines intersected by a transversal. Find the measure of $\angle a$.

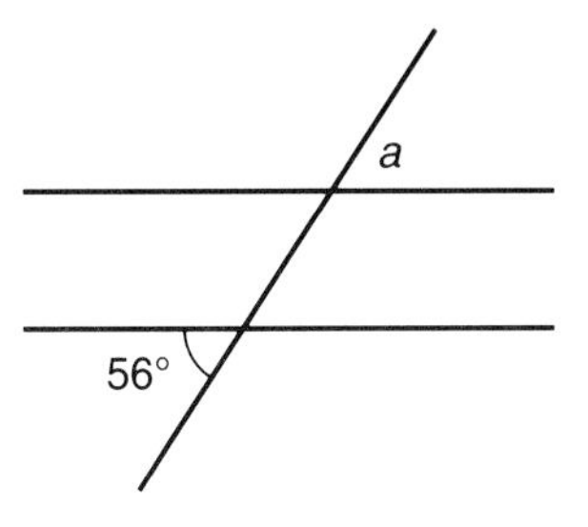

3. How many sides does a pentagon have?

4. How many diagonals does a 6-sided polygon have?

5. Find the perimeter of a rectangle having a width of $4\frac{3}{4}$ inches and a length of 12 inches.

6. Find the area of a rectangle 10.5 centimeters long by 7.4 centimeters wide.

7. In isosceles $\triangle ABC$, $\angle A = 70°$. Find the measures of the other two angles.

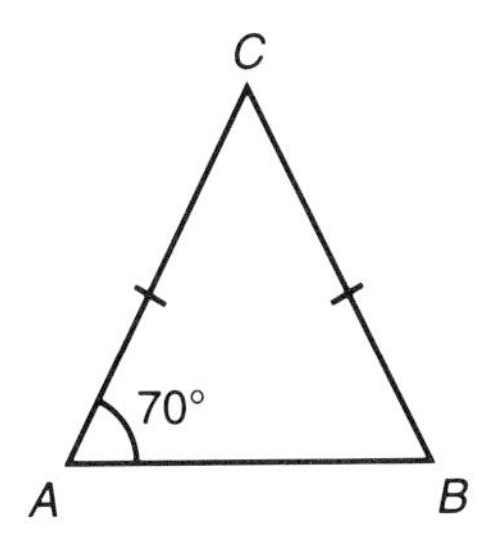

8. Find the area of a triangle with base 14 inches and altitude 8 inches.

9. Find the circumference of a circle having a radius of 4.5 feet. Use $\pi = 3.14$.

10. Find the area of a circle having a radius of $\frac{1}{2}$ meter. Use $\pi = \frac{22}{7}$.

11. Find the volume of a rectangular parallelepiped with length 2.5 feet, width 2.2 feet, and altitude 3.4 feet.

12. Find the volume of a cylinder having a height of $3\frac{1}{2}$ inches and having a circular base with radius $1\frac{3}{4}$ inches. Use $\pi = \frac{22}{7}$.

13. If tile costs $3.50 per square foot, how much will it cost to cover a floor that measures 10 feet by 15 feet?

14. A swimming pool measures 12 feet by 18 feet and is 5 feet deep. If each cubic foot contains about 7.5 gallons, how many gallons of water are needed to fill the pool?

C H A P T E R

5 Concepts and Formulas from Geometry

Geometric designs occur in both nature and manufactured objects.

STUDY HELP

Special features have been placed within this text to help in your studies. The objectives state what is to be learned. A skill check following most examples helps you determine if you understand the concept presented. The margin notes, consisting of informal statements that assist you in understanding the material and also warn of common pitfalls, serve as your own private tutor. Utilizing these important features will maximize your success.

Geometry is based on a process of logical thinking in which conclusions are arrived at in a very structured manner. A thorough presentation of geometry requires more time and study than can be accomplished in one chapter. Our purpose here is to present some of the basic concepts and numerous facts and formulas from geometry and illustrate their usefulness in some everyday situations.

5–1 Angles and Their Measure

OBJECTIVES

Upon completing this section you should be able to:

1. Understand some basic terms used in geometry.
2. Find the complement of an angle.
3. Find the supplement of an angle.
4. Determine the measures of angles formed by intersecting lines.
5. Use a ruler and protractor to measure line segments and angles.

1 Some Basic Terminology

As is true in any logical system, some terms used in geometry cannot be defined because they cannot be expressed in simpler terms (that is, in words that are already familiar to us). For example, **point** is an undefined term. We can describe a point as "an element that has position but no dimension," but this is not a formal definition since *dimension* is defined using the word *point*. **Line** and **plane** are other undefined terms.

Although *point, line,* and *plane* are accepted as undefined, we can form a pictorial model that serves as a representation of these undefined terms. Thus we have an intuitive idea of their meanings and can visualize that they possess certain properties.

We visualize point, line, and plane as follows:

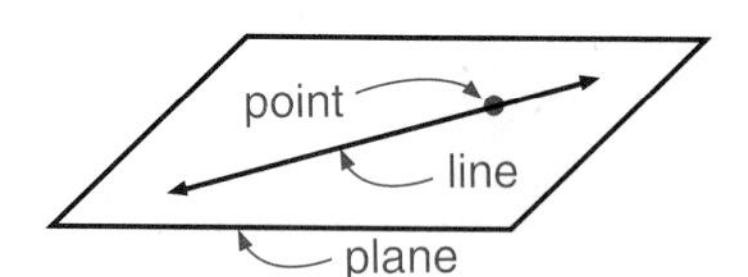

A line is composed of a set of points and may be curved or straight. In our discussion, however, when we use the word *line* we will be referring to a straight line unless otherwise specified. All points on the same straight line are said to be **collinear.**

A line extends indefinitely in opposite directions from any point P on the line.

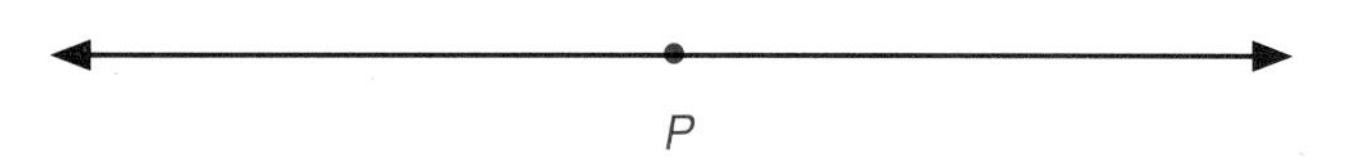

We often say that two points determine a line.

If there are two distinct (different) points A and B, then there is one and only one line AB through these two points.

If three distinct points A, B, and C are not on the same line (noncollinear points), then there is one and only one plane through these three points.

Three noncollinear points determine a plane.

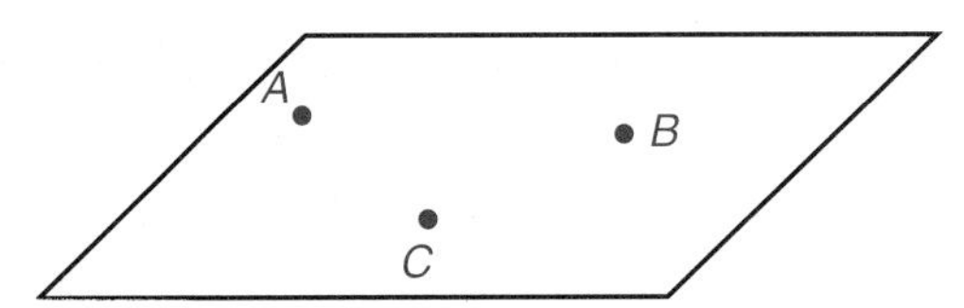

We will now use these undefined terms in some definitions. Notice that the following definition makes use of terms we have accepted as undefined.

DEFINITION A **line segment** consists of two distinct points on a line and all points on the line between them.

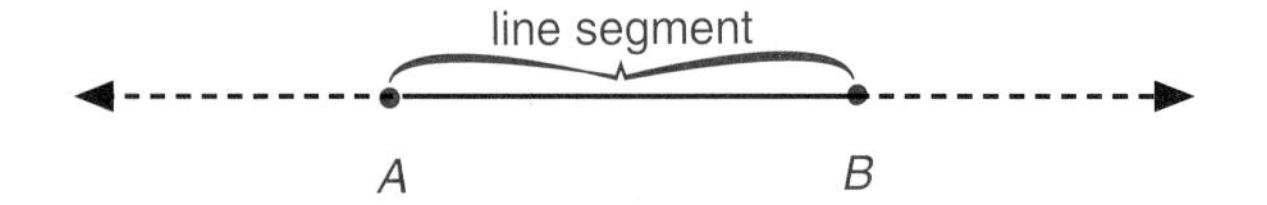

In this figure the line segment contains the point A, the point B, and all points on the line between A and B. The notation for the line segment is $\overline{AB}$. We write the letters indicating the endpoints with a bar over them.

It can be shown that there are infinitely many points between any two points on a line.

The length of a line segment is the distance between its endpoints. Line segments are measured using **linear units** such as inch, centimeter, foot, meter, mile, and so on. The statement $\overline{AB} = 5$ inches means the distance from point A to point B is 5 inches. In this text we will not use a different notation for a line segment and the measure of a line segment.

Sometimes you might see the notation $d(\overline{AB})$ used to mean the distance from A to B.

DEFINITION A **plane figure** is a geometric figure that lies entirely in a plane.

Notice that a line and a line segment are plane figures. We now wish to define another plane figure called a *ray*. We can think of a ray as starting at a point on a line and extending indefinitely in one direction on the line. The formal definition follows.

DEFINITION A **ray** consists of a point A, a point B, all points on the line between A and B, and all points C such that point B is between A and C.

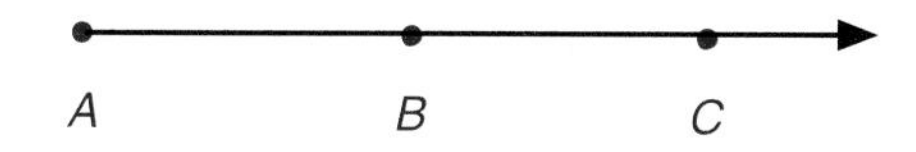

We represent the ray as $\overrightarrow{AB}$. A is called the endpoint or origin of the ray. Notice how this notation differs from the notation for a line segment.

$\overline{AB}$ denotes a line segment where $\overrightarrow{AB}$ denotes a ray with origin A.

We now define an important plane figure called an *angle*. Angles and the measure of angles form a very important part of the study of geometry.

DEFINITION The plane figure formed by two rays having a common endpoint is an **angle**.

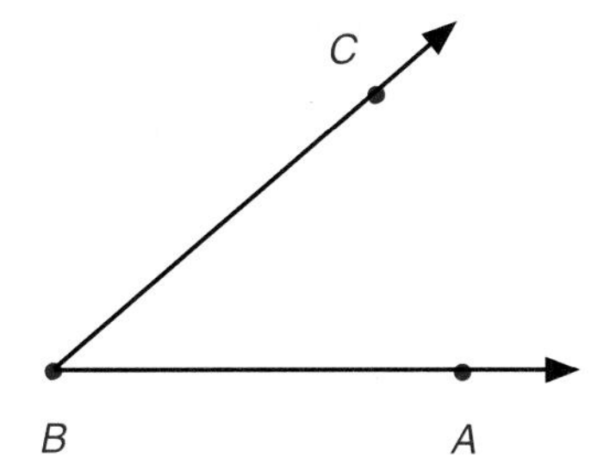

The common endpoint is called the **vertex** of the angle and the two rays are referred to as the **sides** of the angle.

In the above figure B is the vertex. $\overrightarrow{BA}$ and $\overrightarrow{BC}$ are sides.

The symbol used for angle is $\angle$. Angles are usually named using one of three methods.

1. Three capital letters where the first letter represents a point on one side, the second letter represents the vertex, and the third letter represents a point on the other side.

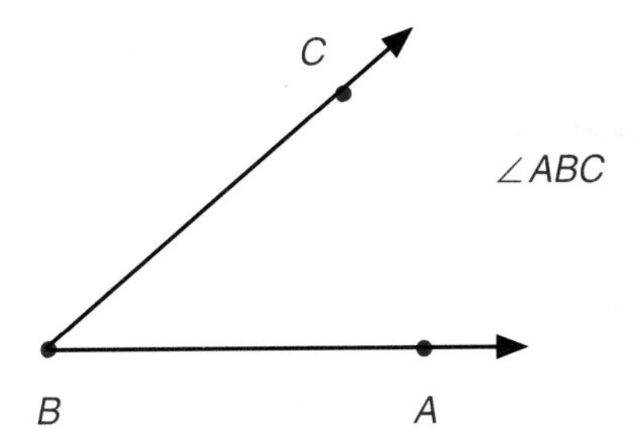

Note the middle letter is the vertex.

2. One capital letter that indicates the vertex.

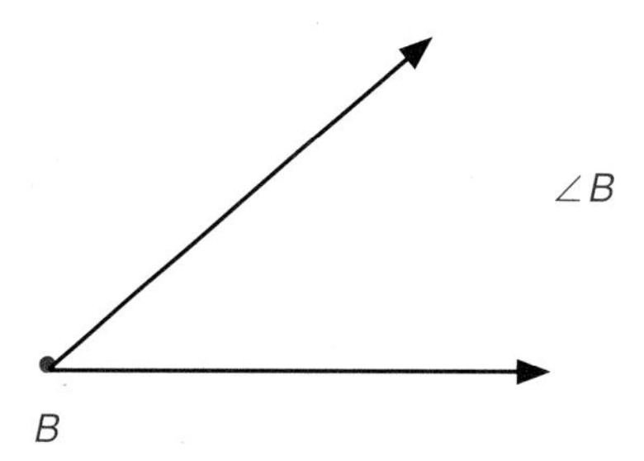

3. A number or lowercase letter, usually a Greek letter.

This is the Greek letter theta.

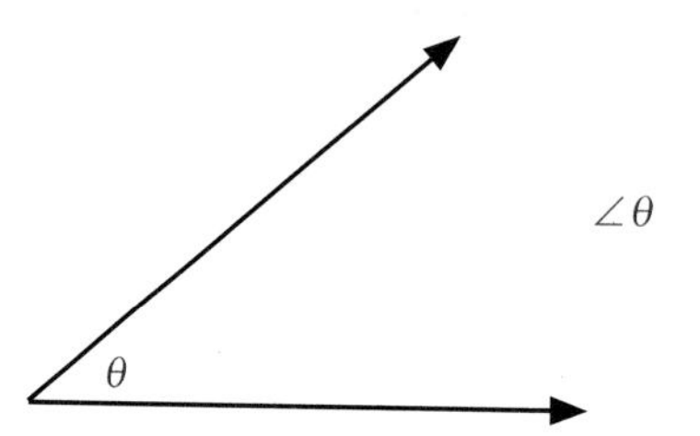

DEFINITION Two angles having a common vertex and a common side between them are called **adjacent angles.**

B is the common vertex and $\overrightarrow{BC}$ is the common side.

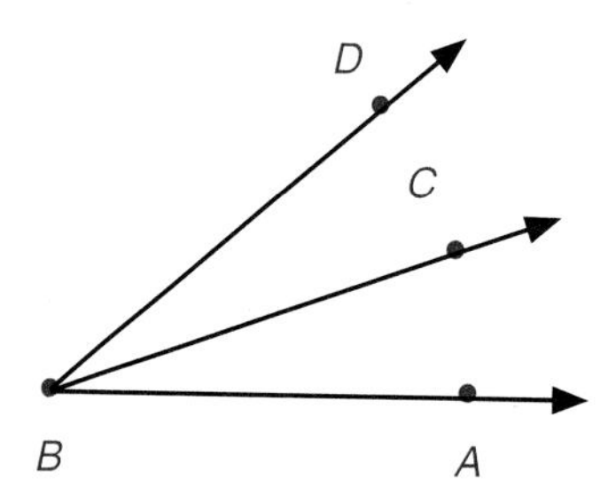

$\angle ABC$ and $\angle CBD$ are adjacent angles.

An angle is measured by the amount of revolution needed to rotate one of its sides about its endpoint so it will coincide with the other side. Degree measurement is commonly used.

DEFINITION **One degree** is $\frac{1}{360}$ of a complete revolution; or one complete revolution is 360 degrees.

The symbol for degree is a small circle (°) usually written after and slightly above a number. (Ten degrees is written as 10°.)

DEFINITION A **right angle** is an angle that is $\frac{1}{4}$ of a revolution and measures 90 degrees.

The symbol ⌝ is often used to indicate a right angle.

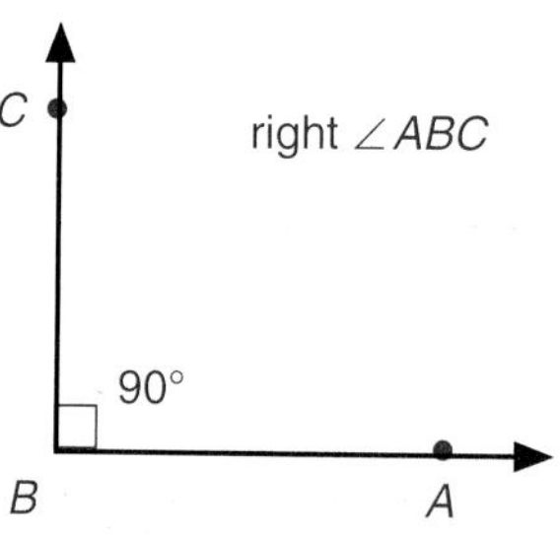

If we want to indicate that the measure of $\angle ABC$ is 90 degrees, we could write $m\,\angle ABC = 90°$. In our discussion however we will simply write $\angle ABC = 90°$ and understand that this means the *measure* of the angle.

Angles that are equal in measure are said to be **congruent angles.**

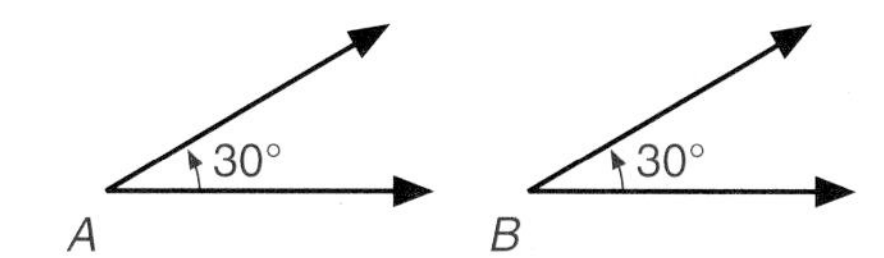

Since $\angle A = 30°$ and $\angle B = 30°$, $\angle A$ and $\angle B$ are congruent. Sometimes the congruent symbol $\cong$ is used and $\angle A \cong \angle B$ indicates that the two angles are congruent. In this text, however, we will simply write $\angle A = \angle B$ to mean that the measures of the two angles are equal and that they are therefore congruent.

2 Complementary Angles

DEFINITION If the sum of the measures of two angles is 90°, they are called **complementary angles.**

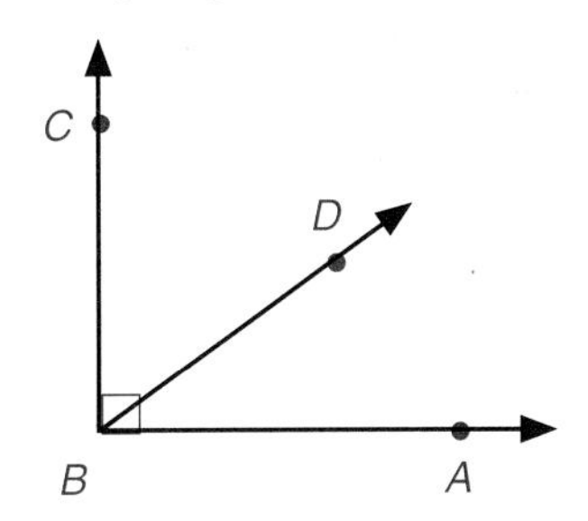

$\angle ABD$ and $\angle DBC$ are complementary since the sum of their measures is a right angle.

Example 1 Find the measure of an angle that is complementary to an angle whose measure is 36°.

Solution The sum of the two measures must be 90° so we must subtract 36° from 90° to obtain our answer of 54°.

$90° - 36° = 54°$

⦿ ***SKILL CHECK 1*** Find the measure of an angle if its complement measures 72°.

DEFINITION An **acute angle** is an angle that measures between 0° and 90°.

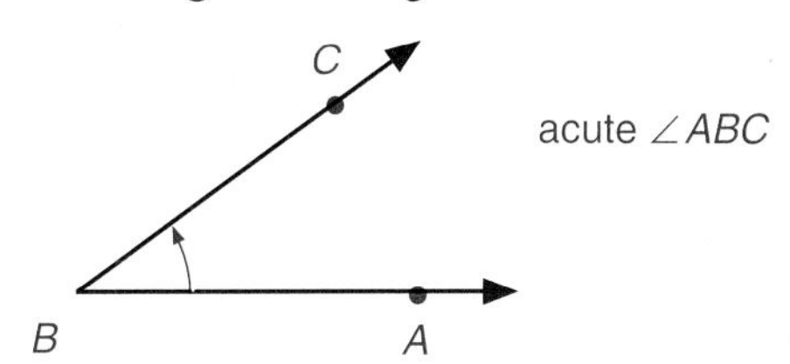

DEFINITION An **obtuse angle** is an angle that measures between 90° and 180°.

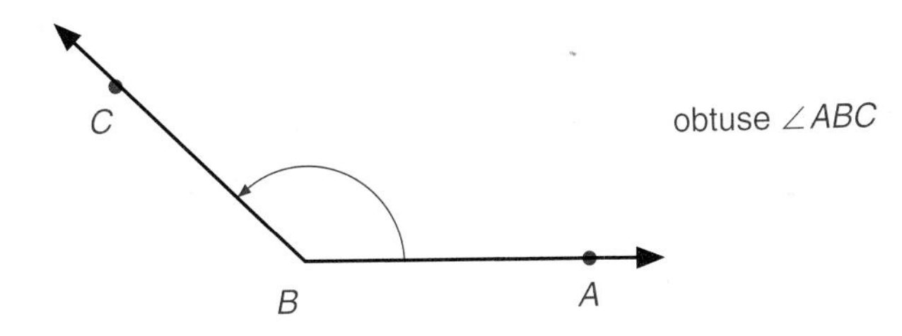

DEFINITION A **straight angle** is an angle that measures 180°.

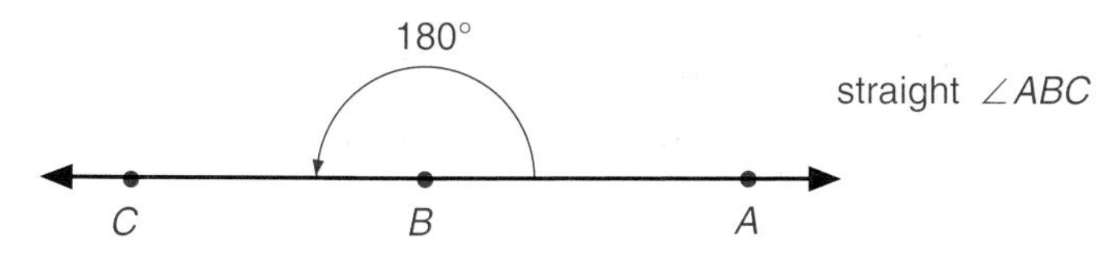

⦿ ***Skill Check Answer***

1. 18°

3 Supplementary Angles

DEFINITION If the sum of the measures of two angles is 180°, the angles are called **supplementary angles.**

If AC is a straight line, then $\angle ABD$ and $\angle DBC$ are supplementary.

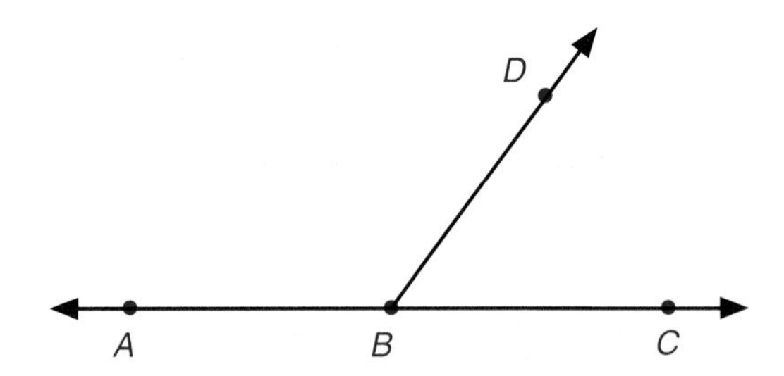

Example 2 Find the measure of an angle that is supplementary to an angle that measures 60°.

Solution Since the sum of the measures must be 180°, we must subtract 60° from 180° to obtain the answer of 120°.

180° − 60° = 120°

⊙ ***SKILL CHECK 2*** Find the measure of an angle if its supplement measures 81°.

DEFINITION If two lines meet at an angle of 90°, they are said to be **perpendicular lines.**

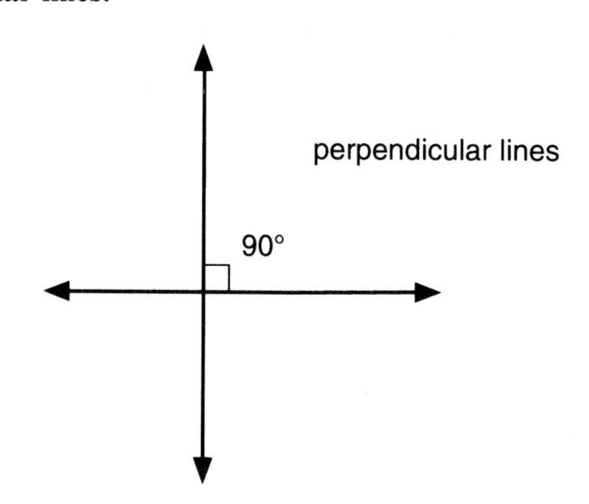

DEFINITION Two lines in the same plane that do not intersect are called **parallel lines.**

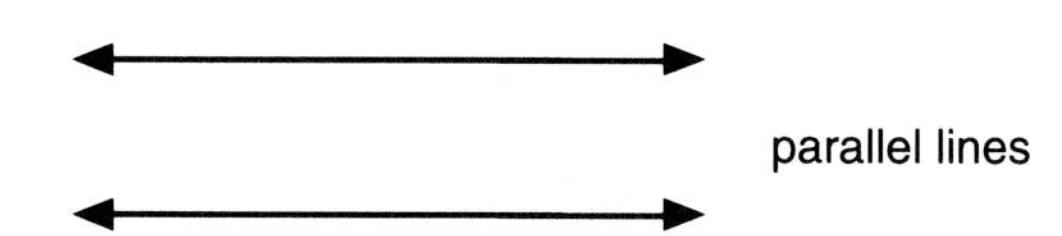

4 Angles Formed by Intersecting Lines

DEFINITION A line intersecting two parallel lines is called a **transversal.**

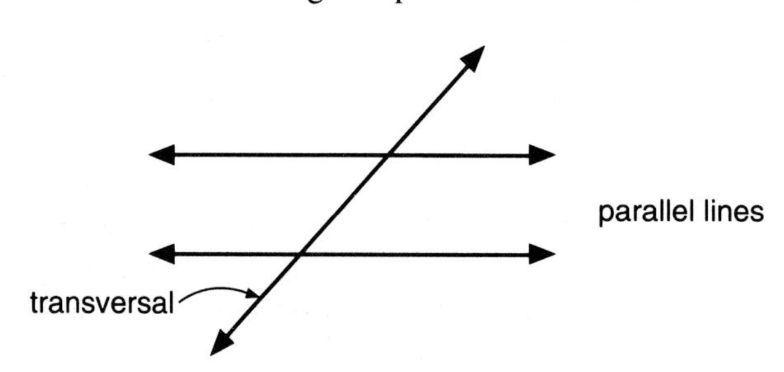

⊙ ***Skill Check Answer***

2. 99°

When parallel lines are intersected by a transversal, the angles formed are given special names according to their positions.

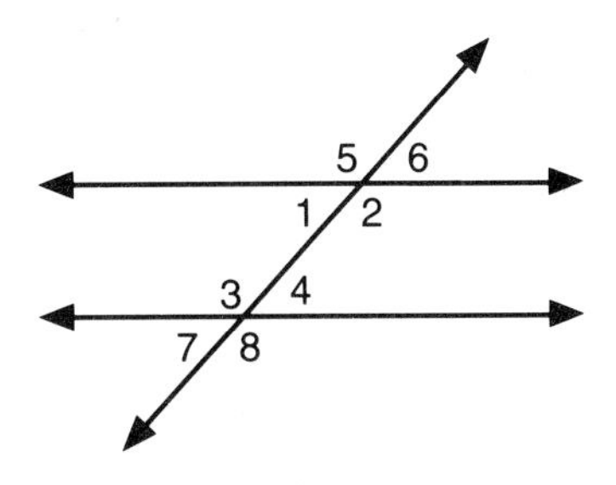

Here we are using numbers to designate angles.

$\angle 1$, $\angle 2$, $\angle 3$, and $\angle 4$ are called **interior angles.**

$\angle 5$, $\angle 6$, $\angle 7$, and $\angle 8$ are called **exterior angles.**

$\angle 1$ and $\angle 4$ are **alternate interior angles.**

$\angle 2$ and $\angle 3$ are **alternate interior angles.**

$\angle 5$ and $\angle 8$ are **alternate exterior angles.**

$\angle 6$ and $\angle 7$ are **alternate exterior angles.**

The following pairs of angles, $\angle 1$ and $\angle 7$, $\angle 2$ and $\angle 8$, $\angle 3$ and $\angle 5$, $\angle 6$ and $\angle 4$ are called **corresponding angles.**

DEFINITION If two lines intersect, the angles opposite each other are called **vertical angles.**

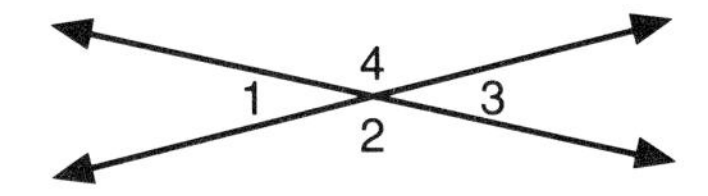

$\angle 1$ and $\angle 3$ are vertical angles.
$\angle 2$ and $\angle 4$ are vertical angles.

Statements that can be proved using previously known facts are an important part of geometry. Such statements are called **theorems.** Theorems are proved using undefined terms, definitions, and previously proved theorems. As stated earlier, we will not attempt a formal presentation of geometry in one short chapter. However, you should be able to informally show why some theorems are true.

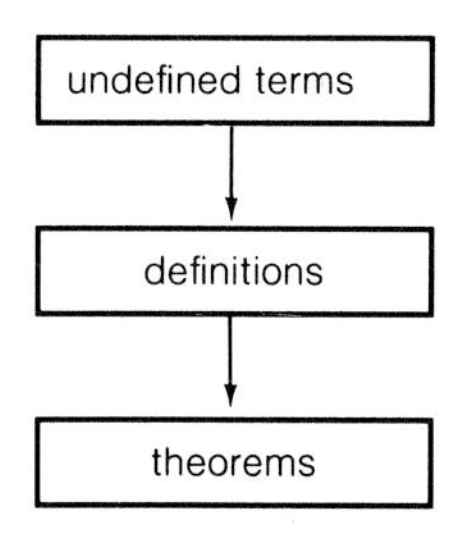

Example 3 Informally show the following theorem to be true.

THEOREM Vertical angles are equal in measure.

Solution Our thinking should proceed in this manner. Since we have the definition of vertical angles, we will sketch a figure and label it for convenience.

Reread this discussion several times until you can take a blank sheet of paper and reason through it on your own.

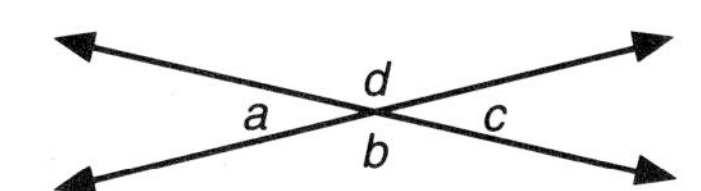

We wish to prove that $\angle a = \angle c$ and $\angle b = \angle d$. We will use the definition of supplementary angles to state that $\angle a + \angle b = 180°$. This is the same as the statement $\angle a = 180° - \angle b$. Using the same definition, we can state that $\angle c = 180° - \angle b$. Now since $\angle a$ and $\angle c$ are both equal to the same quantity $(180° - \angle b)$, they must be equal to each other. So $\angle a = \angle c$. In the same manner we can show that $\angle b = \angle d$. You will be asked to do this in an exercise.

Recall the definition of subtraction.

This is due to the principle of substitution.

Example 4 Referring to the figure, determine the measures of $\angle 1$, $\angle 2$, and $\angle 3$.

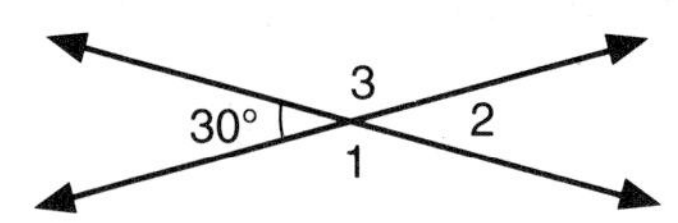

Solution Since we know from example 3 that vertical angles are equal, we can say $\angle 2 = 30°$. Also, since we have straight lines we know that $\angle 1 = 180° - 30°$ or $\angle 1 = 150°$. Since $\angle 1$ and $\angle 3$ are vertical angles we have $\angle 3 = 150°$.

$\angle$ 1 and 30° are supplementary. Vertical angles are equal.

⊙ ***SKILL CHECK 3*** Referring to the figure, determine the measures of $\angle 1$, $\angle 2$, and $\angle 3$.

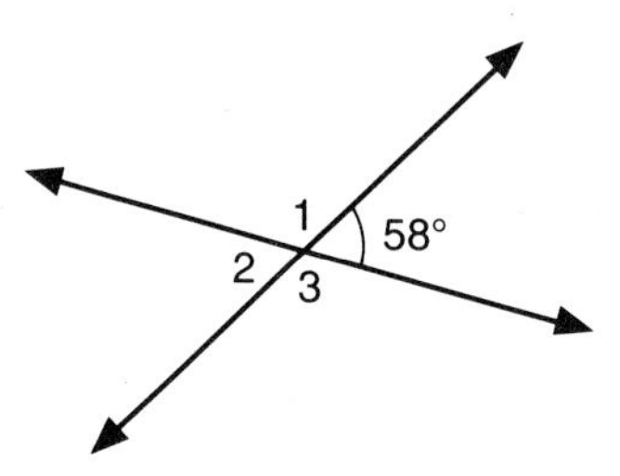

A theorem that we will accept but not prove concerns parallel lines intersected by a transversal. We will use this theorem in future sections.

THEOREM If two parallel lines are intersected by a transversal, the alternate interior angles are equal.

$\angle$ *a* and $\angle$ *b* are alternate interior angles. Also, $\angle$ *c* and $\angle$ *d* are alternate interior angles.

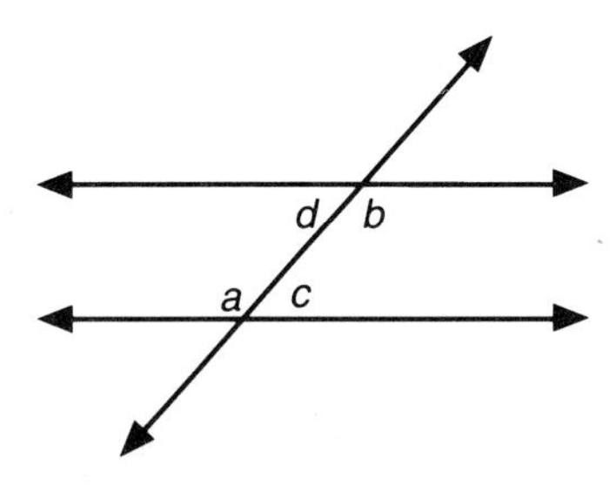

$\angle a = \angle b$ and $\angle c = \angle d$.

Example 5 The following diagram depicts two parallel lines intersected by a transversal. Find the measures of the indicated angles.

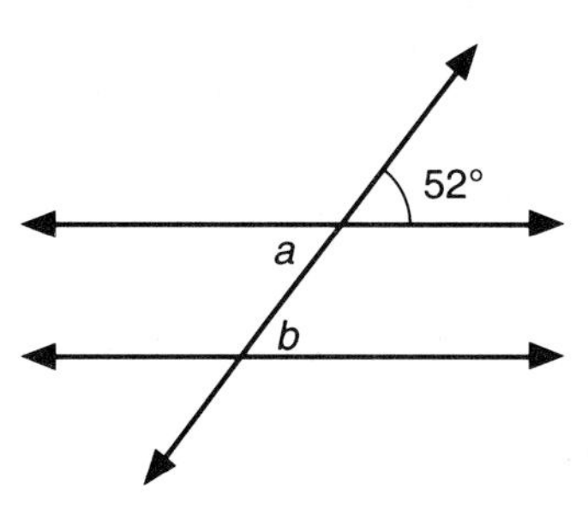

Solution $\angle a$ and $\angle b$ are alternate interior angles and by the above theorem must be equal. $\angle a = 52°$ since vertical angles are equal. Thus $\angle b = 52°$.

Notice that we are using two theorems to solve this problem.

⊙ ***Skill Check Answer***

3. $\angle 1 = 122°$
   $\angle 2 = 58°$
   $\angle 3 = 122°$

⦿ SKILL CHECK 4 The following diagram depicts two parallel lines intersected by a transversal. Find the measures of the indicated angles.

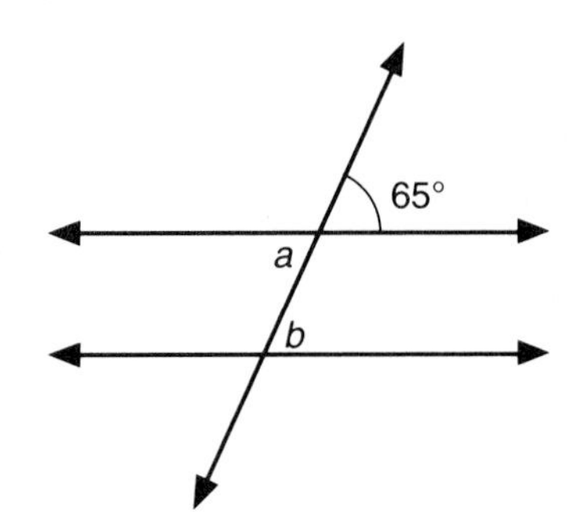

5 Rulers and Protractors

There are two types of measurement that you should be able to perform, linear measurement and angular measurement. Special instruments are used to approximate these measures. Linear measurement involves using a **ruler.** You should be able to use this instrument calibrated in sixteenths of an inch to measure the length of a line segment to the nearest sixteenth of an inch. You should also be able to use a ruler calibrated in millimeters to measure the length of a line segment to the nearest millimeter.

An instrument is only as accurate as its calibration.

Example 6 Use a ruler to measure the length of the following line segment to the nearest sixteenth of an inch.

Solution Placing the end of the ruler on one endpoint, we observe that the other endpoint is closest to the measurement mark for $1\frac{3}{16}$ inches.

Each mark on the ruler represents $\frac{1}{16}$ inch.

Example 7 Use a ruler to measure the length of the following line segment to the nearest millimeter.

Solution Performing the measurement, we see the length is closest to 28 millimeters.

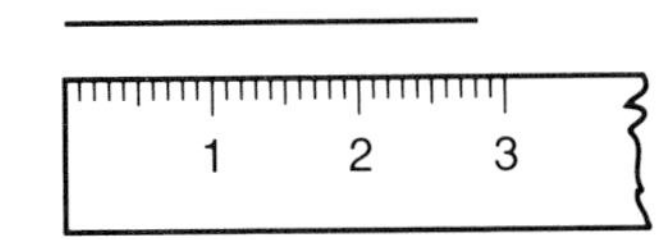

Each mark on the ruler represents 1 mm.

Angular measurement is obtained by using an instrument called a **protractor.** You should be able to use a protractor to measure an angle to the nearest degree.

Example 8 Use a protractor to measure the following angle to the nearest degree.

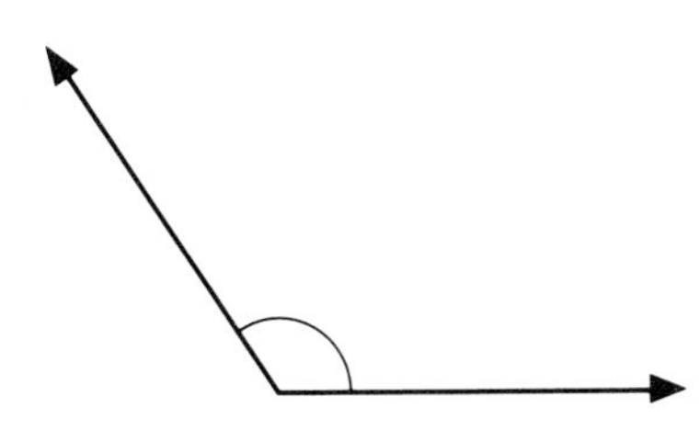

⦿ Skill Check Answer

4. $\angle a = 65°$
 $\angle b = 65°$

Solution Placing the straight edge of the protractor along one side of the angle with the 0-point at the vertex, we see that the other side of the angle passes through the scale closest to the 124° mark.

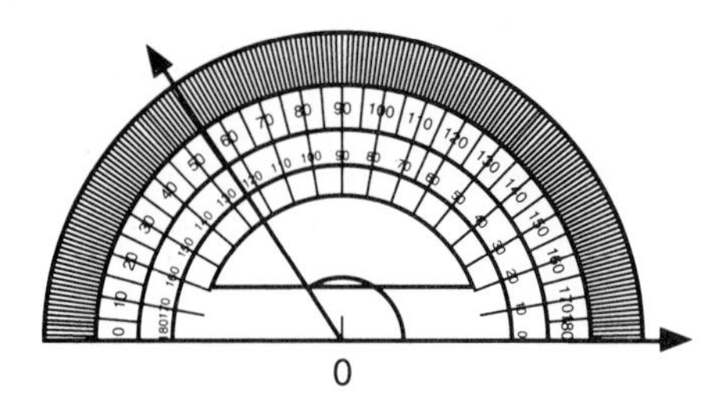

Exercise 5–1–1

1

1. Explain the distinction between a line and a line segment.
2. Explain the distinction between a line segment and a ray.
3. Is an angle that measures 100° an acute angle?
4. What is the name of an angle whose measure is greater than a right angle but less than a straight angle?

2 Find the measure of an angle that is complementary to each of the following.

5. 25°
6. 38°
7. 74°
8. 89°

3 Find the measure of an angle that is supplementary to each of the following.

9. 48°
10. 63°
11. 125°
12. 109°

4 Use the diagram to find the following angle measures.

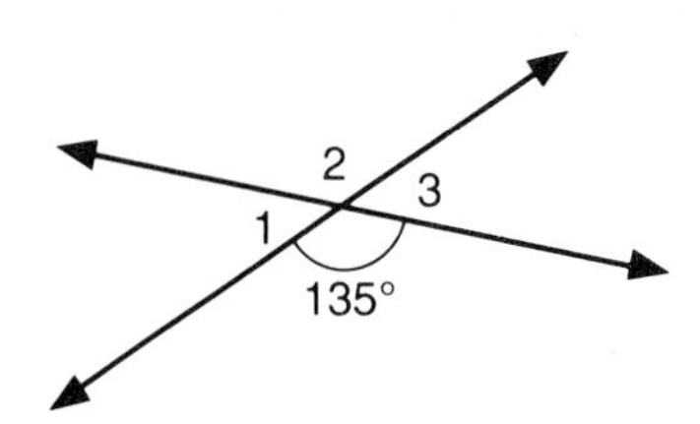

13. ∠1
14. ∠2
15. ∠3

The diagram depicts two parallel lines intersected by a transversal. Find the measures of the indicated angles.

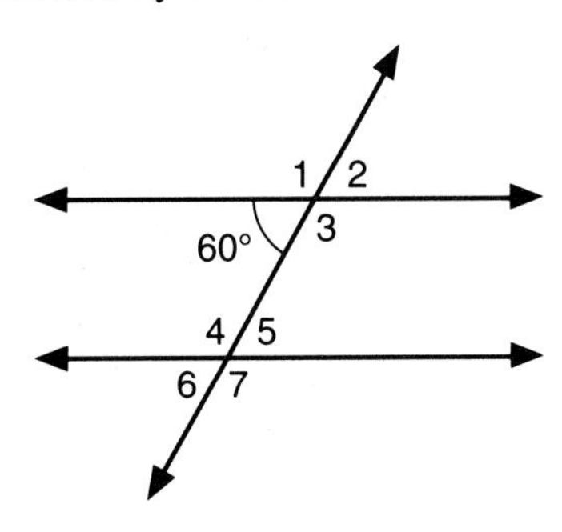

16. $\angle 1$

17. $\angle 2$

18. $\angle 3$

19. $\angle 4$

20. $\angle 5$

21. $\angle 6$

22. $\angle 7$

Use the following diagram to answer the questions.

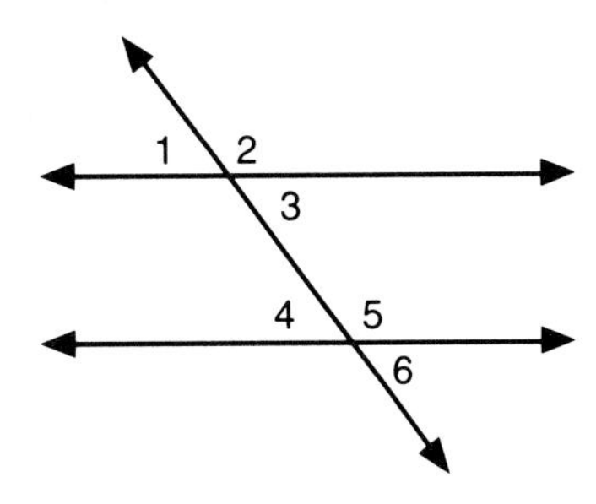

23. Name a pair of vertical angles.

24. Name a pair of alternate interior angles.

25. Name a pair of alternate exterior angles.

26. Name a pair of corresponding angles.

5 Use a ruler to measure the length of each line segment to the nearest sixteenth of an inch.

27. ________

28. ____________________

29. __________________________

30. ____________________

Use a ruler to measure the length of each line segment to the nearest millimeter.

31. ________

32. __________

33. ____________

34. _____________________

Use a protractor to measure each of the indicated angles to the nearest degree.

35.

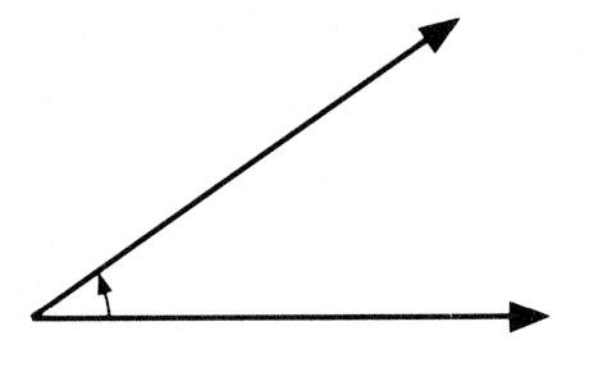

36.

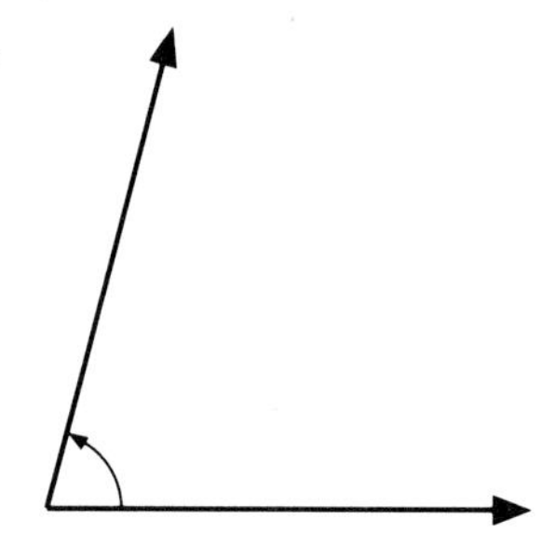

37.

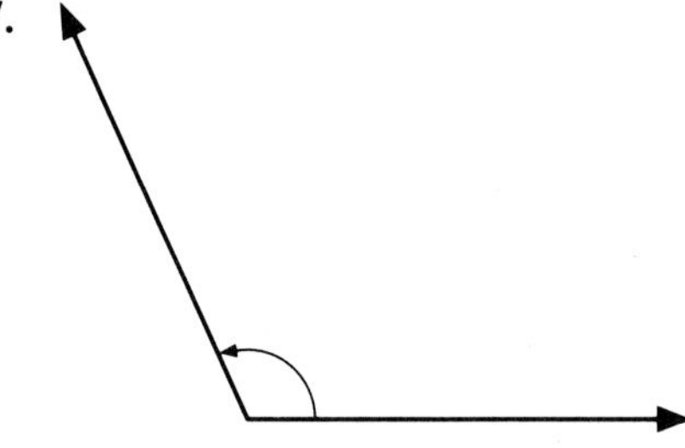

38.

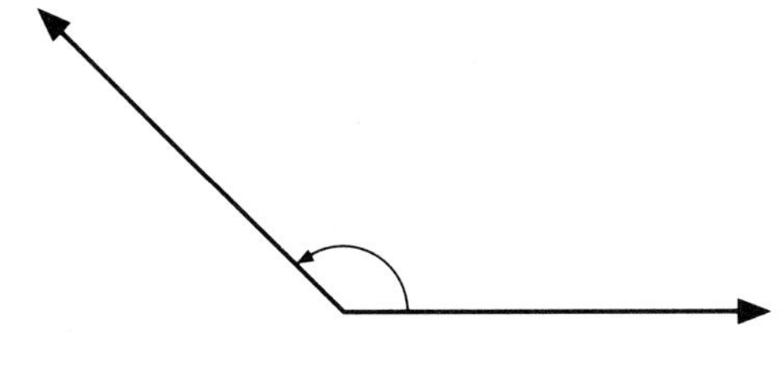

39.

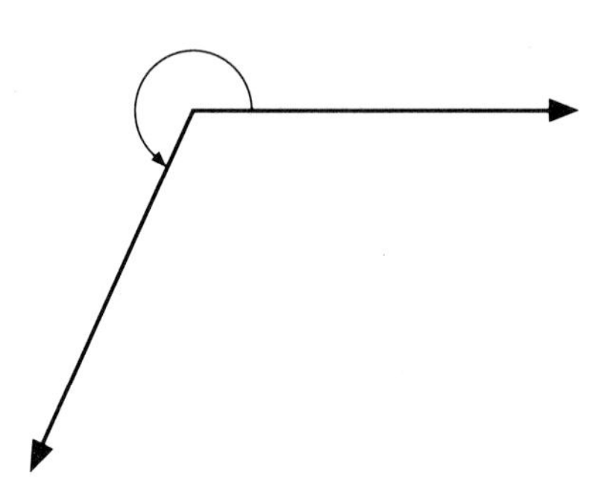

40.

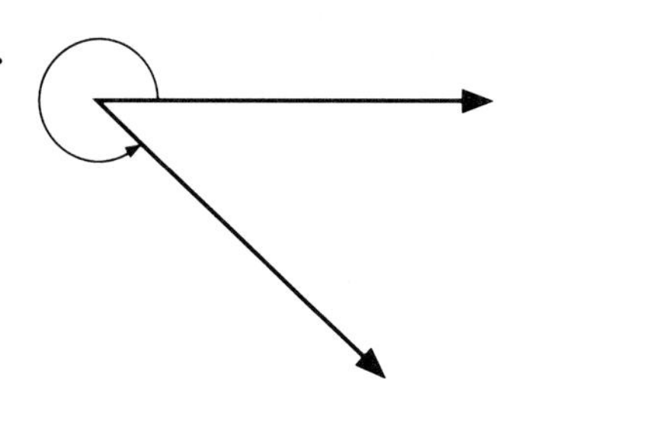

• Concept Enrichment

41. Using the same thought process as in example 3 of this section, informally show that $\angle b = \angle d$.

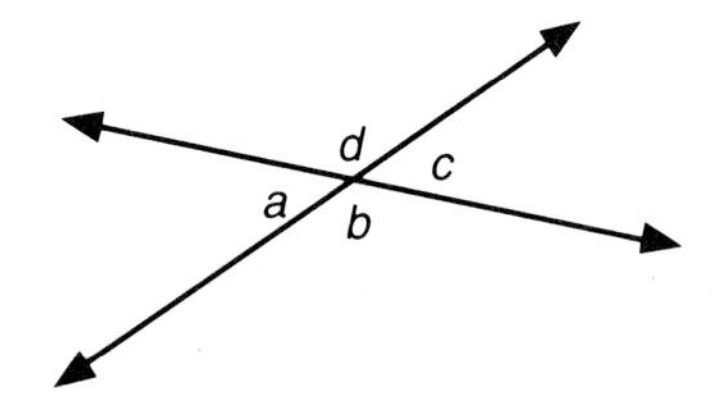

42. The following diagram illustrates two parallel lines intersected by a transversal. Assuming that alternate interior angles are equal, show that corresponding angles are equal. (That is, assuming $\angle 1 = \angle 3$, prove that $\angle 2 = \angle 3$.)

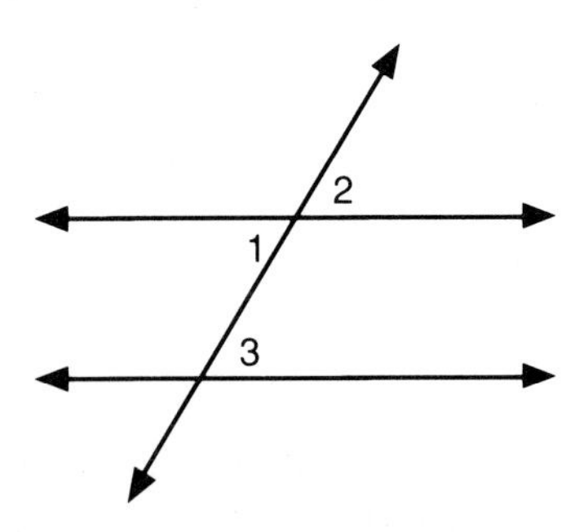

5-2 Polygons

In the previous section we discussed lines, line segments, rays, and angles—all of which are plane figures. In this section we will study some plane figures composed of line segments and angles.

Some definitions of terms are necessary before we can proceed.

OBJECTIVES

Upon completing this section you should be able to:

1. Determine if a plane figure is a polygon.
2. Given the number of sides, calculate the number of diagonals in the polygon.
3. Recognize the names of special polygons.

DEFINITION If a plane figure is sketched by starting at a point and returning to that same point, the figure is a **closed plane figure.**

DEFINITION The portion of the plane enclosed by the closed plane figure is the **interior** of the figure.

Example 1 The following are sketches of some closed plane figures.

You can start at any point and trace along the figure returning to that point.

Example 2 The following plane figures are not closed.

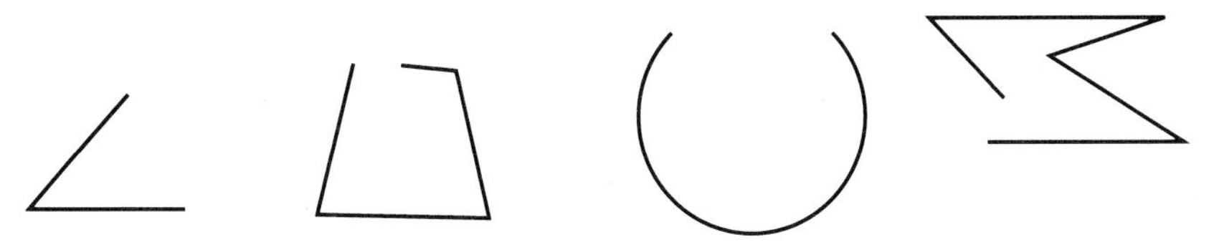

1 Polygons

DEFINITION If each side of a closed plane figure is a line segment, the figure is a **polygon.** The line segments are called **sides** of the polygon. The endpoints of the line segments are called **vertices** of the polygon.

Vertex is singular.

Example 3 The following plane figures are polygons.

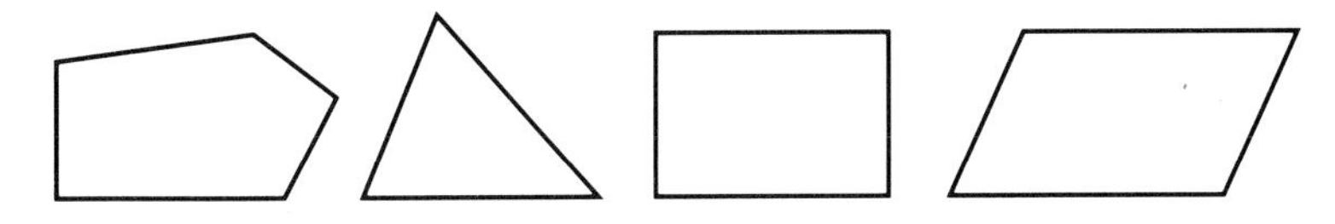

All sides are composed of line segments.

Example 4 The following plane figures are not polygons.

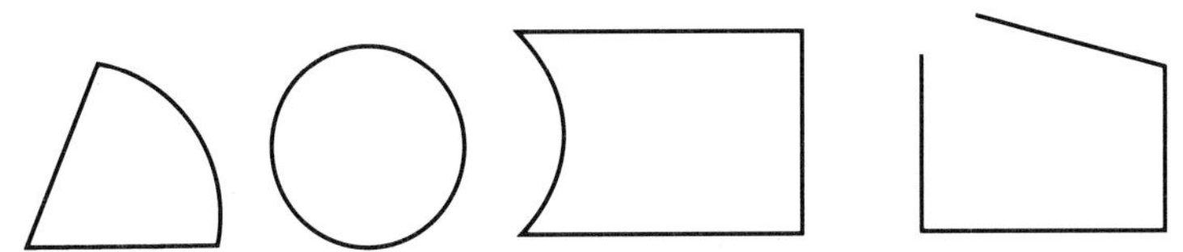

DEFINITION Two vertices that are endpoints of the same side of a polygon are **consecutive vertices.**

Example 5

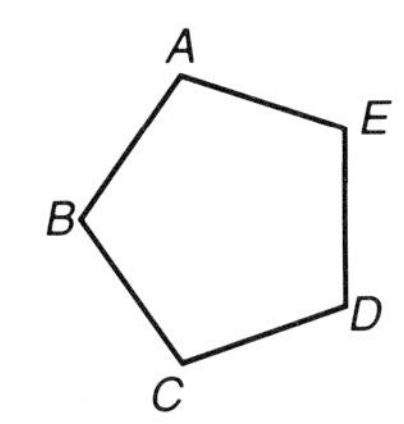

A and B are consecutive vertices. Also B and C, C and D, D and E, E and A.

A and C, for example, are *not* consecutive vertices.

The vertices of a polygon are also the vertices of the **interior angles** of the polygon. Each interior angle is formed by two adjacent sides of the polygon.

2 Diagonals of a Polygon

DEFINITION A line segment that has two nonconsecutive vertices of a polygon as endpoints is a **diagonal** of the polygon.

Example 6 In the following polygons the dashed line segments are diagonals.

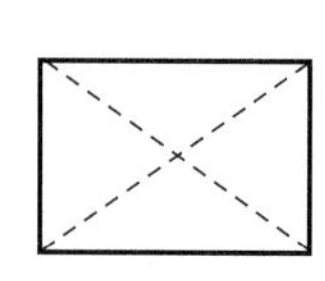
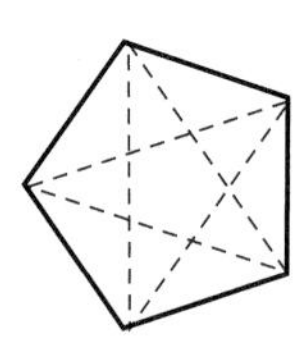
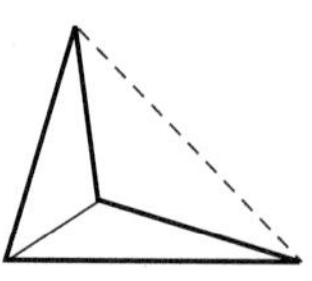

We will now use a formula for the number of diagonals in a polygon. This theorem would be proved in a geometry course but we will state it here without proof.

THEOREM If n represents the number of sides of a polygon, then the number of diagonals (d) is given by the formula $d = \dfrac{n(n-3)}{2}$.

Notice that the number of diagonals depends on the number of sides.

Example 7 Find the number of diagonals in a polygon having 4 sides.

Solution We substitute 4 for n in the formula.

$$d = \frac{n(n-3)}{2}$$
$$= \frac{4(4-3)}{2} = \frac{4(1)}{2} = 2$$

Therefore the polygon has 2 diagonals.

If a polygon has 4 sides, how many diagonals does it have?

Example 8 If a polygon has 8 sides, how many diagonals does it have?

Solution Using $n = 8$ in the formula, we have

$$d = \frac{n(n-3)}{2}$$
$$= \frac{8(8-3)}{2} = \frac{8(5)}{2} = \frac{40}{2} = 20.$$

A polygon with 8 sides has 20 diagonals.

STOP

A stop sign is an example of a polygon having 8 sides.

SKILL CHECK 1 How many diagonals does an 11-sided polygon have?

3 Names of Polygons

There are two main classes of polygons.

DEFINITION If each diagonal of a polygon lies totally in the interior of the polygon, the figure is a **convex polygon.** A polygon that is not convex is a **concave polygon.**

Example 9 This is a sketch of a convex polygon.

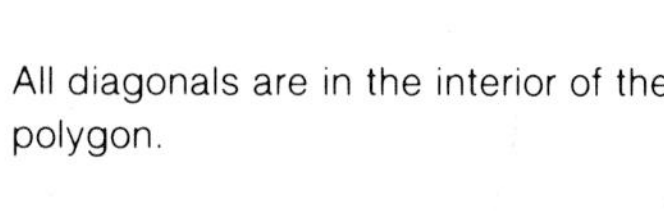
All diagonals are in the interior of the polygon.

Skill Check Answer

1. 44

Example 10 This is a sketch of a concave polygon.

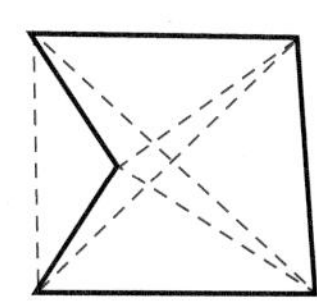

One of the diagonals lies outside the interior of the polygon.

A polygon is a **regular polygon** if its sides are equal and its interior angles are congruent.

Example 11

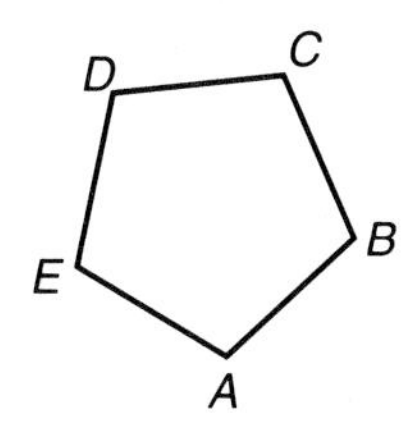

In this figure if $\overline{AB} = \overline{BC} = \overline{CD} = \overline{DE} = \overline{EA}$ and $\angle A = \angle B = \angle C = \angle D = \angle E$, then the polygon is a regular polygon.

Polygons are given names according to the number of sides they have. These names apply to both convex and concave polygons. This list contains the names of some of the more common polygons.

Number of sides	*Name*
3	triangle
4	quadrilateral
5	pentagon
6	hexagon
7	heptagon
8	octagon
9	nonagon
10	decagon
12	dodecagon
n	n-gon

How many sides does the pentagon building in Washington, DC have?

Exercise 5–2–1

1 Identify each plane figure as closed or not closed.

1.

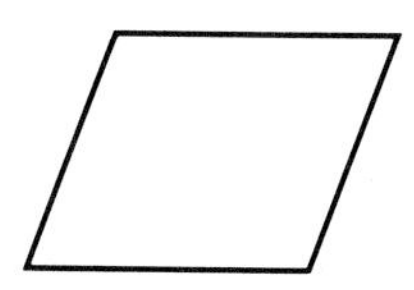

2.

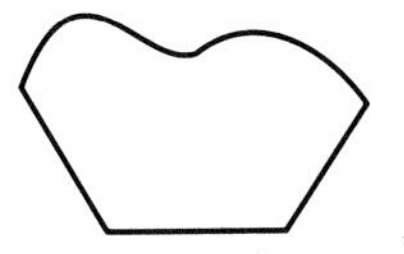

3.

4.

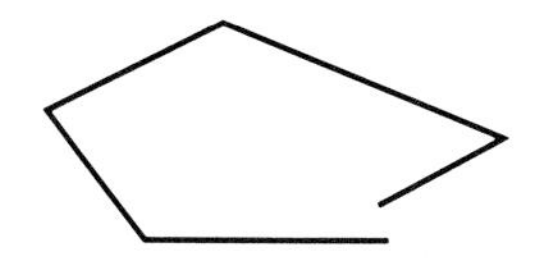

5.

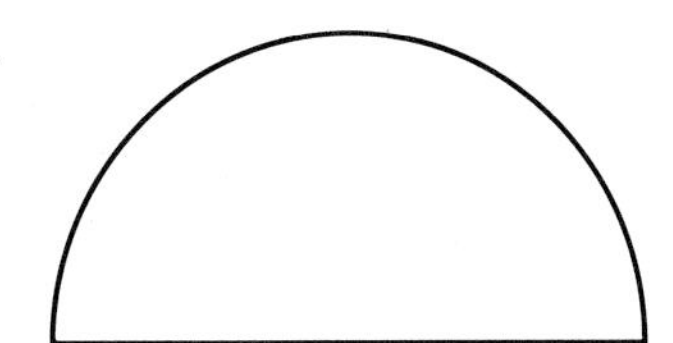

6.

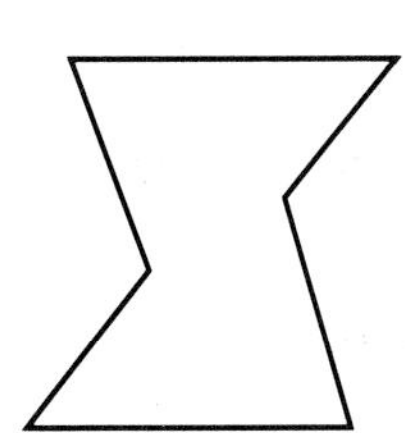

Which of the figures are polygons?

7.

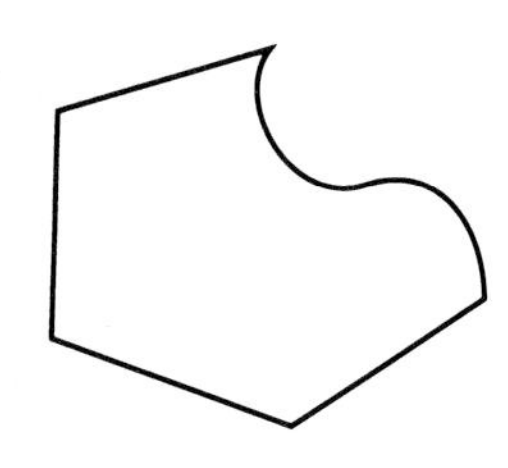

8.

9.

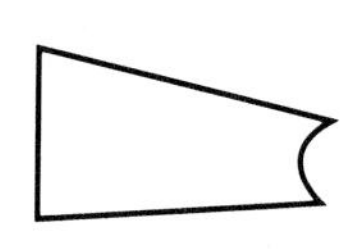

10.

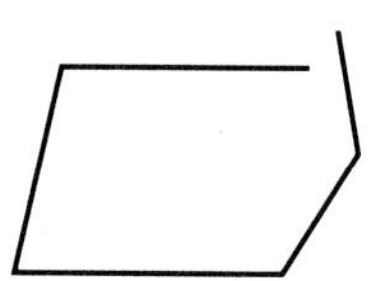

11.

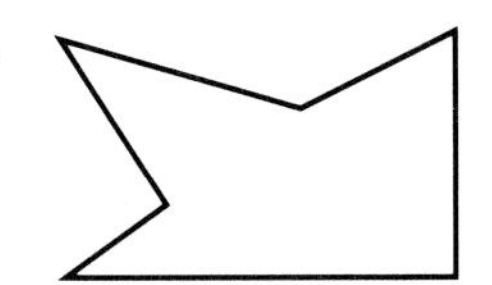

12.

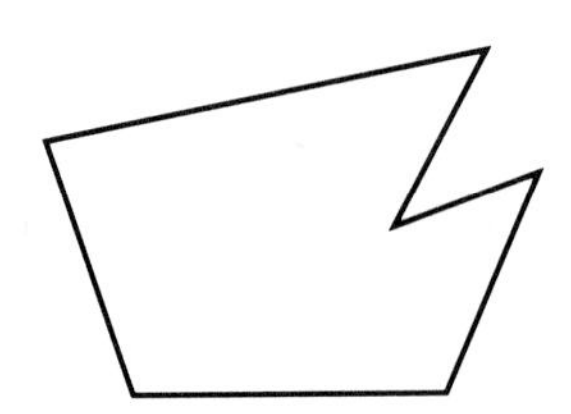

2

13. Sketch a polygon having 5 sides and draw the diagonals. How many diagonals does it have?

14. Sketch a polygon having 6 sides and draw the diagonals. How many diagonals does it have?

15. Use the formula to find the number of diagonals in a polygon having 5 sides.

16. Use the formula to find the number of diagonals in a polygon having 6 sides.

17. Use the formula to find the number of diagonals in a polygon having 7 sides.

18. Use the formula to find the number of diagonals in a polygon having 9 sides.

19. Use the formula to find the number of diagonals in a polygon having 10 sides.

20. Use the formula to find the number of diagonals in a polygon having 12 sides.

3 Classify each polygon as convex or concave.

21.

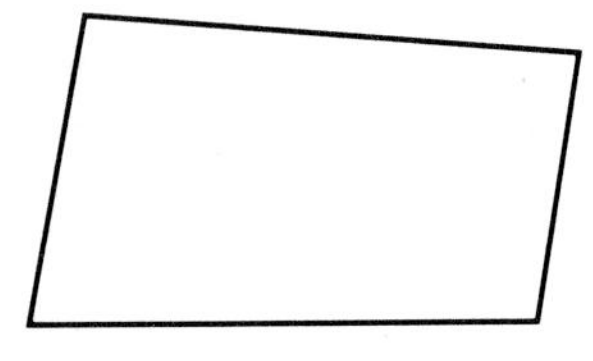

22.

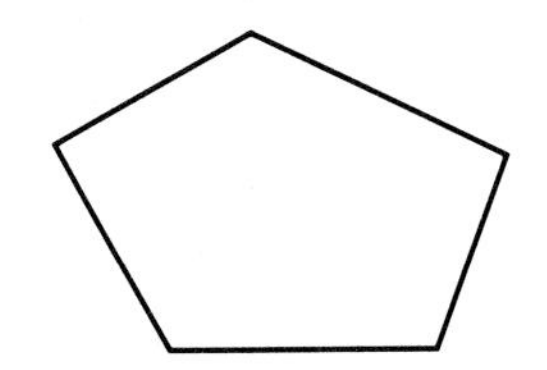

23.

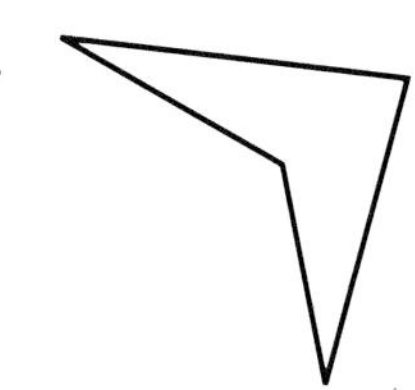

24. 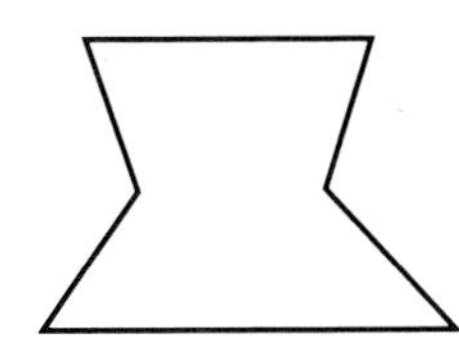

25. 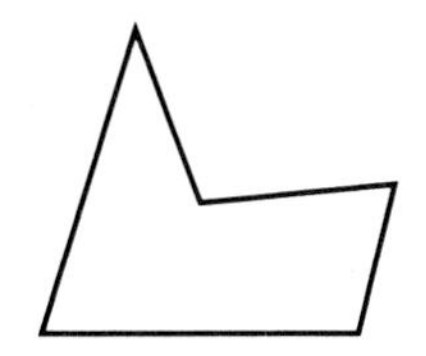

26. 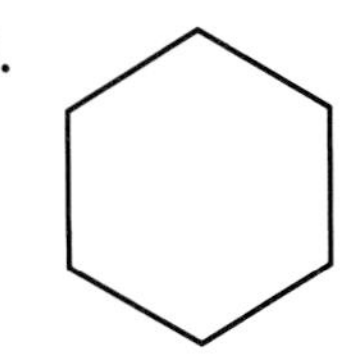

Name each polygon according to the number of sides.

27. 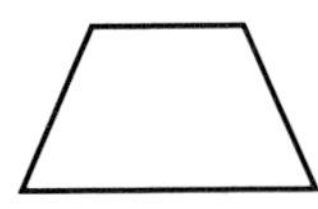

28. 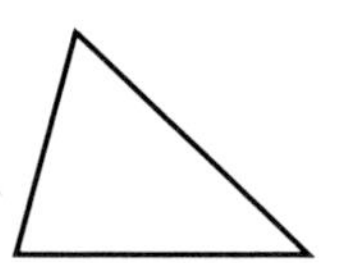

29. 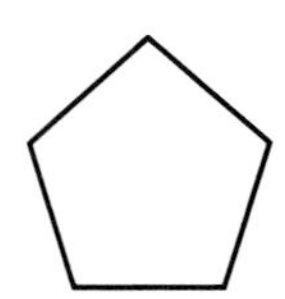

30. 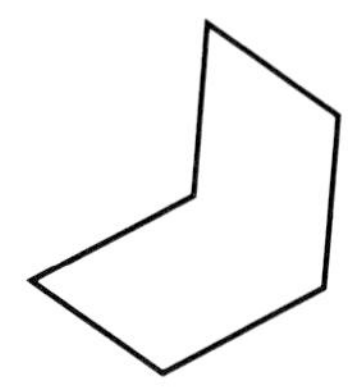

31. 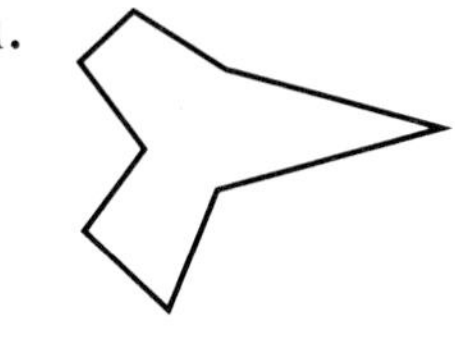

32. 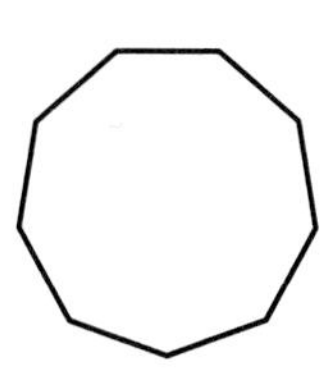

33. How many diagonals does a hexagon have?

34. How many diagonals does an octagon have?

35. How many diagonals does a triangle have?

36. How many diagonals does a quadrilateral have?

37. Draw a sketch of a convex octagon.

38. Draw a sketch of a concave octagon.

39. Draw a sketch of a regular hexagon.

40. Draw a sketch of a regular quadrilateral.

• Concept Enrichment

41. Another way of describing a convex polygon is that the entire polygon lies on one side of any one of its sides extended. Using this description, is the following plane figure convex?

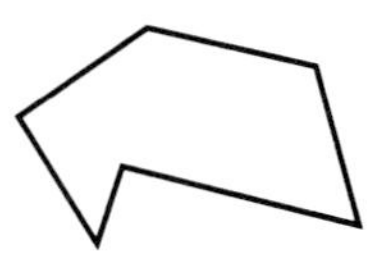

42. Using the description given in exercise 41, is every triangle convex?

5–3 Quadrilaterals—Perimeter and Area

OBJECTIVES

Upon completing this section you should be able to:

1. Identify various types of quadrilaterals.
2. Find the perimeter of a quadrilateral.
3. Find the area of some quadrilaterals.

In section 5–2 we found that polygons are named according to their number of sides. In this section we will discuss a special class of polygons called quadrilaterals and learn some facts and formulas concerning them.

1 Special Types of Quadrilaterals

A polygon having 4 sides is called a **quadrilateral.** Some quadrilaterals are given special names.

DEFINITION A **parallelogram** is a quadrilateral in which opposite sides are parallel.

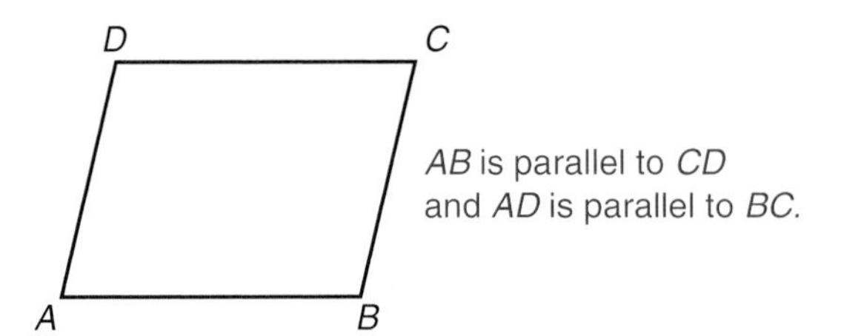

DEFINITION A **rectangle** is a parallelogram in which all four interior angles are right angles.

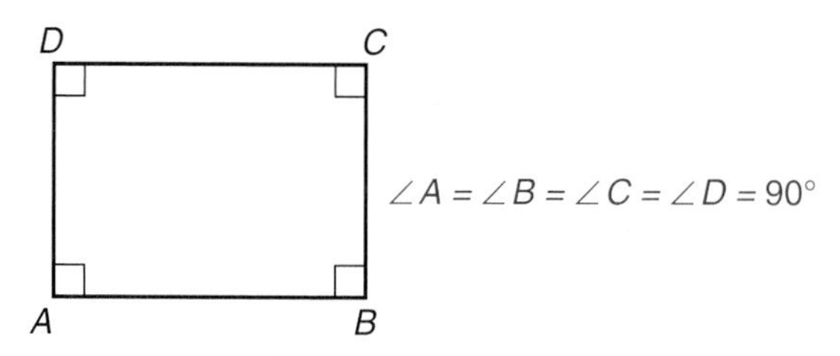

DEFINITION A **square** is a rectangle in which all four sides are equal.

Notice that all squares are rectangles and all rectangles are parallelograms.

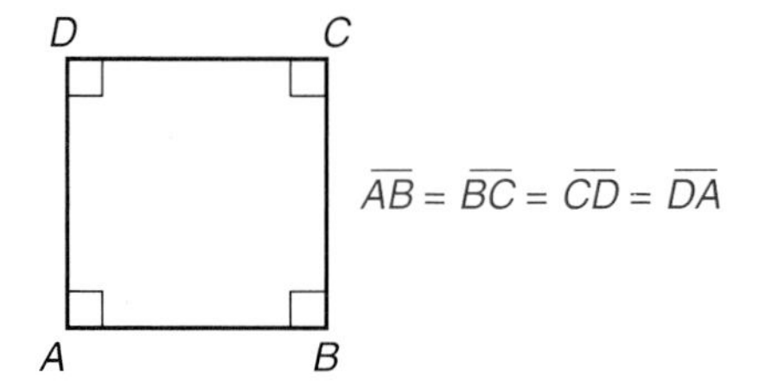

DEFINITION The **altitude** of a parallelogram is the distance, measured along a perpendicular line, between two parallel sides.

DEFINITION The side from which the altitude is measured is called the **base** of the parallelogram.

Lowercase letters are often used to represent the measure of a line segment. Here h represents the altitude.

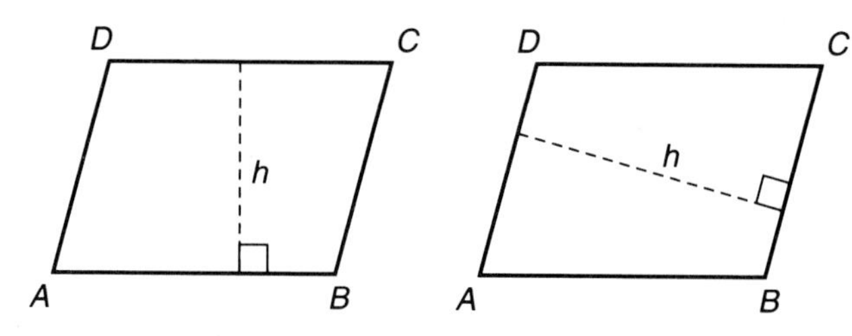

DEFINITION A **trapezoid** is a quadrilateral having only two parallel sides.

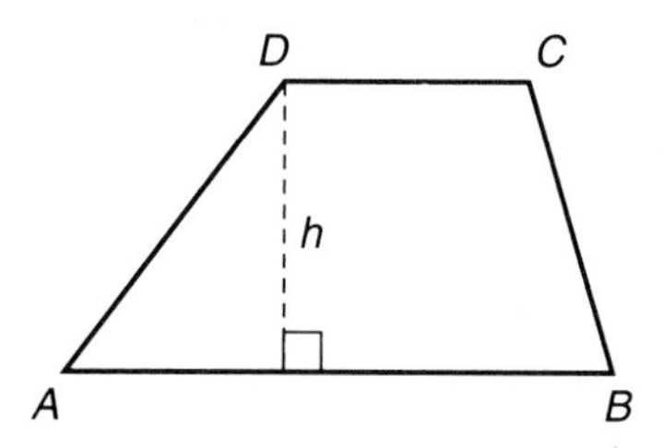

AB is parallel to *CD*.

DEFINITION The two parallel sides of a trapezoid are called the **bases** of the trapezoid.

$\overline{AB}$ and $\overline{CD}$ are the bases in the above figure.

DEFINITION The distance between the parallel sides of a trapezoid is the **altitude** of the trapezoid.

2 Perimeter

We now wish to discuss a measurement called perimeter.

DEFINITION The **perimeter** of a polygon is the sum of the measures of the sides of the polygon.

The sides of a polygon consist of line segments and they are measured using linear units. Thus the perimeter of a polygon is measured in linear units.

Refer to section 5-1.

Example 1 Find the perimeter P of the given figure with measures of sides shown.

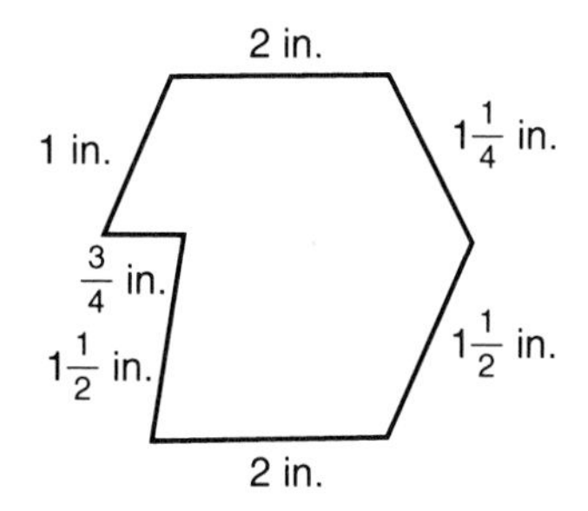

Solution Using the skills learned in previous chapters, we add the measures of the sides, obtaining

$$P = 2 + 1\frac{1}{2} + 1\frac{1}{4} + 2 + 1 + \frac{3}{4} + 1\frac{1}{2} = 10 \text{ inches.}$$

Here we use the skills for adding whole numbers and fractions.

The perimeter of a quadrilateral can be obtained by finding the sum of the measures of its four sides.

Example 2 Find the perimeter of a trapezoid having the indicated measurements for its sides.

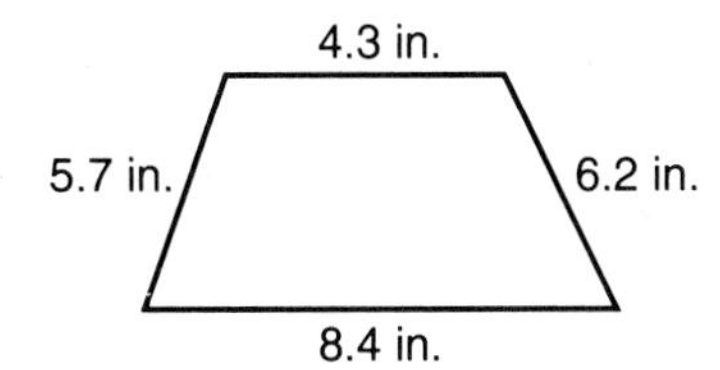

Solution If we let P represent the perimeter, then adding the measures of the four sides, we obtain

$$P = 8.4 + 6.2 + 4.3 + 5.7 = 24.6 \text{ in.}$$

⊙ ***SKILL CHECK 1*** Find the perimeter of a quadrilateral if the four sides measure 3.1 inches, 2.5 inches, 6.3 inches, and 5.4 inches.

⊙ ***Skill Check Answer***

1. 17.3 in.

We will now state two important theorems that give some special facts about parallelograms. These theorems would be proved in a course in geometry but we will state them here without proof.

THEOREM The opposite sides of a parallelogram are equal in measure.

We already know that the opposite sides are parallel.

D C A B

$\overline{AB} = \overline{CD}$

$\overline{AD} = \overline{BC}$

THEOREM The opposite angles of a parallelogram are equal in measure.

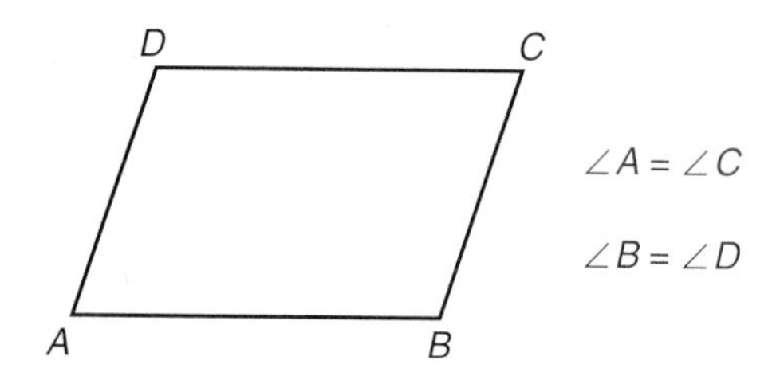

We used formulas in section 2–4.

The perimeter of some closed plane figures can be expressed as a formula. A **formula** in mathematics is a rule or principle that can be stated in algebraic language.

The formula for the perimeter of a parallelogram is found by using the fact that the opposite sides are equal in measure.

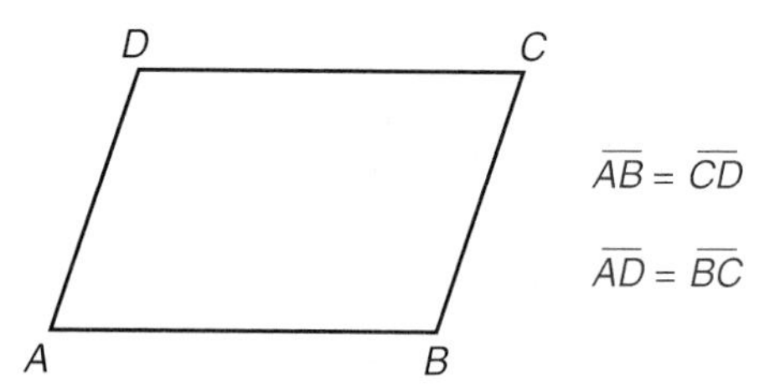

A script or lowercase letter is often used to represent the length of a side of a polygon.

If we use the script letter ℓ to represent the length of the equal sides $\overline{AB}$ and $\overline{CD}$ and if we use the lowercase letter w to represent the length of the equal sides $\overline{AD}$ and $\overline{BC}$, we have the following figure.

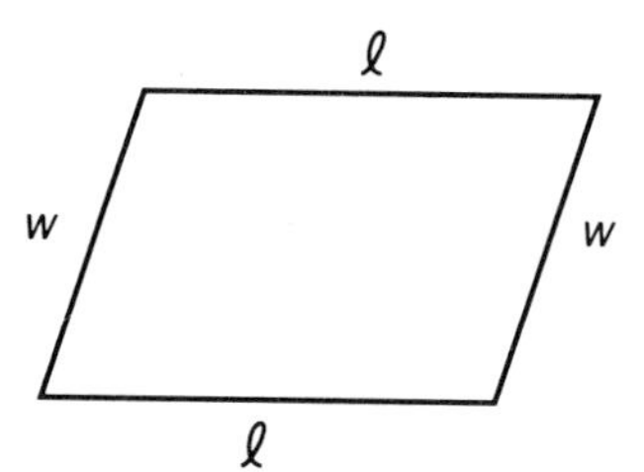

Now using P to represent the perimeter of the parallelogram along with the definition of perimeter, we have the formula

$$P = \ell + w + \ell + w$$

or

$$P = 2\ell + 2w.$$

Example 3 Find the perimeter of a parallelogram if two sides each have a measure of 3.4 meters and the other two sides each have a measure of 2.7 meters.

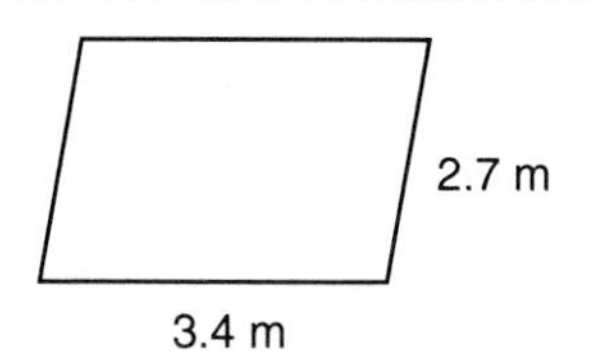

Solution If we let $\ell = 3.4$ and $w = 2.7$, we substitute into the formula obtaining

$$\begin{aligned} P &= 2\ell + 2w \\ &= 2(3.4) + 2(2.7) \\ &= 6.8 + 5.4 \\ &= 12.2 \text{ meters.} \end{aligned}$$

We could also add ℓ and w and then double that sum.
$P = 2(\ell + w)$ (distributive property).

⦿ ***SKILL CHECK 2*** Find the perimeter of a parallelogram if the unequal sides measure 4.8 meters and 9.3 meters.

Since a rectangle is a parallelogram, the formula $P = 2\ell + 2w$ also gives the perimeter of a rectangle.

Example 4 A gardener wishes to build a fence around a vegetable garden. If the garden is rectangular in shape and measures 105 feet long by 52 feet wide, how many feet of fencing material are needed?

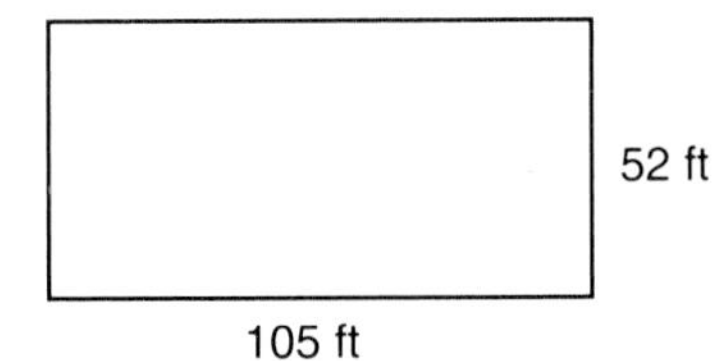

Solution Using $\ell = 105$ ft and $w = 52$ ft we have

$$\begin{aligned} P &= 2\ell + 2w \\ &= 2(105) + 2(52) \\ &= 210 + 104 \\ &= 314 \text{ feet.} \end{aligned}$$

Thus 314 feet of fencing material is needed.

⦿ ***SKILL CHECK 3*** How many feet of fencing are needed if the garden in example 4 measures 75 feet by 34 feet?

⦿ ***Skill Check Answers***

2. 28.2 m
3. 218 ft

Exercise 5–3–1

1

1. What is the name of a quadrilateral whose opposite sides are parallel?

2. What is the name of a quadrilateral having only two parallel sides?

3. Is a square a parallelogram?

4. What property do a trapezoid and a parallelogram have in common?

2 Find the perimeter of each quadrilateral.

5. 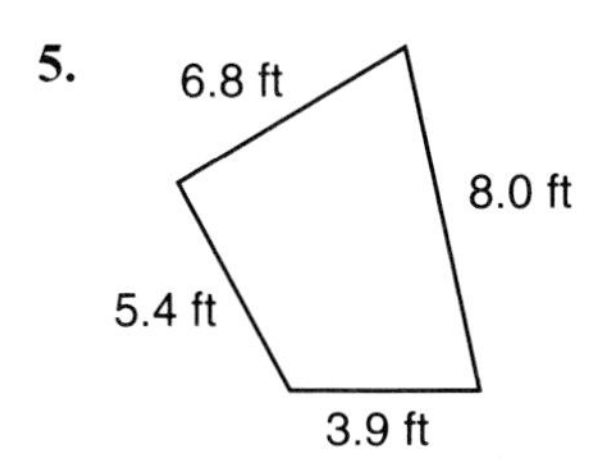

6. 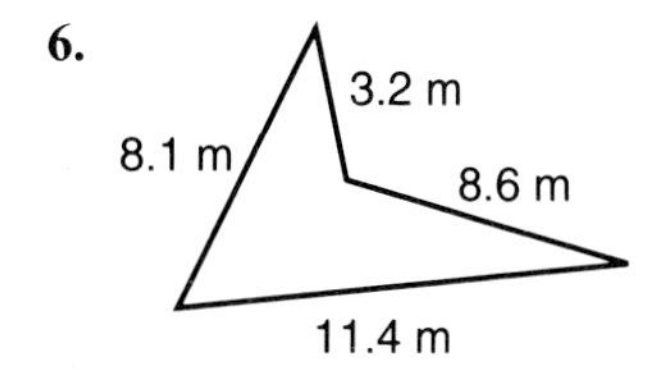

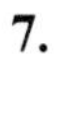

7.

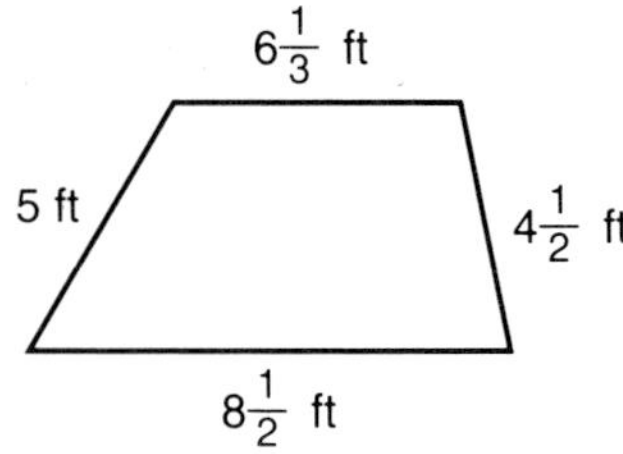

8.

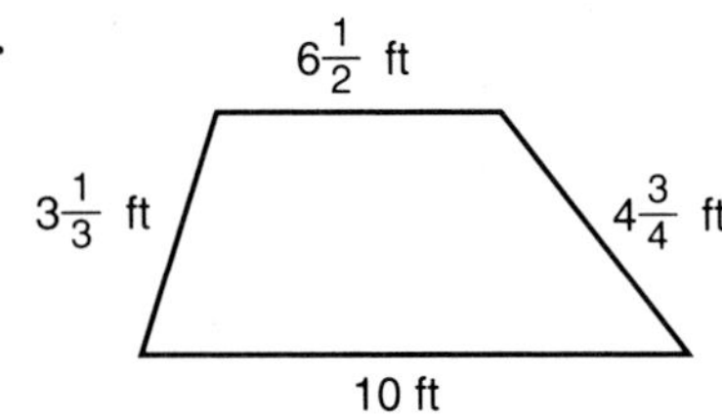

9. Find the perimeter of a parallelogram if two sides each measure 6 feet and the other two sides each measure 4 feet.

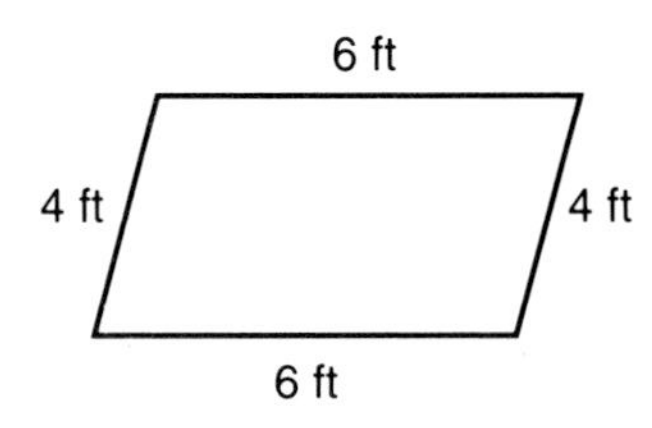

10. Find the perimeter of a parallelogram if two sides each measure 10 meters and the other two sides each measure 7 meters.

11. Find the perimeter of a parallelogram if two sides each measure 4.2 centimeters and the other two sides each measure 2.8 centimeters.

12. Find the perimeter of a parallelogram if two sides each measure $9\frac{1}{2}$ feet and the other two sides each measure $4\frac{2}{3}$ feet.

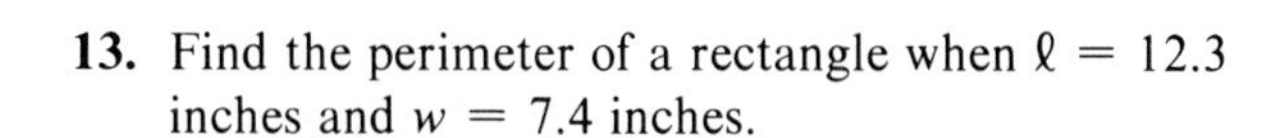

13. Find the perimeter of a rectangle when $\ell = 12.3$ inches and $w = 7.4$ inches.

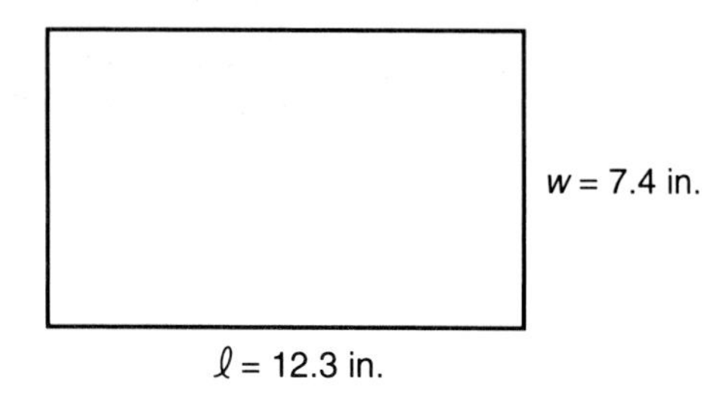

14. Find the perimeter of a rectangle if $\ell = 6.2$ meters and $w = 3.1$ meters.

15. Find the perimeter of a rectangle if the length is $12\frac{1}{2}$ feet and the width is $5\frac{3}{4}$ feet.

16. Find the perimeter of a rectangle if the length is 8.2 centimeters and the width is 2.7 centimeters.

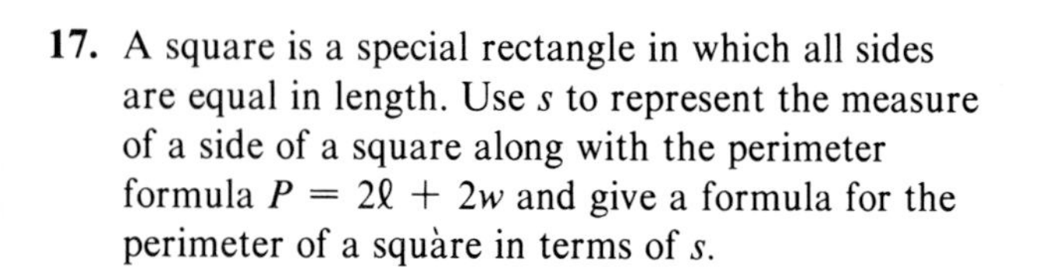

17. A square is a special rectangle in which all sides are equal in length. Use s to represent the measure of a side of a square along with the perimeter formula $P = 2\ell + 2w$ and give a formula for the perimeter of a square in terms of s.

18. Using the formula derived in question 17, find the perimeter of a square having a side that measures 3 inches.

19. Find the perimeter of a square if $s = 8$ feet.

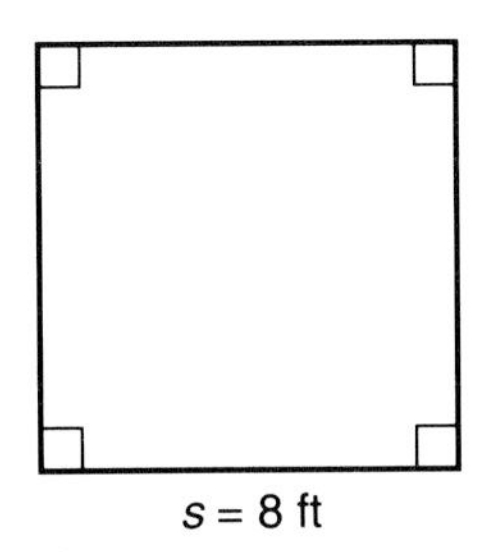

20. Find the perimeter of a square if $s = 3\frac{2}{3}$ inches.

21. Find the perimeter of a square having a side of length 5.4 centimeters.

22. Find the perimeter of a square having a side of length 8.5 meters.

23. A rectangular piece of land is 120 feet by 85 feet. Find the perimeter of the land.

24. How much edging is needed to go around a square garden having a side of length $12\frac{3}{8}$ feet?

3 Area

We now consider another measurement called area.

DEFINITION The **area** of a polygon is the measure of the interior of the polygon.

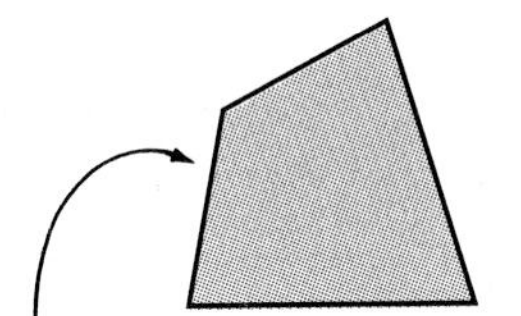
The shaded portion is the interior of the polygon.

Measurement of area is based on square units. Earlier in this section we defined a square. Using that definition we will define a *square unit*.

DEFINITION If the measure of the side of a square is one unit, then the area of the square is **one square unit.**

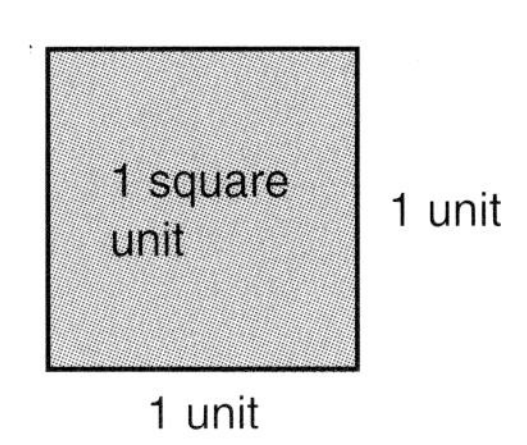

Example 5 A square with a side measuring 1 inch has an area of 1 square inch.

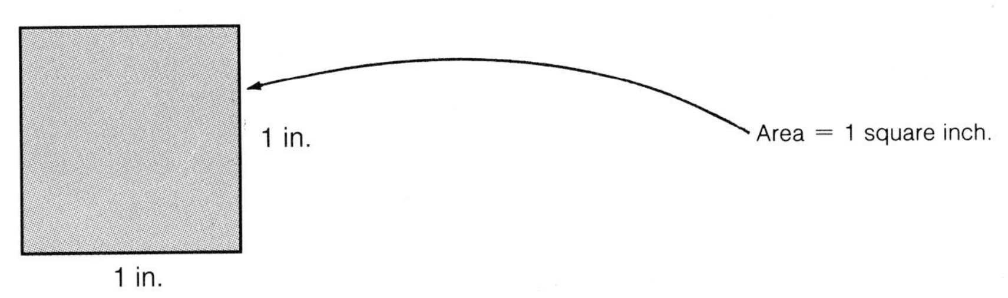

Example 6 A square whose side measures 1 centimeter has an area of 1 square centimeter.

The area of a polygon is thus the sum of the areas of squares that would completely cover the interior of the polygon. Area is measured in square units and these units depend on the linear units used to measure the sides of the polygon.

Square is abbreviated sq.

Example 7 5 square inches can be written 5 sq in.

Example 8 12 square centimeters can be written 12 sq cm.

An exponent of 2 is sometimes used to abbreviate square units.

Example 9 5 square inches can be written 5 in.^{2}.

Example 10 12 square centimeters can be written 12 cm^2.

We will now explore an example that will help us informally develop a formula for the area of a parallelogram.

Example 11 Find the number of square units in a rectangle if the length of the base is 7 units and the length of the altitude is 4 units.

Solution We will first sketch a figure to help in solving the problem.

Sketching a figure is usually helpful in solving geometric problems.

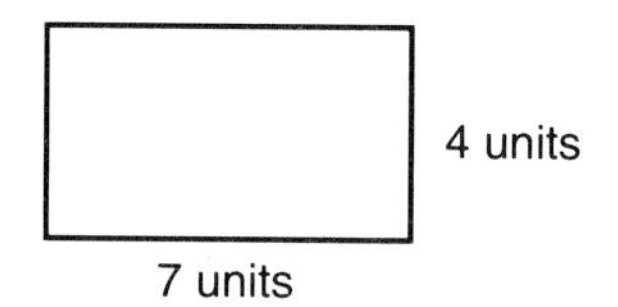

Our thinking can now proceed in this manner. We can draw 7 columns, each of which is 1 unit wide.

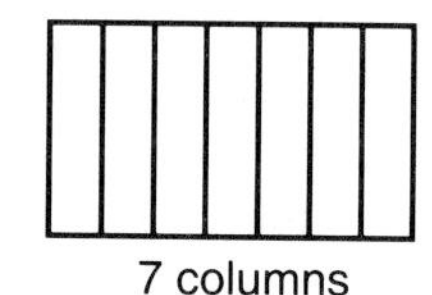

We can then draw 4 rows, each of which is 1 unit high.

How many square units are in this rectangle?

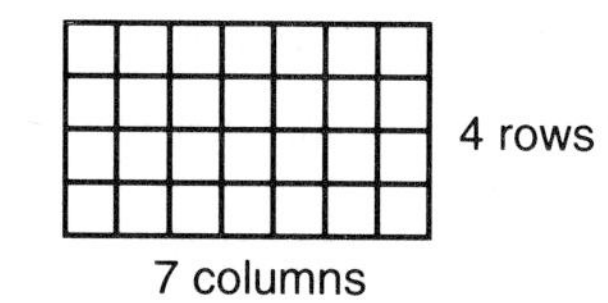

We have now covered the interior of the rectangle with square units. Since we have 7 columns each with 4 square units, we have $(7)(4) = 28$ square units.

This can be shown in a course in geometry.

From this example we see that the area is the product of the base and altitude. The same will be true for any parallelogram. The preceding discussion is not a proof but it does give plausibility to the following theorem.

THEOREM The area A of a parallelogram is the product of the base b and the altitude h.

$$A = bh$$

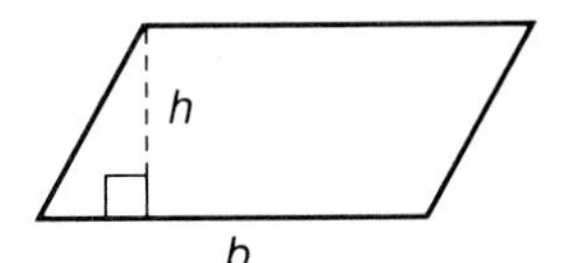

Example 12 Find the area of a parallelogram if the length of the base is 12 inches and the altitude is 7 inches.

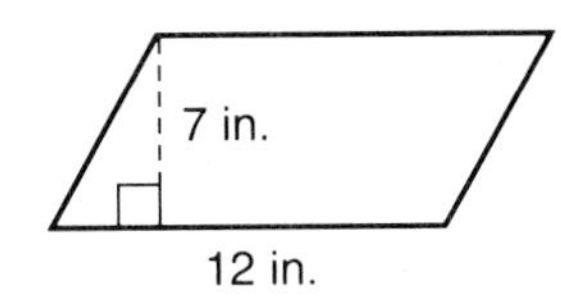

Always draw and label a sketch so that you will make the correct substitutions.

Solution We are asked to find A in the formula when $b = 12$ in. and $h = 7$ in.

$$\begin{aligned} A &= bh \\ &= (12)(7) = 84 \text{ in.}^2 \end{aligned}$$

Note the units used.

⦿ ***SKILL CHECK 4*** Find the area of a parallelogram if the base measures 18 inches and the altitude is 9 inches.

The formula for a rectangle (which is a parallelogram) is sometimes written as $A = \ell w$ since the length ℓ of the rectangle is the same as the base b and the width w is the same as the altitude h.

Recall that the width of a rectangle is perpendicular to its length.

Example 13 Find the area of a rectangle if the length measures 40 yards and the width measures 5 yards.

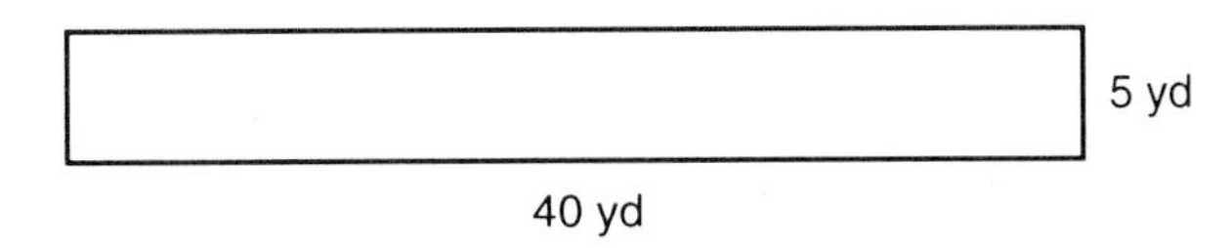

Solution Using the formula $A = \ell w$, we have

$$\begin{aligned} A &= \ell w \\ &= (40)(5) = 200 \text{ yd}^2 \end{aligned}$$

$\ell = 40$ yd
$w = 5$ yd

⦿ ***SKILL CHECK 5*** Find the area of a rectangle if the length measures 13 yards and the width measures 6 yards.

In a course in geometry we could also establish a formula for the area of a trapezoid. We will state this formula here without proof.

THEOREM The area A of a trapezoid having altitude h and bases measuring b_1 and b_2 is found by using the formula $A = \frac{1}{2}h(b_1 + b_2)$.

These are read "b sub one" and "b sub two."

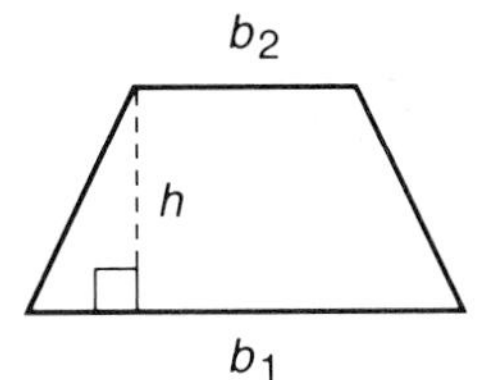

⦿ ***Skill Check Answers***

4. 162 in.^{2}
5. 78 yd^2

Example 14 Find the area of a trapezoid if $b_1 = 12.2$ cm, $b_2 = 7.6$ cm, and $h = 6$ cm.

Again, draw and label a sketch.

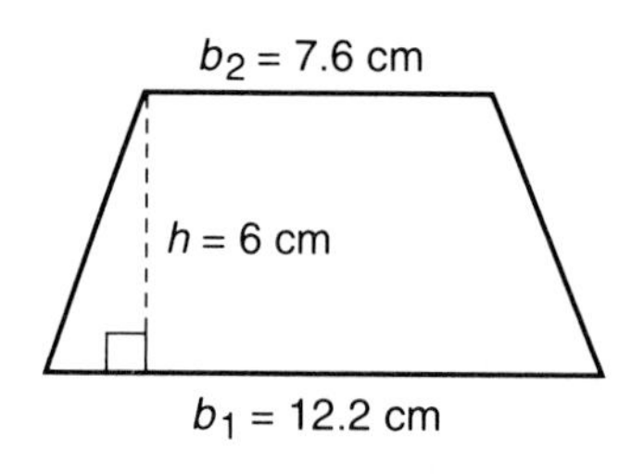

Solution

$$A = \frac{1}{2}h(b_1 + b_2)$$
$$= \frac{1}{2}(6)(12.2 + 7.6)$$
$$= 59.4 \text{ cm}^2$$

⦿ ***SKILL CHECK 6*** Find the area of a trapezoid if $b_1 = 15.5$ cm, $b_2 = 8.7$ cm, and $h = 7$ cm.

⦿ ***Skill Check Answer***

6. 84.7 cm²

Exercise 5–3–2

3

1. Determine the formula for the area of a square in terms of the length of a side s. (Recall that a square is a special rectangle and substitute s for b and h in the formula for its area.)

2. Using the formula derived in question 1, find the area of a square whose side measures 5 in.

3. Find the area of a square if $s = 2.3$ ft. Round answer to one decimal place.

4. Find the area of a square if $s = 3.4$ m. Round answer to one decimal place.

5. Find the area of a parallelogram if $b = 3.4$ ft and $h = 2.5$ ft.

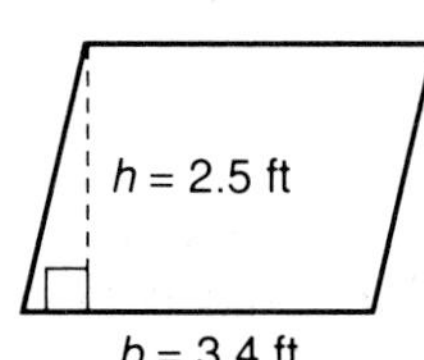

6. Find the area of a parallelogram if $b = 5\frac{1}{2}$ cm and $h = 3$ cm.

7. Find the area of a rectangle if $\ell = 4\frac{2}{3}$ ft and $w = 3\frac{3}{4}$ ft.

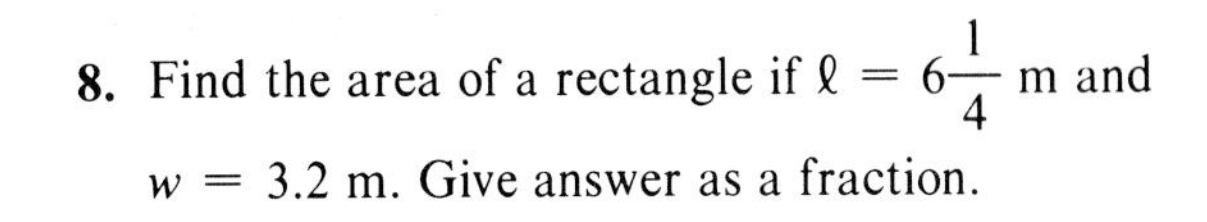

8. Find the area of a rectangle if $\ell = 6\frac{1}{4}$ m and $w = 3.2$ m. Give answer as a fraction.

9. Find the area of a trapezoid if $b_1 = 8$ cm, $b_2 = 4$ cm, and $h = 3$ cm.

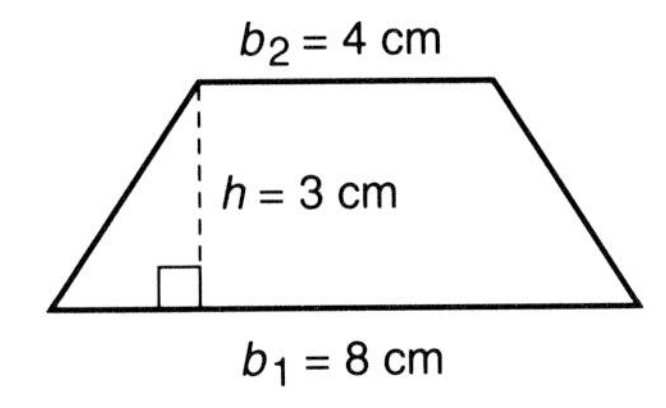

10. Find the area of a trapezoid if $b_1 = 5.4$ m, $b_2 = 4.7$ m, and $h = 2.6$ m. Round answer to one decimal place.

11. A **rhombus** is a parallelogram having all four sides equal in length. Find the area of a rhombus having a side that measures 8 in. and an altitude of 6 in.

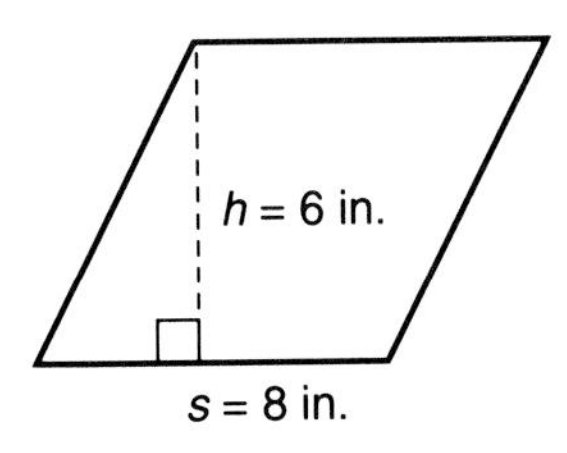

12. Find the area of a rhombus having a side that measures 8.4 cm and an altitude of 6.5 cm.

13. Find the area of a square whose side measures 13.6 cm. Round the answer to one decimal place.

14. Find the area of a parallelogram having an altitude of 8.4 m and a base that measures 11.2 m. Round the answer to one decimal place.

15. Find the area of a rectangle having sides that measure $4\frac{1}{2}$ m and $3\frac{2}{3}$ m.

16. Find the area of a parallelogram having an altitude of $3\frac{1}{2}$ in. and a base that measures $5\frac{3}{4}$ in.

17. Find the area of a square garden having a side that measures $4\frac{1}{2}$ yd.

18. Find the area of a rectangular piece of cloth 12.5 ft long and 3.5 ft wide. Round the answer to one decimal place.

19. Find the area of a trapezoid having bases that measure 7 m and 13 m and an altitude of $\frac{4}{5}$ m.

20. Find the area of a trapezoid having bases of lengths 6.2 in. and 4.8 in. and an altitude of 5.4 in.

• Concept Enrichment

21. Why is perimeter measured in linear units and area in square units?

22. Find the area of a square if its perimeter is 14.00 cm.

5–4 Triangles—Perimeter and Area

OBJECTIVES

Upon completing this section you should be able to:

1. Identify certain properties of triangles.
2. Find the perimeter of a triangle.
3. Find the area of a triangle.

In this section we will discuss another special class of polygons called triangles. The study of triangles and their properties is a very important part of the study of geometry. We will begin by restating the definition of a triangle as described in section 5–2.

1 Properties of Triangles

A polygon having three sides is called a **triangle.**

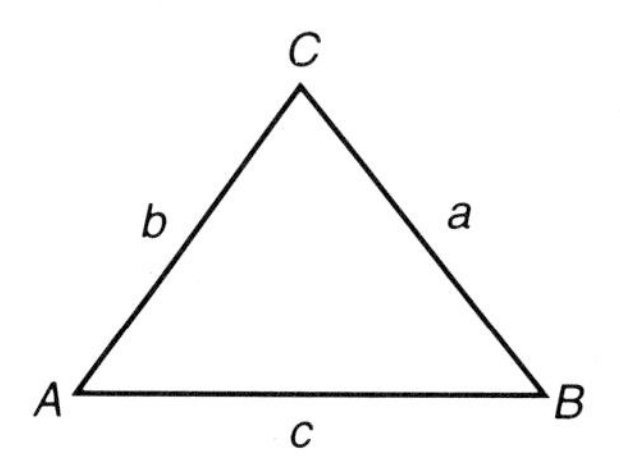

Proper labeling is very important in geometry.

The *vertices* of the triangle diagrammed above are represented by A, B, and C. The lowercase letters a, b, and c represent the lengths of sides $\overline{BC}$, $\overline{AC}$, and $\overline{AB}$, respectively. Note that the lowercase letters are opposite the vertex of the same uppercase letter. That is, a represents the length of the side opposite $\angle A$, and so on.

A triangle is named by using its three vertices. The symbol $\triangle$ is used to indicate a triangle.

We usually read counterclockwise around the figure.

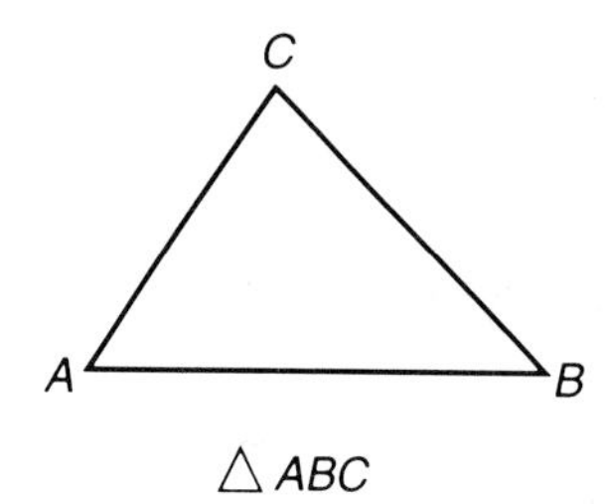

$\triangle ABC$

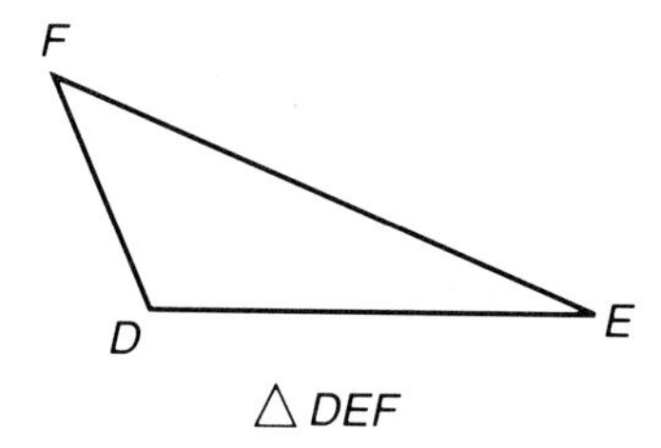

$\triangle DEF$

Triangles are classified according to the size of their angles.

DEFINITION An **acute triangle** has all three angles measuring less than 90° each.

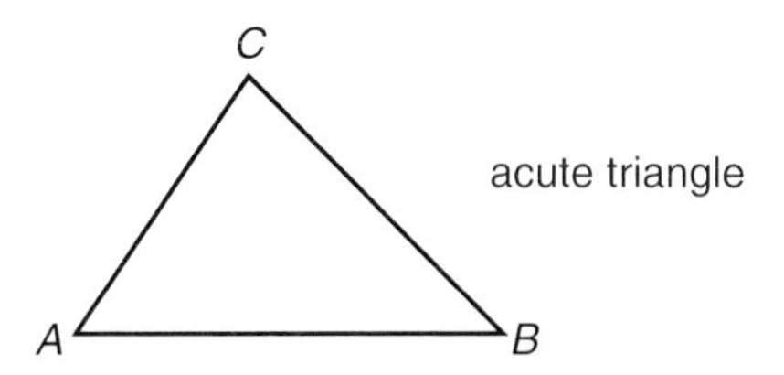

$\angle A$, $\angle B$, and $\angle C$ are each less than 90° in measure.

DEFINITION A **right triangle** has one angle whose measure is 90°.

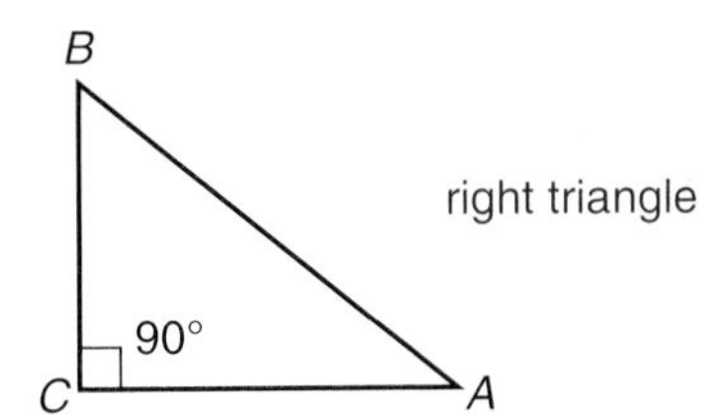

DEFINITION In a right triangle the side opposite the 90° angle is called the **hypotenuse.**

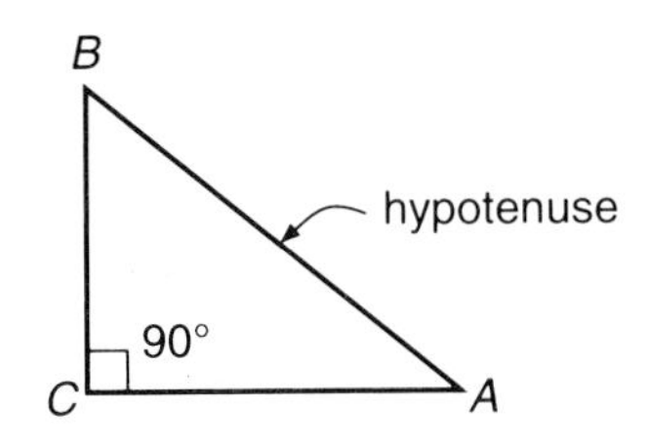

The other two sides are sometimes referred to as the *legs* of the right triangle.

DEFINITION An **obtuse triangle** has one angle that measures greater than 90°.

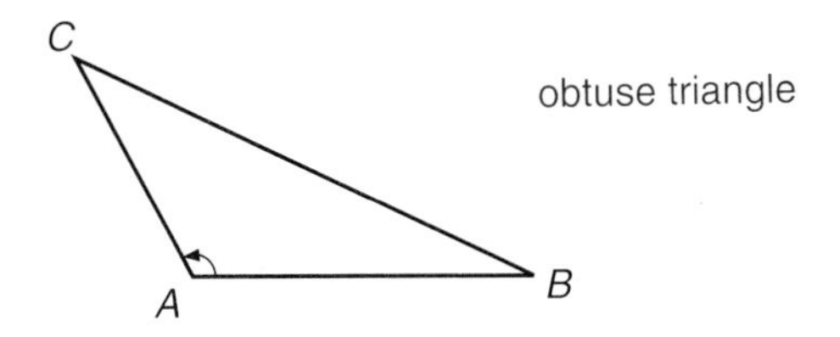

$\angle A$ measures greater than 90°.

DEFINITION If all three angles of a triangle have the same measure, the triangle is **equiangular.**

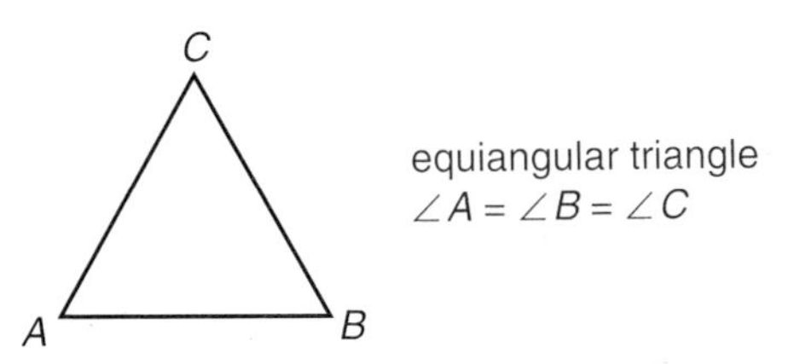

Triangles are also classified according to the relative lengths of their sides.

DEFINITION If two sides of a triangle are equal in length, the triangle is **isosceles.**

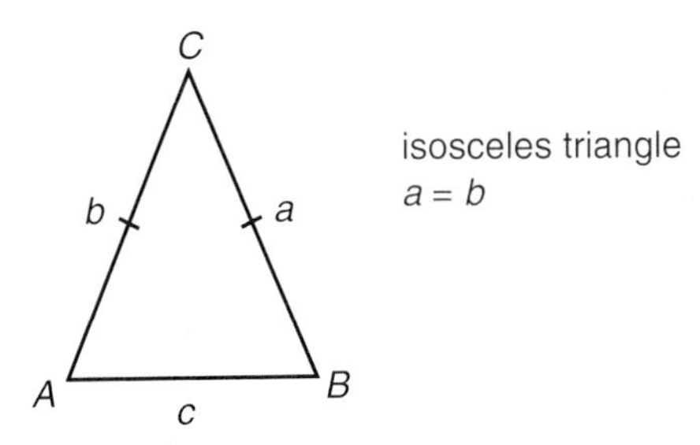

The hash marks on sides $\overline{AC}$ and $\overline{BC}$ indicate equal lengths.

DEFINITION If all three sides of a triangle are equal in length, the triangle is **equilateral.**

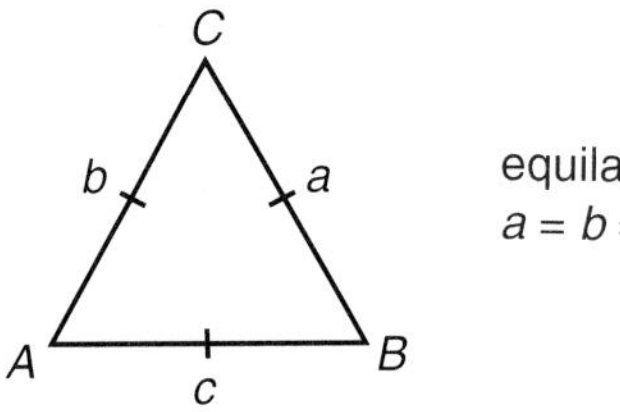

equilateral triangle
$a = b = c$

DEFINITION If no two sides of a triangle are equal in length, the triangle is **scalene.**

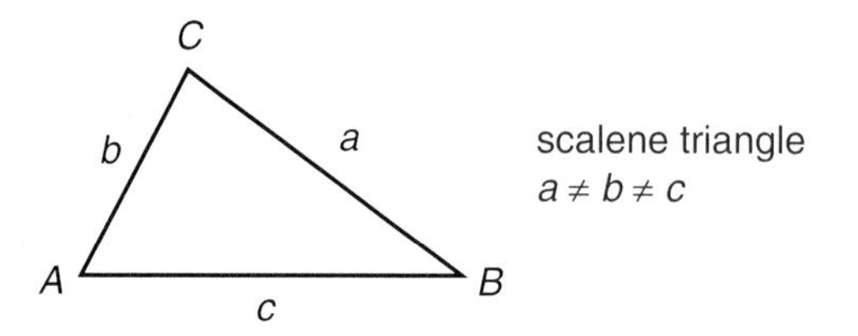

scalene triangle
$a \neq b \neq c$

Suppose we have two triangles illustrated as follows.

A' is read as "A prime."

$\angle A = \angle A'$
$\angle B = \angle B'$
$\angle C = \angle C'$

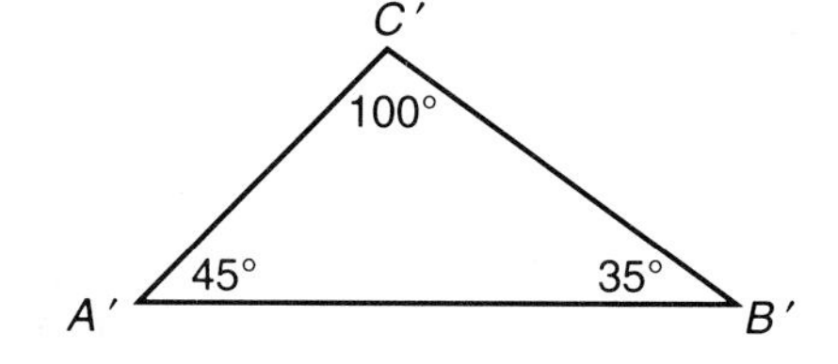

Notice that the three angles of one triangle are equal, respectively, to the three angles of the other triangle. This illustrates a special class of triangles.

$\triangle ABC \sim \triangle A'B'C'$ means the two triangles are similar.

DEFINITION If the three angles of one triangle are equal in measure to the three angles of another triangle, the triangles are called **similar triangles.**

Also, $\overline{AC}$ and $\overline{A'C'}$ are corresponding sides as are $\overline{BC}$ and $\overline{B'C'}$.

The sides opposite the equal angles of two similar triangles are said to be **corresponding sides.** For example, in the figure above we see that sides $\overline{AB}$ and $\overline{A'B'}$ are corresponding sides.

Note that the corresponding sides of similar triangles do not necessarily need to be the same length. If the corresponding sides of two similar triangles *are* equal in length, the triangles are said to be **congruent triangles.**

$\triangle ABC \cong \triangle A'B'C'$ means the two triangles are congruent.

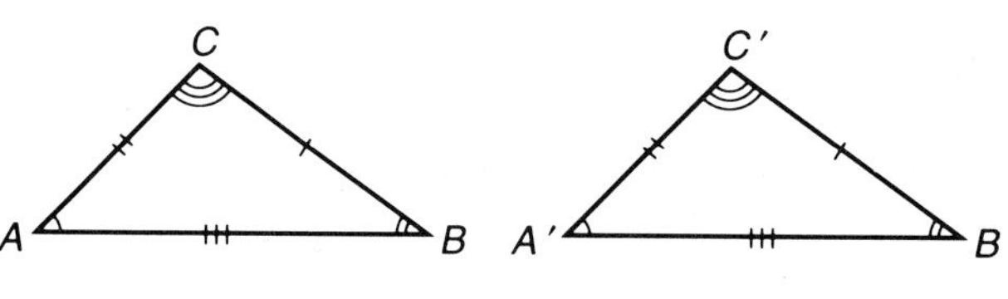

If $\angle A = \angle A'$, $\angle B = \angle B'$, $\angle C = \angle C'$, $\overline{AB} = \overline{A'B'}$, $\overline{BC} = \overline{B'C'}$, and $\overline{AC} = \overline{A'C'}$, then the triangles are congruent.

If two triangles are congruent, their interiors will have the same measure (area).

See section 8-3.

We will discuss similar triangles again later when we study proportions.

DEFINITION The **altitude** of a triangle is the distance measured along a perpendicular from any vertex to the opposite side extended if necessary.

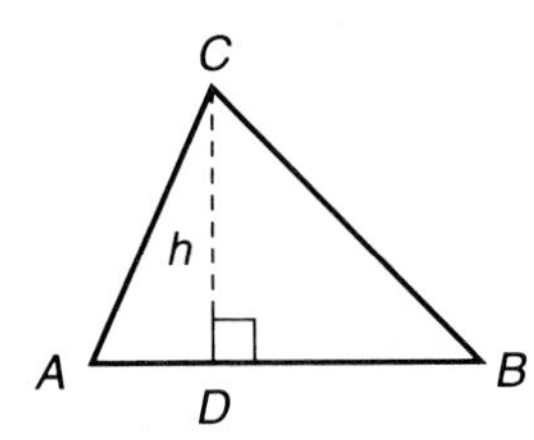

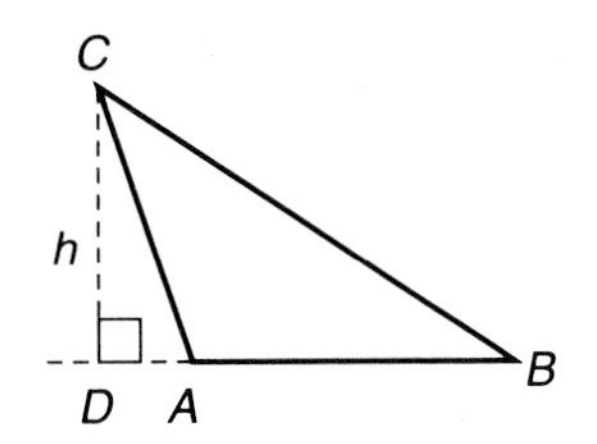

The distance from a point to a line is the length of the perpendicular line segment drawn from the point to the line.

$\overline{CD}$ is the perpendicular drawn from vertex C to the opposite side $\overline{AB}$. h represents the length of $\overline{CD}$.

DEFINITION The side to which an altitude is measured is called the **base** of the triangle.

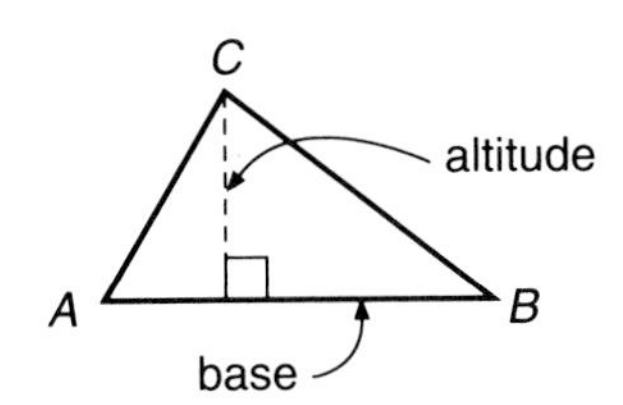

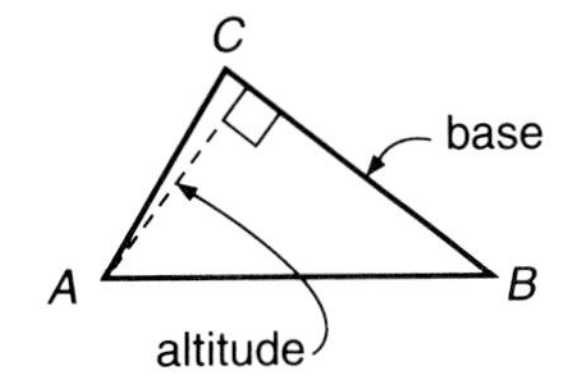

An altitude can be drawn from any vertex to the opposite side.

As mentioned in section 5–1, theorems are statements that can be proved to be true. We will state but not prove some important theorems concerning triangles.

THEOREM In an isosceles triangle the angles opposite the equal sides are equal in measure.

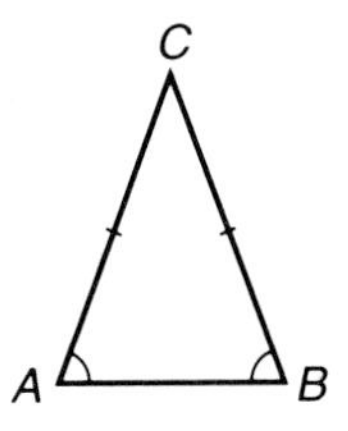

This diagram shows $\overline{AC} = \overline{BC}$.

In $\triangle ABC$ if $\overline{AC} = \overline{BC}$, then $\angle A = \angle B$.

THEOREM In an equilateral triangle the three angles are equal in measure.

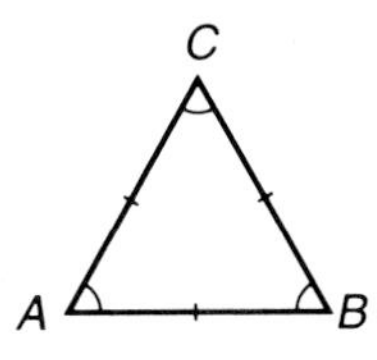

Thus the triangle is also equiangular.

If $\triangle ABC$ is equilateral, then $\angle A = \angle B = \angle C$.

Using the definitions and theorems from this section and the previous section, we can give an informal proof for the following theorem.

Example 1 Prove that the sum of the measures of the interior angles of any triangle is 180°.

Solution We first sketch a triangle and name it $\triangle ABC$.

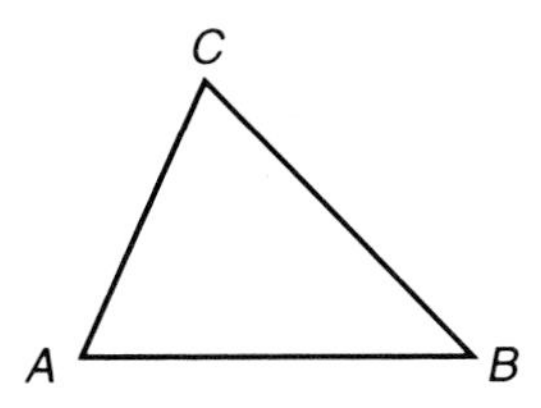

Since the theorem holds for any triangle, we sketch a general triangle.

We next draw a line through the vertex C parallel to the line that contains $\overline{AB}$.

ℓ_1 is parallel to ℓ_2. We label the angles for easy reference.

Again, notice the use of previous definitions and theorems in our thinking.

From section 5–1 we know that the measures of $\angle 4 + \angle 3 + \angle 5 = 180°$ since we have a straight line. Also we know that $\angle 1 = \angle 4$ and $\angle 2 = \angle 5$ because they are alternate interior angles of parallel lines. Now if we use the principle of substitution and replace $\angle 4$ with $\angle 1$ and $\angle 5$ with $\angle 2$, we obtain $\angle 1 + \angle 2 + \angle 3 = 180°$.

Notice the discussion did not depend on the size or shape of the triangle, so we can conclude that the sum of the measures of the interior angles of any triangle is 180°.

◉ ***SKILL CHECK 1*** If two angles of a triangle measure 35° and 87°, find the measure of the third side.

Now that we have shown that the sum of the measures of the interior angles of a triangle is 180°, we can find a formula for the sum of the measures of the interior angles of any polygon.

Example 2 Find the sum of the measures of the interior angles of a pentagon.

Solution We first sketch a pentagon. Then choosing any point in the interior of the pentagon, we draw a line segment from that point to each vertex.

How many triangles are formed?

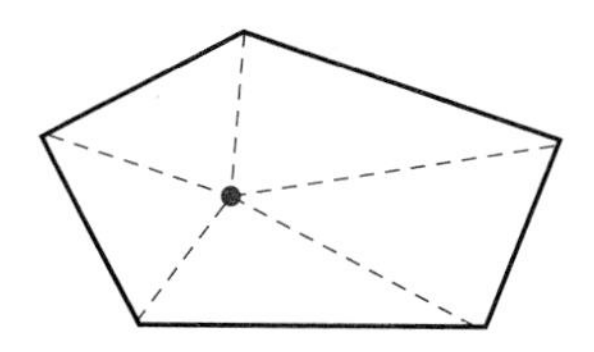

Note that one triangle is formed for each of the 5 sides of the pentagon. Since we know that the sum of the interior angles of each triangle is 180°, then the sum of all angles is $5(180°) = 900°$. Also note that the sum of the angles around the interior point is one revolution or 360°. If we subtract 360° from 900°, we find that the sum of the interior angles of the pentagon is $900° - 360° = 540°$. Note that we multiplied 5 by 180° and then subtracted 2 times 180°. This gives $5(180°) - 2(180°) = (5 - 2)180°$.

Distributive property.

Generalizing the above example gives us the following theorem.

THEOREM The sum S of the measures of the interior angles of a polygon having n sides is found by using the formula
$S = (n - 2)180°$.

Example 3 Find the sum of the measures of the interior angles of an octagon.

Solution Since an octagon has 8 sides we let $n = 8$, obtaining

$$\begin{aligned} S &= (n - 2)180° \\ &= (8 - 2)180° \\ &= (6)180° = 1{,}080°. \end{aligned}$$

◉ ***SKILL CHECK 2*** Find the sum of the measures of the interior angles of a decagon.

◉ ***Skill Check Answers***

1. 58°
2. 1,440°

Exercise 5-4-1

1 In triangle ABC the measures of two angles are given. Find the measure of the third angle and classify the triangle as acute, right, or obtuse.

1. $\angle A = 35°, \angle B = 55°$

2. $\angle A = 68°, \angle B = 22°$

3. $\angle A = 30°, \angle C = 70°$

4. $\angle B = 54°, \angle C = 59°$

5. $\angle B = 43°, \angle C = 38°$

6. $\angle A = 35°, \angle B = 49°$

Classify each triangle as isosceles, equilateral, or scalene given the measures of sides a, b, and c.

7. $a = 3$ in., $b = 3$ in., $c = 5$ in.

8. $a = 4$ ft, $b = 4$ ft, $c = 4$ ft

9. $a = 2$ ft, $b = 4$ ft, $c = 5$ ft

10. $a = 6$ cm, $b = 4$ cm, $c = 6$ cm

11. $a = 3.2$ cm, $b = 3.2$ cm, $c = 3.2$ cm

12. $a = 5\frac{1}{2}$ in., $b = 3\frac{1}{4}$ in., $c = 7\frac{1}{2}$ in.

13. In a right triangle ABC if $\angle A = 38°$ and $\angle C = 90°$, find the measure of $\angle B$.

14. In a right triangle ABC if $\angle B = 52°$ and $\angle C = 90°$, find the measure of $\angle A$.

15. In isosceles $\triangle ABC$, $\angle C = 50°$. Find the measures of the other two angles.

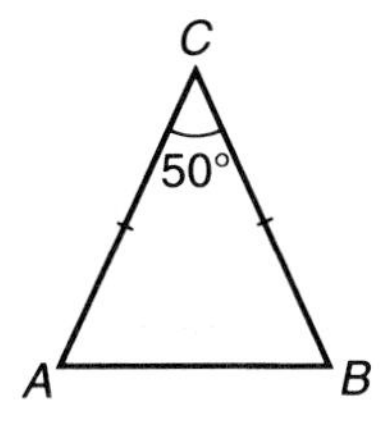

16. In isosceles $\triangle ABC$, $\angle C = 65°$. Find the measures of the other two angles.

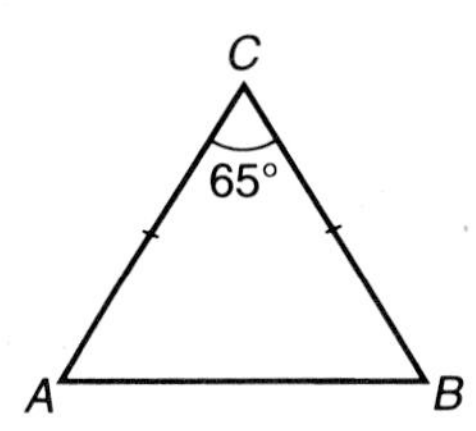

17. In isosceles $\triangle ABC$, $\angle A = 53°$. Find the measures of the other two angles.

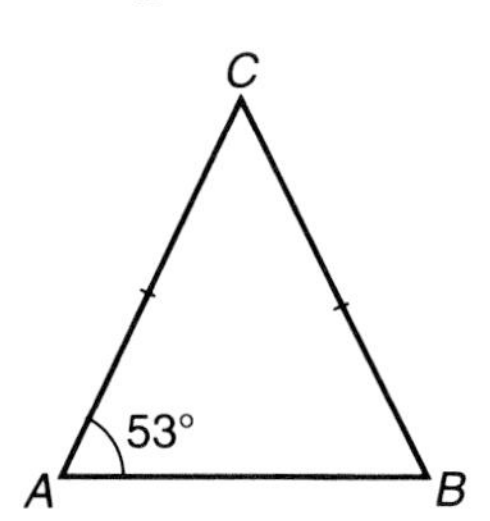

18. One angle of a right triangle measures 30°. Find the measures of the other two angles.

19. If a right triangle is isosceles, find the measures of its angles.

20. What is the measure of each angle of an equilateral triangle?

21. Is an equilateral triangle an isosceles triangle?

22. Find the sum of the measures of the interior angles of a quadrilateral.

23. Find the sum of the measures of the interior angles of a hexagon.

24. Find the measure of one angle of a regular pentagon.

25. Find the measure of one angle of a regular decagon.

• Concept Enrichment

26. Give an informal proof of the theorem that states the three angles of an equilateral triangle are equal in measure. (*Hint:* Use the theorem that states the angles opposite the equal sides of an isosceles triangle are equal in measure.)

27. Prove that the two acute angles of a right triangle are complementary.

28. In the following figure $\overline{DE}$ is parallel to $\overline{AB}$. Prove that $\triangle ABC$ and $\triangle DEC$ are similar triangles. (*Hint:* Use the result established in question 42 of exercise 5–1–1.)

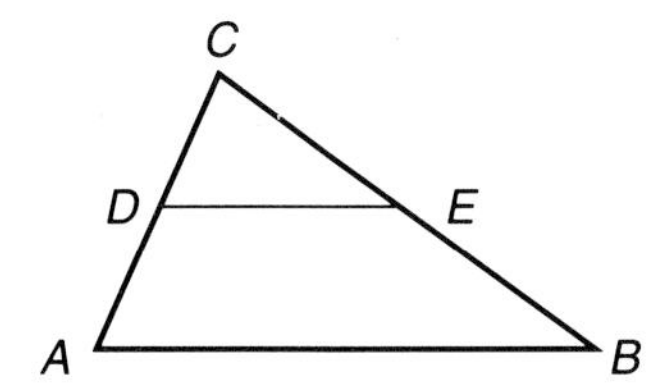

We now wish to develop formulas for finding the perimeter and area of triangles.

2 Perimeter of a Triangle

We have defined the perimeter of a polygon as the sum of the measures of the sides of the polygon. The perimeter P of a triangle with sides measuring a, b, and c is expressed by the formula

See section 5–3.

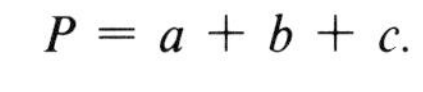

$$P = a + b + c.$$

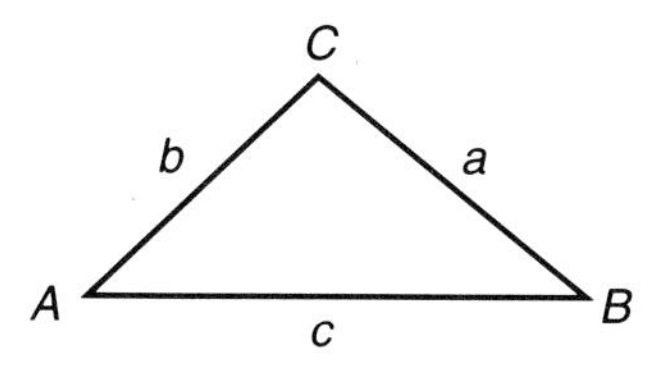

Example 4 Find the perimeter of the following triangle.

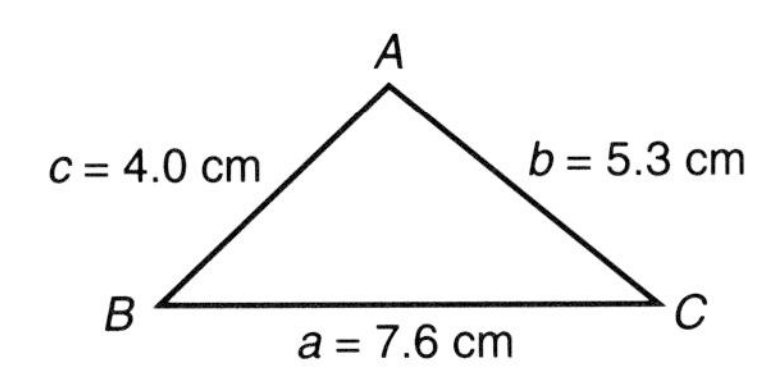

Solution Substituting the values for a, b, and c in the formula, we obtain

$$\begin{aligned} P &= a + b + c \\ &= 7.6 + 5.3 + 4.0 \\ &= 16.9 \text{ cm.} \end{aligned}$$

⦿ ***SKILL CHECK 3*** Find the perimeter of a triangle whose sides measure 3.7 cm, 8.6 cm, and 10.2 cm.

Example 5 Find the perimeter of the following isosceles triangle.

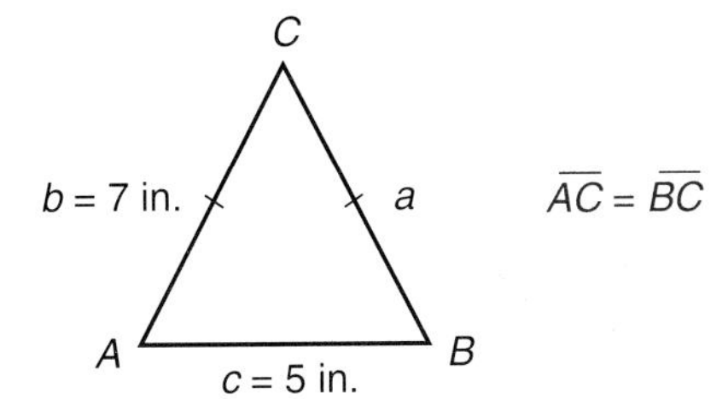

$\overline{AC} = \overline{BC}$

An isosceles triangle has two sides of the same length.

⦿ ***Skill Check Answer***

3. 22.5 cm

Solution Since $\overline{AC} = \overline{BC}$, $a = 7$ in. Thus

$$\begin{aligned} P &= a + b + c \\ &= 7 + 7 + 5 \\ &= 19 \text{ in.} \end{aligned}$$

⦿ ***SKILL CHECK 4*** Find the perimeter of the following isosceles triangle.

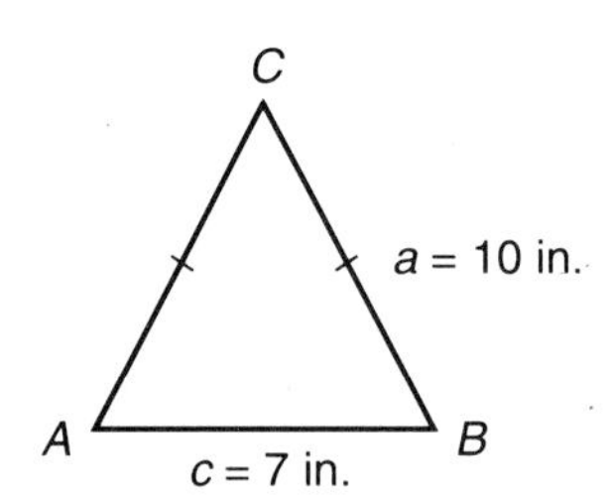

3 Area of a Triangle

What is the formula for the area of a parallelogram?

In section 5–3 we defined area of a polygon and established formulas for finding the areas of several quadrilaterals. We now wish to derive a formula for finding the area of a triangle. First we sketch a triangle ABC having altitude h to side $\overline{AC}$.

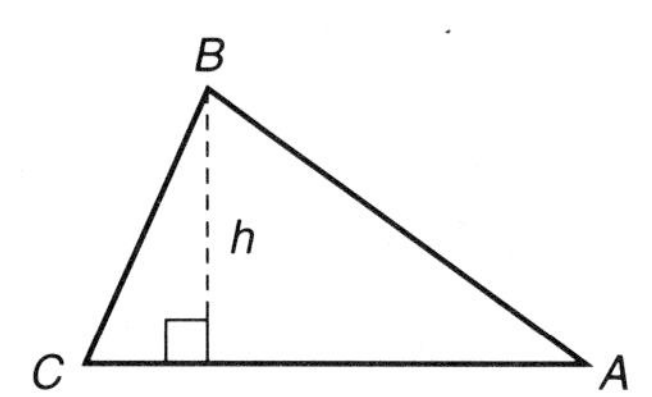

Quadrilateral $AC'BC$ is a parallelogram since opposite sides are parallel.

Next we draw a line through vertex B, parallel to the base $\overline{AC}$. Likewise we draw another line through vertex A parallel to side $\overline{BC}$. We label the intersection of the two drawn lines C'. This forms the parallelogram $AC'BC$. We also draw a line segment from vertex A perpendicular to side BC'. Its length is also h.

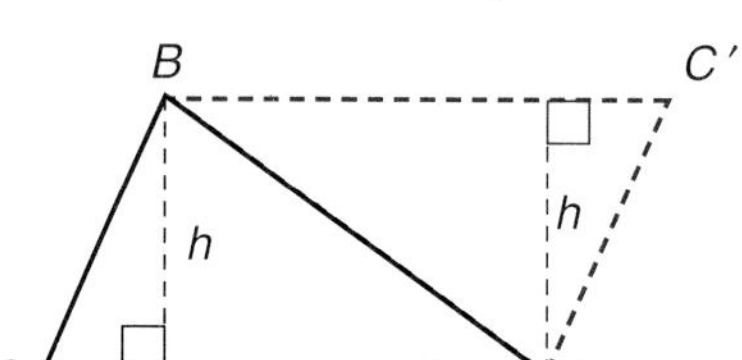

Parallel lines cut by a transversal. Refer to section 5–1.

See section 5–3. $\overline{AC} = \overline{BC'}$ and $\overline{BC} = \overline{AC'}$. $\overline{AB}$ is a common side to both triangles.

We now have two triangles, $\triangle ABC$ and $\triangle BAC'$. Notice that $\overline{AC}$ is parallel to $\overline{BC'}$, thus $\angle CAB = \angle ABC'$. Also since $\overline{BC}$ is parallel to $\overline{AC'}$, $\angle ABC = \angle BAC'$. Since opposite angles of a parallelogram are equal in measure, it follows that $\angle C = \angle C'$. Thus the two triangles are similar. But since the opposite sides of a parallelogram are equal in measure, we see that the corresponding sides of the two triangles are equal in measure and therefore the two triangles are congruent. Thus each has the same area. In other words, the area of each triangle is one-half the area of the parallelogram. Thus $\triangle ABC$ has an area that is one-half the area of the parallelogram.

This discussion leads to the following theorem.

THEOREM The area of a triangle is equal to one-half the product of the measure of the base and the altitude to the base.

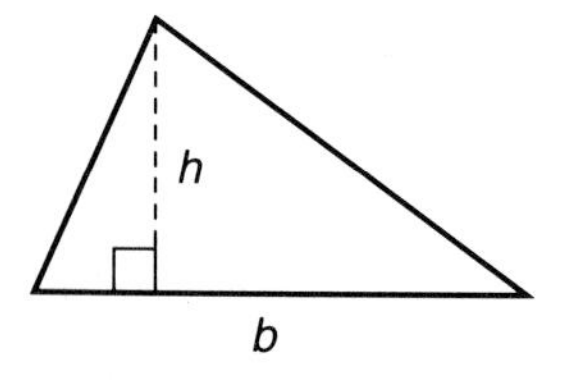

Sometimes the altitude is called the height.

The formula derived from this theorem is written

$$A = \frac{1}{2}bh,$$

where A is the area in square units, b is the measure of the base, and h is the altitude to the base. It is important that the measures of b and h are in the same units.

⦿ ***Skill Check Answer***

4. 27 in.

Example 6 Find the area of a triangle having an altitude of 4 inches and a base that measures 7 inches.

Solution First sketch the triangle and label the known parts.

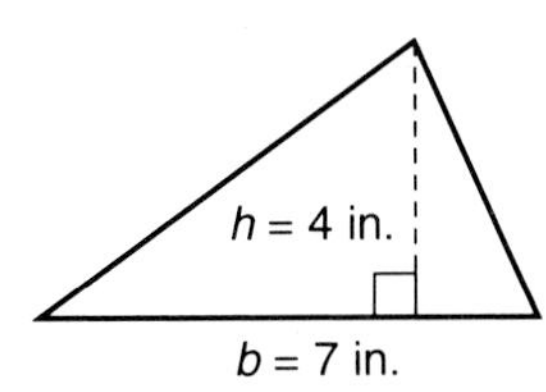

Always draw and label a sketch so that you will make the correct substitutions.

Next substitute $b = 7$ and $h = 4$ in the formula for the area of a triangle obtaining

$$\begin{aligned} A &= \frac{1}{2}bh \\ &= \frac{1}{2}(7)(4) \\ &= 14 \text{ in.}^2 \end{aligned}$$

Note the answer is in square units.

⊙ ***SKILL CHECK 5*** Find the area of a triangle having an altitude of 12 inches and a base that measures 9 inches.

⊙ ***Skill Check Answer***

5. 54 in.^{2}

Exercise 5–4–2

2

1. Find the perimeter of a triangle if $a = 6$ cm, $b = 4$ cm, and $c = 8$ cm.

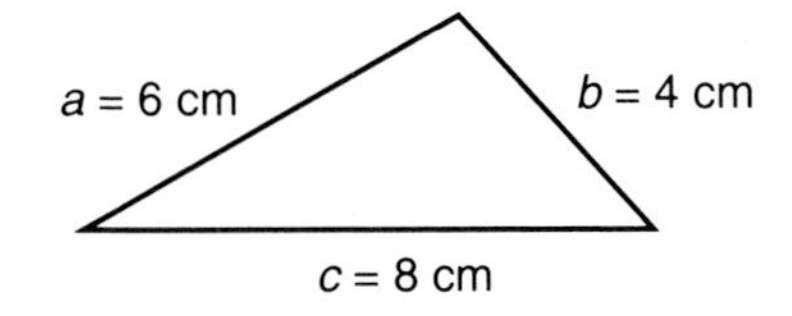

2. Find the perimeter of a triangle if $a = 3.2$ in., $b = 5.3$ in., and $c = 6.9$ in.

3. Find the perimeter of a triangle having sides of lengths 14.5 in., 10.6 in., and 17.8 in.

4. Find the perimeter of a triangle having sides of lengths $12\frac{1}{2}$ ft, $7\frac{3}{4}$ ft, and 6 ft.

3

5. Find the area of a triangle if $b = 6$ in. and $h = 5$ in.

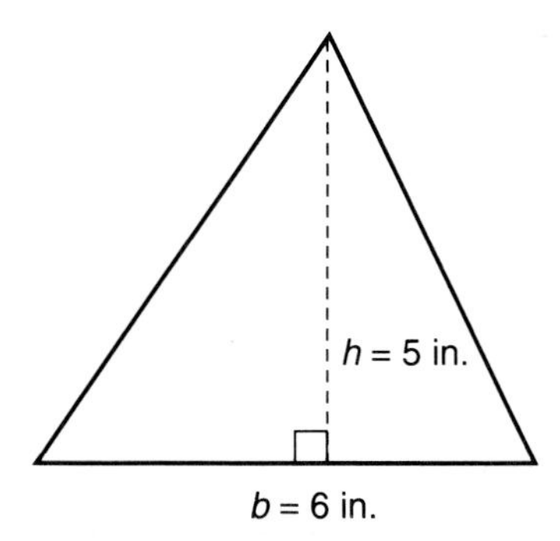

6. Find the area of a triangle if $b = 7$ cm and $h = 10$ cm.

7. Find the area of a triangle having a base of 6.4 cm and an altitude of 3.5 cm.

8. Find the area of the right triangle shown here.

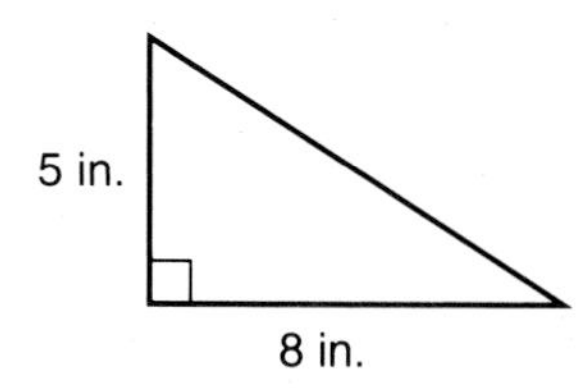

9. Find the area of a triangle having a base of 5.4 in. and an altitude of $4\frac{1}{2}$ in. Give the answer as a decimal rounded to one decimal place.

10. A fenced enclosure in the form of a triangle has sides of 120.2 ft, 142.5 ft, and 87.6 ft. What is the perimeter of the triangle?

11. Seventy-two feet of fencing are to be used to enclose a plot of ground in the shape of an equilateral triangle. What is the length of each side if all the fencing is used?

• Concept Enrichment

12. Find the perimeter and area of the following figure.

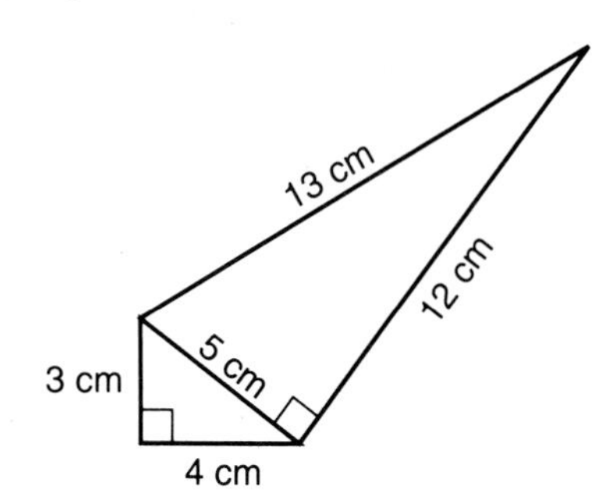

13. Determine the area of the shaded portion of the trapezoid.

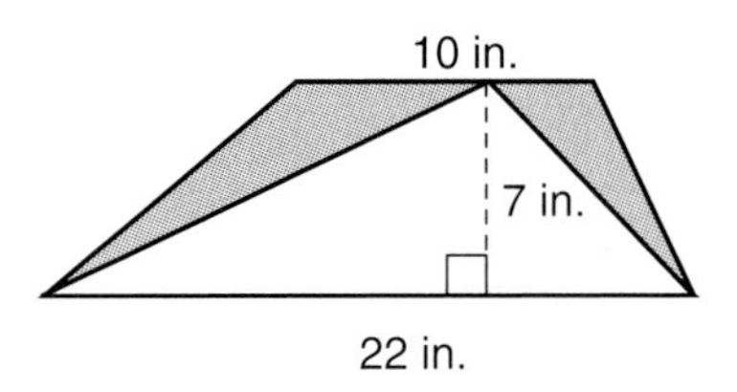

14. You now have enough information to derive the formula for the area of a trapezoid,

$$A = \frac{1}{2}h(b_1 + b_2).$$

First draw a sketch of a trapezoid $ABCD$ and label the bases b_1 and b_2 and altitude h. Now draw the diagonal $\overline{BD}$. You now have the trapezoid separated into two triangles. Draw the perpendicular from vertex B to side $\overline{DC}$ extended and label the altitude h.

Now using the formula for the area of a triangle, represent the area of $\triangle ABD$ and the area of $\triangle BCD$. Indicate the sum of the areas of these two triangles and use the distributive property to put the formula in its final form.

5-5 Circles—Circumference and Area

The closed plane figures discussed in previous sections were all composed of straight line segments. In this section we are concerned with a closed plane figure, the circle, that involves a curved line. We will first define a circle and some important terms associated with it. Then we will establish formulas for finding the circumference and area of a circle.

OBJECTIVES

Upon completing this section you should be able to:

1. Identify some terms and properties involving circles.
2. Find the circumference of a circle.
3. Find the area of a circle.

1 Properties of Circles

DEFINITION A **circle** is the set of all points on a plane that are an equal distance from a fixed point called the center of the circle.

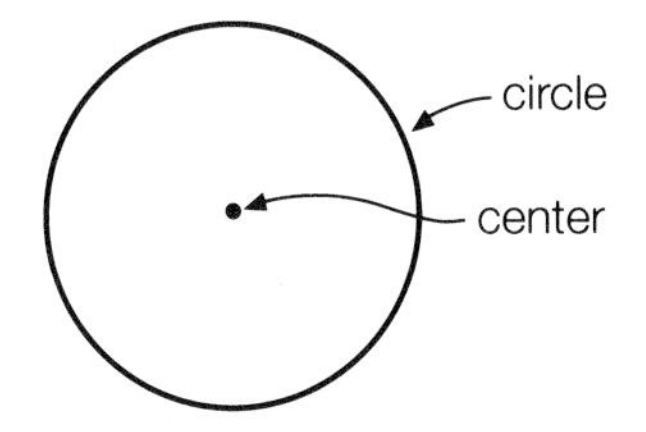

DEFINITION The distance from the center to the circle is the **radius** and is usually designated by the letter r. The plural of radius is radii.

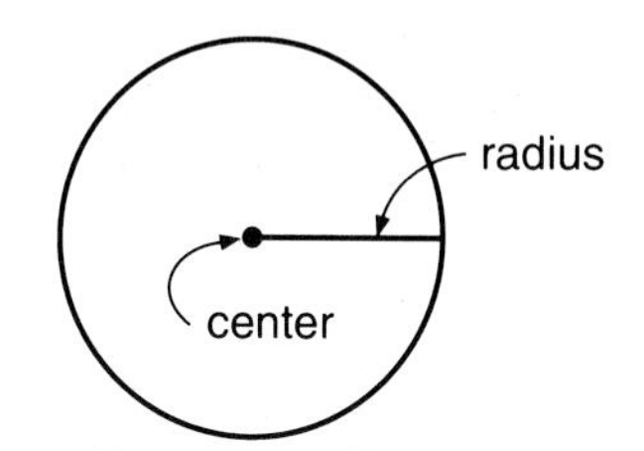

DEFINITION A line that intersects a circle in two points is called a **secant** of the circle.

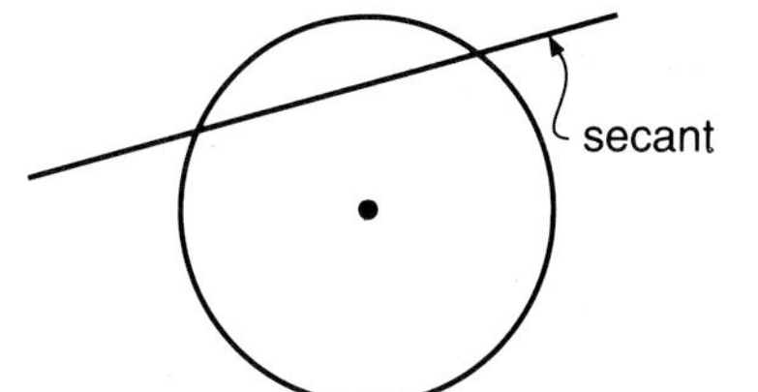

DEFINITION A line that intersects a circle in exactly one point is called a **tangent** to the circle.

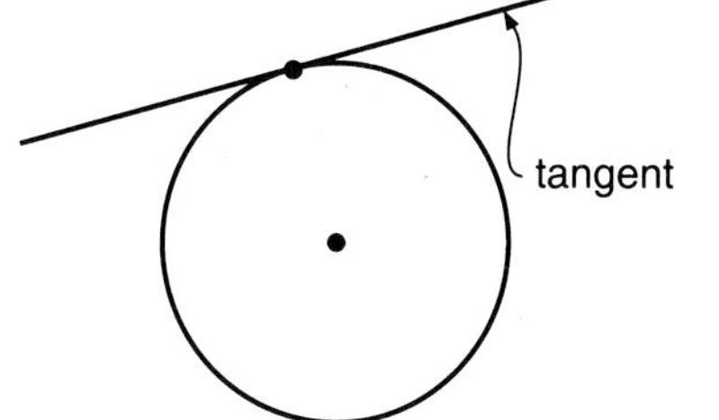

DEFINITION A line segment between two points on a circle is a **chord** of the circle.

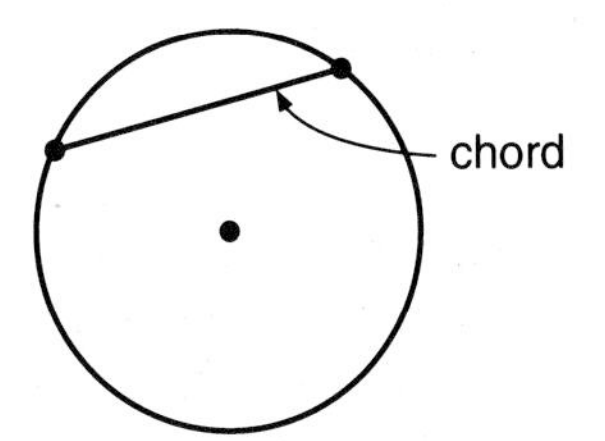

DEFINITION A chord that passes through the center of a circle is a **diameter** of the circle and is usually designated by the letter d.

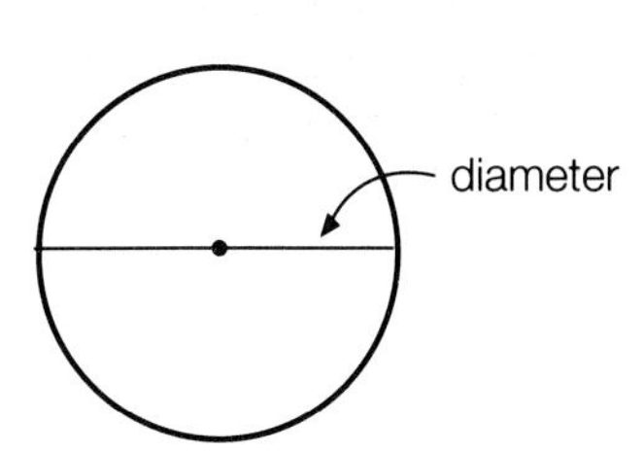

The length of the diameter of a circle is twice the length of the radius ($d = 2r$).

SKILL CHECK 1 If the length of the radius of a circle is 10 inches, find the length of the diameter.

Skill Check Answer

1. 20 in.

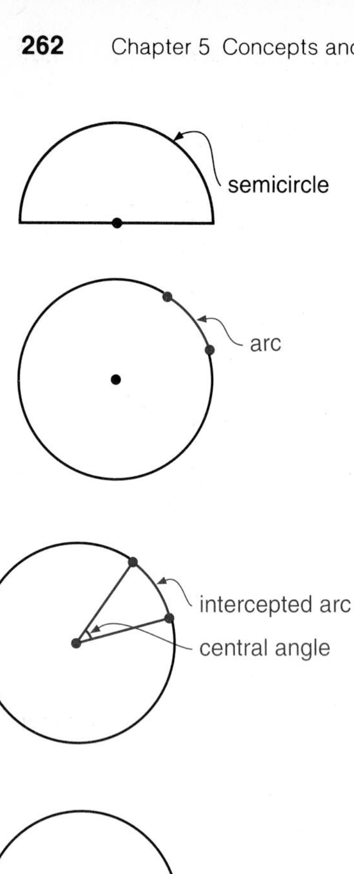

DEFINITION A **semicircle** is that portion of a circle cut off by a diameter.

DEFINITION Any part of a circle bounded by two points on the circle is an **arc** of the circle.

DEFINITION An angle with vertex at the center of a circle and whose sides are radii is a **central angle.** The arc bounded by the sides of the angle is called the **intercepted arc.**

The measure of a central angle is the same as the amount of rotation in its intercepted arc.

Example 1 A central angle that measures 30° intercepts an arc of 30°.

◉ ***SKILL CHECK 2*** What is the measure of an arc intercepted by a central angle of 46°?

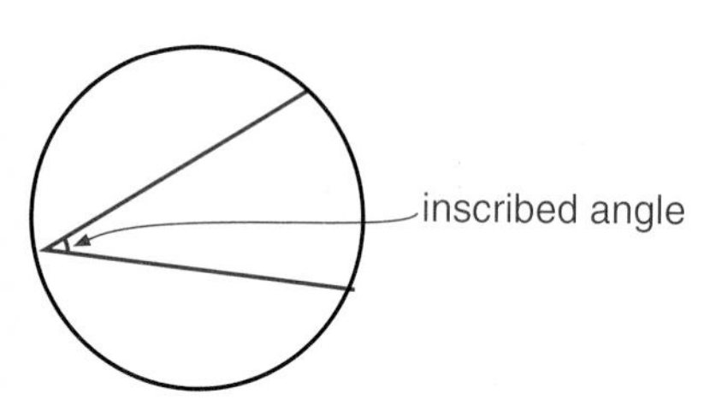

DEFINITION An angle formed by two chords that intersect on a circle is an **inscribed angle.**

The measure of an inscribed angle is equal to one-half the rotation in the intercepted arc.

Example 2 An inscribed angle measuring 30° intercepts an arc measuring 60°.

30°

arc measures
60° rotation

◉ ***SKILL CHECK 3*** What is the measure of an arc intercepted by an inscribed angle of 46°?

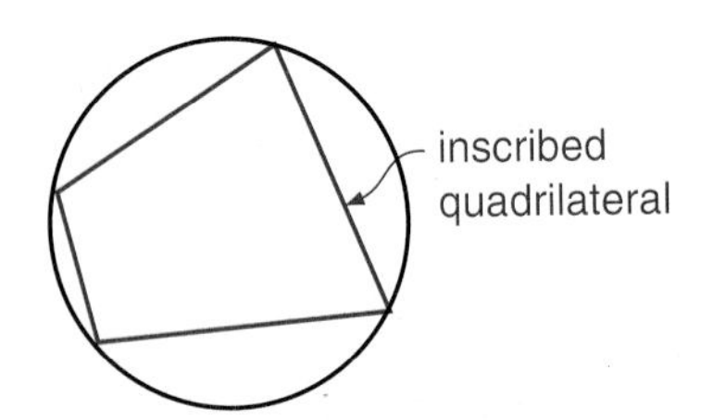

DEFINITION A polygon that has every vertex on a circle is an **inscribed polygon.**

◉ ***Skill Check Answers***

2. 46°
3. 92°

DEFINITION A **sector** of a circle is that part of the circle bounded by two radii.

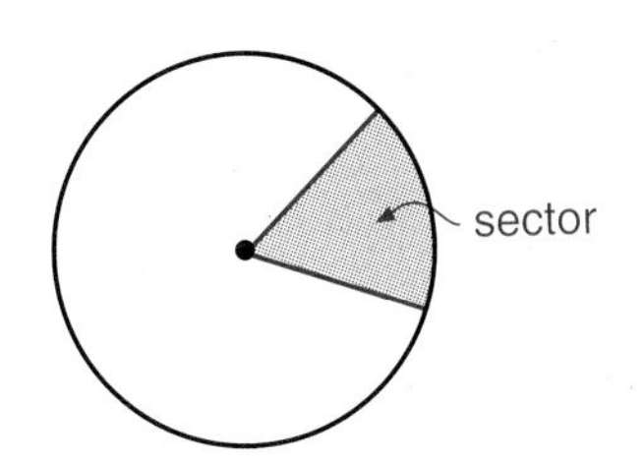

DEFINITION A **segment** of a circle is the area of the region bounded by a chord and the circle.

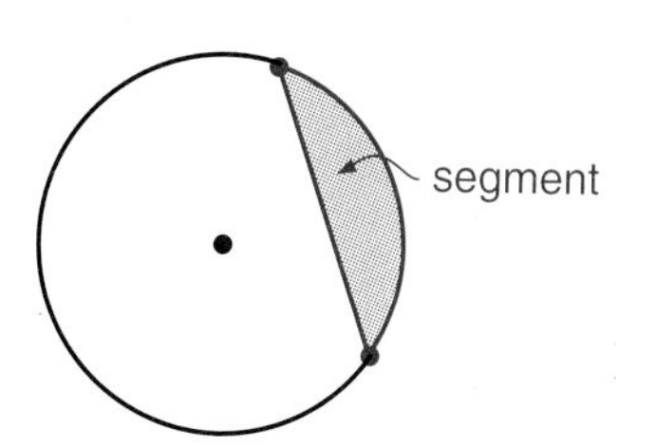

2 Circumference of a Circle

DEFINITION The perimeter of a circle is called the **circumference** of the circle and is usually designated by the letter C.

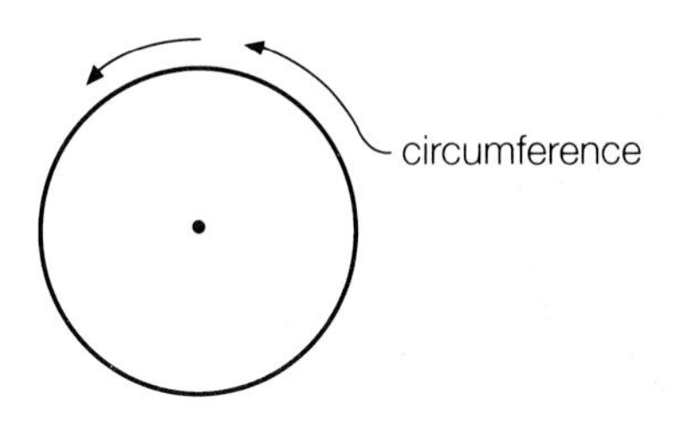

An interesting relationship exists between the diameter and circumference of a circle. If the diameter is measured as a decimal number, the circumference will not measure as a decimal number. Likewise, if the measure of the circumference is given as a decimal number, the diameter's measure is not a decimal number. This type of number that cannot be expressed as a decimal number is called an **irrational number.** We will discuss this type of number again in the next chapter.

See section 6-10.

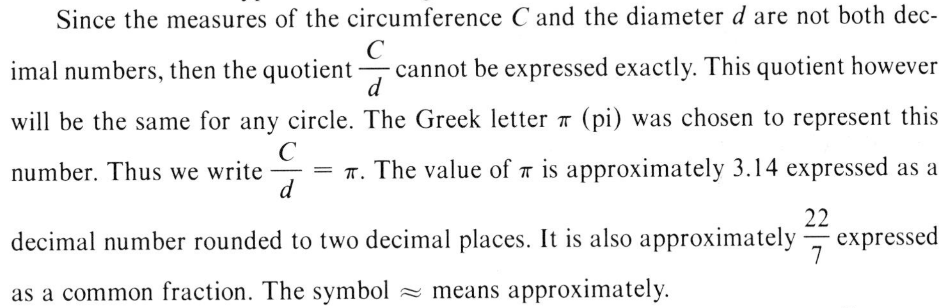

Since the measures of the circumference C and the diameter d are not both decimal numbers, then the quotient $\frac{C}{d}$ cannot be expressed exactly. This quotient however will be the same for any circle. The Greek letter π (pi) was chosen to represent this number. Thus we write $\frac{C}{d} = \pi$. The value of π is approximately 3.14 expressed as a decimal number rounded to two decimal places. It is also approximately $\frac{22}{7}$ expressed as a common fraction. The symbol $\approx$ means approximately.

$\pi \approx 3.14$

$\pi \approx \frac{22}{7}$

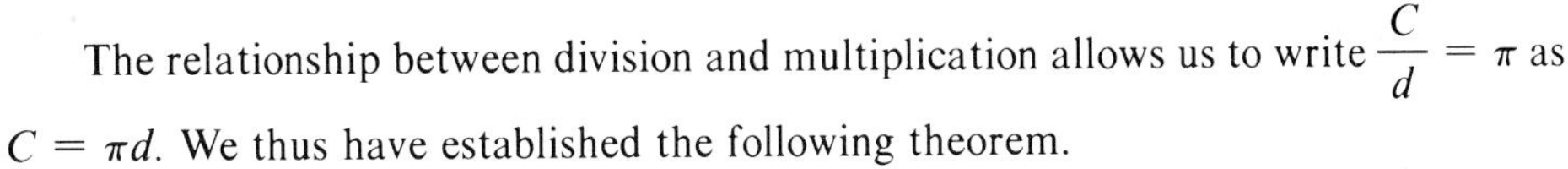

The relationship between division and multiplication allows us to write $\frac{C}{d} = \pi$ as $C = \pi d$. We thus have established the following theorem.

See section 2-2.

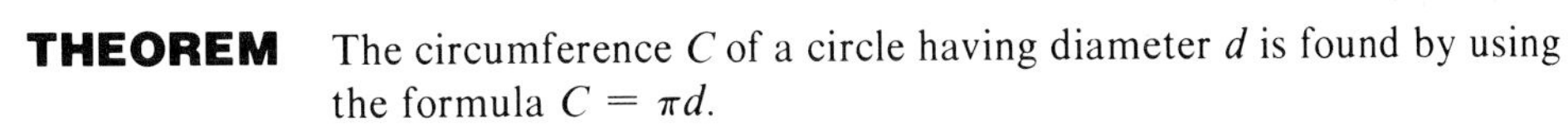

THEOREM The circumference C of a circle having diameter d is found by using the formula $C = \pi d$.

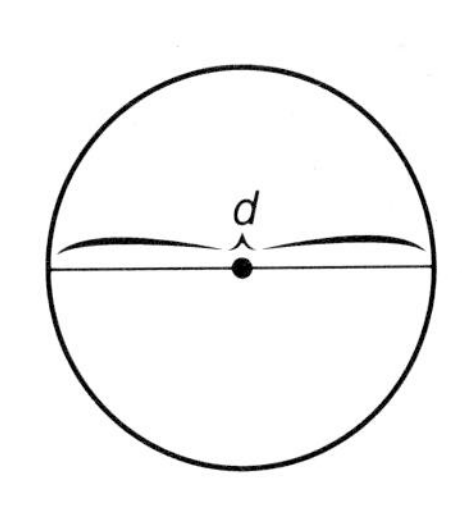

Example 3 Find the circumference of a circle if the diameter measures $5\frac{1}{2}$ inches. Use $\pi = \frac{22}{7}$.

Solution We use the formula for circumference, obtaining

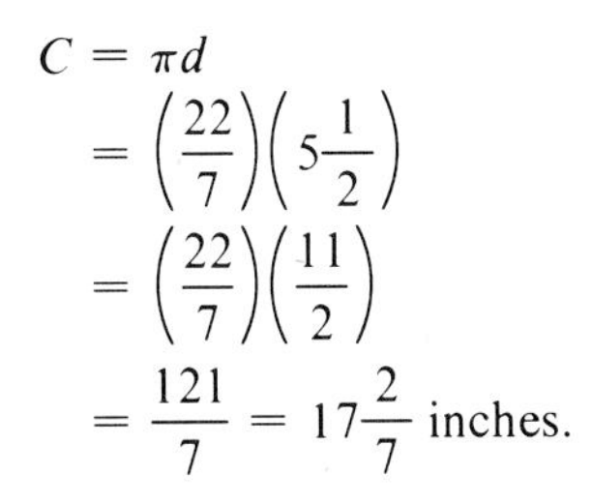

$$\begin{aligned} C &= \pi d \\ &= \left(\frac{22}{7}\right)\left(5\frac{1}{2}\right) \\ &= \left(\frac{22}{7}\right)\left(\frac{11}{2}\right) \\ &= \frac{121}{7} = 17\frac{2}{7} \text{ inches.} \end{aligned}$$

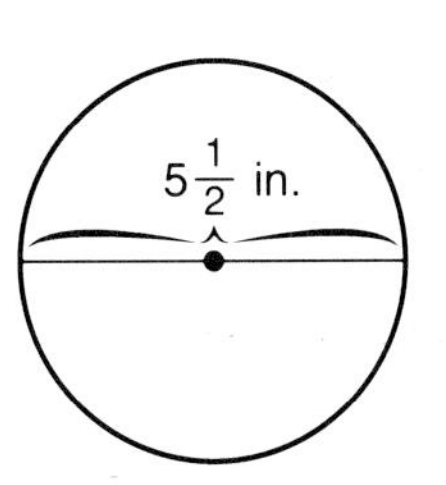

SKILL CHECK 4 Find the circumference of a circle if the diameter measures $3\frac{1}{2}$ inches. Use $\pi = \frac{22}{7}$.

Skill Check Answer

4. 11 in.

$d = 2r$, so $C = \pi d$
$= \pi(2r)$
$= 2\pi r$

Since we know the measure of the diameter of a circle is twice that of the radius, we can rewrite the circumference formula as $C = 2\pi r$.

Example 4 Find the circumference of a circle if $r = 2.4$ inches. Round the answer to one decimal place.

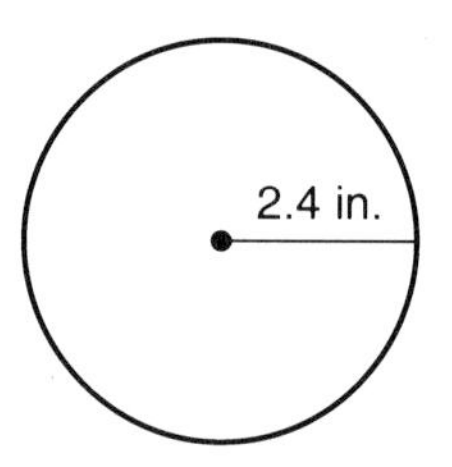

It is important to emphasize that 3.14 and $\frac{22}{7}$ are only approximations for π.

Solution Since r is given as a decimal, we will use $\pi = 3.14$. If r were given as a fraction, we would use $\pi = \frac{22}{7}$.

$$\begin{aligned} C &= 2\pi r \\ &= 2(3.14)(2.4) \\ &= 15.1 \text{ inches} \end{aligned}$$

Answer rounded to one decimal place.

⦿ ***SKILL CHECK 5*** Find the circumference of a circle if the radius measures 3.6 inches. Round the answer to one decimal place.

3 Area of a Circle

The following discussion will give reasonableness to the formula for the area of a circle.

We begin by cutting a circle into many small equal pieces as shown.

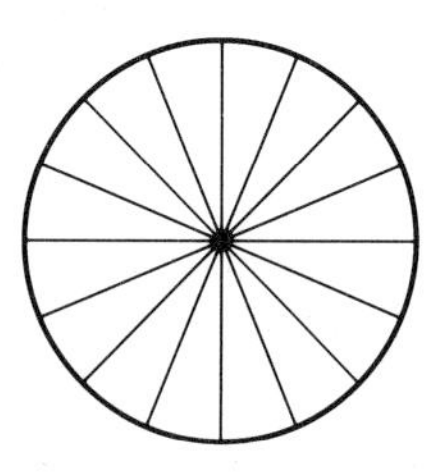

We next arrange the pieces as shown.

These arcs add up to one-half the circumference.

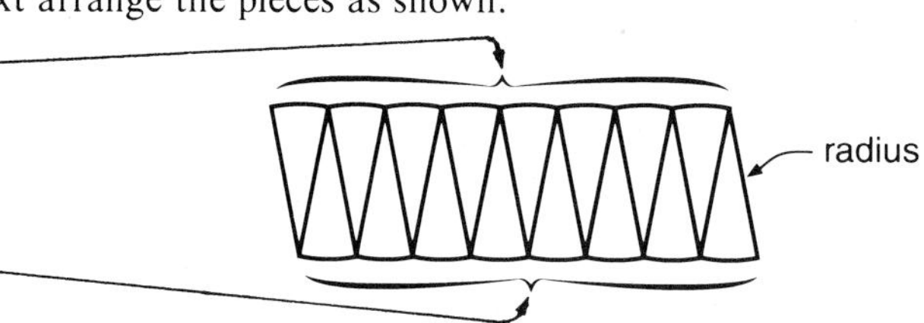

These arcs add up to one-half the circumference.

Note that if the pieces were small enough, the base of the figure would *almost* be a straight line segment and the altitude would *almost* be equal to the radius of the circle. In other words, the figure would be close to that of a rectangle having a base that measures one-half the circumference of the circle and an altitude equal to the radius of the circle.

Since $C = 2\pi r$, one-half the circumference is $\left(\frac{1}{2}\right)2\pi r = \pi r$.

$h = r$

$b = \pi r$

⦿ ***Skill Check Answer***

5. 22.6 in.

Using the formula for the area of a rectangle, we have

$$\begin{aligned} A &= bh \\ &= (\pi r)(r) \\ &= \pi r^2. \end{aligned}$$

Thus the following theorem is plausible.

THEOREM The area A of a circle having radius r is given by the formula $A = \pi r^2$.

Remember that area is always measured in square units.

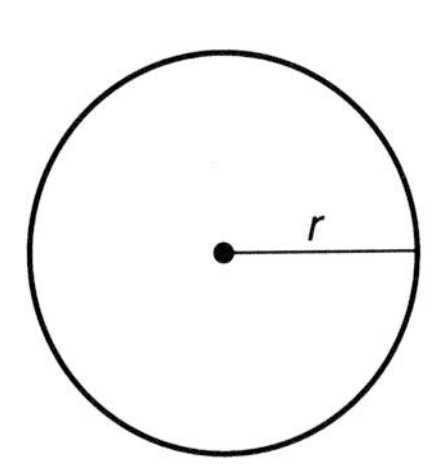

Example 5 Find the area of a circle if $r = 3.2$ m. Round the answer to one decimal place.

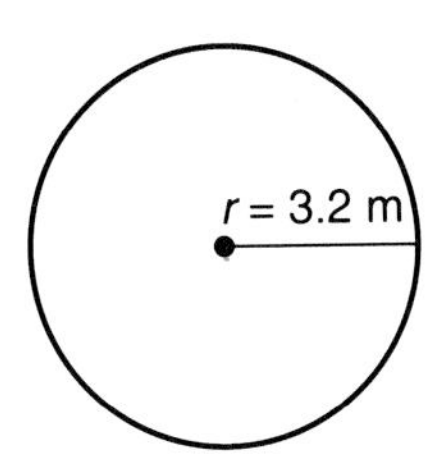

Solution

$$\begin{aligned} A &= \pi r^2 \\ &= (3.14)(3.2)^2 \\ &= 32.2 \text{ m}^2 \end{aligned}$$

Since the radius was given as a decimal, we use $\pi = 3.14$.

⦿ ***SKILL CHECK 6*** Find the area of a circle if $r = 4.6$ m. Round the answer to one decimal place.

Example 6 Find the area of a circle if $r = 4\frac{3}{4}$ inches.

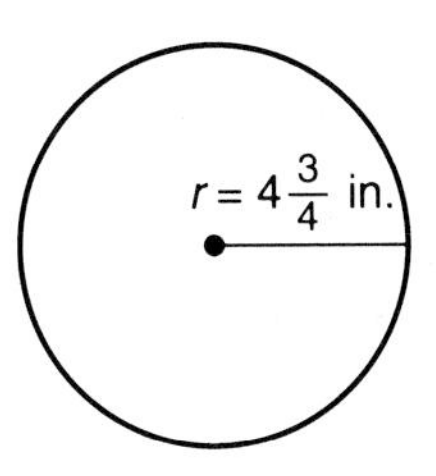

Solution

$$\begin{aligned} A &= \pi r^2 \\ &= \left(\frac{22}{7}\right)\left(4\frac{3}{4}\right)^2 \\ &= 70\frac{51}{56} \text{ in.}^2 \end{aligned}$$

Since r was given as a fraction, we use $\pi = \frac{22}{7}$.

⦿ ***SKILL CHECK 7*** Find the area of a circle if $r = 2\frac{1}{3}$ in.

⦿ ***Skill Check Answers***

6. 66.4 m^2
7. $17\frac{1}{9} \text{ in.}^2$

Exercise 5–5–1

1

1. Find the diameter of a circle having a radius of 4 inches.

2. Find the radius of a circle having a diameter of 21.0 centimeters.

3. What is the measure of a central angle that intercepts an arc of 90°?

4. What is the measure of a central angle that intercepts an arc of 130°?

5. What is the measure of an inscribed angle that intercepts an arc of 66°?

6. What is the measure of an inscribed angle that intercepts an arc of 152°?

2

7. Find the circumference of a circle when $r = 3\frac{1}{2}$ inches. Use $\pi = \frac{22}{7}$.

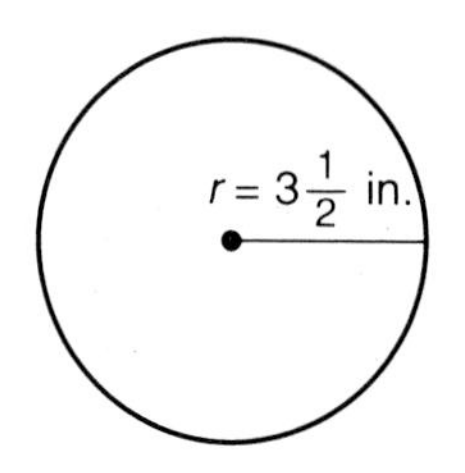

8. Find the circumference of a circle when $r = 2.5$ centimeters. Use $\pi = 3.14$.

9. Find the circumference of a circle whose diameter measures 14 centimeters. Use $\pi = \frac{22}{7}$.

10. Find the circumference of a circle having a diameter that measures $12\frac{1}{2}$ feet. Use $\pi = \frac{22}{7}$.

11. Find the circumference of a semicircle having a radius of 6.2 feet. Use $\pi = 3.14$ and round the answer to one decimal place.

12. Find the circumference of a semicircle having a radius of $3\frac{1}{2}$ meters. Use $\pi = \frac{22}{7}$.

13. How much edging is needed to go around a circular tablecloth having a radius of 1.5 meters? Use $\pi = 3.14$. Round the answer to one decimal place.

14. How many meters of fencing is needed to surround a circular garden having a radius of 10 meters? Use $\pi = 3.14$.

3

15. Find the area of a circle if $r = 2$ inches. Use $\pi = \frac{22}{7}$.

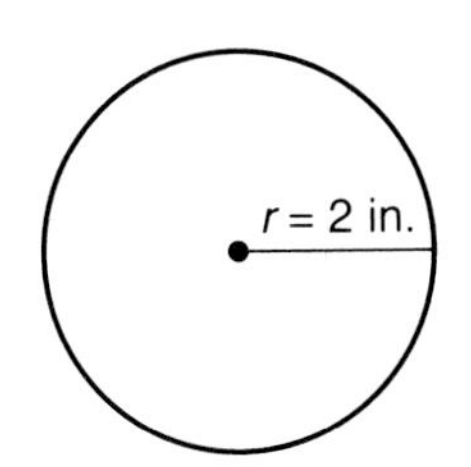

16. Find the area of a circle if $r = 3.1$ centimeters. Use $\pi = 3.14$. Round the answer to one decimal place.

17. Find the area of a semicircle having a diameter of 14 inches. Use $\pi = \frac{22}{7}$.

18. Find the area of a semicircle having a diameter of 8.0 meters. Use $\pi = 3.14$ and round the answer to one decimal place.

19. Find the area of a circle whose radius is 8.5 meters. Round the answer to one decimal place. Use $\pi = 3.14$.

20. Find the area of a circle having a radius of 3.80 inches. Round the answer to the nearest hundredth of a square inch. Use $\pi = 3.14$.

• Concept Enrichment

Find the area of the shaded region. Use $\pi = 3.14$ and round the answers to one decimal place.

21.

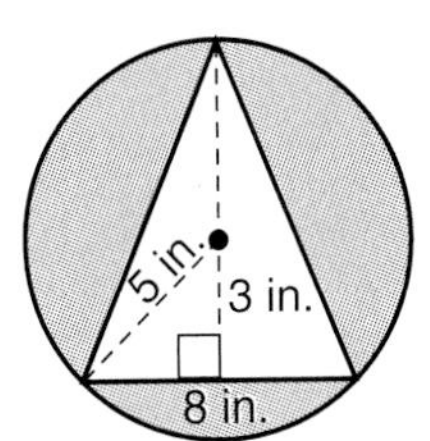

22.

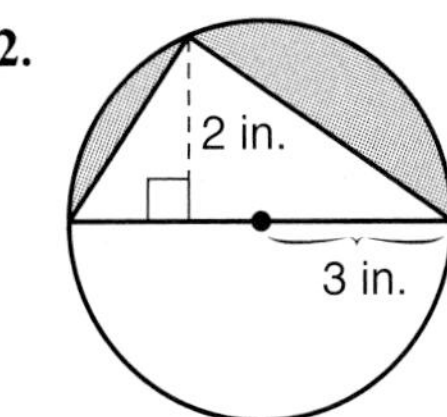

23.

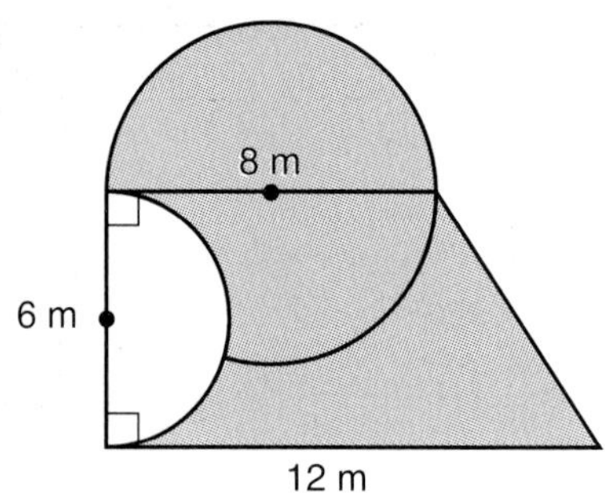

24

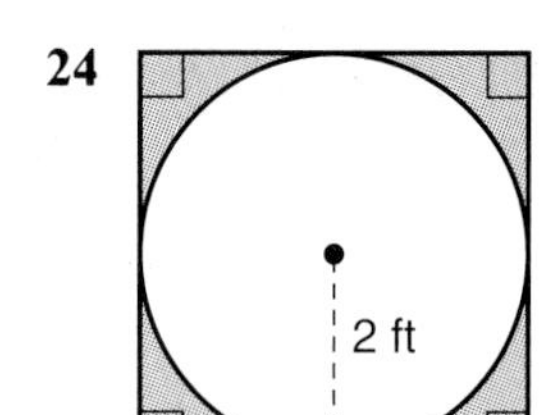

25. State a formula for the circumference of a semicircle having a radius r.

26. An arithmetic book written by R. C. Smith and published in 1858 states, "To obtain the area of a circle, multiply one-half the circumference by one-half the diameter." In formula form this would be $A = \left(\frac{C}{2}\right)\left(\frac{d}{2}\right)$. Use the substitution principle to substitute $2\pi r$ for C and $2r$ for d and show that this formula is the same as $A = \pi r^2$.

5–6 Volume of Some Geometric Figures

OBJECTIVE

Upon completing this section you should be able to:

1. Find the volume of some solid geometric figures.

In previous sections we have discussed perimeter and area of plane geometric figures. These topics were from the study of plane geometry. **Solid geometry** is the study of figures that have three dimensions. The space in which we live is three-dimensional space. We refer to the three dimensions as length, width, and height. In this section we will discuss some three-dimensional figures called solid geometric figures and give formulas for their volumes.

1 Volume

DEFINITION The **volume** of a solid geometric figure is the measure of the space enclosed by the figure.

Note three units are being multiplied together. For example (in.)(in.)(in.) = in.^{3}.

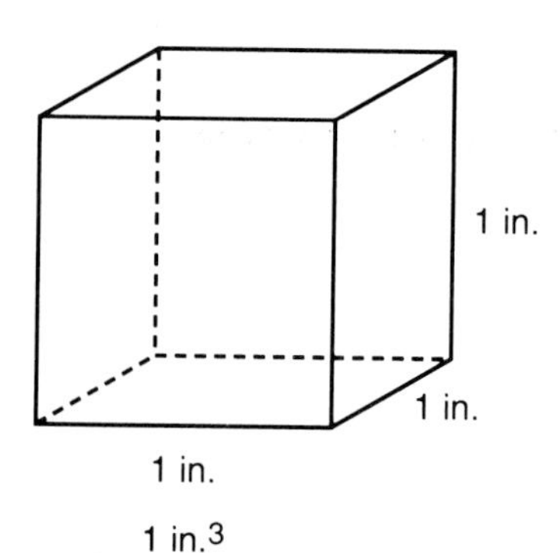

1 in.^{3}

We saw that perimeter is measured in linear units and area in square units. Volume is measured in **cubic units.** Five cubic inches can be abbreviated as 5 cu in. or 5 in.^{3}.

Recall that the area of a plane figure is the sum of the areas of squares that would completely cover the interior of the figure. The volume of a solid figure is the sum of the cubes that would completely fill the figure.

There are many types of solid geometric figures. In this section we will state the definitions and formulas for a few of the figures that are most commonly encountered in practical applications.

RECTANGULAR PARALLELEPIPED

A rectangular parallelepiped is a solid geometric figure, such as a box or a room, in which each face is a rectangle. The volume V of a rectangular parallelepiped with length ℓ, width w, and height h is given by the formula

$$V = \ell wh.$$

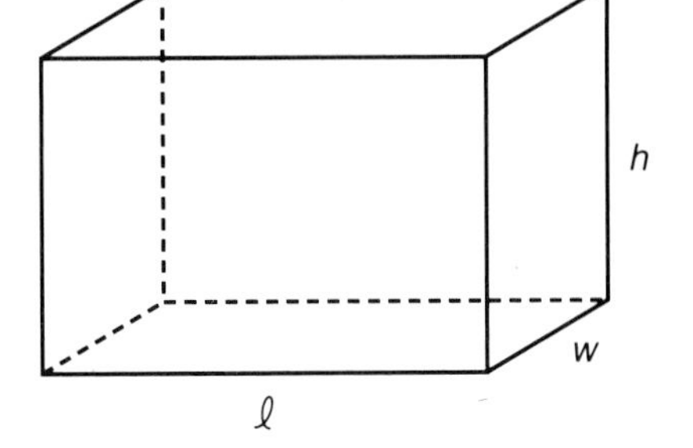

Example 1 Find the volume of a rectangular parallelepiped if $\ell = 4$ feet, $w = 2$ feet, and $h = 3$ feet.

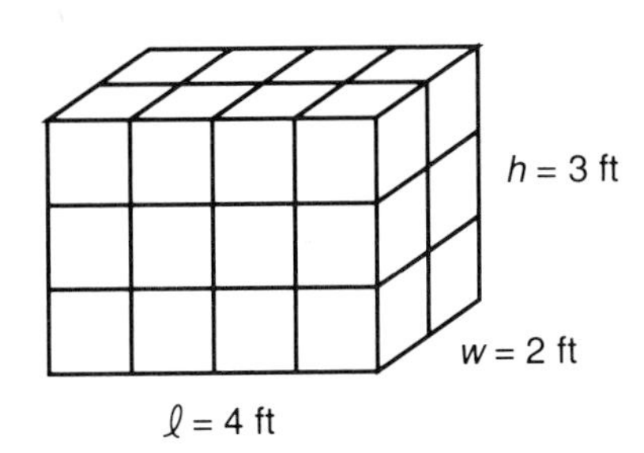

Note that we can find the volume of a figure only if it is three-dimensional. For example, we cannot find the volume of a rectangle since it is only two-dimensional.

Solution

$$\begin{aligned} V &= \ell wh \\ &= (4)(2)(3) \\ &= 24 \text{ ft}^3 \end{aligned}$$

Again, we use the principle of substitution.

⊙ SKILL CHECK 1 Find the volume of a rectangular parallelepiped if $\ell = 5$ feet, $w = 3$ feet, and $h = 4$ feet.

CUBE

A cube can be thought of as a rectangular parallelepiped in which the length, width, and height are equal. The volume of a cube with edge s is given by the formula

$$V = s^3.$$

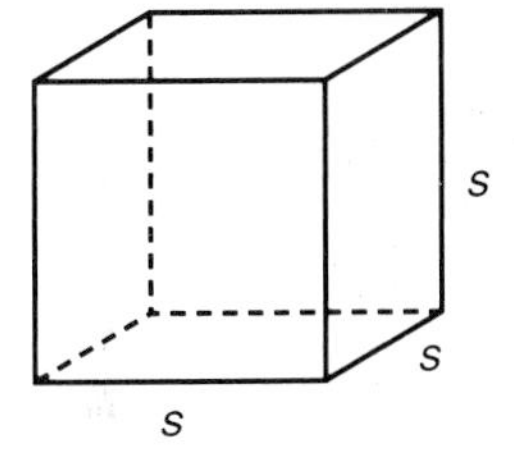

Example 2 Find the volume of a cube whose edge measures 3.2 inches. Round the answer to one decimal place.

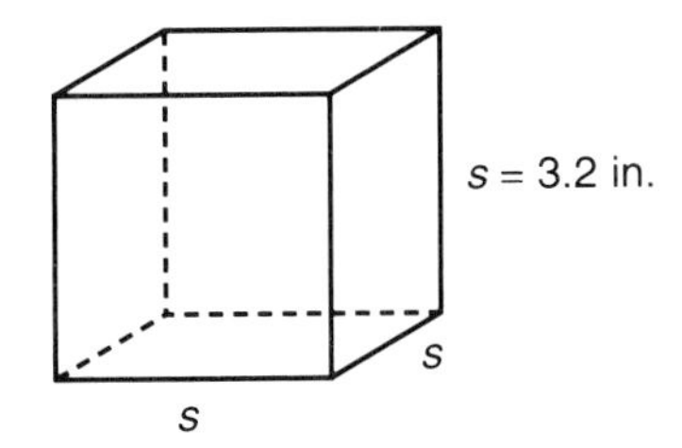

How many faces does a cube have?

Solution

$$\begin{aligned} V &= s^3 \\ &= (3.2)^3 \\ &= 32.8 \text{ in.}^3 \end{aligned}$$

⊙ SKILL CHECK 2 Find the volume of a cube whose edge measures 5.4 inches. Round the answer to one decimal place.

⊙ Skill Check Answers

1. 60 ft^3
2. 157.5 in.^3

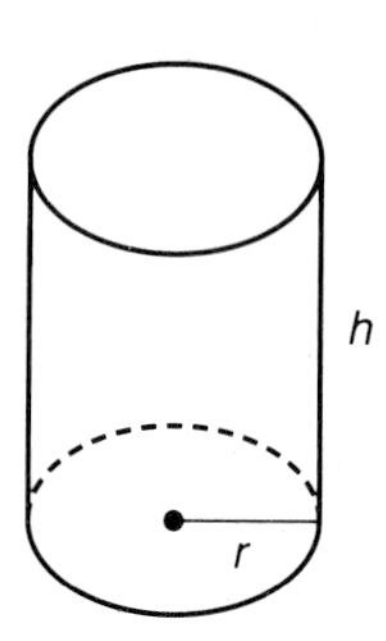

CIRCULAR CYLINDER A circular cylinder is a solid geometric figure having two congruent circles on parallel planes as bases. The volume of a circular cylinder with height h and a base with radius r is given by the formula

$$V = \pi r^2 h.$$

Example 3 Find the volume of a circular cylinder if the radius of the base measures 2 inches and the height measures 7 inches. Use $\pi = \frac{22}{7}$.

Determine the volume of a can of soup.

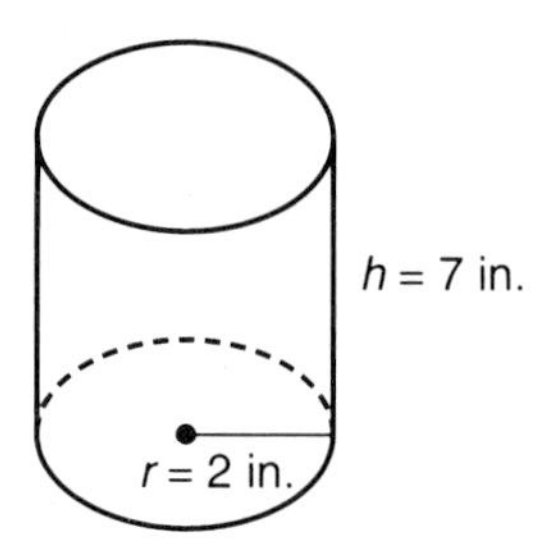

Solution

$$\begin{aligned} V &= \pi r^2 h \\ &= \left(\frac{22}{7}\right)(2)^2(7) = 88 \text{ in.}^3 \end{aligned}$$

⊙ SKILL CHECK 3 Find the volume of a circular cylinder if the radius of the base measures 3 inches and the height measures 14 inches. Use $\pi = \frac{22}{7}$.

At this point it may be noted that to find the volume of a rectangular parallelepiped, a cube, or a circular cylinder we use the same basic formula. This basic formula is $V = Bh$, where B represents the area of the base and h the altitude of the figure.

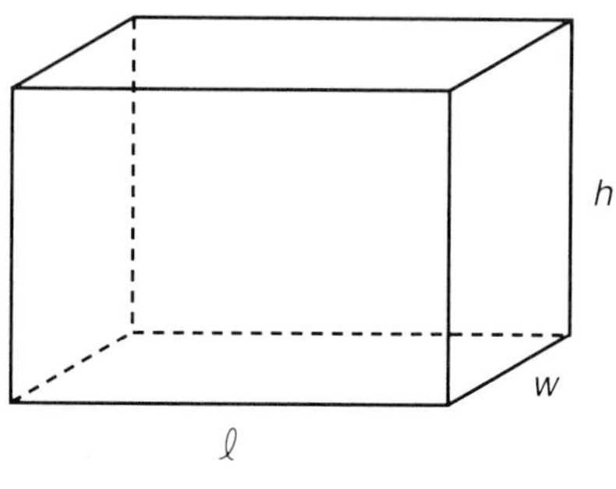

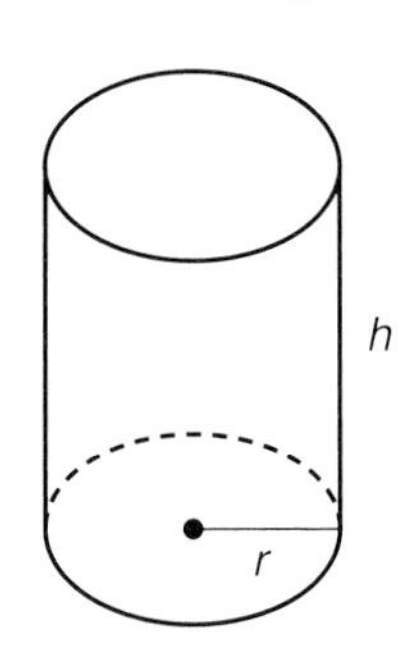

Base is a rectangle (ℓw).	Base is a square (s^2).	Base is a circle (πr^2).
$V = Bh$	$V = Bh$	$V = Bh$
$= (\ell w)h$	$= (s^2)s$	$= (\pi r^2)h$
$= \ell wh$	$= s^3$	$= \pi r^2 h$

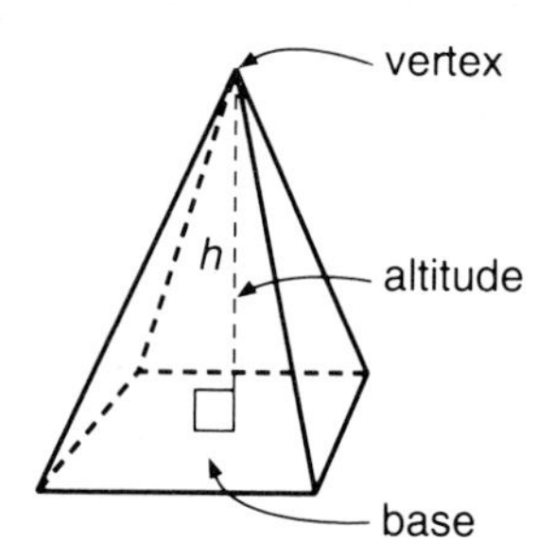

PYRAMID A pyramid is a solid geometric figure in which one face is a polygon (called the base) and the other faces are triangles having a common vertex.

The volume V of a pyramid is $V = \frac{1}{3}Bh$, where B is the area of the base and h is the altitude (the perpendicular distance from the vertex to the plane of the base).

⊙ Skill Check Answer

3. 396 in.^{3}

Example 4 Find the volume of a pyramid having a square base if one side of the base measures 3 inches and the altitude is 5 inches.

Solution Since the base is a square, its area $B = s^2 = (3)^2 = 9$ in.^{2}. Also $h = 5$ in. Thus

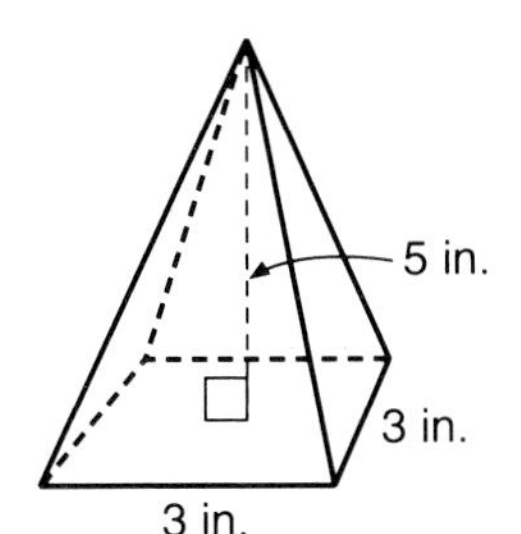

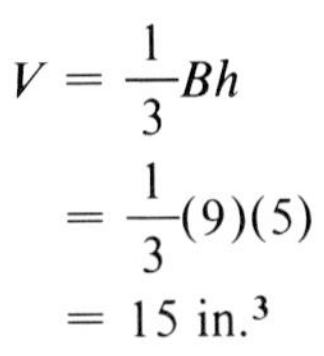

$$
\begin{aligned}
V &= \frac{1}{3}Bh \\
&= \frac{1}{3}(9)(5) \\
&= 15 \text{ in.}^3
\end{aligned}
$$

⦿ SKILL CHECK 4 Find the volume of a pyramid if its square base measures 5 inches on a side and the altitude is 12 inches.

CIRCULAR CONE If, instead of a polygon, the base is a circle and we draw line segments to every point on the circle from the vertex, the figure thus formed is a circular cone. The volume of a circular cone is also $V = \frac{1}{3}Bh$, but since the base is a circle

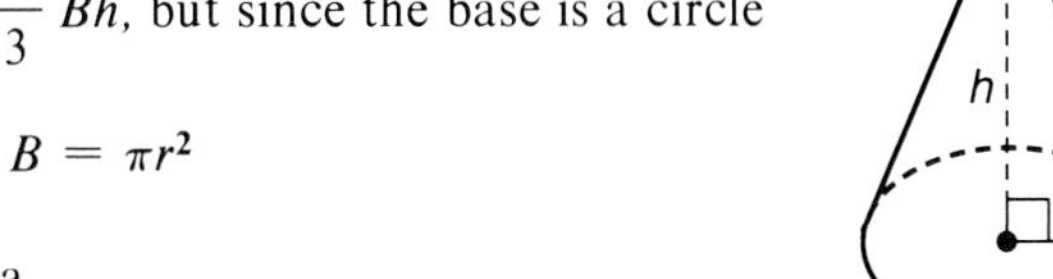

$$B = \pi r^2$$

and we have the formula

$$V = \frac{1}{3}\pi r^2 h.$$

Example 5 Find the volume of a circular cone when $r = 6$ centimeters and $h = 14$ centimeters. Use $\pi = \frac{22}{7}$.

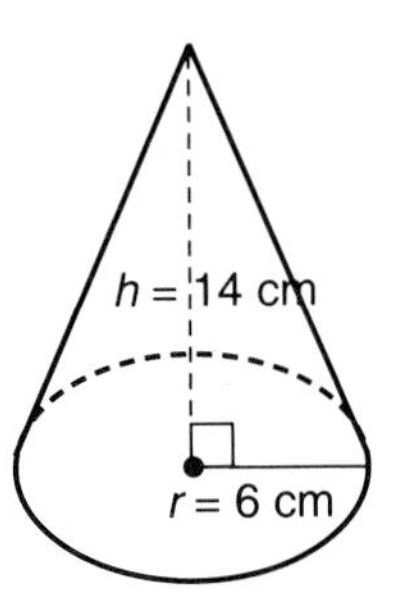

Solution

$$
\begin{aligned}
V &= \frac{1}{3}\pi r^2 h \\
&= \frac{1}{3}\left(\frac{22}{7}\right)(6^2)(14) \\
&= 528 \text{ cm}^3
\end{aligned}
$$

Perform the computation here in the margin.

⦿ SKILL CHECK 5 Find the volume of a circular cone when $r = 7$ cm and $h = 9$ cm. Use $\pi = \frac{22}{7}$.

⦿ *Skill Check Answers*

4. 100 in.^{3}
5. 462 cm^3

Note that the volume of a circular cone is one-third the volume of a circular cylinder having the same base and altitude.

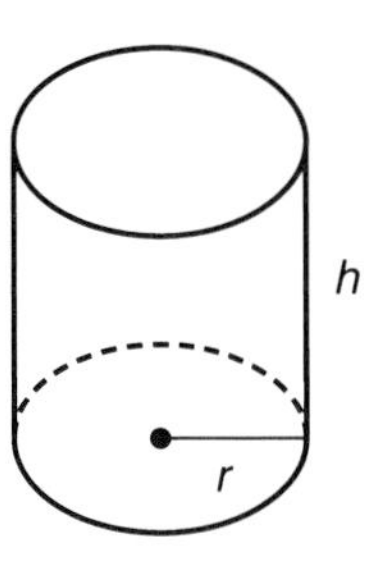

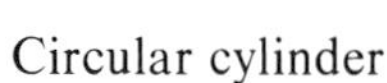
Circular cylinder

$$V = \pi r^2 h$$

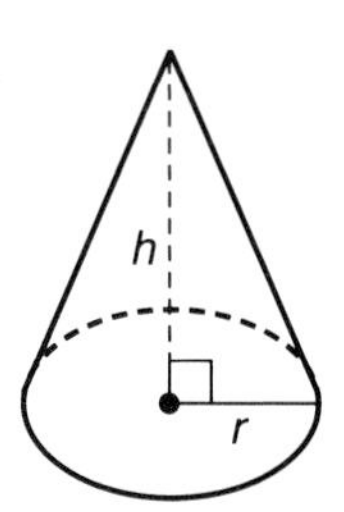

Circular cone

$$V = \frac{1}{3}\pi r^2 h$$

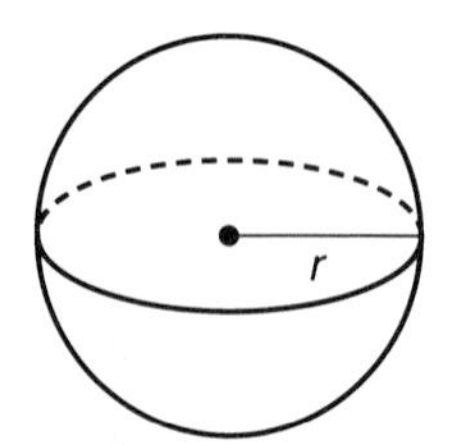

SPHERE A sphere is a solid geometric figure made up of all points in space that are at a given distance from a fixed point called the center of the sphere. The given distance from the center to the sphere is called the radius of the sphere. The volume of a sphere with radius r is given by the formula

$$V = \frac{4}{3}\pi r^3.$$

Example 6 Find the volume of a sphere if the radius measures 4 inches. Use $\pi = \frac{22}{7}$.

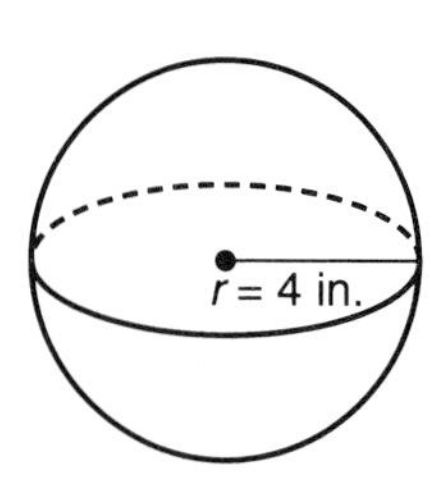

Solution

$$\begin{aligned} V &= \frac{4}{3}\pi r^3 \\ &= \left(\frac{4}{3}\right)\left(\frac{22}{7}\right)(4)^3 \\ &= 268\frac{4}{21} \text{ in.}^3 \end{aligned}$$

Perform the computation here in the margin.

⊙ **SKILL CHECK 6** Find the volume of a sphere if the radius measures $1\frac{1}{2}$ feet.

⊙ ***Skill Check Answer***

6. $14\frac{1}{7}$ ft^3

Exercise 5-6-1

A

1. Find the volume of a rectangular parallelepiped if $\ell = 7$ feet, $w = 5$ feet, and $h = 3$ feet.

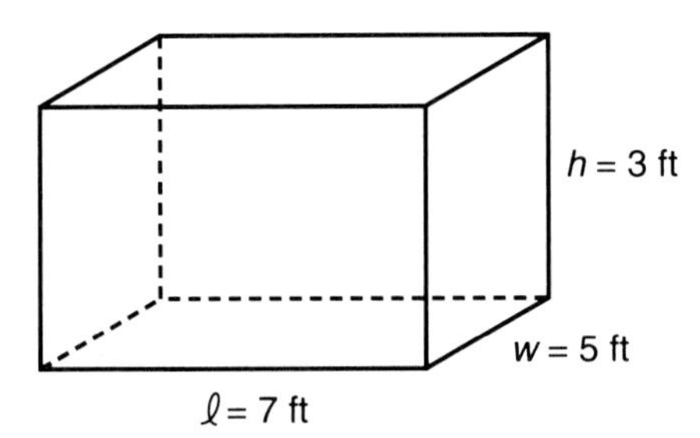

2. Find the volume of a rectangular parallelepiped if $\ell = 6.3$ centimeters, $w = 3.2$ centimeters, and $h = 4.5$ centimeters.

3. Find the volume of a rectangular parallelepiped if the edges are $5\frac{1}{2}$ feet by $6\frac{2}{3}$ feet by $2\frac{1}{4}$ feet.

4. Find the volume of a rectangular parallelepiped if the edges measure 15 meters by $7\frac{3}{4}$ meters by 6.8 meters. Give the answer as a mixed number.

5. Find the volume of a cube if $s = 3$ centimeters.

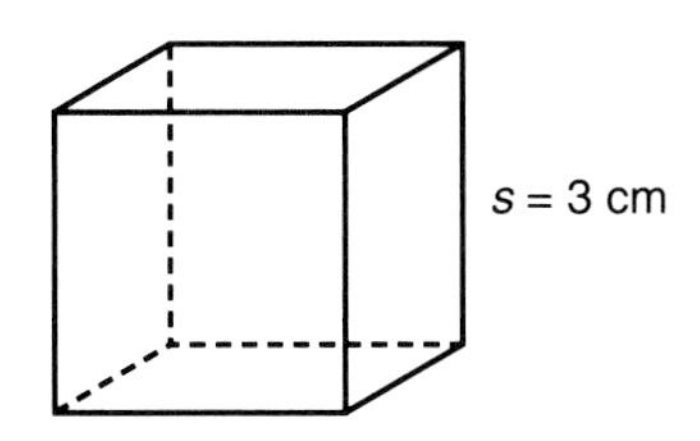

6. Find the volume of a cube if $s = 2.3$ inches.

7. Find the volume of a circular cylinder if $r = 3$ meters and $h = 14$ meters.

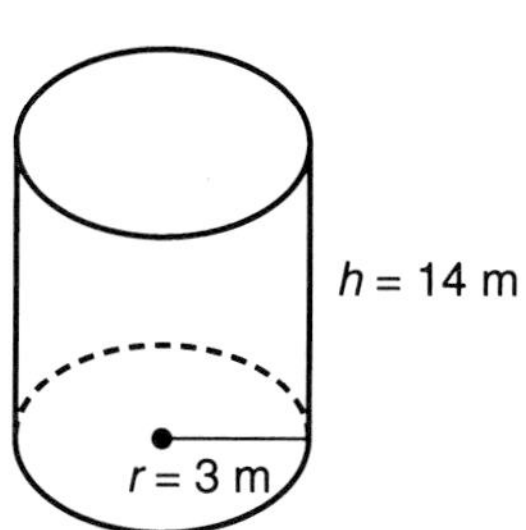

8. Find the volume of a circular cylinder if the radius of the base is $3\frac{1}{2}$ feet and the height is 5 feet.

9. Find the volume of a sphere whose radius is 6 centimeters.

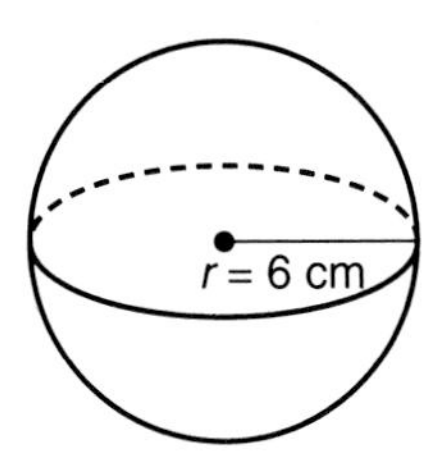

10. Find the volume of a sphere having a radius of $1\frac{1}{2}$ feet.

11. Find the volume of a cube having an edge of length 3.5 inches. Round the answer to one decimal place.

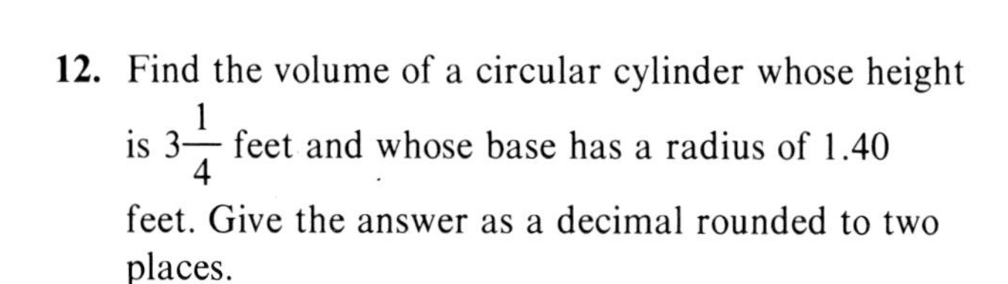

12. Find the volume of a circular cylinder whose height is $3\frac{1}{4}$ feet and whose base has a radius of 1.40 feet. Give the answer as a decimal rounded to two places.

13. Find the volume of a pyramid if the base is a parallelogram having an area of 27 square inches and the altitude of the pyramid is 10 inches.

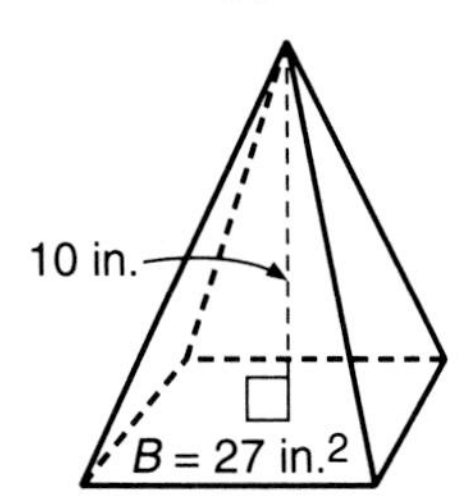

14. Find the volume of a pyramid if the base is a right triangle whose legs measure 3 feet and 6 feet and the altitude of the pyramid is 5 feet.

15. Find the volume of a circular cone when $\pi = \frac{22}{7}$, $r = 3$ inches, and $h = 7$ inches.

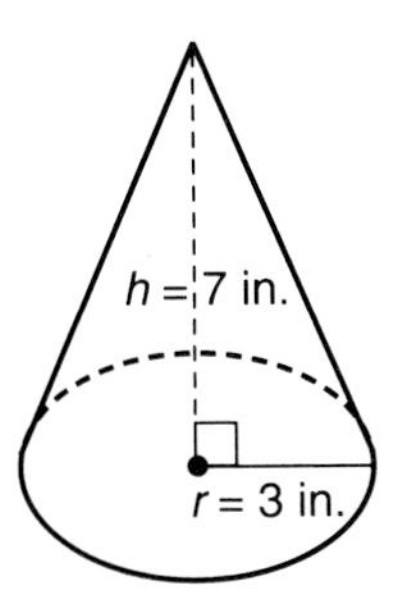

16. Find the volume of a circular cone whose base has a radius of 3.5 centimeters and the altitude of the cone is 6.0 centimeters. Use $\pi = \frac{22}{7}$.

17. Find the volume of a cube in which each edge measures 15 centimeters.

18. Calculate the volume of a rectangular room 6.2 meters long, 4.1 meters wide, and 2.8 meters high. Round the answer to one decimal place.

19. Find the volume of a sphere whose radius is 3.5 inches. Use $\pi = 3.14$ and round the answer to one decimal place.

20. Find the volume of a sphere whose radius is 2.1 feet. Use $\pi = \frac{22}{7}$. Round the answer to one decimal place.

21. A rectangular room 15 feet by 20 feet has a ceiling 10 feet high. What is the volume of the room?

22. How many cubic meters of concrete are needed to fill a rectangular hole 0.4 meters deep, 2.1 meters wide, and 4.7 meters long? Round the answer to the nearest tenth of a cubic meter.

• Concept Enrichment

23. Find the combined volume of the circular cone and half a sphere (hemisphere). Use $\pi = \frac{22}{7}$.

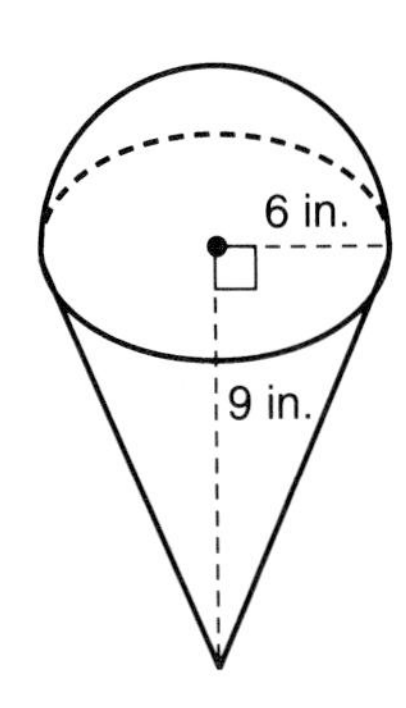

24. Find the combined volume of the rectangular parallelepiped and pyramid.

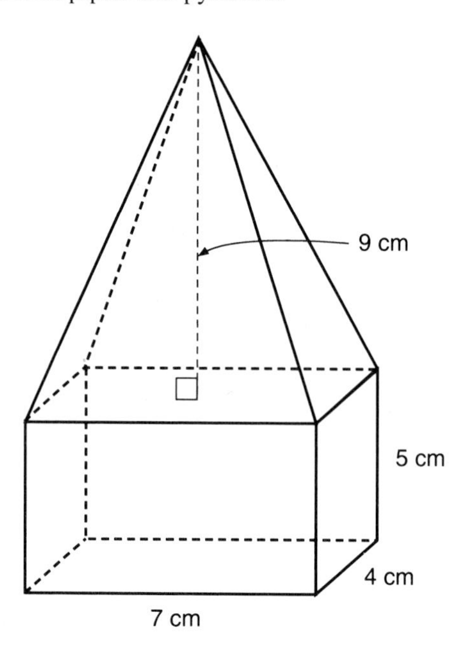

5–7 Applied Problems Involving Geometric Formulas

OBJECTIVE

Upon completing this section you should be able to:

1. Use geometric formulas to solve some practical problems.

Many problems encountered in practical situations require the use of geometric formulas in their solutions. In this section we will give examples and exercises in which geometric formulas together with facts you have learned in previous chapters are used to solve such problems.

1 Applications

Example 1 Mrs. Smith wishes to carpet her bedroom. She found the carpet she wishes to buy and was told, "We will furnish the carpet, padding, and installation for \$26.50 per square yard." If her rectangular-shaped bedroom measures 4 yards by 5 yards, how much will the job cost?

Solution We must first find the number of square yards in the bedroom.

Don't forget to draw a sketch.

4 yd

5 yd

Using the formula for the area of a rectangle, we have

$$\begin{aligned} A &= \ell w \\ &= (5)(4) \\ &= 20 \text{ yd}^2 \end{aligned}$$

Next we must find the cost of 20 square yards at \$26.50 per square yard. To do this we multiply 20 by \$26.50.

$$(20)(26.50) = 530$$

So the cost is \$530.00

SKILL CHECK 1 How much would it cost Mrs. Smith to install the same carpet in her rectangular living room if it measures 5 yards by 7 yards?

Example 2 Mike had an aluminum gas tank with rectangular faces custom made for his boat. It was $2\frac{1}{2}$ feet long, $1\frac{1}{2}$ feet wide, and $\frac{2}{3}$ foot high. If the capacity of one cubic foot is about $7\frac{1}{2}$ gallons, how many gallons will his tank hold?

Since the measurements are given in fraction form, we will give the answer using the same form.

Solution We must first find the volume of the tank in cubic feet. We will use the formula for the volume of a rectangular parallelepiped.

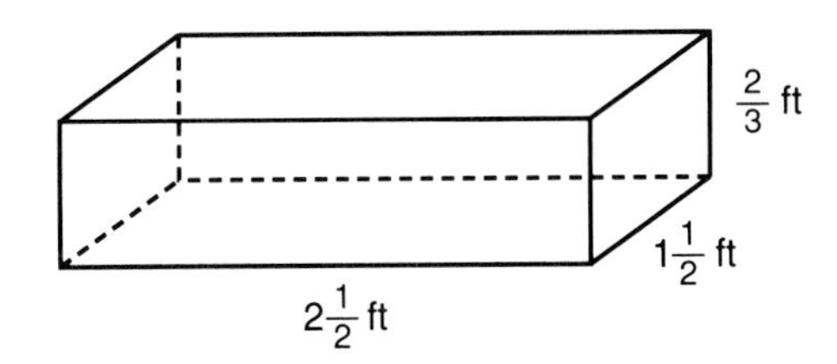

Again, first draw a sketch.

$$\begin{aligned} V &= \ell wh \\ &= \left(2\frac{1}{2}\right)\left(1\frac{1}{2}\right)\left(\frac{2}{3}\right) \\ &= 2\frac{1}{2}\ \text{ft}^3 \end{aligned}$$

$$\left(2\frac{1}{2}\right)\left(1\frac{1}{2}\right)\left(\frac{2}{3}\right) = \left(\frac{5}{2}\right)\left(\frac{3}{2}\right)\left(\frac{2}{3}\right) = 2\frac{1}{2}$$

Since each cubic foot contains about $7\frac{1}{2}$ gallons, we will multiply $2\frac{1}{2}$ ft³ by $7\frac{1}{2}$.

$$\left(7\frac{1}{2}\right)\left(2\frac{1}{2}\right) = 18\frac{3}{4}$$

$$\left(\frac{15}{2}\right)\left(\frac{5}{2}\right) = \frac{75}{4} = 18\frac{3}{4}$$

The gas tank will thus hold about $18\frac{3}{4}$ gallons.

SKILL CHECK 2 If the price of gasoline is \$1.29 per gallon, how much will it cost Mike to fill the tank? Round the answer to the nearest cent.

Example 3 A circular rose garden has a radius of 8 feet. A walkway around the garden is 3 feet wide. Find the area of the walkway. Round the answer to two decimal places.

Skill Check Answers

1. \$927.50
2. \$24.19

Solution

What is the radius of the larger circle?

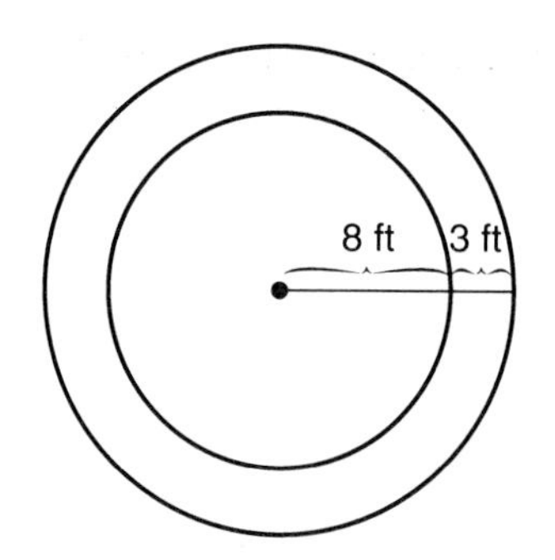

From the figure we see that we are asked to find the difference of the areas of the larger circle and the smaller circle. The radius of the smaller circle is 8 feet. When we add the width of the walkway we obtain the radius of the larger circle, which is 11 feet.

We use $\pi = 3.14$.

Area of smaller circle	Area of larger circle
$A = \pi r^2$	$A = \pi r^2$
$= (3.14)(8)^2$	$= (3.14)(11)^2$
$= 200.96 \text{ ft}^2$	$= 379.94 \text{ ft}^2$

Subtracting the smaller area from the larger, we have

$$379.94 - 200.96 = 178.98.$$

Thus the area of the walkway is 178.98 ft².

⊙ ***SKILL CHECK 3*** Suppose in example 3 the walkway was 4 feet wide and the radius of the garden was reduced to 7 feet. What is the area of the walkway?

⊙ ***Skill Check Answer***

3. 226.08 ft²

Exercise 5–7–1

1

1. Mr. Jones wishes to cover his dining room floor with a type of floor covering that is sold in squares one foot by one foot. He has measured the floor and found it to be 15 feet long and 12 feet wide. How many pieces of floor covering will he need?

2. Suppose Mr. Jones (exercise 1) has chosen a covering that sells for $2.45 per piece. What is the cost of the materials needed?

3. A farmer wishes to enclose a rectangular piece of land for a small pasture. He measured the land and found it to be 680 feet long and 425 feet wide. How many feet of fence will he need?

4. The cost of the fence in exercise 3 is $5.50 per foot. What will it cost for the materials to fence the field?

5. A rectangular tablecloth measures 60 inches by 72 inches. What is the cost for the material to edge the tablecloth if the edging costs 5 cents per inch?

6. A woman wishes to put a border around her small rectangular vegetable garden that measures 5 feet by 20 feet. The price of the border is 59 cents per foot. How much will it cost to buy enough material to border the garden?

7. A farmer wishes to enclose a square field with an electric fence. If the field measures 220 meters on a side, and if the wire for the fence costs 75 cents per meter, what will be the cost of the wire for the electric fence?

8. A rectangular piece of land is 186 meters by 250 meters. If the price of fencing is 6.50 per meter including labor and materials, what is the cost of enclosing the land with fencing?

9. A square tile 30.5 cm on each side costs 28 cents. What is the cost to tile a rectangular floor 488 cm by 732 cm?

10. If carpet costs $12.98 per square meter, what is the cost of carpeting a rectangular room 5.5 m by 6.5 m? Round answer to nearest cent.

11. A fenced enclosure in the form of a triangle has sides of 120 feet, 135 feet, and 98 feet. If the cost of fencing is $4.50 per foot, what is the total cost of fencing the enclosure?

12. A rectangular lawn measures 200 feet by 90 feet. A bag of "weed and feed" costs $5.00 and covers 1,500 square feet of lawn. How much will it cost to "weed and feed" the entire lawn?

13. The price of a certain type of carpet is $19.95 per square meter. How much would it cost to carpet a hallway 1.6 m wide and 8.5 m long?

14. A parking lot is 46.2 m long and 18 m wide. Find the cost of fencing the lot if the price of fencing is $12.95 per meter.

- A **circular cone** is formed by using a circle as its base and from a point in space drawing line segments to each point on the circle.
- A **sphere** is a solid geometric figure formed by all points in space that are at a given distance from a fixed point.

Procedures

Section 5–1

- To find the complement of a given angle subtract the measure of the given angle from 90°.
- To find the supplement of a given angle subtract the measure of the given angle from 180°.
- Linear measurement is performed using a ruler.
- Angular measurement is performed using a protractor.

Section 5–2

- The number of diagonals d in a polygon having n sides can be found by using the formula $d = \dfrac{n(n-3)}{2}$.
- Some polygons are given special names according to the number of sides they have.

Section 5–3

- To find the perimeter of a quadrilateral find the sum of the measures of its four sides.
- The perimeter P of a parallelogram having sides ℓ and w is found by using the formula

$$P = 2\ell + 2w.$$

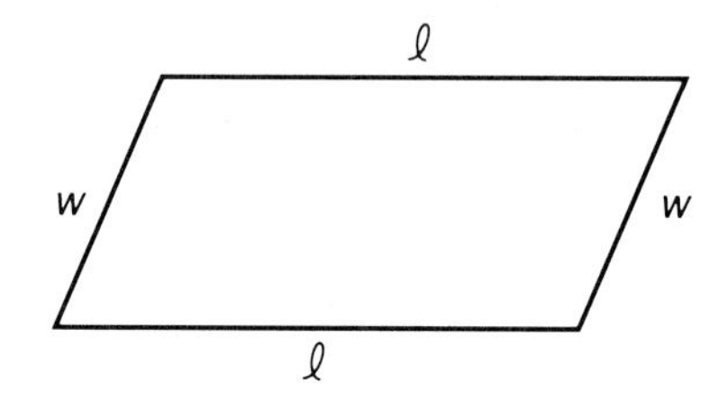

- The area A of a parallelogram having base b and altitude h is found by using the formula

$$A = bh.$$

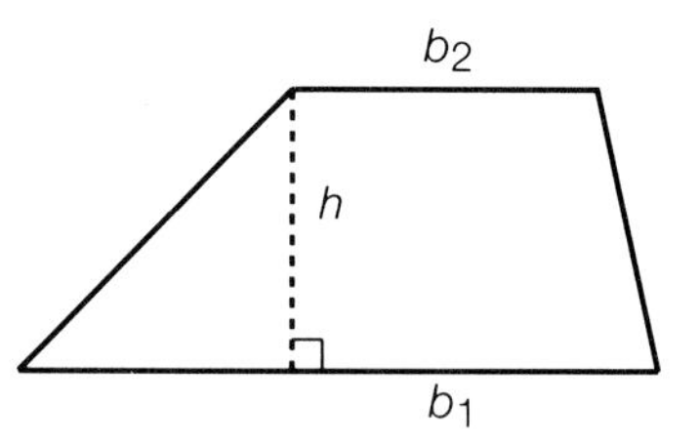

- The area of a trapezoid having altitude h and bases b_1 and b_2 is found by using the formula

$$A = \frac{1}{2}h(b_1 + b_2).$$

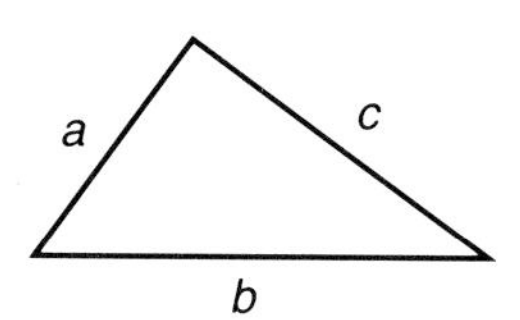

Section 5–4

- The sum S of the measures of the interior angles of a polygon having n sides is found by using the formula $S = (n - 2)180°$.
- The perimeter P of a triangle with sides measuring a, b, and c is found by using the formula

$$P = a + b + c.$$

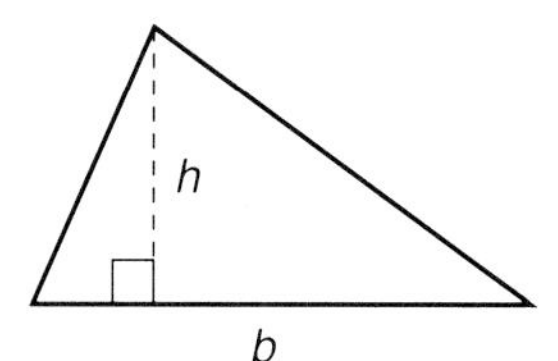

- The area of a triangle with base b and altitude h is found by using the formula

$$A = \frac{1}{2}bh.$$

Section 5–5

- The measure of a central angle is the same as the amount of rotation in its intercepted arc.
- The measure of an inscribed angle is equal to one-half the rotation in its intercepted arc.
- The circumference of a circle having diameter d is found by using the formula

$$C = \pi d.$$

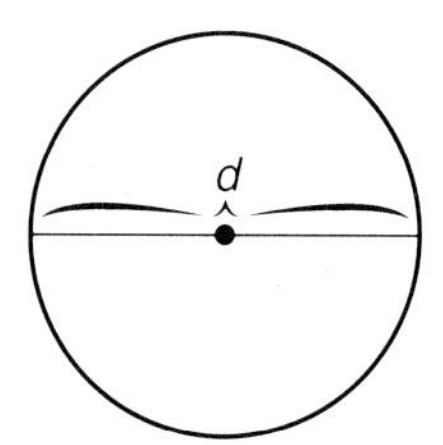

- The area of a circle having radius r is found by using the formula

$$A = \pi r^2.$$

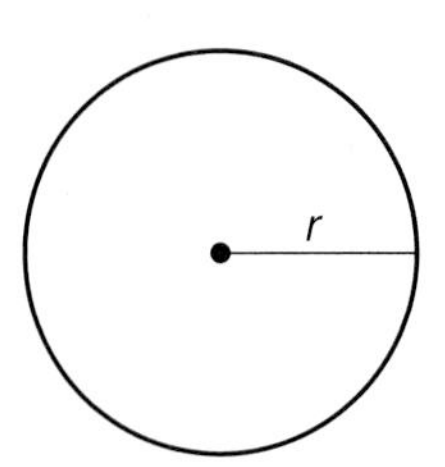

Section 5–6

- The volume of a rectangular parallelepiped with length ℓ, width w, and altitude h is found by using the formula

$$V = \ell wh.$$

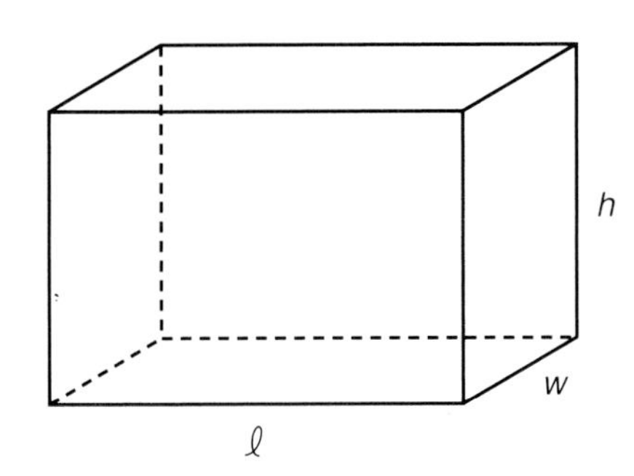

- The volume of a cube with edge s is found by using the formula

$$V = s^3.$$

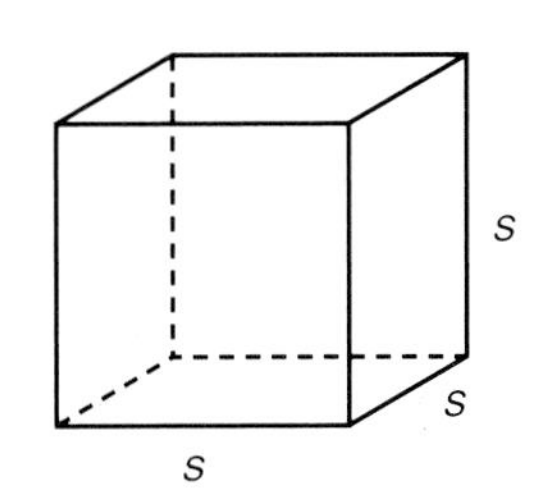

- The volume of a circular cylinder with height h and having a base with radius r is found by using the formula

$$V = \pi r^2 h.$$

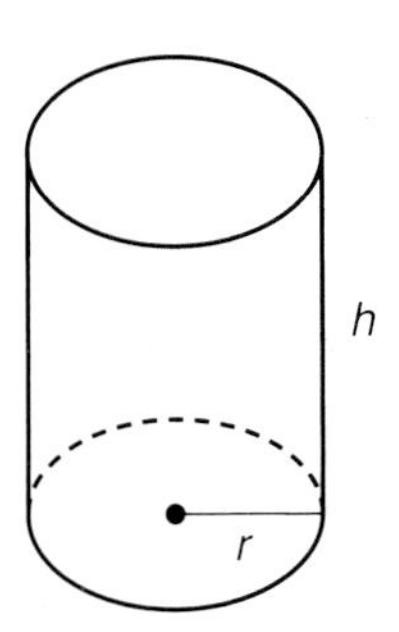

- The common formula for the volume of a rectangular parallelepiped, a cube, or a circular cylinder is $V = Bh$ where B is the area of the base and h is the altitude.
- The volume of a pyramid having an altitude h and a base of area B is found by using the formula

$$V = \frac{1}{3}Bh.$$

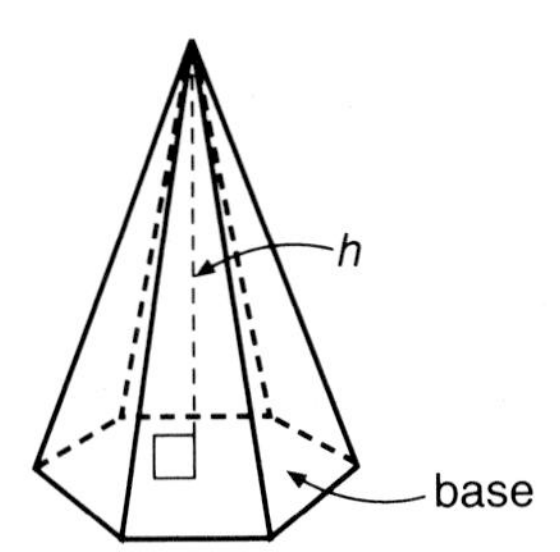

- The volume of a circular cone is also found by using the formula $V = \frac{1}{3}Bh$, where B is the area of the circular base and h is the altitude of the cone.

$$V = \frac{1}{3}Bh = \frac{1}{3}\pi r^2 h.$$

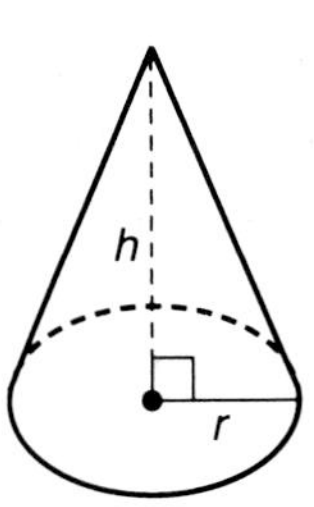

- The volume of a sphere with radius r is found by using the formula

$$V = \frac{4}{3}\pi r^3.$$

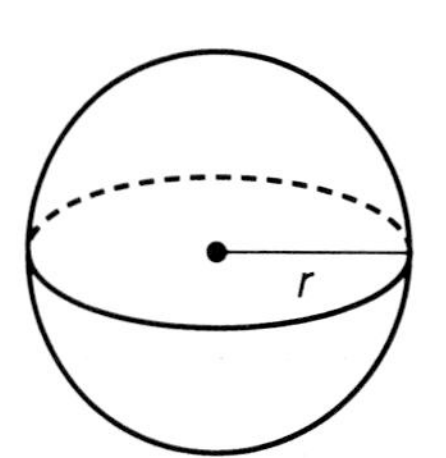

CHAPTER 5 REVIEW

1. Find the measure of an angle whose complement measures 24°.

2. Find the measure of an angle whose complement measures 55°.

3. Find the measure of an angle whose supplement measures 82°.

4. Find the measure of an angle whose supplement measures 39°.

Use the following diagram of two parallel lines intersected by a transversal to find the measures of the indicated angles.

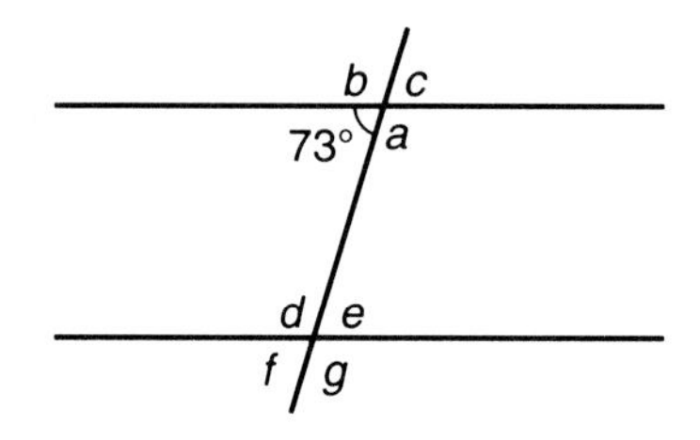

5. $\angle c$

6. $\angle b$

7. $\angle a$

8. $\angle d$

9. $\angle e$

10. $\angle g$

11. Draw a sketch of a pentagon.

12. Draw a sketch of a hexagon.

13. How many sides does an octagon have?

14. How many sides does a decagon have?

15. How many sides does a quadrilateral have?

16. How many sides does a dodecagon have?

17. How many diagonals does a 9-sided polygon have?

18. How many diagonals does a 7-sided polygon have?

19. How many diagonals does an octagon have?

20. How many diagonals does a decagon have?

Draw a sketch to help solve each of the following.

21. A rectangular garden is 30 feet long and 20 feet wide. What is the area of the garden?

22. If the garden in exercise 21 is to be enclosed by a fence, how many feet of fencing are needed?

23. A rectangular room measures 18 feet by 14 feet. How many feet of baseboard molding are needed to go around the room?

24. What is the area of the floor of the room in exercise 23?

25. In right $\triangle ABC$ if $\angle A = 50°$ and $\angle C = 90°$, find the measure of $\angle B$.

26. In right $\triangle ABC$ if $\angle B = 63°$ and $\angle C = 90°$, find the measure of $\angle A$.

27. In isosceles $\triangle ABC$, $\angle C = 120°$. Find the measures of the other two angles.

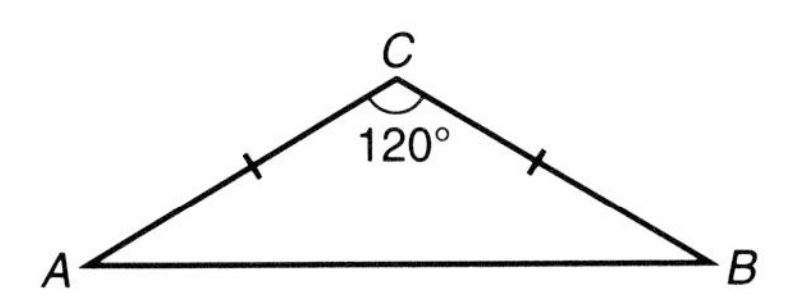

28. In isosceles $\triangle ABC$, $\angle B = 44°$. Find the measures of the other two angles.

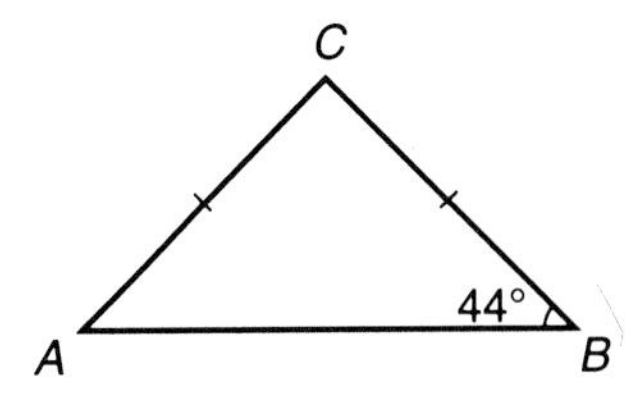

29. A circular fish pond has a diameter of 12 feet. What is the circumference of the pool? Use $\pi = 3.14$.

30. What is the area of the pond in exercise 29?

31. Find the volume of a cube having a side that measures 4 inches.

32. Find the volume of a rectangular solid whose dimensions are $2\frac{1}{2}$ feet by $5\frac{1}{4}$ feet by $3\frac{1}{3}$ feet.

33. A sail on a boat is in the form of a triangle with a base of 10 feet and a height of 17 feet. What is the area of the sail?

34. A triangular platform has one side that measures 20 feet. The altitude to the opposite vertex is 15 feet. Find the area of the platform.

35. Find the volume of a circular cylinder having a height of 14 centimeters and whose base has a radius of 5 centimeters.

36. Find the volume of a sphere whose diameter is 7 centimeters. Give the answer in fraction form.

37. A square-shaped garden is to be fenced. The length of one side is 16 feet. How many feet of fencing are needed?

38. The perimeter of a triangular-shaped planter is 40 feet. If two sides measure 10 feet and 16 feet, respectively, what is the length of the third side?

39. A circular landing pad for a helicopter has a radius of 65 feet. What is the area of the pad? Use $\pi = 3.14$.

40. How much fencing is needed to enclose the pad in exercise 39?

41. Find the volume of a right circular cone having an altitude of 7 centimeters and a base with a radius of 3 centimeters. Use $\pi = \frac{22}{7}$.

42. Find the volume of a circular-cylinder-shaped water tank having a height of 21 feet and a base with a diameter of 18 feet.

43. A side of a square-shaped flower garden measures 24 feet. The garden is surrounded by a 3-feet-wide wooden walkway. Find the area of the walkway.

44. Inside the garden in exercise 43 is a circular fountain having a diameter of 10 feet. Find the area of the garden excluding the fountain. Use $\pi = 3.14$.

45. A rectangular-shaped swimming pool measuring 15 feet by 30 feet has a one-foot-wide tile border around it. Find the area of the border.

46. What is the cost of the border in exercise 45 if the price of a one-foot-square tile is $1.15?

47. A rectangular floor measures 18 feet by 15 feet. How much will it cost to carpet the floor if the price is $22.50 per square yard? (One square yard = 9 square feet.)

48. The level of a rectangular building lot 120 meters by 240 meters must be raised by a height of 0.4 meter to meet flood insurance requirements for building a house. How many cubic meters of dirt are needed to accomplish this?

CHAPTER 5 TEST

1. Find the measure of an angle whose supplement measures 36°.

2. Find the measure of $\angle a$.

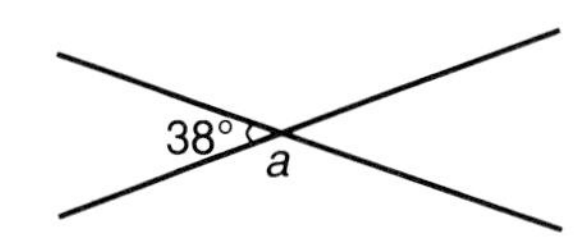

3. How many sides does a decagon have?

4. How many diagonals does an 8-sided polygon have?

5. Find the perimeter of a parallelogram whose sides measure 3.2 meters and 7.4 meters.

6. Find the area of a trapezoid having bases of $7\frac{1}{2}$ feet and $3\frac{1}{2}$ feet and whose altitude is 6 feet.

7. In isosceles $\triangle ABC$, $\angle C = 68°$. Find the measures of the other two angles.

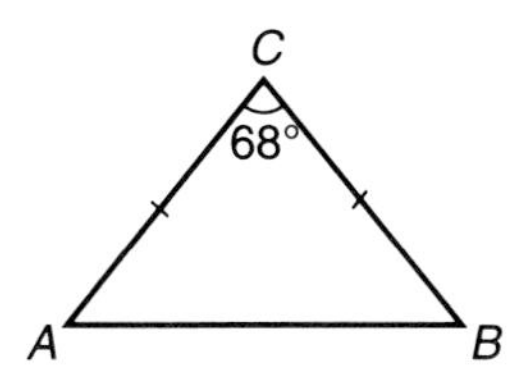

8. Find the area of a triangle whose base is 14 centimeters and whose altitude is 8 centimeters.

9. Find the circumference of a circle having a diameter of $2\frac{1}{2}$ feet.

10. Find the area of a circle whose diameter is 20 inches. Use $\pi = 3.14$.

11. Find the volume of a circular cylinder having a height of $3\frac{1}{2}$ centimeters and whose base has a radius of $2\frac{1}{2}$ centimeters. Give the answer as a mixed number.

12. Find the volume of a sphere whose diameter is 3 meters. Use $\pi = \frac{22}{7}$.

13. If a certain carpet is priced at $27.50 per square yard installed, how much will it cost to carpet a room 5 yards wide by 7 yards long?

14. A circular fountain has a radius of 10 feet. There is a 4-feet wide walkway around the outside border of the fountain. Find the area of the walkway. Use $\pi = 3.14$ and round the answer to two decimal places.

CHAPTERS 1–5 CUMULATIVE TEST

1. Add: $196 + 2{,}508 + 791 + 8$

2. Evaluate $2x^2 + xz - 3y$ if $x = 3$, $y = 5$, and $z = 4$.

3. Reduce: $\dfrac{6x^2y}{10xy^3}$

4. Round 18.93451 to the nearest thousandth.

5. Multiply: $\begin{array}{r} 5.64 \\ \times\ 3.5 \\ \hline \end{array}$

6. Simplify: $3xy + 2x^2 - 2xy + 5x^2$

7. Simplify: $5a^2 + 6a + b - 3a^2 - 4a - b$

8. Find the least common multiple of 6, 14, and 21.

9. Write 0.34 as a common fraction in reduced form.

10. Add: $2\frac{3}{8} + 5\frac{1}{6}$

11. Find the measure of an angle if its supplement measures 78°.

12. Find the prime factorization of 297.

13. Find the volume of a cube having an edge that measures 5 centimeters.

14. Multiply: $\left(\frac{3}{5}\right)\left(\frac{20}{27}\right)$

15. Divide: $41\overline{)50{,}416}$

16. Find the area of a triangle if the measure of the base is 6 inches and the altitude is 5 inches.

17. Subtract: $500 - 392$

18. Subtract: $\frac{5}{9} - \frac{4}{15}$

19. How many terms are in the expression $x^3 + 3x^2 - 4x + 1$?

20. Multiply: $\begin{array}{r} 362 \\ \underline{\times\ 47} \end{array}$

21. Divide: $1.4\overline{)45.64}$

22. Find the area of a circle if its diameter measures 7 feet. Use $\pi = \frac{22}{7}$.

23. Divide: $\frac{2}{3} \div \frac{9}{4}$

24. A rectangular garden is 15.0 feet long and 9.5 feet wide. What is the area of the garden?

25. If the price of bananas is $0.28 per pound, what is the cost of 4.5 pounds?

C H A P T E R

6 Pretest

Answer as many of the following questions as you can before starting this chapter. When you finish the chapter, take the test at the end and compare your scores to see how much you have learned.

1. Find the opposite of -5.9.

2. Find the opposite of $\frac{3}{7}$.

3. Simplify: $-|-13|$

4. Use the number line to show the addition of $(-6) + (+9)$.

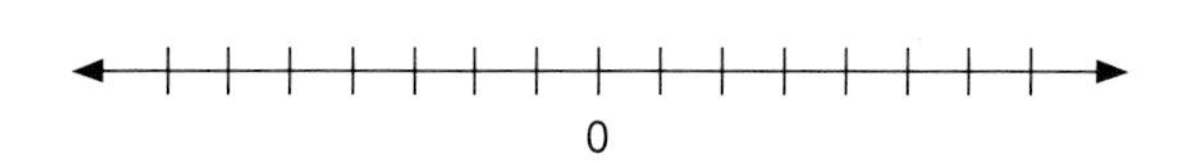

5. Add: $(+15) + (-28)$

6. Add: $(-6) + (-16)$

7. Subtract: $(-9) - (+20)$

8. Subtract: $(-5) - (-19)$

9. Simplify: $-8 - 2 + 13 - (-5) + 1$

10. Remove parentheses and simplify: $16 - (12 - 31) - 6$

11. Combine like terms: $9x^2y + 4x^2y - 6x^2y$

12. Simplify: $6 - [3x - (7 + 5x)]$

13. Subtract the difference of 11 and 7 from 12.

14. Find the product: $(+5)(-14)$

15. Find the product: $(-6)(-8)$

16. Find the product: $(-2)(-3)(6)(-1)$

17. Find the quotient: $\dfrac{-36}{-9}$

18. Find the quotient: $\dfrac{2}{3} \div \left(-\dfrac{4}{9}\right)$

19. Evaluate: $\dfrac{(-14)(-6)}{-21}$

20. Evaluate: $\dfrac{-6 - 15}{-7}$

CHAPTER

6 Operations on Signed Numbers

Negative numbers were so feared and misunderstood that they were not readily accepted in mathematics until the sixteenth century.

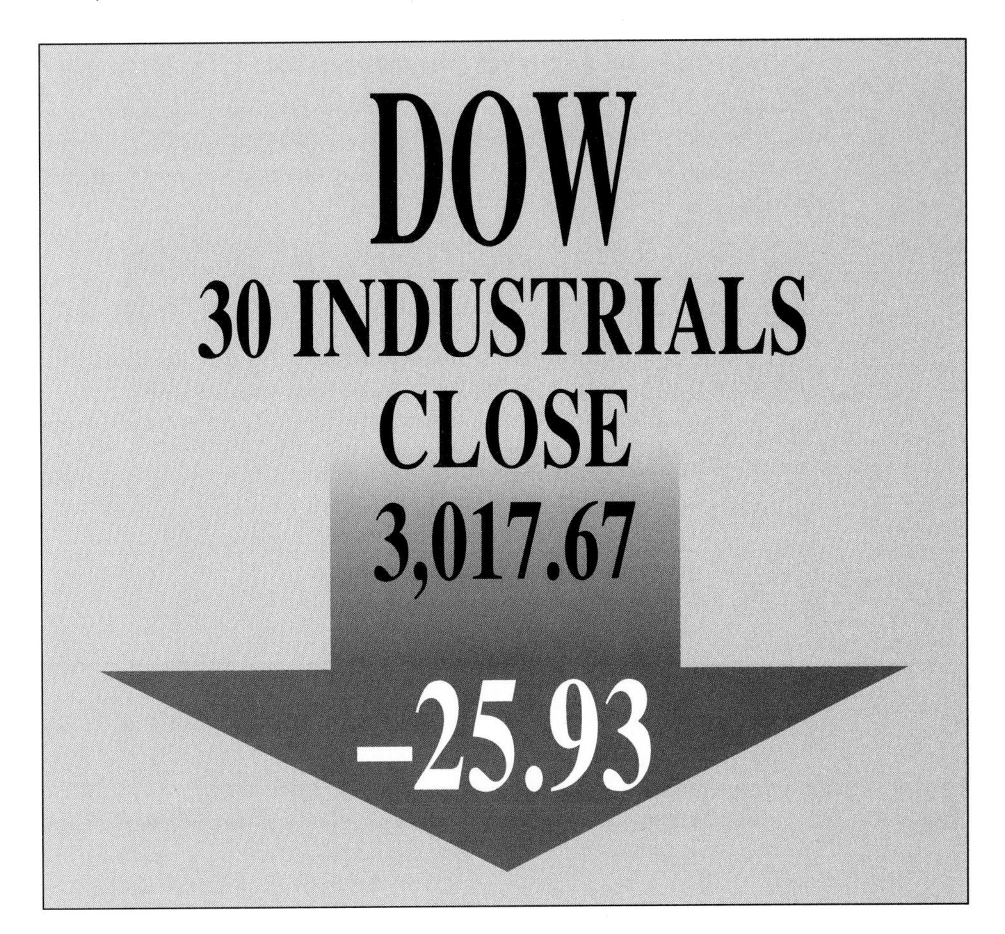

STUDY HELP

Effective review for a test is a problem for some students. The chapter review, chapter test, and cumulative test provide an opportunity for you to check your understanding of concepts. The keyed answers to these tests direct you to the proper sections for further study.

In previous chapters we dealt with the operations on a set of numbers called the **numbers of arithmetic** and solved problems involving these numbers. In this chapter we will introduce **signed numbers** and establish the four basic operations on this set.

6–1 Signed Numbers

OBJECTIVES

Upon completing this section you should be able to:

1. Locate signed numbers on the number line.
2. Use the symbol for inequality to compare two signed numbers.
3. Find the opposite of a given number.
4. Find the absolute value of a number.

In chapter 1 we introduced the number line and used it to discuss the order of the whole numbers. We also used the inequality symbols to indicate that one number was less than or greater than another. Our discussion included numbers that were greater than or equal to zero. On the number line these numbers included zero and numbers to the right of zero.

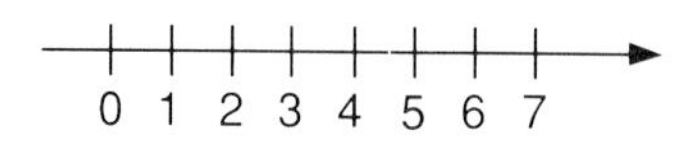

1 Locating Signed Numbers on the Number Line

We now wish to extend our thinking to include numbers to the left of zero. Such numbers are called **negative numbers** and are written with a $-$ sign preceding them. Numbers such as -1, -3, and -5 are read "negative one," "negative three," and "negative five."

The arrows indicate that the line continues indefinitely in either direction.

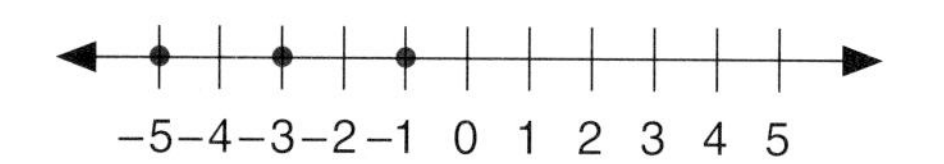

The whole numbers that we have studied up to this point and the negative whole numbers together make up the set of numbers called **integers.**

The set of numbers that can be expressed as the quotient of two integers is called the set of **rational numbers.** Fractions, such as $\frac{1}{2}$, $-\frac{2}{3}$, $\frac{8}{3}$, $-\frac{5}{2}$, and so on, or their decimal forms will fall proportionally on the number line.

-3.75 $-\frac{5}{2}$ $-\frac{2}{3}$ 0.5 $\frac{8}{3}$ 4.5

-5 -4 -3 -2 -1 0 1 2 3 4 5

It should be noted that the integers are also rational numbers since, for instance, $+5$ could be written $\frac{5}{1}$, $\frac{10}{2}$, and so on.

The numbers to the right of zero are called **positive numbers** and are written with a $+$ sign. If a nonzero number is written without a sign, it is assumed to be a positive number. Hence $+7$ or simply 7 would represent positive seven. Also note that "to the right" is a positive direction and "to the left" is a negative direction. Zero is neither positive nor negative.

These positive and negative numbers are sometimes referred to as **signed numbers** or **directed numbers.**

Example 1 Represent the number $+5$ on the number line.

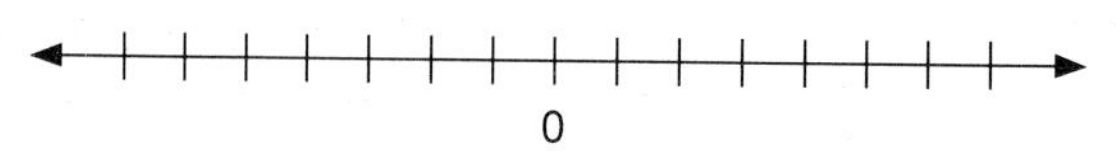

Solution Since $+5$ indicates 5 units to the right of zero on the number line, we place a dot at this location and label it $+5$.

Make sure you count the correct number of spaces.

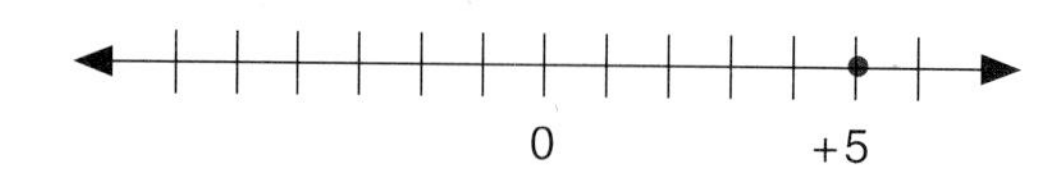

Example 2 Represent the number -2 on the number line.

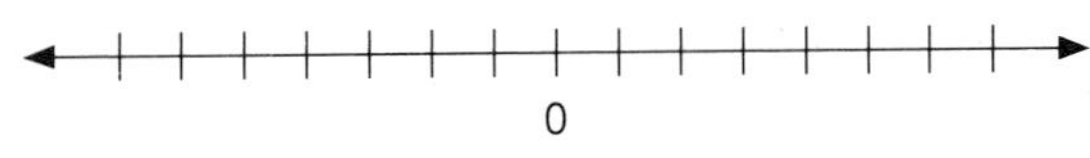

Solution

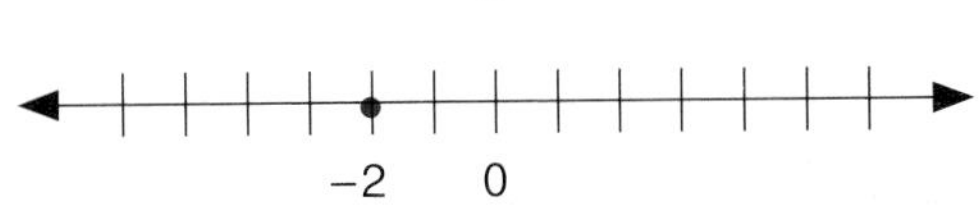

⦿ ***SKILL CHECK 1*** Represent the number -4 on the number line.

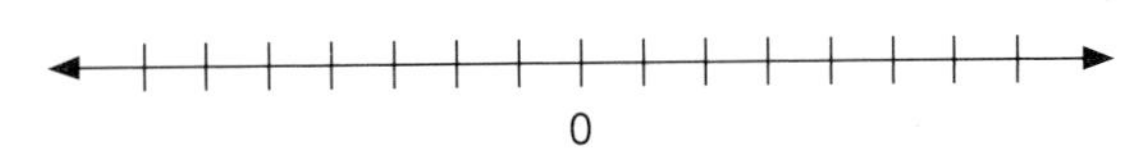

2 Comparing the Relative Size of Two Signed Numbers

We can use the inequality symbol to compare the relative values of two signed numbers.

Example 3 Replace the question mark in $-6 \ ? \ 3$ with the inequality symbol positioned to indicate a true statement.

Solution Since -6 is to the left of 3 on the number line we write $-6 < 3$.

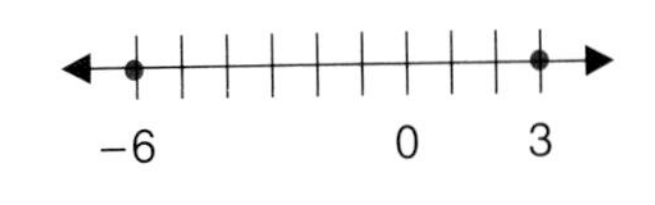

Example 4 Replace the question mark in $-\frac{1}{2} \ ? \ -7$ with the inequality symbol positioned to indicate a true statement.

Solution Since $-\frac{1}{2}$ is to the right of -7 on the number line we write $-\frac{1}{2} > -7$.

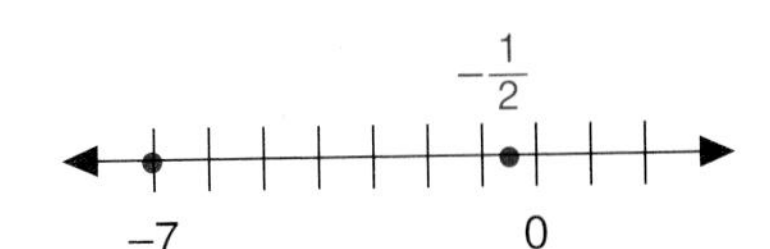

⦿ ***SKILL CHECK 2*** Replace the question mark in $-8 \ ? \ -2$ with the inequality symbol positioned to indicate a true statement.

Most likely you have encountered signed numbers many times in your newspaper or on TV. For instance, if the temperature is ten degrees below zero, it is usually written $-10°$. If the stock market falls four points, it is usually written -4. You can probably think of other examples you have encountered.

Example 5 If a football team gained three yards on a play, how would you represent this using a signed number? How would you represent a loss of five yards?

Solution $+3$ would represent a gain of three yards.
-5 would represent a loss of five yards.

Example 6 Jack had a bank balance of \$23 and wrote a check for \$30. Represent his balance with a signed number.

Solution Since he spent seven dollars more than he had, his balance is $-\$7$ and the bank is probably not too happy.

⦿ ***SKILL CHECK 3*** If you had a bank balance of \$65 and wrote a check for \$93, represent your balance with a signed number.

⦿ ***Skill Check Answers***

1.

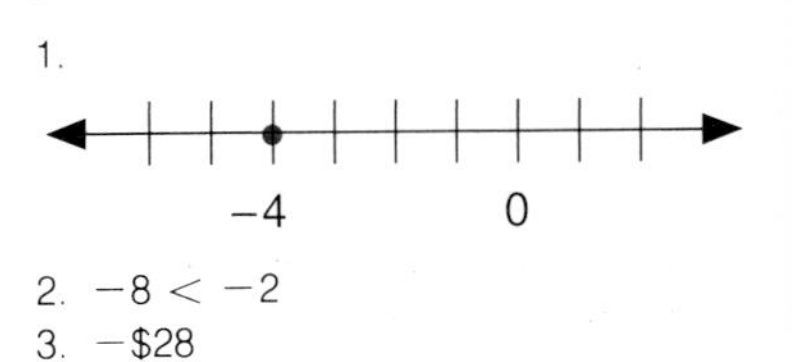

2. $-8 < -2$
3. $-\$28$

Exercise 6–1–1

1 Identify the signed numbers represented by the following points.

1. *A*
2. *B*
3. *C*
4. *D*

2 Locate each number on the number line and replace the question mark with $<$ or $>$ to indicate a true statement.

5. $6 \,?\, 10$

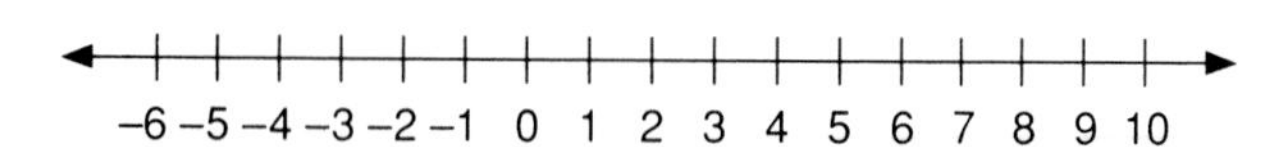

6. $-6 \,?\, -10$

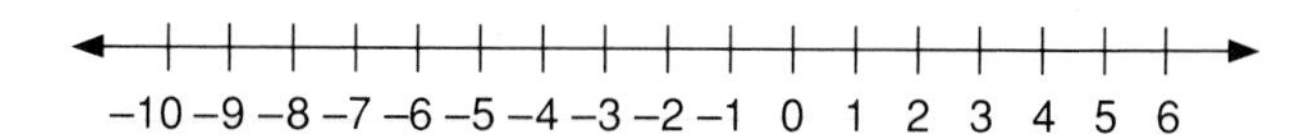

7. $-3 \,?\, 3$

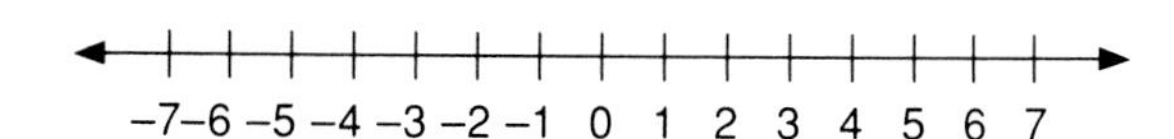

8. $-4 \,?\, -1$

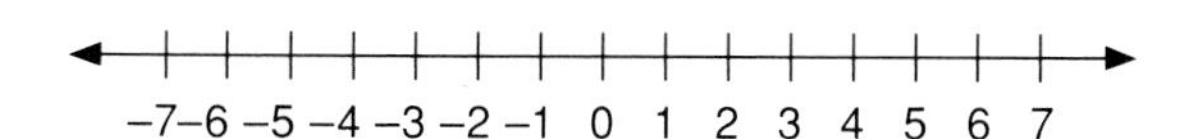

9. $4 \,?\, 1.5$

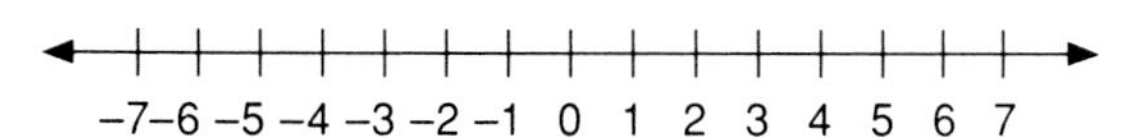

10. $1.75 \,?\, -3$

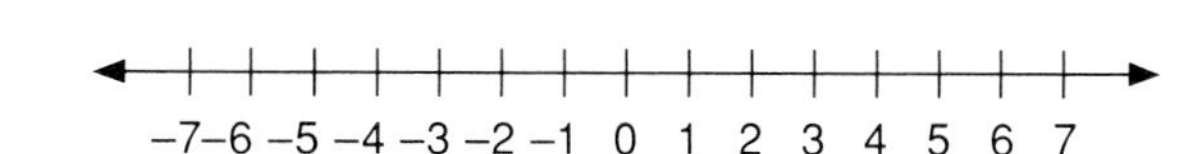

11. $-2 \,?\, -\frac{5}{2}$

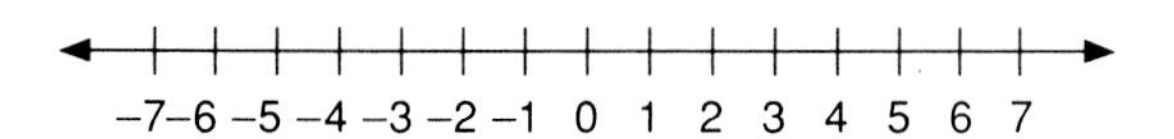

12. $-\frac{1}{2} \,?\, -3$

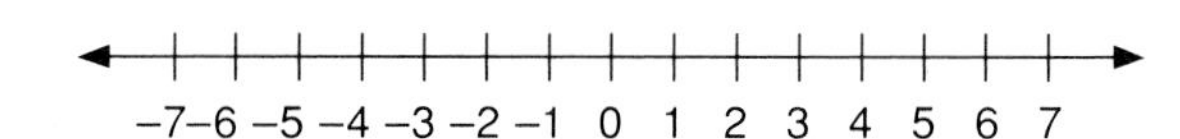

13. $0 \,?\, 7$

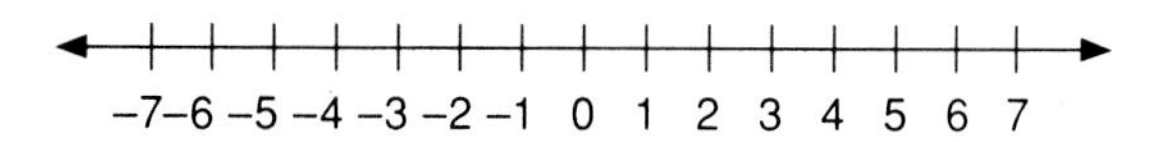

14. $0 \,?\, -3$

−7 −6 −5 −4 −3 −2 −1 0 1 2 3 4 5 6 7

15. If $+20$ represents a temperature of twenty degrees above zero, use a signed number to represent a temperature of twenty degrees below zero.

16. If -10 represents a drop of ten points on the stock market, use a signed number to represent a gain of ten points.

17. A football team loses five yards on a play. Use a signed number to indicate this loss.

18. A news reporter stated that the Dow-Jones Average dropped thirteen points. Indicate this drop using a signed number.

19. A person's weight was recorded two weeks ago at 165 pounds. This week the person's weight is 159 pounds. Indicate the change from last week to this week using a signed number.

20. Last month a club had 38 members. This month the membership is 43. Indicate the change in membership from last month to this month using a signed number.

3 The Opposite of a Number

To be successful in using operations on signed numbers it is important to understand what is meant by the *opposite* of a number.

> The **opposite** of a number is that number with the opposite sign.

The opposite of a number is sometimes referred to as the "negative of a number."

Example 7 The opposite of $+5$ is -5.

⦿ ***SKILL CHECK 4*** Find the opposite of $+12$.

Example 8 The opposite of -3.2 is $+3.2$.

⦿ ***SKILL CHECK 5*** Find the opposite of $-\frac{2}{3}$.

Example 9 The opposite of $+x$ is $-x$.

Example 10 The opposite of $-x$ is $+x$.

We will agree that the opposite of zero is zero.

Remember, zero has no sign.

The negative sign is often used as a symbol for "the opposite of." Therefore, $-(-10)$ means "the opposite of" -10.

Example 11 $-(+7) = -7$

This is the sign of the number.
This means "the opposite of."

⦿ ***SKILL CHECK 6*** Simplify: $-(+9)$

Example 12 $-\left(-\frac{1}{2}\right) = +\frac{1}{2}$

⦿ ***SKILL CHECK 7*** Simplify: $-(-5.3)$

4 Absolute Value

Another concept often used is that of **distance.** The distance from $+3$ to 0 on the number line is 3 units. The distance from -3 to 0 is also 3 units.

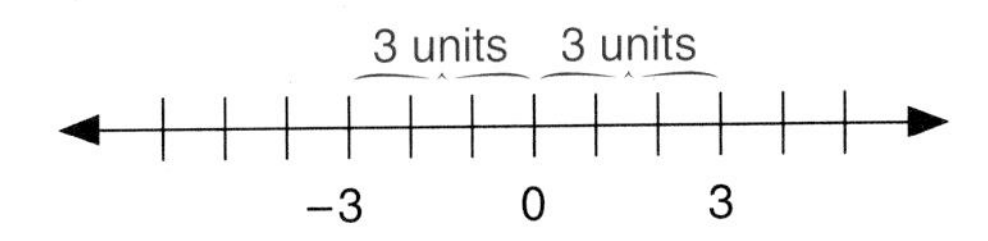

Distance is always expressed as a nonnegative number, that is, always positive or zero.

In determining distance, direction is not important.

⦿ ***Skill Check Answers***

4. -12 5. $+\frac{2}{3}$ 6. -9
7. $+5.3$

> The distance of any number from zero on the number line is defined to be the **absolute value** of that number. The absolute value of a number x is written $|x|$.

From this definition we see that the absolute value of a number will always be a positive number or zero.

Example 13 Find: $|+6|$

Solution We see that $+6$ is 6 units from zero on the number line.

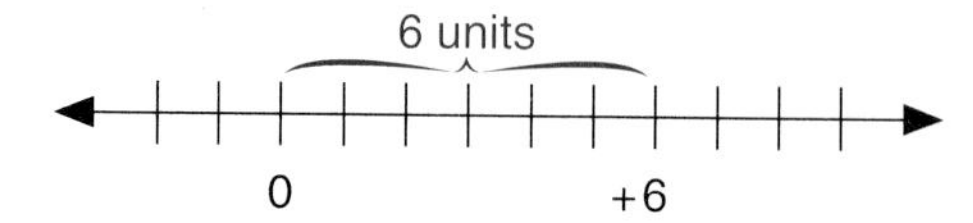

Therefore we write

$$|+6| = +6.$$

◉ ***SKILL CHECK 8*** Find: $|+10|$

Example 14 Find: $|-5|$

Solution We see that -5 is 5 units from zero on the number line.

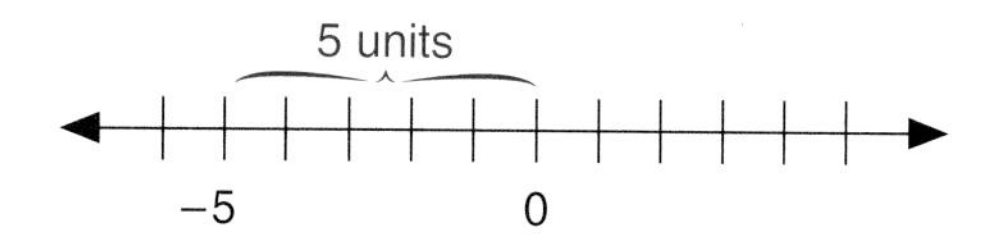

Therefore we write

$$|-5| = 5.$$

Remember that a positive number can be written without the positive sign.

Example 15 Find: $|-2.14|$

Solution

$$|-2.14| = 2.14$$

◉ ***SKILL CHECK 9*** Find: $\left|-\frac{1}{3}\right|$

WARNING

Be careful when the negative sign precedes the absolute value symbol.

Example 16 Find: $-|-9|$

Solution We are asked to find the opposite of the absolute value of -9. $|-9| = 9$ and the opposite of 9 is -9 so we write

$$-|-9| = -9.$$

◉ ***SKILL CHECK 10*** Find: $-|-15|$

◉ ***Skill Check Answers***

8. $+10$
9. $+\frac{1}{3}$
10. -15

Exercise 6–1–2

3 Find the opposite of each number.

1. $+5$ **2.** $+16$ **3.** -6 **4.** -28

5. 14

6. 3.1

7. $\frac{1}{2}$

8. $-4\frac{1}{3}$

9. 0

10. 100

11. $5\frac{3}{4}$

12. -7.38

Simplify.

13. $-(+4)$

14. $-(21)$

15. $-(-15)$

16. $-(-8)$

17. $-(+4.25)$

18. $-(-2.05)$

19. $-\left(\frac{1}{2}\right)$

20. $-\left(-\frac{3}{4}\right)$

4 Find each value.

21. $|+9|$

22. $|13|$

23. $|-4|$

24. $|-17|$

25. $|0|$

26. $\left|-\frac{3}{4}\right|$

27. $\left|\frac{2}{3}\right|$

28. $|-(+4)|$

29. $|-3.16|$

30. $|+8.25|$

31. $-|+5|$

32. $-|-14|$

• Concept Enrichment

33. What is the only number that does not have a sign?

34. Distinguish between a negative number and the negative of a number.

35. Why is the statement "the absolute value of any number is always positive" not correct?

36. What value(s) of x would make the statement $|x| = 10$ true?

6–2 Using the Number Line to Add Signed Numbers

OBJECTIVES

Upon completing this section you should be able to:

1. Understand how signed numbers show direction.
2. Use the number line to add signed numbers.

In the previous section we established the meaning of signed numbers. We next wish to consider the four basic operations on signed numbers. The first operation we will discuss is **addition.** The symbol + is used to indicate both the operation of addition and a positive number. We must be careful to distinguish the meaning of this sign.

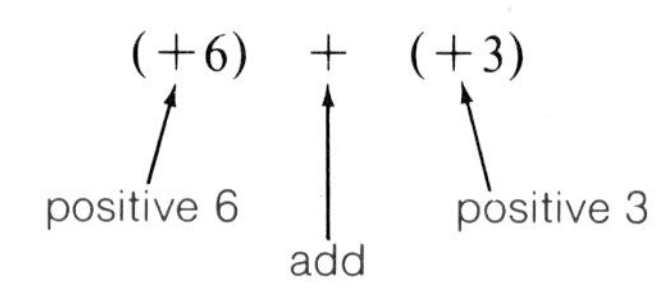

We do not write two such symbols together without parentheses separating them.

1 Signed Numbers Show Direction

Consider the following three examples that relate to the operation of addition on signed or directed numbers.

Example 1 If you start at a point and move six steps north, then move three more steps north, where will you be in relation to your starting point?

Solution The answer to this problem is nine steps north. We can think of the answer as the sum of the two movements. If "steps north" is represented as positive, then the sum of six steps north followed by three steps north equals nine steps north could be represented by writing $(+6) + (+3) = +9$.

Remember, signed numbers show direction.

SKILL CHECK 1 Use a signed number to indicate the result of moving five steps north and then nine steps north.

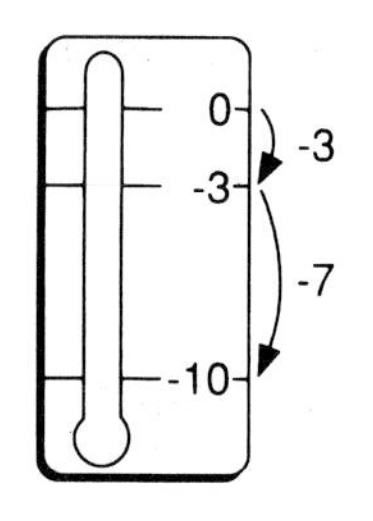

Example 2 The temperature at 6:00 P.M. is zero degrees. From 6:00 P.M. until midnight the temperature falls three degrees, and from midnight until 4:00 A.M. it falls seven more degrees. What is the temperature at 4:00 A.M.?

Solution The result is a sum of changes in temperature. If a fall in temperature is represented by a negative sign, then a fall of 3° followed by a fall of 7° can be written

$$(-3°) + (-7°) = -10°.$$

SKILL CHECK 2 Use a signed number to represent the result of a drop in temperature of 7 degrees followed by a drop of 6 degrees.

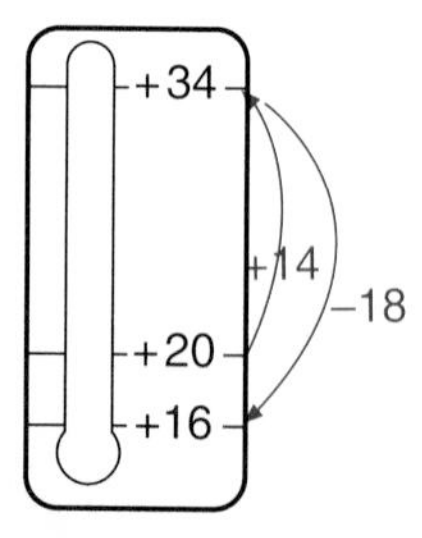

Example 3 At 9:00 A.M. the temperature is +20° and from 9:00 A.M. until 1:00 P.M. it rises 14°. Then from 1:00 P.M. until 5:00 P.M. it falls 18°. What is the temperature at 5:00 P.M.?

Solution Starting at +20°, a rise of 14° takes the temperature to +34°. A fall of 18° then takes it to +16°. If a rise in temperature is positive and a fall in temperature is negative, then we can write

$$(+20°) + (+14°) + (-18°) = +16°.$$

Skill Check Answers

1. +14 2. −13°

⦿ ***SKILL CHECK 3*** Use a signed number to represent the result of rise in temperature of 8 degrees followed by a drop of 12 degrees.

⦿ ***Skill Check Answer***

3. $-4°$

Exercise 6–2–1

1 Express each result as a signed number.

1. If we consider north as positive and south as negative, find the result of ten steps north, followed by three steps south.

2. Find the result of five steps south, followed by two steps north.

3. Find the result of four steps north, followed by six steps south.

4. Find the result of three steps south, followed by six steps south.

5. Find the result of eight steps south, followed by ten steps north.

6. Find the result of a profit of $10 combined with a profit of $12.

7. Find the result of a profit of $16 combined with a loss of $9.

8. Find the result of a loss of $20 combined with a profit of $12.

9. Find the result of a loss of $6 combined with a loss of $18.

10. Find the result of a profit of $30 combined with a loss of $36.

11. Find the result of a gain of two yards in a football game, followed by a loss of six yards.

12. Find the result of a loss of two yards, followed by a loss of five yards.

13. Find the result of a loss of eight yards, followed by a gain of twelve yards.

14. Find the result of a gain of twelve yards, followed by a loss of fifteen yards.

15. Find the result of a fall of six degrees in temperature, followed by a fall of two degrees.

16. Find the result of a rise of eight degrees in temperature, followed by a fall of two degrees.

17. If at 10:00 P.M. the temperature was 65° and from 10:00 P.M. until 4:00 A.M. it fell 12 degrees, what was the temperature at 4:00 A.M.?

18. If the temperature was 50° at 6:00 A.M. and from 6:00 A.M. to 2:00 P.M. it rose fifteen degrees, then from 2:00 P.M. to 7:00 P.M. it fell nine degrees, what was the temperature at 7:00 P.M.?

19. Ten steps north, followed by six steps south, followed by nine steps north, followed by twelve steps south equals

20. A temperature starts at 10°. It falls fifteen degrees, then rises eight degrees, then falls three degrees. What is the final temperature?

2 Using the Number Line

We will use the number line discussed in section 6–1 as an aid in establishing rules for adding signed numbers. If a positive number represents a direction to the right and a negative number represents a direction to the left, then we can illustrate addition of signed numbers as a sequence of directed movements on the number line.

Do you see why signed numbers are sometimes referred to as *directed numbers?*

WARNING

Unless the problem states otherwise, always start at zero for the first move.

Example 4 Use a number line to indicate the sum $(+3) + (+4)$.

Solution We start at zero and move 3 units to the right. From that point we then move 4 units to the right.

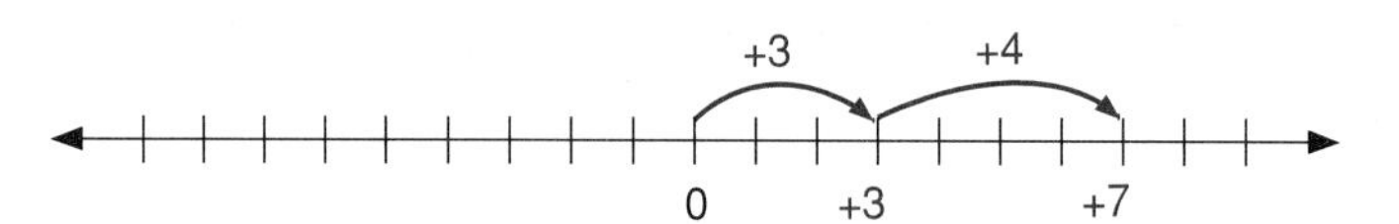

The result is a movement of 7 units to the right. So $(+3) + (+4) = +7$.

Example 5 Use a number line to indicate the sum $(-2) + (-4)$.

Solution Starting at zero we move 2 units to the left and then 4 more units to the left.

Both movements are in the negative direction.

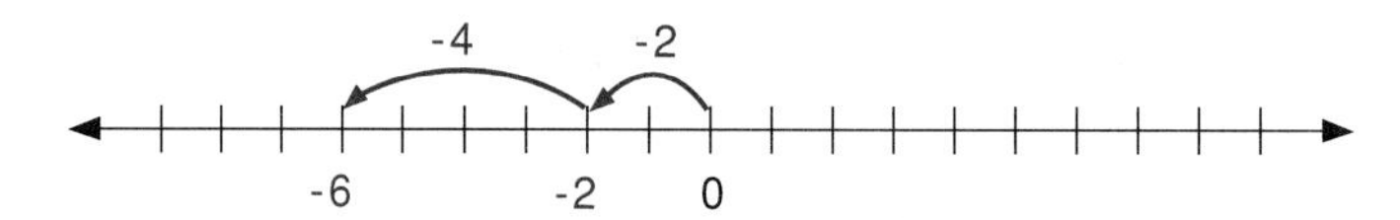

The result is a movement of 6 units to the left. So $(-2) + (-4) = -6$.

◉ ***SKILL CHECK 4*** Use the number line to show the sum $(-3) + (-2)$.

We will now use the number line to add two numbers having unlike signs.

Example 6 Use the number line to find the sum $(-6) + (+8)$.

Solution Starting at zero we first move 6 units to the left and then 8 units to the right.

Notice here we are moving in opposite directions.

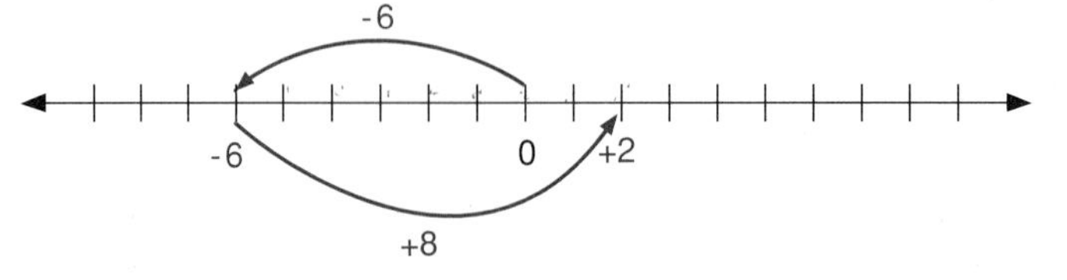

The result is the same as a movement of 2 units to the right. So $(-6) + (+8) = +2$.

◉ ***Skill Check Answer***

4.

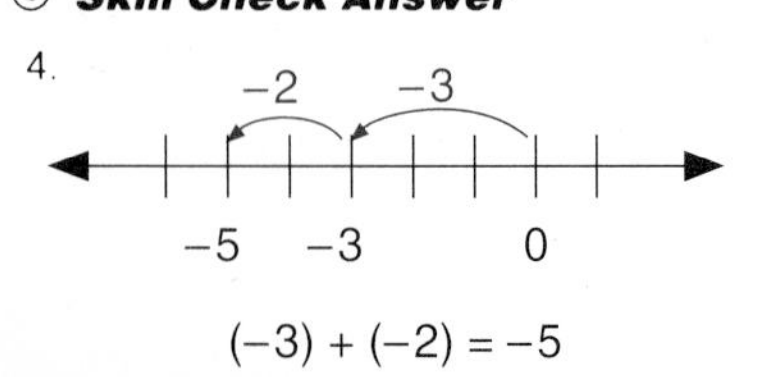

$(-3) + (-2) = -5$

Example 7 Use the number line to find the sum $(-7) + (+3)$.

Solution Starting at zero we move 7 units to the left and then 3 units to the right.

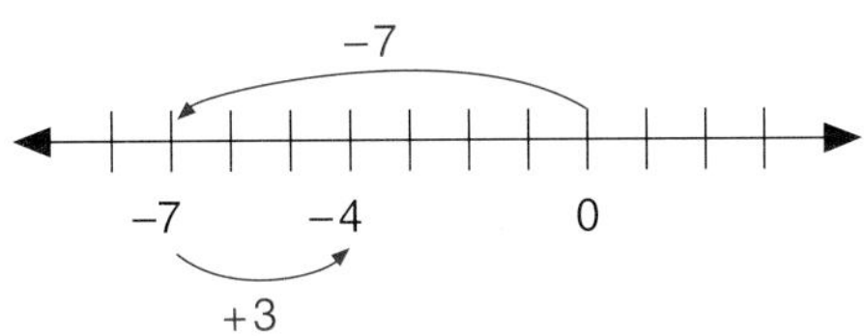

In which direction do we move the greatest distance?

The result is the same as a movement of 4 units to the left. Thus $(-7) + (+3) = -4$.

If we move more units to the right than to the left, the end result is to the right. If we move more units to the left than to the right, the end result will be to the left.

◉ ***SKILL CHECK 5*** Use the number line to find the sum $(+3) + (-8)$.

We can also use the number line to find the indicated sum of several signed numbers.

Example 8 Add: $(+3) + (-6) + (+4) + (-5)$

Solution Using the number line and starting at zero, first move 3 units to the right.

Remember, the sign of the number indicates the direction.

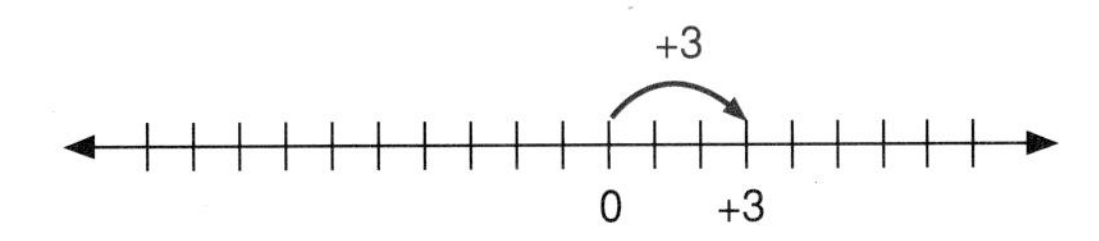

Then move 6 units to the left.

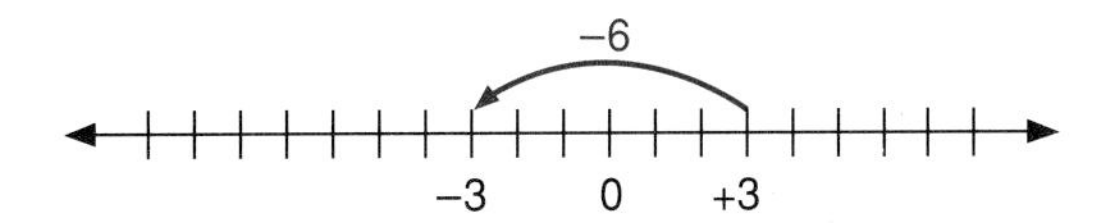

Then move 4 units to the right.

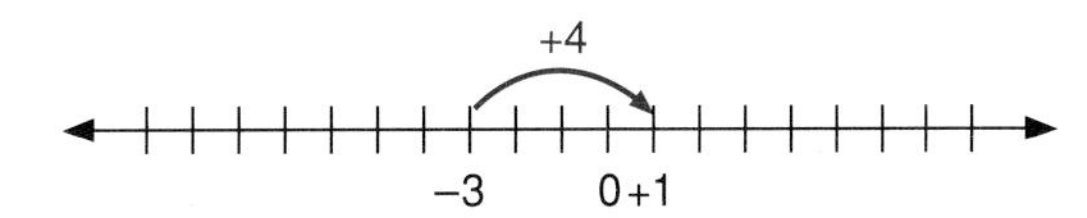

Finally move 5 units to the left.

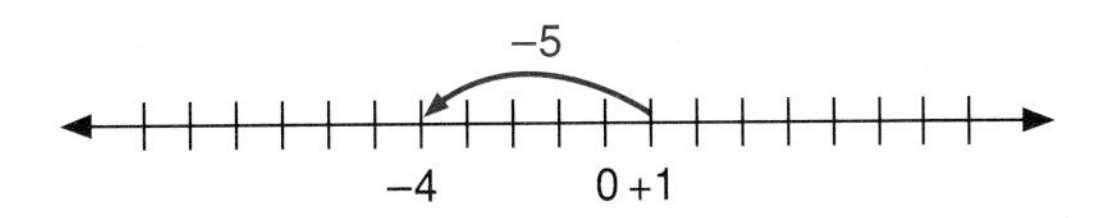

The final location is the solution.

The final location is at -4.
Therefore, $(+3) + (-6) + (+4) + (-5) = -4$.

◉ ***Skill Check Answer***

5.

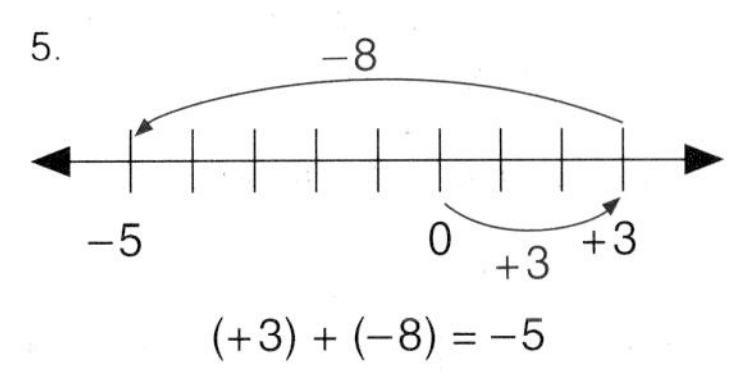

Example 9 After losing nine yards on the first down, the Browns completed a pass for a twelve-yard gain on the second down. What was the total gain or loss?

Solution First write the number statement $(-9) + (+12) =$

Using the number line, we obtain

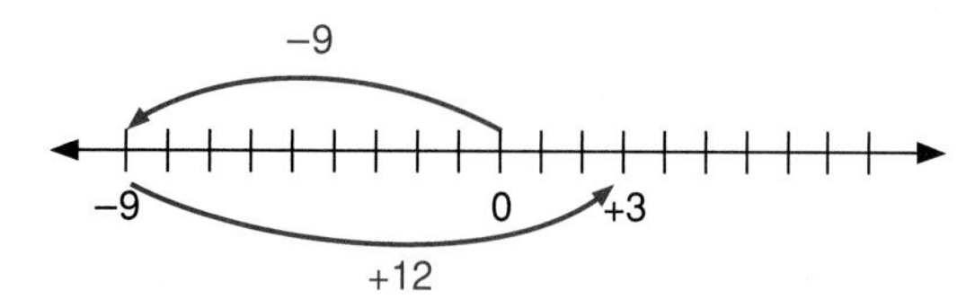

Thus $(-9) + (+12) = +3$ (a three-yard gain).

Exercise 6–2–2

Use a number line to add.

1. $(+3) + (+5) =$

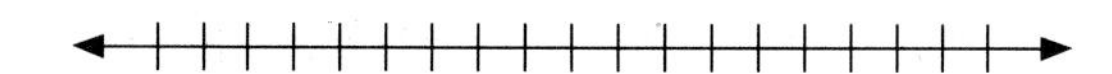

2. $(-3) + (-4) =$

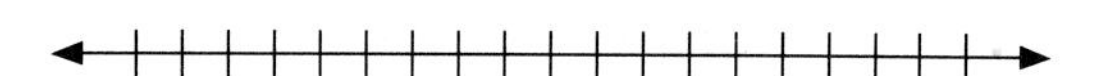

3. $(-4) + (+10) =$

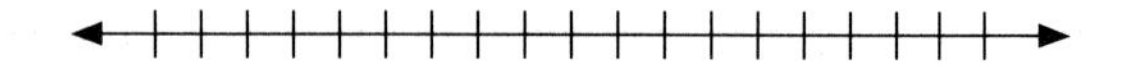

4. $(+9) + (-6) =$

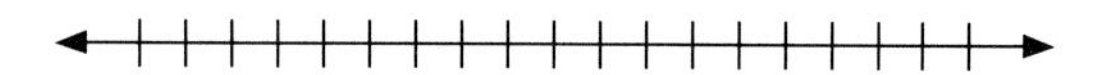

5. $(-2) + (+7) =$

6. $(-11) + (+5) =$

7. $(+8) + (-8) =$

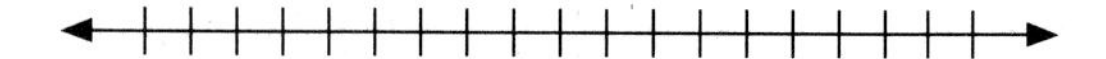

8. $(+13) + (-20) =$

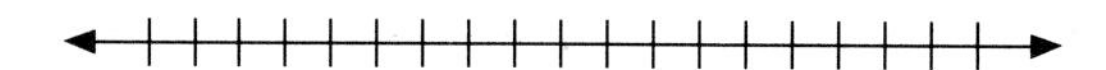

9. $(+9) + (-15) =$

10. $(-11) + (+18) =$

11. $(+8) + (-17) =$

12. $(+4) + (-2) + (+7) =$

13. $(-6) + (+13) + (-7) =$

14. $(-4) + (+13) + (-9) =$

15. $(-5) + (+8) + (-4) + (+1) =$

16. $(+3) + (-8) + (+5) + (-4) =$

17. $(+3) + (-9) + (+6) + (-14) =$

18. $(-14) + (+3) + (-9) + (+6) =$

19. $(-8) + (+6) + (-5) + (+7) =$

20. $(-5) + (+6) + (-8) + (+7) =$

21. The stock market gained ten points in the morning, then lost six points in the afternoon. What was the net gain or loss for the day?

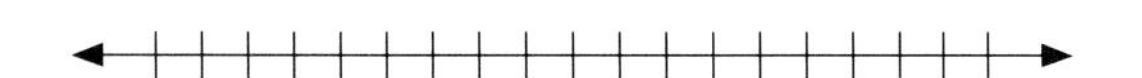

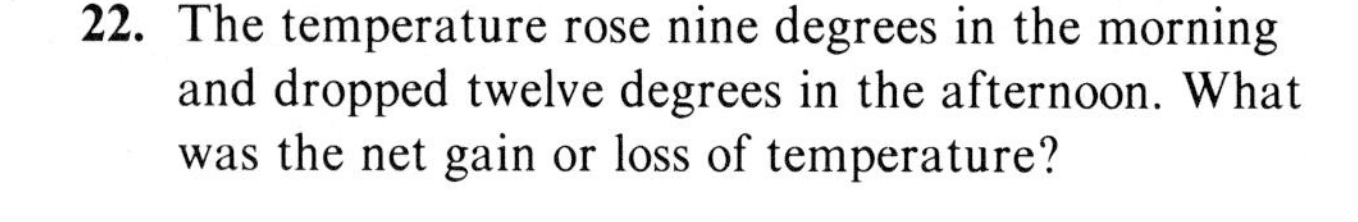

22. The temperature rose nine degrees in the morning and dropped twelve degrees in the afternoon. What was the net gain or loss of temperature?

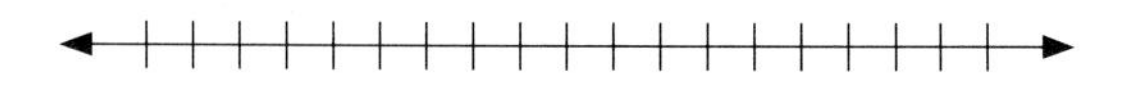

23. A worker earned \$25 in the morning and spent \$18 in the afternoon. What was the net gain or loss for the day?

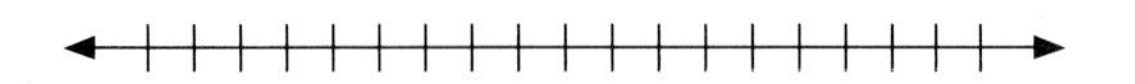

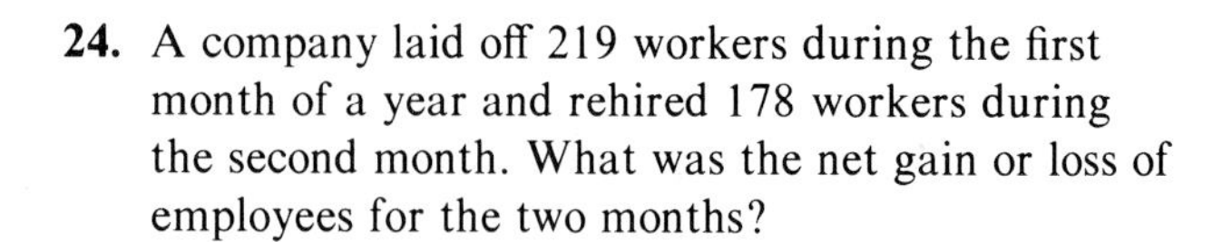

24. A company laid off 219 workers during the first month of a year and rehired 178 workers during the second month. What was the net gain or loss of employees for the two months?

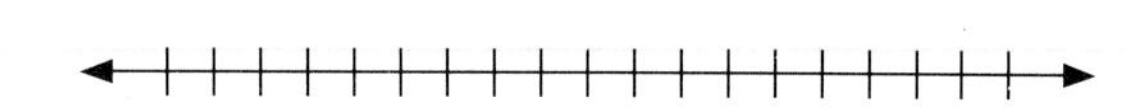

• Concept Enrichment

25. Why are signed numbers sometimes called directed numbers?

26. Explain what each of the three signs in the expression $(+5) + (-8)$ means.

27. A weight-watcher lost five pounds the first week, gained two pounds the second week, lost three pounds the third week, and lost two pounds the fourth week. What was the net gain or loss for the four weeks?

28. The temperature in Denver was 28° C at 1:00 P.M. During the next hour the temperature rose four degrees. During the second hour the temperature dropped three degrees, and during the third hour it dropped another six degrees. What was the temperature at 4:00 P.M.?

6–3 Rules for Adding Signed Numbers

OBJECTIVE

Upon completing this section you should be able to:

1 Use the rules for adding signed numbers.

In the preceding section we used the meaning of a signed number and the number line to add these numbers. We are now going to establish rules for adding signed numbers that will be consistent with the results obtained by using the number line.

1 Rules for Addition

We will use the concept of absolute value that was discussed in section 6–1 in stating the rules for addition. We first state a rule for adding two numbers having the same sign.

To add two numbers with like signs add their absolute values. The sum will have the same sign as the numbers being added.

Recall that absolute value refers to distance without regard to direction.

Example 1 Add: $(+3) + (+4)$

Solution Using the rule, we take the absolute value of each.

$$|+3| = 3$$
$$|+4| = 4$$

Adding these numbers, we obtain

$$3 + 4 = 7.$$

Since both numbers are positive the answer is positive. Thus we have

$$(+3) + (+4) = +7.$$

Is the result the same as when using the number line?

Of course we could simply write this as $3 + 4 = 7$, since any number except zero written without a sign is assumed to be positive.

⦿ ***SKILL CHECK 1*** Add: $(+5) + (+6)$

Example 2 Add: $(-2) + (-4)$

We could write this as $-2 + (-4)$, but notice that -4 must have parentheses since we cannot write two signs together without parentheses separating them.

Solution

$$|-2| = 2$$
$$|-4| = 4$$
$$2 + 4 = 6$$

Since both numbers are negative the answer must be negative. So we write

$$(-2) + (-4) = -6.$$

⦿ ***SKILL CHECK 2*** Add: $(-3) + (-7)$

The following rule for adding numbers with unlike signs is also consistent with the values obtained using the number line.

To add two numbers with unlike signs find the difference of their absolute values. The answer will have the sign of the number having the greater absolute value.

Example 3 Add: $(-6) + (+8)$

We could also write this as $-6 + 8$.

Solution

$$|-6| = 6$$
$$|+8| = 8$$

The difference of 6 and 8 is 2. Since $+8$ has the greater absolute value the answer is positive. Therefore

8 is greater than 6. In symbols, $8 > 6$.

$$(-6) + (+8) = +2.$$

Or $-6 + 8 = 2$.

⦿ ***SKILL CHECK 3*** Add: $(+10) + (-6)$

⦿ ***Skill Check Answers***

1. $+11$
2. -10
3. $+4$

Example 4 Add: $(-7) + (+3)$

Solution

$$|-7| = 7$$
$$|+3| = 3$$

7 is greater than 3. In symbols, $7 > 3$.

The difference of 7 and 3 is 4. Since -7 has the greater absolute value the answer is negative. Therefore

$$(-7) + (+3) = -4.$$

⦿ ***SKILL CHECK 4*** Add: $(+8) + (-14)$

Example 5 Add: $\left(\frac{2}{3}\right) + \left(-\frac{3}{4}\right)$

See section 3-4 for comparing the size of fractions.

Solution

$$\left|\frac{2}{3}\right| = \frac{2}{3} = \frac{8}{12}$$
$$\left|-\frac{3}{4}\right| = \frac{3}{4} = \frac{9}{12}$$

The difference of $\frac{8}{12}$ and $\frac{9}{12}$ is $\frac{1}{12}$.

Since $-\frac{3}{4}$ has the greater absolute value the answer is negative.

Therefore $\left(\frac{2}{3}\right) + \left(-\frac{3}{4}\right) = -\frac{1}{12}$.

⦿ ***SKILL CHECK 5*** Add: $\left(+\frac{5}{8}\right) + \left(-\frac{2}{3}\right)$

Addition is a *binary* operation. See section 1-4.

We noted in chapter 1 that only two numbers can be added at a time. We also used the associative and commutative properties of addition to rearrange and regroup numbers to be added. These same properties hold true for signed numbers.

This could also be written as $6 + (-10) + 3$.

Example 6 Add: $(+6) + (-10) + (+3)$

Solution We can first add $(+6)$ and (-10) to obtain -4. Then add (-4) and $(+3)$ to obtain -1. So

$$(+6) + (-10) + (+3) = -1.$$

We could also use the commutative property to rewrite $(+6) + (-10) + (+3)$ as $(+6) + (+3) + (-10)$. Then add $(+6)$ and $(+3)$ to obtain $+9$. Next add $(+9)$ and (-10) to obtain -1.

Example 7 Add: $(+3) + (-4) + (-6) + (+8)$

Solution We can add the numbers as they occur from left to right, obtaining

$$\begin{aligned}(+3) + (-4) + (-6) + (+8) &\\ &= (-1) + (-6) + (+8)\\ &= (-7) + (+8)\\ &= +1\end{aligned}$$

Try this method here in the margin.

You might prefer to first find the sum of the positive numbers, then the sum of the negative numbers, and finally find the sum of those two resulting numbers.

⦿ ***SKILL CHECK 6*** Add: $(+6) + (-5) + (-6) + (+1)$

⦿ ***Skill Check Answers***

4. -6 5. $-\frac{1}{24}$ 6. -4

In chapter 2 we learned rules for removing grouping symbols. Numbers within a set of grouping symbols should be added before the symbol is removed.

See section 2–3.

Example 8 $(+12) + [(+5) + (-7)] = (+12) + [-2]$
$= +10$

We first add +5 and −7.

Example 9 $[(-6) + (-2)] + [(-4) + (+9)] = [-8] + [+5]$
$= -3$

⊙ ***SKILL CHECK 7*** Add: $[(+3) + (-10)] + [(-5) + (+6)]$

Recall that when we simplify an expression containing grouping symbols within grouping symbols, we remove the innermost set of symbols first.

Example 10 $10 + \{+3 + [(+5) + (-21)]\} = 10 + \{+3 + [-16]\}$
$= 10 + \{-13\}$
$= -3$

⊙ ***SKILL CHECK 8*** Add: $6 + \{(-5) + [(-6) + (+9)]\}$

⊙ ***Skill Check Answers***

7. −6
8. 4

Exercise 6–3–1

1 Add.

1. $(+5) + (+7)$

2. $(+13) + (28)$

3. $(+8) + (+9)$

4. $(+7) + (+6)$

5. $(-8) + (-15)$

6. $(-3) + (-18)$

7. $(-15) + (-16)$

8. $(-18) + (-14)$

9. $(+8) + (+13) + (+4)$

10. $(+5) + (+7) + (+11)$

11. $(-4) + (-8) + (-16)$

12. $(-6) + (-15) + (-9)$

13. $(+8) + (-3)$

14. $(+11) + (-7)$

15. $(+14) + (-20)$

16. $(+6) + (-8)$

17. $(-13) + (+7)$

18. $(-28) + (+14)$

19. $(+1) + (-1)$

20. $(+18) + (-18)$

21. $(+5) + (-7) + (-2)$

22. $(-8) + (+9) + (-6)$

23. $(-14) + (+7) + (+4)$

24. $(+15) + (-23) + (+6)$

25. $\left(+\frac{3}{7}\right) + \left(-\frac{5}{7}\right)$

26. $\left(-\frac{2}{5}\right) + \left(+\frac{4}{5}\right)$

27. $(+12.7) + (-21.3)$

28. $(-4.2) + (+17.1)$

29. $\left(-\frac{2}{3}\right) + \left(+\frac{1}{2}\right) + \left(-\frac{3}{4}\right)$

30. $\left(+\frac{4}{5}\right) + \left(-\frac{1}{3}\right) + \left(-\frac{5}{6}\right)$

31. $(+5) + [(+9) + (-12)]$

32. $(-7) + [(+3) + (-8)]$

33. $-8 + [7 + (-9)]$

34. $11 + [-25 + (+7)]$

35. $[(-2) + (+8)] + [(+5) + (-12)]$

36. $[(+7) + (-12)] + [(-15) + (-2)]$

37. $4 + \{-13 + [(-6) + (+4)]\}$

38. $8 + \{10 + [(-21) + (+3)]\}$

39. $[-8 + 3] + \{-11 + [-6 + 4]\}$

40. $[16 + (-20)] + \{9 + [-14 + 2]\}$

Write as an addition statement and find the value.

41. A gain of 2 yards in a football game, followed by a loss of 6 yards.

42. A gain of 12 yards, followed by a loss of 15 yards.

43. If the temperature was 50 degrees at 6:00 A.M. and from 6:00 A.M. to 2:00 P.M. it rose 15 degrees, then from 2:00 P.M. to 7:00 P.M. it fell 9 degrees, what was the temperature at 7:00 P.M.?

44. A temperature starts at 10 degrees. It falls 15 degrees, then rises 8 degrees, then falls 3 degrees. What is the final temperature?

• Concept Enrichment

45. What property allows us to write $[(-3) + (-5)] + (+3) = [(-5) + (-3)] + (+3)$?

46. What property allows us to write $[(-5) + (-3)] + (+3) = (-5) + [(-3) + (+3)]$?

6–4 Subtracting Signed Numbers

OBJECTIVES

Upon completing this section you should be able to:

1. Write a subtraction problem as an addition problem.
2. Add and subtract signed numbers without writing parentheses.
3. Add and subtract several signed numbers at a time.
4. Add and subtract signed numbers within grouping symbols.

The second operation we will consider on signed numbers is **subtraction.** The symbol $-$ is used to indicate both the operation of subtraction and a negative number. Again we must be careful to distinguish between these two meanings.

$$(+5) - (-4)$$

negative 4 (the second − sign)

subtract (the first − sign)

1 Subtraction Defined as Addition

A simple illustration will help lead us to the definition of subtraction. In a previous example we noted that if a check for \$30 was written on an account containing \$23, the balance in the account would be $-\$7$. This could be written

$$23 - 30 = -7.$$

The same answer would be obtained in the addition problem $23 + (-30) = -7$.

This illustrates the relationship between subtraction and addition. Every subtraction problem can be written as an addition problem using the following rule.

$$a - b = a + (-b)$$

This rule states that to subtract a number add the opposite of that number.

Example 1 Subtract: $(+5) - (+8)$

Solution We are subtracting $+8$. The opposite of $+8$ is -8. So using the above rule, we write

$$(+5) - (+8) = (+5) + (-8).$$

Then using the rule for addition, we obtain

$$(+5) + (-8) = -3.$$

The opposite of a number is sometimes referred to as the *additive inverse* of that number.

SKILL CHECK 1 Subtract: $(+3) - (+7)$

Skill Check Answer

1. -4

Example 2 Subtract: $(-7) - (-2)$

Solution Using the subtraction rule, we write

$$(-7) - (-2) = (-7) + (+2) = -5.$$

⦿ ***SKILL CHECK 2*** Subtract: $(-2) - (-5)$

Example 3 Subtract -8 from 5.

Solution The statement "subtract -8 from 5" means $5 - (-8)$. We can then write

Remember, $5 + 8$ means $(+5) + (+8)$.

$$5 - (-8) = 5 + 8 = 13.$$

⦿ ***SKILL CHECK 3*** Subtract -6 from 9.

Example 4 Subtract -0.58 from -2.06.

Solution

$$-2.06 - (-0.58) = -2.06 + 0.58 = -1.48$$

⦿ ***SKILL CHECK 4*** Subtract 3.05 from -1.82.

⦿ ***Skill Check Answers***

2. $+3$
3. 15
4. -4.87

Exercise 6–4–1

1 Use the rule for subtraction to rewrite each as an addition problem and determine the answer.

1. $(+6) - (+4)$ **2.** $(+10) - (+3)$ **3.** $(+9) - (+12)$

4. $(+6) - (+8)$ **5.** $(+13) - (-2)$ **6.** $(+11) - (-5)$

7. $(-16) - (+7)$ **8.** $(-14) - (+6)$ **9.** $(-13) - (-9)$

10. $(-8) - (-15)$ **11.** $\left(+\frac{2}{3}\right) - \left(+\frac{3}{4}\right)$ **12.** $\left(+\frac{1}{3}\right) - \left(+\frac{3}{8}\right)$

13. $(-3.25) - (-6.72)$ **14.** $(-12.50) - (-4.68)$ **15.** $24 - (+24)$

16. $18 - (+18)$ **17.** $-11 - (-1)$ **18.** $-25 - (-17)$

19. $14 - 27$ **20.** $8 - 21$ **21.** Subtract 12 from 8.

22. Subtract -4 from 9.

23. Subtract -23 from -18.

24. Subtract -14 from -8.

• Concept Enrichment

25. What is meant by "the additive inverse of a number"?

26. What is the additive inverse of zero?

27. Explain in words what is meant by "$8 - 3$" in terms of addition.

28. Explain in words what is meant by "$(-15) - (-4)$" in terms of addition.

2 Adding and Subtracting Mentally

With a little practice you should be able to add and subtract signed numbers without writing all of the parentheses and positive and negative signs. For instance, you should be able to determine the answer to $8 - 15$ without writing it as $(+8) + (-15)$. You can mentally think of that expression without actually writing it.

You should be able to write $8 - 15 = -7$.

Example 5 Simplify: $6 - 11$

Solution We mentally think of $6 - 11$ as adding -11 to $+6$ and write

$$6 - 11 = -5.$$

Example 6 $-9 + 12 = 3$

Example 7 $-14 + 8 = -6$

⦿ ***SKILL CHECK 5*** Add: $-21 + 9$

Example 8 $16 - 23 = -7$

Example 9 $-5 - 9 = -14$

⦿ ***SKILL CHECK 6*** Add: $-4 - 18$

⦿ ***Skill Check Answers***

5. -12
6. -22

Exercise 6–4–2

2 Simplify without writing parentheses and without rewriting a subtraction problem as an addition problem.

1. $8 - 2$

2. $9 - 5$

3. $4 - 9$

4. $11 - 20$

5. $-2 + 5$

6. $-6 + 8$

7. $7 - 13$

8. $9 - 21$

9. $-6 + 1$ **10.** $-3 + 2$ **11.** $-9 + 12$ **12.** $-6 + 14$

13. $-1 - 8$ **14.** $-3 - 6$ **15.** $-3 - 24$ **16.** $-12 - 7$

17. $23 - 17$ **18.** $24 - 19$ **19.** $13 - 45$ **20.** $16 - 25$

21. $-5.4 + 2.3$ **22.** $-3.8 + 2.6$ **23.** $23 - 41$ **24.** $13 - 52$

25. $\frac{3}{7} - \frac{5}{7}$ **26.** $\frac{1}{5} - \frac{3}{5}$ **27.** $5 - 6.4$ **28.** $-3.81 + 4$

29. $\frac{3}{8} - \frac{1}{2}$ **30.** $\frac{2}{3} - \frac{5}{6}$

3 Adding and Subtracting Several Numbers

If an expression contains several numbers, we must remember that the rules for adding and subtracting signed numbers apply to only two numbers at a time.

Example 10 Simplify: $-6 + 4 + 8 - 9$

Solution We can rewrite the expression in the form $(-6) + (+4) + (+8) + (-9)$, but, as mentioned earlier, try to look at the original expression and mentally picture the expression this way. We first add -6 and $+4$ to obtain -2.

Think: $-6 + 4 = (-6) + (+4) = -2$.

$$\boxed{-6 + 4} + 8 - 9 = \boxed{-2} + 8 - 9$$

We next add -2 and $+8$ to obtain $+6$.

Think: $-2 + 8 = (-2) + (+8) = +6$.

$$\boxed{-2 + 8} - 9 = \boxed{6} - 9$$

Next we add $+6$ and -9 to obtain the final result.

Think: $6 - 9 = (+6) + (-9) = -3$.

$$6 - 9 = -3$$

⊙ ***SKILL CHECK 7*** Simplify: $-8 + 6 + 5 - 7$

As we discussed in the previous section, you could have solved the problem in example 10 by rearranging the numbers to find the sum of the positive numbers, then the sum of the negative numbers, and finally add those two sums to obtain the final result.

We could write the problem as $-6 - 9 + 4 + 8 = -15 + 12 = -3$.

⊙ ***Skill Check Answer***

7. -4

Example 11 Simplify: $6 - 3 - 8 - 7 + 4$

Solution We rearrange numbers and write

$$\begin{aligned} 6 - 3 - 8 - 7 + 4 &= 6 + 4 - 3 - 7 - 8 \\ &= 10 - 10 - 8 \\ &= -8. \end{aligned}$$

Work the example by adding the numbers in order from left to right.

Note in this last example that this order gives easier combinations than combining left to right. The choice of order is left to the student since the answer will be the same either way.

Exercise 6-4-3

3 Simplify.

1. $3 - 6 + 9$

2. $2 - 7 + 5$

3. $-5 + 8 - 2$

4. $-11 + 3 - 2$

5. $5 - 6 + 4$

6. $6 - 8 - 4$

7. $-3 + 9 - 5$

8. $3 - 9 + 7$

9. $5 + 9 - 5$

10. $21 - 10 - 12$

11. $13 - 17 + 5 - 4$

12. $-9 + 10 - 6 + 3$

13. $-7 + 5 + 7 + 9$

14. $9 + 20 - 9 + 5$

15. $27 - 59 - 27 + 50$

16. $12 - 7 + 4 - 5$

17. $-20 + 17 - 5 + 25$

18. $8 - 7 + 2 - 4 + 9$

19. $34 - 16 + 9 + 16 - 3$

20. $6 - 4 + 8 + 16 - 9 - 8$

4 Grouping Symbols

Subtraction problems may also involve grouping symbols. Again, remember to simplify the terms within the grouping symbol before removing the symbol.

Example 12

$$\begin{aligned} 15 - (4 - 9) &= 15 - (-5) \\ &= 15 + 5 \\ &= 20 \end{aligned}$$

You should now be able to determine $4 - 9 = -5$ without writing all the parentheses and signs.

⊙ ***SKILL CHECK 8*** Simplify: $17 - (5 - 8)$

We may also have grouping symbols within grouping symbols.

Remember to remove the innermost set of grouping symbols first and to perform one step at a time.

Example 13

$$\begin{aligned} 14 - [-5 - (6 - 29)] &= 14 - [-5 - (-23)] \\ &= 14 - [-5 + 23] \\ &= 14 - [+18] \\ &= 14 + (-18) \\ &= -4 \end{aligned}$$

⊙ ***SKILL CHECK 9*** Simplify: $8 - [-5 - (3 - 9)]$

Example 14 Subtract the sum of 5 and -2 from 10.

Always analyze what the words are saying.

Solution We interpret the statement as follows: "First find the sum of 5 and -2. Then subtract the result from 10." We can write this statement in symbols as $10 - [(+5) + (-2)]$. Then simplifying, we have

$$\begin{aligned} 10 - [(+5) + (-2)] &= 10 - [+3] \\ &= 10 + (-3) \\ &= 7. \end{aligned}$$

⊙ ***SKILL CHECK 10*** Subtract the sum of -16 and 9 from 4.

⊙ ***Skill Check Answers***

8. 20
9. 7
10. 11

Exercise 6–4–4

A Simplify.

1. $7 + (8 - 12)$

2. $5 + (2 - 6)$

3. $6 - (4 + 8)$

4. $9 - (3 + 11)$

5. $5 + (9 - 12)$

6. $7 + (3 - 8)$

7. $10 - (2 + 5)$

8. $18 - (6 + 4)$

9. $13 - (2 - 1)$

10. $11 - (8 - 10)$

11. $(15 - 17) - (4 - 10)$

12. $(23 - 18) - (7 - 11)$

13. $(7 - 12) + (-5 - 4)$

14. $(3 - 13) + (-6 - 2)$

15. $(9 - 24) - (-5 + 9)$

16. $(6 - 14) - (-10 + 12)$

17. $4 + [13 - (6 + 4)]$

18. $8 + [10 - (3 + 4)]$

19. $11 + [-6 - (3 + 8)]$

20. $14 + [-8 - (4 + 7)]$

21. $8 - [4 - (3 - 2)]$

22. $9 - [5 - (4 - 7)]$

23. $-3 - [7 + (4 - 8)]$

24. $-6 - [4 - (9 - 4)]$

25. $10 - \{8 + [5 - (9 - 2)]\}$

26. $12 - \{10 - [4 - (5 - 9)]\}$

27. Subtract the sum of 5 and -11 from 16.

28. Subtract the sum of -6 and -2 from -18.

29. Subtract the difference of 7 and 3 from 10.

30. Subtract the difference of 9 and -4 from 12.

6–5 Adding and Subtracting Like Terms Involving Signed Numbers

OBJECTIVES

Upon completing this section you should be able to:

1. Add and subtract like terms involving signed numbers.
2. Remove grouping symbols to simplify expressions having like terms.

In chapter 1 we defined like terms as terms that have exactly the same literal factors. We also noted that only like terms can be added or subtracted. We will now extend those rules to terms involving signed numbers.

1 Adding and Subtracting Like Terms

Example 1 Simplify: $-3x - 7x$

Solution Recall that to add these like terms we must add the coefficients of x, in this case -3 and -7.

$$\begin{aligned}-3x - 7x &= (-3x) + (-7x)\\ &= [(-3) + (-7)]x\\ &= -10x\end{aligned}$$

◉ ***SKILL CHECK 1*** Simplify: $-5a - 3a$

Example 2 Simplify: $-5x + 3x - 4x$

Solution

$$\begin{aligned}-5x + 3x - 4x &= (-5 + 3 - 4)x\\ &= -6x\end{aligned}$$

◉ ***SKILL CHECK 2*** Simplify: $6x - 8x + 3x$

Example 3 Simplify: $12x^2 + 3y - 4x^2 - 4y$

Solution:

$$\begin{aligned}12x^2 + 3y - 4x^2 - 4y &= (12 - 4)x^2 + (3 - 4)y\\ &= 8x^2 - y\end{aligned}$$

We first rearrange the terms and then add the coefficients.

◉ ***SKILL CHECK 3*** Simplify: $10x^2 - 7y - 8x^2 + 2y$

2 Removing Grouping Symbols

Now consider the expression $4x - 3y - (3x + 2 - 5y)$. Since an expression enclosed in grouping symbols is treated as a single number, a grouping symbol preceded by a subtraction symbol indicates we are to subtract each term within the grouping symbol. We can rewrite the expression as

$$\begin{aligned}4x - 3y - (3x + 2 - 5y) &= 4x - 3y - (+3x) - (+2) - (-5y)\\ &= 4x - 3y - 3x - 2 + 5y\\ &= (4 - 3)x + (-3 + 5)y - 2\\ &= x + 2y - 2.\end{aligned}$$

◉ ***Skill Check Answers***

1. $-8a$
2. x
3. $2x^2 - 5y$

You should now be able to mentally rearrange the terms.

Example 4 Simplify: $2x^2 + 3x - 1 - (x^2 - 5x - 6)$

Solution

$$2x^2 + 3x - 1 - (x^2 - 5x - 6)$$
$$= 2x^2 + 3x - 1 - x^2 + 5x + 6$$
$$= x^2 + 8x + 5$$

⦿ ***SKILL CHECK 4*** Simplify: $3a^2 - 4a + 1 - (2a^2 - a + 4)$

Remember when working with grouping symbols within grouping symbols to remove the innermost set of symbols first.

Remember to simplify within the grouping symbols before removing them.

Example 5

$$5x - [3x - (2x + 3)] = 5x - [3x - 2x - 3]$$
$$= 5x - [x - 3]$$
$$= 5x - x + 3$$
$$= 4x + 3$$

⦿ ***SKILL CHECK 5*** Simplify: $9x - [4x - (3x - 1)]$

⦿ ***Skill Check Answers***

4. $a^2 - 3a - 3$
5. $8x - 1$

Exercise 6–5–1

1 Simplify.

1. $8x + 5x$

2. $5a - 3a$

3. $6x - 9x$

4. $11x^2 + 7x^2$

5. $8xy - 12xy$

6. $11a - 9b$

7. $17x^3 + 3x^2$

8. $9abc - 3abc$

9. $-7ab^2c - 5abc$

10. $3x^2y^2 - 2x^2y^2$

11. $4x + 6x - 10x$

12. $8y - 10y - 6y$

13. $3ab - 9ab + 4ab$

14. $3x - (7x - 5x)$

15. $16a^2b - 31a^2b + 5ab^2$

16. $2ab + 5bc - 6ac$

2

17. $7ab - (6ab - 3ac)$

18. $4xy + 13a - (7xy + 5a)$

19. $8xy - (5x^2y - 2xy) - 10xy + 5x^2y$

20. $21xyz + 15xy - (17xyz - 7xy^2)$

21. $2x + [3x - (2x + 1)]$

22. $4x - [3x - (x + 1)]$

23. $2x - [5x - (3x + 1)]$

24. $2x - [1 - (3x + 4)]$

25. $5 - [x + (8 - 3x)]$

26. $[3a - (4 - 2a)] - 3a$

27. $8x + [3x - 4 - (x + 5)]$

28. $5a - [-4 - (2a - 3)]$

29. $2x - 3 + [4x - (x - 7)]$

30. $7a + 4 - [3a - (5 - 2a)]$

31. $6a + 3b + [a + (3a - 4b)]$

32. $10x - y - [3x - (2x - 5y)]$

33. $a + b - [a - b - (a + b)]$

34. $u - v - [2u + 7v - (3u + 2v)]$

35. $6x + \{3x - [4x - 3 - (x - 1)]\}$

36. $2x^2 - 3x + \{x - 3x^2 - [4x - (2x^2 - x)]\}$

37. Subtract $(4x - 3y + 1)$ from $(8x - y - 4)$.

38. Subtract $(5x^2y + 3xy - y^2)$ from $(2x^2y - xy + 5y^2)$.

• Concept Enrichment

39. What property allows us to add or subtract like terms?

40. Explain what is wrong with this statement: $5x + 4x^2 = 9x^3$

6–6 Multiplying and Dividing Signed Numbers

OBJECTIVES

Upon completing this section you should be able to:

1. Multiply two numbers having opposite signs.
2. Apply the distributive property of multiplication over addition.
3. Multiply two negative numbers.
4. Divide signed numbers.

In previous sections we established rules for adding and subtracting signed numbers. In this section we will study the two other basic operations, **multiplication** and **division.**

1 Product of Two Numbers with Opposite Signs

Remember from chapter 2 that multiplication can be thought of as "shortcut addition." For instance, if we wish to multiply $(7)(5)$, we can think of this as "seven fives" or $5 + 5 + 5 + 5 + 5 + 5 + 5$ and obtain the result of 35. Or we can think of it as "five sevens" or $7 + 7 + 7 + 7 + 7$ and also obtain the same result of 35.

Applying this same technique to signed numbers will lead to one of the rules for their multiplication.

Think of $(+7)(-5)$ as seven (-5)s or

$$(-5) + (-5) + (-5) + (-5) + (-5) + (-5) + (-5) = -35.$$

Thus $(+7)(-5) = -35$.

This illustrates the following rule.

> The product of a positive number and a negative number yields a negative number.

Remember that multiplying any number by zero always gives a product of zero.

$$(5)(0) = 0$$
$$(-4)(0) = 0$$

Example 1 $(-2)(5) = -10$

Example 2 $(+3)(-12) = -36$

Example 3 $\left(+\frac{2}{3}\right)\left(-\frac{3}{4}\right) = -\frac{1}{2}$

◉ ***SKILL CHECK 1*** Multiply: $(-2.4)(3.5)$

From arithmetic we know the following rule.

> The product of two positive numbers is positive.

For this reason multiplication is called a *binary* operation.

In each of these rules be careful to note that the word *two* is very important. Since we can only multiply two numbers at a time, we only need rules for two numbers.

Example 4 Find the product: $(3)(-2)(7)$

Solution There are several ways this could be done. We could first multiply $(3)(-2)$, obtaining

$$\begin{aligned}(3)(-2)(7) &= (-6)(7)\\ &= -42.\end{aligned}$$

This is because multiplication is *commutative.* That is

$$(-2)(7) = (7)(-2).$$

We could also first multiply $(3)(7)$, obtaining

$$\begin{aligned}(3)(-2)(7) &= (3)(7)(-2)\\ &= (21)(-2)\\ &= -42.\end{aligned}$$

This is because multiplication is *associative.* That is $[(3)(-2)](7) = (3)[(-2)(7)]$.

Still another approach would be to first multiply $(-2)(7)$, obtaining

$$\begin{aligned}(3)(-2)(7) &= (3)(-14)\\ &= -42.\end{aligned}$$

◉ ***Skill Check Answer***

1. -8.4

It should be clear that the order in which the numbers are multiplied does not matter. The result will be the same.

◉ ***SKILL CHECK 2*** Find the product: $(5)(-3)(4)$

◉ ***Skill Check Answer***

2. -60

Exercise 6–6–1

1 Find the products.

1. $(+8)(-5)$

2. $(+9)(-4)$

3. $(-5)(+8)$

4. $(-4)(+9)$

5. $(7)(-3)$

6. $(6)(-7)$

7. $(-4)(0)$

8. $(-8)(0)$

9. $14(-13)$

10. $6(-11)$

11. $\left(-\frac{1}{2}\right)(8)$

12. $\left(-\frac{1}{3}\right)(12)$

13. $\left(\frac{2}{3}\right)\left(-\frac{9}{10}\right)$

14. $\left(\frac{4}{5}\right)\left(-\frac{3}{4}\right)$

15. $(-3)(2.5)$

16. $(-4)(4.2)$

17. $(5.1)(-2.3)$

18. $(4.0)(-6.2)$

19. $\left(\frac{2}{3}\right)\left(-\frac{3}{8}\right)$

20. $\left(-\frac{4}{5}\right)\left(\frac{15}{16}\right)$

21. $\left(-\frac{1}{3}\right)(6.3)$

22. $\left(-\frac{1}{2}\right)(4.6)$

23. $(5)(-2)(4)$

24. $(2)(-4)(3)$

25. $(-3)(2)(8)$

26. $(-6)(3)(10)$

27. $(5)(3)(-4)$

28. $(6)(2)(-1)$

29. $(-10)\left(\frac{1}{2}\right)(7)$

30. $(-25)(4)\left(\frac{1}{5}\right)$

31. $(-16)(121)(0)$

32. $(14)(0)(-15)$

33. $(4)\left(-\frac{1}{2}\right)(3)$

34. $\left(\frac{2}{3}\right)\left(-\frac{1}{4}\right)(6)$

35. $\left(-\frac{2}{7}\right)(14)\left(2\frac{3}{8}\right)$

36. $(1.2)(-3.1)(5.0)$

37. $(18)(2.3)(-4)$

38. $\left(-\frac{3}{4}\right)(18)\left(1\frac{5}{9}\right)$

39. $(5.4)\left(-3\frac{4}{5}\right)(0)$

40. $\left(\frac{1}{2}\right)\left(\frac{1}{4}\right)\left(-\frac{1}{16}\right)$

2 Using the Distributive Property

To establish a rule for the product or quotient of two negative numbers we will need to use a property discussed in section 2–1 called the distributive property of multiplication over addition. In symbols the distributive property of multiplication over addition is

Notice that a is a factor, that is, it is multiplying the quantity $(b + c)$.

$$a(b + c) = ab + ac.$$

We may express this property in words by saying "to multiply terms enclosed in parentheses by a number we must multiply each term in the parentheses by that number."

Also, since $b - c = b + (-c)$, then

$$\begin{aligned} a(b - c) &= a[b + (-c)] \\ &= ab + a(-c) \\ &= ab - ac \end{aligned}$$

WARNING

In an expression such as $2 + (x + 3)$ many students are tempted to multiply the expression in the parentheses by 2. But 2 is *not* a factor. The number directly multiplying the parentheses is $(+1)$.

Example 5 Simplify: $5(4 + 3)$

Solution We could just add $4 + 3$ and obtain

$$\begin{aligned} 5(4 + 3) &= 5(7) \\ &= 35. \end{aligned}$$

If we use the distributive property, we have

$$\begin{aligned} 5(4 + 3) &= 5(4) + 5(3) \\ &= 20 + 15 \\ &= 35. \end{aligned}$$

Example 6 Simplify: $-3(2x + 5)$

Solution

$$\begin{aligned} -3(2x + 5) &= -3(2x) + (-3)(+5) \\ &= -6x - 15 \end{aligned}$$

⦿ SKILL CHECK 3 Simplify: $-2(4x + 3)$

Example 7 Simplify: $3x + 8y - 5(x + 2y)$

Solution We can rewrite this as

$$\begin{aligned} 3x + 8y - 5(x + 2y) &= 3x + 8y + (-5)(x + 2y) \\ &= 3x + 8y - 5x - 10y \\ &= -2x - 2y. \end{aligned}$$

Subtracting $5(x + 2y)$ is the same as adding $(-5)(x + 2y)$.

⦿ SKILL CHECK 4 Simplify: $2a + 5b - 4(2a + b)$

Example 8 Simplify: $8x - (3x + 2)$

Solution In section 6–5 we removed grouping symbols preceded by a subtraction symbol by subtracting each term within the grouping symbols. So we would have

$$\begin{aligned} 8x - (3x + 2) &= 8x - 3x - 2 \\ &= 5x - 2. \end{aligned}$$

We can also approach this type of problem by remembering that the number directly in front of the parentheses is understood to be 1. In other words,

Remember from section 1–6 that if no coefficient is written it is assumed to be 1.

$$8x - (3x + 2) = 8x - 1(3x + 2).$$

Then using the rule for subtraction, we have

$$= 8x + (-1)(3x + 2).$$

Next using the distributive property, we have

$$= 8x - 3x - 2.$$

Finally adding like terms, we obtain

$$= 5x - 2.$$

We see that the result is the same by using either approach. The method used is left to the preference of the student.

⦿ SKILL CHECK 5 Simplify: $7x - (5x + 3)$

⦿ *Skill Check Answers*

3. $-8x - 6$
4. $-6a + b$
5. $2x - 3$

Exercise 6–6–2

Simplify.

1. $5(x + 4)$

2. $2(x - 9)$

3. $-2(3x + 4)$

4. $-3(2x + 5)$

5. $14(3x - 2)$

6. $-8(2x + 5)$

7. $3(2x^2 + 4x + 2)$

8. $5(2x^2 - x + 3)$

9. $-2(x^2 + 3x + 1)$

10. $-4(3a + 2b + c)$

11. $2x - 3(x + 5)$

12. $7x - 2(x + 3)$

13. $9 + (12x - 7)$

14. $-4(2x + 3) + 15$

15. $-3(4a + 1) + 12a$

16. $6(4a - 3) - 10a + 18$

17. $3(2x - y) + x - 2y$

18. $-4(3x + 2y) + 9x + 10y$

19. $7x - (4x + 6)$

20. $5x - (9x + 3)$

21. $11a - 3b - (a + 7b)$

22. $16u - 13v - (12u + 3v)$

23. $21w - 13 - 4(6w + 19)$

24. $7m - 12n - 5(2m + 9n)$

25. $3(5x - 4) - 2(x + 3)$

26. $-2(3x + 7) - (5x + 12)$

27. $3(2a - b) - 3(a + 4b + 6)$

28. $3(x - 4) - 2(x^2 + 2x + 1)$

3 The Product of Two Negative Numbers

We now wish to establish a rule for multiplying two negative numbers.

When the sum of two numbers is zero we say that one number is the *additive inverse* of the other.

By using our rules for adding signed numbers we have worked problems such as $(-15) + (+15) = -15 + 15 = 0$.

We also know that multiplying by zero gives zero as a result. For instance, $(-3)(0) = 0$.

Since $(5) + (-5) = 0$ we say 5 is the additive inverse of (-5).

Now consider this problem. Find the result of $(-3)[(5) + (-5)]$. If we note that $(5) + (-5) = 0$, then we have

$$\begin{aligned}(-3)[(5) + (-5)] &= (-3)(0)\\ &= 0.\end{aligned}$$

Remember, the distributive property tells us to multiply each term inside the grouping symbols by the factor preceding the grouping symbols.

However, if we first use the distributive property, we obtain

$$\begin{aligned}(-3)[(5) + (-5)] &= (-3)(5) + (-3)(-5)\\ &= -15 + (-3)(-5).\end{aligned}$$

We know this result must be zero. That is, $-15 + (-3)(-5) = 0$. Therefore $(-3)(-5) = +15$ since $+15$ is the only number that can be added to -15 to obtain a result of zero.

The choice of numbers in this discussion would not change the conclusion. Hence we have the following rule.

The product of two negative numbers is positive.

WARNING

Adding two negative numbers yields a negative number. *Multiplying* two negative numbers yields a positive number. Don't confuse these two rules.

Example 9 $(-2)(-4) = 8$

Example 10 $(-4)(-6) = 24$

Example 11 $\left(-\frac{2}{3}\right)\left(-\frac{3}{4}\right) = \frac{1}{2}$

⊙ ***SKILL CHECK 6*** Multiply: $(-2.6)(-3.5)$

Once more we must realize the importance of the word *two*. This rule, as all others, must be applied to only *two* numbers at a time.

Example 12 Find the product of $(-3)(-2)(-5)$.

Solution If we first multiply $(-3)(-2)$, we obtain $(+6)$. Then multiplying $(+6)(-5)$, we obtain the final result.

$$\begin{aligned}(-3)(-2)(-5) &= (+6)(-5)\\ &= -30\end{aligned}$$

Find this product by first multiplying $(-2)(-5)$.

⊙ ***SKILL CHECK 7*** Find the product of $(-4)(-7)(-2)$.

Example 13 Find the product of $(-3)(4)(-5)$.

Solution

$$\begin{aligned}(-3)(4)(-5) &= (-12)(-5)\\ &= 60\end{aligned}$$

Find this product by first multiplying $(-3)(-5)$.

⊙ ***SKILL CHECK 8*** Find the product of $(-2)(5)(-6)$.

Example 14 Find the product of $(-2)(5)(-3)(-1)(4)$.

Solution

$$\begin{aligned}(-2)(5)(-3)(-1)(4) &= (-10)(-3)(-1)(4)\\ &= (+30)(-1)(4)\\ &= (-30)(4)\\ &= -120\end{aligned}$$

⊙ ***SKILL CHECK 9*** Find the product of $(-3)(2)(-1)(6)(-4)$.

⊙ ***Skill Check Answers***

6. 9.1
7. −56
8. 60
9. −144

Exercise 6–6–3

3 Find the product.

1. $(-2)(-3)$
2. $(-4)(-3)$
3. $(-1)(-8)$
4. $(-1)(-5)$
5. $(-3)(-7)$
6. $(-8)(-10)$
7. $(-5)(-9)$
8. $(-6)(-4)$
9. $\left(-\frac{1}{2}\right)(-30)$

10. $\left(-\frac{1}{3}\right)(-9)$

11. $\left(-\frac{1}{3}\right)\left(-\frac{3}{5}\right)$

12. $\left(-\frac{2}{3}\right)\left(-\frac{1}{4}\right)$

13. $(-5)(+8)$

14. $(-3)(+5)$

15. $7(-3)$

16. $9(-4)$

17. $(-1.2)(-5.0)$

18. $(-3.2)(-1.5)$

19. $(6)(-2)(-5)$

20. $(4)(-9)(-3)$

21. $(-1)(-3)(-4)$

22. $(-2)(-8)(5)$

23. $(-5)(6)(-2)$

24. $(-3)(4)(-5)$

25. $(-1)(-5)\left(-\frac{1}{5}\right)$

26. $\left(-\frac{1}{3}\right)(-12)\left(-\frac{1}{2}\right)$

27. $(-3)(5)(0)$

28. $(-7)(-14)(0)$

29. $(3)(-1)(-5)(10)$

30. $(-4)(3)(-11)(2)$

31. $(-1)(-5)(-2)(-3)$

32. $(-3)(-1)(-4)(-6)$

33. $(-2)(-7)(3)(4)$

34. $(3)(-2)(-1)(-7)$

35. $(-10)(10)(-1)(-3)$

36. $(-8)(4)(-2)(10)$

37. $(-2)(-1)(-10)(4)(1)$

38. $\left(-\frac{1}{2}\right)(-14)\left(-\frac{3}{7}\right)(-2)$

39. $(27)\left(-\frac{1}{9}\right)\left(-\frac{1}{3}\right)(-37)$

40. $(49)(-72)(-104)(0)(-23)$

- **Concept Enrichment**

41. Several nonzero numbers are to be multiplied. An even number of them are negative numbers. Determine the sign of the product.

42. Several nonzero numbers are to be multiplied and an odd number of them are negative numbers. Determine the sign of the product.

4 Dividing Signed Numbers

Division is defined as "multiplying by the inverse." The **multiplicative inverse** of a number is sometimes referred to as the **reciprocal** of the number.

In a problem such as $12 \div 6$ we are dividing by 6. Using the definition, the problem is the same as multiplying by the multiplicative inverse (or reciprocal) of 6, which is $\frac{1}{6}$.

Thus: $$12 \div 6 = 12\left(\frac{1}{6}\right) = 2.$$

Similarly, in the problem $12 \div (-6)$ the multiplicative inverse of -6 is $-\frac{1}{6}$, so

$$12 \div (-6) = 12\left(-\frac{1}{6}\right) = -2.$$

This relationship ties division and multiplication together so that the rules for division are the same as the rules for multiplication. For convenience we will restate the rules as a single rule.

Examples:

The multiplicative inverse of $\frac{2}{3}$ is $\frac{3}{2}$.

The multiplicative inverse of 8 is $\frac{1}{8}$.

The multiplicative inverse of $-\frac{3}{5}$ is $-\frac{5}{3}$.

> The product or quotient of two numbers having like signs is positive and the product or quotient of two numbers having unlike signs is negative.

Example 15 $(-12) \div (-6) = +2$

⦿ ***SKILL CHECK 10*** Divide: $(-8) \div (-2)$

Example 16 $(-12) \div (+6) = -2$

Example 17 $(+12) \div (-6) = -2$

⦿ ***SKILL CHECK 11*** Divide: $(+35) \div (-5)$

Example 18 $\frac{-8}{-2} = +4$

Example 19 $\frac{-8}{+2} = -4$

Example 20 $$\left(-\frac{2}{3}\right) \div \left(-\frac{3}{4}\right) = \left(-\frac{2}{3}\right)\left(-\frac{4}{3}\right) = +\frac{8}{9}$$

Note that the multiplicative inverse of $-\frac{3}{4}$ is $-\frac{4}{3}$.

⦿ ***Skill Check Answers***

10. $+4$
11. -7

⦿ ***SKILL CHECK 12*** Divide: $\left(-\frac{3}{4}\right) \div \left(-\frac{5}{6}\right)$

Example 21 $\frac{(-6)(-7)}{(-3)} = (2)(-7) = -14$

⦿ ***SKILL CHECK 13*** Simplify: $\frac{(-10)(+9)}{(-6)}$

⦿ ***Skill Check Answers***

12. $+\frac{9}{10}$
13. $+15$

Exercise 6–6–4

Find the quotient.

1. $\frac{-6}{-2}$
2. $\frac{-14}{-2}$
3. $\frac{-6}{+2}$
4. $\frac{-14}{+2}$
5. $\frac{-15}{-3}$
6. $\frac{-8}{-4}$
7. $\frac{+15}{-3}$
8. $\frac{-18}{-3}$
9. $\frac{-20}{+5}$
10. $\frac{+18}{-3}$
11. $\frac{-42}{-6}$
12. $\frac{+24}{-6}$
13. $(-20) \div (-5)$
14. $(-56) \div (-7)$
15. $(+81) \div (-3)$
16. $(+20) \div (-4)$
17. $(-28) \div (+4)$
18. $(-100) \div (+20)$
19. $\left(-\frac{1}{2}\right) \div \left(-\frac{1}{8}\right)$
20. $\left(\frac{3}{4}\right) \div \left(-\frac{9}{16}\right)$
21. $\left(-\frac{2}{5}\right) \div \left(\frac{3}{7}\right)$
22. $\left(-\frac{4}{5}\right) \div \left(\frac{5}{8}\right)$
23. $\left(\frac{3}{8}\right) \div \left(-\frac{3}{16}\right)$
24. $\left(\frac{2}{7}\right) \div \left(-\frac{4}{5}\right)$
25. $(-3) \div \left(-\frac{6}{7}\right)$
26. $-9 \div \left(-\frac{3}{4}\right)$
27. $\left(-\frac{5}{7}\right) \div 15$

28. $\left(-\frac{3}{8}\right) \div 12$

29. $\frac{(-5)(-3)}{(-6)}$

30. $\frac{(-2)(-5)}{-10}$

31. $\frac{(-3)(-6)}{-9}$

32. $\frac{(-2)(-9)}{-3}$

33. $\frac{(+4)(-6)}{(-2)}$

34. $\frac{3(-10)}{-6}$

35. $\frac{(-6)(+4)}{12}$

36. $\frac{(-5)(+4)}{-10}$

37. $\frac{3 - 15}{-6}$

38. $\frac{5 - 33}{-7}$

39. $\frac{9 - 27}{6}$

40. $\frac{18 - 4}{-7}$

CHAPTER 6 SUMMARY

Key Words

Section 6–1

- **Signed numbers** or **directed numbers** are preceded by a positive or negative sign.
- **Positive numbers** are numbers preceded by a positive sign.
- **Negative numbers** are numbers preceded by a negative sign.
- The **opposite** of a number is that number with the opposite sign.
- The **absolute value** of a number is the distance of that number from zero on the number line.

Section 6–2

- **Addition** of signed numbers can be accomplished by using the number line.

Section 6–3

- **Addition** of signed numbers can be accomplished by using a rule involving the absolute values of the numbers to be added.

Section 6–4

- Every **subtraction** problem can be written as an addition problem.

Section 6–5

- **Like terms** can be added and subtracted by using the rules for adding and subtracting signed numbers.

Section 6–6

- **Multiplication** and **division** use the same rule for obtaining the sign of the answer.

Procedures

Section 6–1

- To find the opposite of a number change its sign.
- To find the absolute value of a number determine its distance from zero on the number line.

Section 6–2

- If a positive number indicates a direction to the right and a negative number indicates a direction to the left, then the number line can be used to add signed numbers.

Section 6–3

- To add two signed numbers with like signs add their absolute values. The sum will have the same sign as the numbers being added.
- To add two numbers with unlike signs find the difference of their absolute values. The answer will have the sign of the number having the greater absolute value.

Section 6–4

- To subtract a number use the rule

$$a - b = a + (-b).$$

Section 6–5

- To add and subtract like terms add or subtract the coefficients to obtain the coefficient of the result.
- A grouping symbol preceded by a subtraction symbol indicates we are to subtract each term within the grouping symbol.

Section 6–6

- The product or quotient of two numbers having like signs is positive.
- The product or quotient of two numbers having unlike signs is negative.

CHAPTER 6 REVIEW

Locate the numbers on the number line and replace the question mark with $<$ or $>$ to indicate a true statement.

1. 5 ? 2

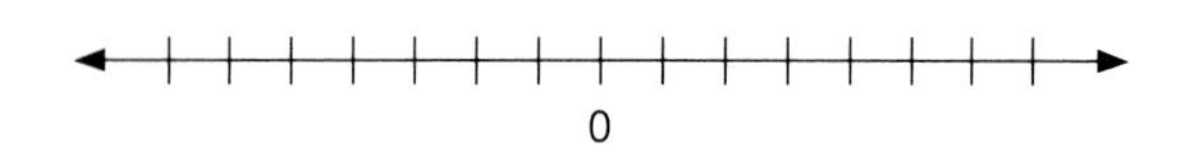

2. 7 ? −2

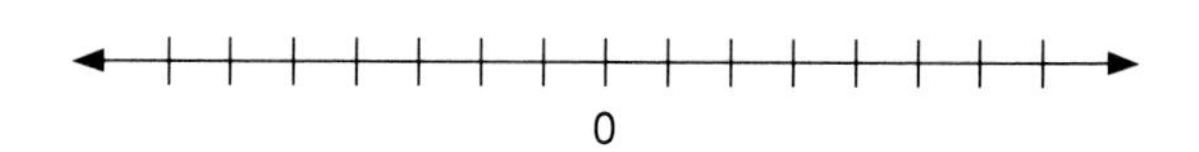

3. −5 ? −2

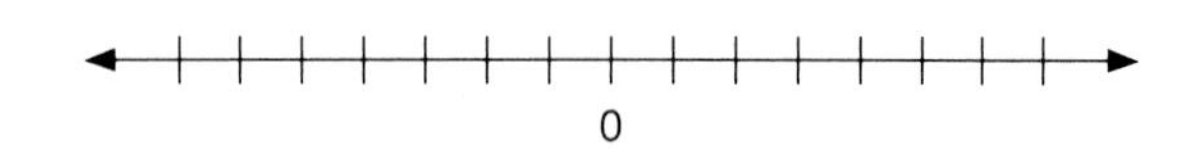

4. −3 ? 0

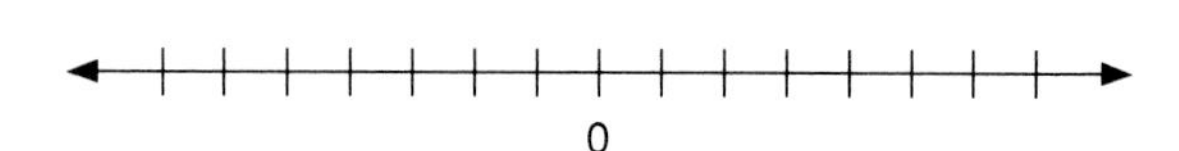

Find the opposite.

5. $+10$

6. $+\frac{2}{3}$

7. $-\frac{3}{8}$

8. -5.1

Find the value.

9. $|+4|$

10. $|-12|$

11. $-|-3|$

12. $-|+6|$

Use a number line to add.

13. $(+5) + (-8)$

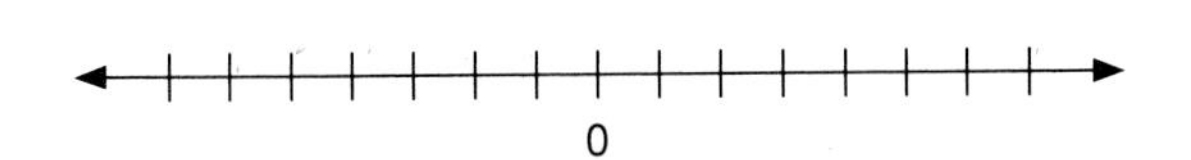

14. $(-3) + (+7)$

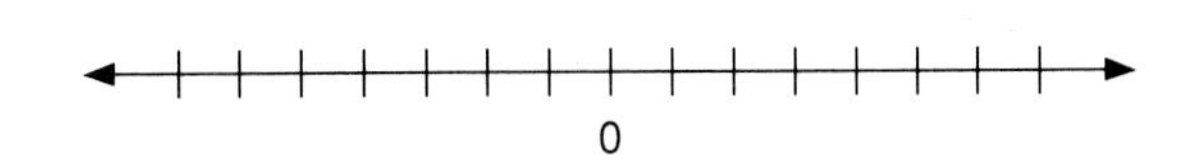

15. $(+2) + (-5) + (+1)$

0

16. $(-4) + (+6) + (-2)$

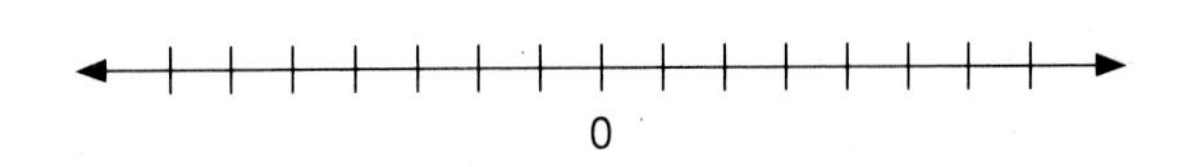

Simplify.

17. $(+7) + (-9)$

18. $(-17) + (-5)$

19. $\left(+\frac{5}{9}\right) + \left(-\frac{2}{9}\right)$

20. $\left(-\frac{1}{2}\right) + \left(+\frac{2}{3}\right)$

21. $(-5) + (+8) + (-10)$

22. $(+6) + (-15) + (+4)$

23. $5 - 8 + 2$

24. $4 - (-6) - 1$

25. $-3 - (-4) + 9$

26. $-13 + 5 - (-16) - 2$

27. $-4 + 9 + 2 - 8 - 7 + 6$

28. $4 - (9 - 2)$

29. $5 - (-6 - 9) + 2$

30. $7 - (5 - 16) - 4$

31. $(-4)(+12)$

32. $(-6)\left(+\frac{1}{3}\right)$

33. $\left(-\frac{3}{4}\right)\left(-\frac{2}{7}\right)$

34. $(-2.3)(-3.5)$

35. $(8)(-5)(-7)$

36. $(-4)(3)(10)$

37. $\frac{-36}{-3}$

38. $\frac{-32}{+8}$

39. $\left(\frac{4}{9}\right) \div \left(-\frac{2}{3}\right)$

40. $(-15) \div \left(-\frac{3}{5}\right)$

41. $\frac{(-6)(5)}{-10}$

42. $\frac{(-8)(-6)}{+12}$

43. $(-3)(-4)(-5)$

44. $(-3)(-1)(5)(-10)\left(-\frac{2}{5}\right)(3)$

45. $\frac{(-4)(-9)}{-6}$

46. $\frac{-18 + 3}{-5}$

47. $3x^2 + 8x^2 - 11x^2$

48. $-ab + 4ab - 5ab$

49. $3x - (2x + 5) - 7$

50. $10a - (9 - 4a) + 16$

51. $8 - (3x + 5) + 12x$

52. $15 + (5x - 7) - (9x - 2)$

53. $14 - 3(2x - 1)$

54. $5(3a - 2) - 4(3a + 1)$

55. $3(a - 2) - 5(2a - 4)$

56. $8(3x - 4) + 3x - 2(6x + 5)$

CHAPTER 6 TEST

1. Find the opposite of $-\frac{5}{8}$

2. Find the opposite of 6.9

3. Simplify: $-|-28|$

4. Use the number line to show the addition of $(+5) + (-11)$.

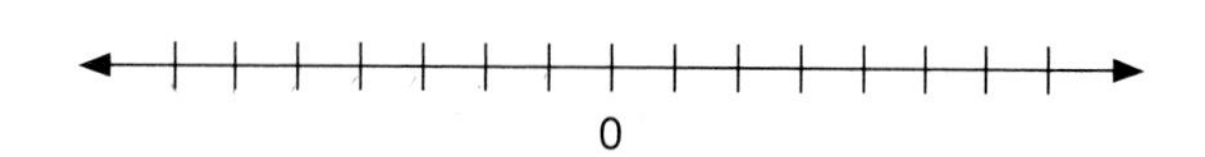

5. Add: $(-32) + (+18)$

6. Add: $(-13) + (-26)$

7. Subtract: $(-6) - (+14)$

8. Subtract: $(-19) - (-23)$

9. Simplify: $4 - 11 - 8 - (-3) + 5$

10. Remove parentheses and simplify: $11 - (-2 - 8) + 15$

11. Combine like terms: $16ab + 9a^2 - 12ab - 6a^2$

12. Simplify: $8 - [3a - (5a + 2)]$

13. Subtract the sum of 13 and 6 from 10.

14. Find the product: $(-11)(-5)$

15. Find the product: $(-8)(+12)$

16. Find the product: $(-3)(8)(-4)(-2)$

17. Find the quotient: $\frac{-26}{-2}$

18. Find the quotient: $\left(-\frac{4}{5}\right) \div 8$

19. Evaluate: $\frac{(-8)(-9)}{-6}$

20. Evaluate: $\frac{-21 + 5}{-2}$

CHAPTERS 1–6 CUMULATIVE TEST

1. What is the place value of the digit 6 in the number 621,435?

2. Multiply: $\begin{array}{r} 2{,}314 \\ \times\ \ 265 \\ \hline \end{array}$

3. Add: $\frac{2}{3} + \frac{1}{4} + \frac{5}{6}$

4. Divide: $3.05\overline{)65.27}$

5. Find: $-|-3|$

6. Multiply: $\begin{array}{r} 2.43 \\ \times 0.28 \\ \hline \end{array}$

7. Simplify: $3a + 2b - a - b$

8. Find the least common multiple of 12, 15, and 20.

9. Divide: $\frac{12}{35} \div \frac{21}{10}$

10. Evaluate: $6 + 2[3 + (7 - 4)]$

11. Find the area of a triangle if the measure of the base is 5.2 feet and the altitude is 3 feet.

12. Round 6.80713 to the nearest hundredth.

13. Find the measure of an angle if its complement is 63°.

14. Replace the question mark in 18 ? 24 with the inequality symbol positioned to indicate a true statement.

15. Round 120,504 to the nearest thousand.

16. Multiply: $\left(\frac{3x^2}{11y^2}\right)\left(\frac{22y}{6x}\right)$

17. Write 0.063 as a common fraction.

18. Add: $(-5) + (+9) + (-6)$

19. Simplify: $7 - 4(2x - 3) + 2(3x - 5)$

20. Find the perimeter of a square if a side measures 13.45 meters.

21. Simplify: $-3 + 5 + 7 - 8 - 5$

22. Find the prime factorization of 300.

23. Simplify: $\frac{(-3)(-10)}{-6}$

24. A tank contains $4\frac{1}{2}$ liters of a solution. If $2\frac{3}{4}$ liters are poured out, how many liters remain in the tank?

25. A rectangular floor measures 7.0 yards by 5.5 yards. If the cost of carpeting and padding is $22.50 per square yard, what will it cost to carpet the floor?

CHAPTER

7 Pretest

Answer as many of the following questions as you can before starting this chapter. When you finish the chapter, take the test at the end and compare your scores to see how much you have learned.

1. Simplify: $(-3)(2) - (5)(-1)$

2. Simplify: $8 - 2(6 - 10) - 10 \div 2$

3. Simplify: $(-5)^3$

4. Simplify: $(-3)^3(-2)^2$

5. Simplify: $-3x(x^2 - 2x + 4)$

6. Simplify: $\dfrac{-12a^2b}{3ab}$

7. Simplify: $7x - 5[x - 2(3x - 5)]$

8. Evaluate $2a^3 - 4a^2b - b^2$ when $a = -5$ and $b = -2$.

9. Write 21.6×10^4 without exponents.

10. Write 0.00000319 in scientific notation.

11. Write 3.02×10^{-5} without exponents.

12. Find the square roots of 121.

13. Find: $\sqrt{169}$

14. The length of one leg of a right triangle is 8 inches and the hypotenuse is 10 inches. Find the length of the other leg.

15. Which property of the real numbers is illustrated by

$$[(2)(-4)]\left(-\frac{1}{4}\right) = 2\left[(-4)\left(-\frac{1}{4}\right)\right]?$$

CHAPTER

7 Using Signed Numbers

Signed numbers are used in many areas of mathematics. One such use is in representing extremely large and extremely small numbers.

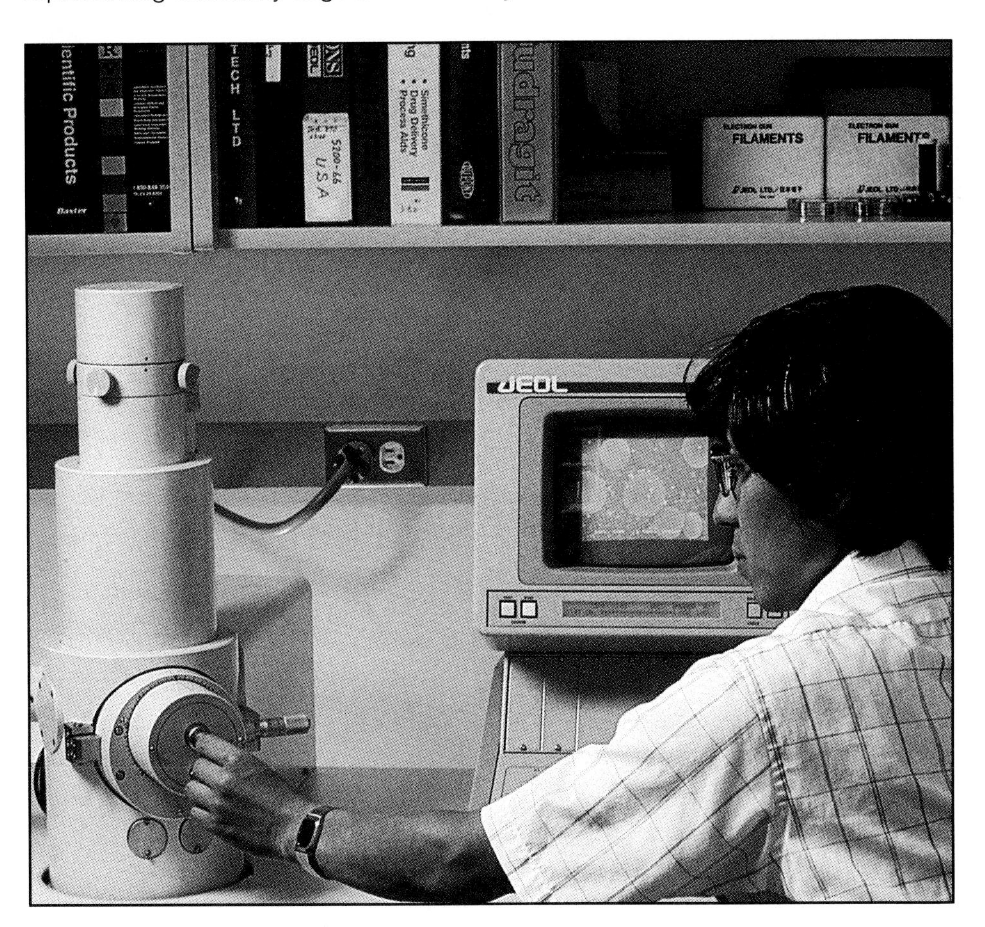

STUDY HELP

Taking a test is an unpleasant task for many students. A systematic approach to taking a test can greatly reduce your anxiety. A good strategy is to first take time to scan the entire test. By doing this you will see many problems that appear easy to you. Work these problems before attempting the more difficult ones. In this way your time will be better utilized and valuable time will not be spent on a problem that is probably not worth more points than any other question.

In chapter 6 we learned to perform the four basic operations of addition, subtraction, multiplication, and division on signed numbers. In this chapter we will work various exercises involving signed numbers and discover important uses of this set of numbers. Square roots and the Pythagorean theorem will be discussed. Finally we will summarize the sets of numbers we have introduced and establish the set of real numbers.

7–1 Exercises Using Signed Numbers

OBJECTIVES

Upon completing this section you should be able to:

1. Simplify expressions involving signed numbers.
2. Use the distributive property to remove grouping symbols.
3. Evaluate algebraic expressions involving signed numbers.

In chapter 2 we discussed grouping symbols and the order of operations. There we worked problems involving only the numbers of arithmetic. We can now work the same types of problems using signed numbers.

1 Simplifying Expressions

Recalling the order of operations (section 2–3), we can simplify expressions that involve all four basic operations.

Example 1 Simplify: $12 + (3)(-4) - 2(-1) - 5$

Solution First we perform the operation of multiplication from left to right.

$$\begin{aligned} &12 + (3)(-4) - 2(-1) - 5 \\ &\quad = 12 + (-12) - (-2) - 5 \end{aligned}$$

Remember, first multiplication and division, then addition and subtraction.

We then perform the operations of addition and subtraction from left to right.

$$\begin{aligned} 12 + (-12) - (-2) - 5 &= 0 - (-2) - 5 \\ &= 0 + 2 - 5 \\ &= 2 - 5 \\ &= -3 \end{aligned}$$

SKILL CHECK 1 Simplify: $6 + (5)(-2) - 4(-3) - 14$

Example 2 Simplify: $16 - 2(3 - 10) - 6 \div (-2)$

Simplify parentheses before using the order of operations.

Solution Here we first simplify the expression within the parentheses and then follow the order of operations.

$2(-7) = -14$ and $6 \div (-2) = -3$.

$$\begin{aligned} &16 - 2(3 - 10) - 6 \div (-2) \\ &\quad = 16 - 2(-7) - 6 \div (-2) \\ &\quad = 16 - (-14) - (-3) \\ &\quad = 16 + 14 + 3 \\ &\quad = 30 + 3 \\ &\quad = 33 \end{aligned}$$

SKILL CHECK 2 Simplify: $20 - 5(6 - 10) + 8 \div (-2)$

We should also be able to simplify expressions involving powers of signed numbers.

Example 3 Simplify: $(-2)^3$

See section 1–3.

Solution Recall the meaning of an exponent.

$$\begin{aligned} (-2)^3 &= (-2)(-2)(-2) \\ &= (+4)(-2) \\ &= -8 \end{aligned}$$

SKILL CHECK 3 Simplify: $(-5)^3$

Skill Check Answers

1. -6
2. 36
3. -125

Example 4 Simplify: $(2)^3(-5)^2$

Solution Remembering that $(2)^3 = (2)(2)(2) = 8$ and $(-5)^2 = (-5)(-5) = 25$, we have

$$(2)^3(-5)^2 = (8)(25) = 200.$$

Notice that we must follow the order of operations. See section 2–3.

⊙ ***SKILL CHECK 4*** Simplify: $(-3)^2(-1)^3$

In chapter 2 we discussed multiplying monomials. This concept can be extended to include monomials having negative as well as positive coefficients.

Example 5 Simplify: $(-2x)(5x^2)$

Solution Recalling the rule for multiplying monomials, we write

See section 2–6.

$$\begin{aligned}(-2x)(5x^2) &= (-2)(5)(x)(x^2)\\ &= -10x^3.\end{aligned}$$

⊙ ***SKILL CHECK 5*** Simplify: $(-3a)(-2a^2)$

An expression consisting of the sum or difference of one or more monomials is called a **polynomial.** $2x^2 + 3x - 5$ is an example of a polynomial. We can use the distributive property to multiply a monomial and another polynomial.

Example 6 Simplify: $-2x^2(3x^2 - 4x + 6)$

Solution Each term within the parentheses must be multiplied by $-2x^2$.

$$\begin{aligned}&-2x^2(3x^2 - 4x + 6)\\ &= (-2x^2)(3x^2) + (-2x^2)(-4x) + (-2x^2)(6)\\ &= (-6x^4) + (+8x^3) + (-12x^2)\\ &= -6x^4 + 8x^3 - 12x^2\end{aligned}$$

We are using the distributive property.

⊙ ***SKILL CHECK 6*** Simplify: $-5x^2(x^2 + 3x - 4)$

Example 7 Simplify: $x(2x^2 - 4x + 5) - 3x^2(3x + 4)$

Solution We use the distributive property and write

$$\begin{aligned}&x(2x^2 - 4x + 5) - 3x^2(3x + 4)\\ &= x(2x^2) + x(-4x) + x(5) + (-3x^2)(3x) + (-3x^2)(4)\\ &= 2x^3 + (-4x^2) + 5x + (-9x^3) + (-12x^2)\\ &= -7x^3 - 16x^2 + 5x\end{aligned}$$

⊙ ***SKILL CHECK 7*** Simplify: $2a(3a^2 - 2a - 1) - a^2(4a - 3)$

Example 8 Simplify: $\dfrac{-18a^2b}{+3ab^3}$

Solution In chapter 3 we worked problems of this type with positive coefficients. The procedure here is the same only we must remember the rules for dividing signed numbers.

See section 3–1.

$$\frac{-18a^2b}{+3ab^3} = \frac{-(2)(\cancel{3})(3)(\cancel{a})(a)(\cancel{b})}{+(\cancel{3})(\cancel{a})(\cancel{b})(b)(b)} = \frac{-6a}{b^2} = -\frac{6a}{b^2}$$

In example 8 we note that the numerator of the fraction is negative and the denominator is positive. Thus a negative number divided by a positive number yields a negative result and we can write the negative sign in front of the fraction to indicate the resulting fraction is negative.

$\dfrac{-6a}{b^2}$, $\dfrac{6a}{-b^2}$, and $-\dfrac{6a}{b^2}$ are all equivalent.

⊙ ***Skill Check Answers***

4. -9 5. $6a^3$
6. $-5x^4 - 15x^3 + 20x^2$
7. $-2a^3 - a^2 - 2a$

⊙ ***Skill Check Answer***

8. $-\frac{8y}{x}$

⊙ ***SKILL CHECK 8*** Simplify: $\frac{24xy^2}{-3x^2y}$

Exercise 7–1–1

1 Simplify.

1. $3 + (-2)(4)$

2. $8 + (-3)(5)$

3. $(-1)(-5) + (-4)(3)$

4. $(3)(-2) + (-1)(-4)$

5. $(6)(-2) - (3)(-8)$

6. $(-2)(-5) - (-3)(2)$

7. $6 - (4)(5) + (-3)(2) - 4$

8. $13 + (3)(-6) - (-2)(8) - 7$

9. $12 - 3(7 - 10) - 5(-3)$

10. $17 - 5(2 - 9) - 6(-3)$

11. $10 + (-6) - (-2) - 2(9 - 13)$

12. $14 - (-15) - (-3) - (7 - 11)$

13. $10 - 2(1 - 5) - 4 \div (-1)$

14. $8 + 3(5 - 7) + 15 \div (-3)$

15. $-6(3 - 5) - 12 \div 2 - 1$

16. $-4(1 - 3) - 6 \div (-2) - 8$

17. $(-2)^5$

18. $(-5)^2$

19. $-(-2)^2$

20. $-(-3)^3$

21. $(2)^3(-3)^2$

22. $(-4)^3(-1)^5$

23. $(-a)(-2a)$

24. $(x)(-3x)$

25. $(2x)(-5x^2)$

26. $(-2a)(-3a^2)$

27. $(-2x)(4x^3)$

28. $(-3x^2)(-4x^3)$

29. $-x(2x - 5)$

30. $-a(3a - 4)$

31. $-3a(3a^2 - 4a - 1)$

32. $-2x(x^2 - 2x - 3)$

33. $-2a^2(3a^2 - 3a + 4)$

34. $-3x^3(-2x^2 + 5x - 3)$

35. $\frac{-6xy}{-2x}$

36. $\frac{-15ab}{-3b}$

37. $\frac{-a^2b}{3ab}$

38. $\frac{4xy^2}{-xy}$

39. $\frac{20a^3b}{-4a}$

40. $\frac{-24xy^3}{6x^2y}$

41. $\frac{-21xy^3}{15x^2y^2}$

42. $\frac{-35ab^3}{-7a^3b^2}$

43. $a(3a - 1) - a(-2a + 5)$

44. $x(-2x - 3) - x(x - 4)$

45. $x(3x^2 - 4x + 1) - 2x(x^2 + 5x - 7)$

46. $3a^2(2a - 4) - 4a(a^2 + 3a - 2)$

• Concept Enrichment

47. Consider the expression $(-1)^n$, where n represents a whole number. For what value(s) of n will the value of the expression be positive?

48. Consider the expression given in problem 47. For what value(s) of n will the value of the expression be negative?

49. What property allows us to write $-x(2x^2 + x - 5) = -2x^3 - x^2 + 5x$?

50. Explain the distinction between -2^2 and $(-2)^2$.

2 Grouping Symbols

In section 6–5 we worked with grouping symbols within grouping symbols that involved addition and subtraction. Use of the distributive property may also be involved.

Example 9 Simplify: $3x - 2[5(x - y) + 8y]$

Solution

$$\begin{aligned} 3x - 2[5(x - y) + 8y] & \\ &= 3x - 2[5x - 5y + 8y] && \text{Multiply } 5(x - y). \\ &= 3x - 2[5x + 3y] && \text{Add } (-5y) + (+8y). \\ &= 3x - 10x - 6y && \text{Multiply } (-2)[5x + 3y]. \\ &= -7x - 6y && \text{Add } (+3x) + (-10x). \end{aligned}$$

◉ ***SKILL CHECK 9*** Simplify: $8x - 3[2(2x - y) - 3y]$

Example 10 Simplify: $3x^2 - x[2x - 5(x - 1)]$

Solution

$$\begin{aligned} 3x^2 - x[2x - 5(x - 1)] & \\ &= 3x^2 - x[2x - 5x + 5] \\ &= 3x^2 - x[-3x + 5] \\ &= 3x^2 + 3x^2 - 5x \\ &= 6x^2 - 5x \end{aligned}$$

WARNING

To avoid errors perform only one step at a time.

◉ ***SKILL CHECK 10*** Simplify: $4a^2 - a[5a - 3(2a - 4)]$

3 Evaluating Expressions

In chapter 2 we evaluated algebraic expressions using only the numbers of arithmetic. We are now able to evaluate expressions that involve signed numbers.

Example 11 Evaluate (a) x^2, (b) $-x^2$, and (c) $(-x)^2$ when $x = -2$.

Solution In each case we must replace x with -2 and evaluate the expression.

a. $x^2 = (-2)^2 = 4$
b. $-x^2 = -(-2)^2 = -4$
c. $(-x)^2 = [-(-2)]^2 = [+2]^2 = 4$

Note the distinction between *b* and *c*.
$-x^2 = -(x)(x)$
$(-x)^2 = (-x)(-x)$

◉ ***SKILL CHECK 11*** Evaluate $-x^2$ when $x = -3$.

Example 12 Evaluate $x^2 - 4x + 6$ when $x = -3$.

Solution Replacing x with -3, we obtain

$$\begin{aligned} x^2 - 4x + 6 &= (-3)^2 - 4(-3) + 6 \\ &= 9 + 12 + 6 \\ &= 27. \end{aligned}$$

Remember that $(-3)^2$ means $(-3)(-3)$.

◉ ***SKILL CHECK 12*** Evaluate $a^2 + 4a - 1$ when $a = -2$.

Example 13 Evaluate $3x^2 - 2y + x$ when $x = 1$ and $y = -3$.

Solution We replace x with 1 and y with -3, obtaining

$$\begin{aligned} 3x^2 - 2y + x &= 3(1)^2 - 2(-3) + (1) \\ &= 3 + 6 + 1 \\ &= 10. \end{aligned}$$

◉ ***SKILL CHECK 13*** Evaluate $2x^2 - 5y + 2x$ when $x = -1$ and $y = -4$.

◉ ***Skill Check Answers***

9. $-4x + 15y$
10. $5a^2 - 12a$
11. -9
12. -5
13. 20

Exercise 7–1–2

2 Simplify.

1. $4x - 3[(x + 1) - 3x]$

2. $a - [a - 5(2a - 3)]$

3. $2x + 5[y - 4(2x - 3y)]$

4. $3a + 4[4b - 2(2b - a)]$

5. $x - 3[2x - (3 + 3x)]$

6. $9 - 2[x - 3(2 - 4x)]$

7. $6a - 3[4 - 5(a - 2)]$

8. $6[-2(x - 5) - 3(4 - 2x)]$

9. $2[5x^2 - x(2x - 1)]$

10. $3[2a^2 - a(3a + 4)]$

11. $x^2 - 2x[3x + 2(x + 4)]$

12. $5a^2 - 3a[a + 2(a - 2)]$

13. $3x(x + 2y) - [2x(2x + y) - 6y]$

14. $2x(3x - y) - [x(x + y) - y]$

15. $2[x - y(3y - 5)] - 7[3x + 2y(y - 8)]$

16. $8a^2 - 3[3ax + a(2a + 4x) - 6ay]$

3 Evaluate.

17. x, when $x = -2$

18. $-x$, when $x = -2$

19. $3x + 5$, when $x = -7$

20. $16 - 2x$, when $x = -3$

21. x^2, when $x = 5$

22. $-x^2$, when $x = 5$

23. x^2, when $x = -6$

24. $-x^2$, when $x = -6$

25. $(-x)^2$, when $x = -5$

26. $(-x)^3$, when $x = -1$

27. $3x^3$, when $x = -2$

28. $(3x)^3$, when $x = -2$

29. $2x - 3y$, when $x = 4$ and $y = -2$

30. $3xy - 2y^2$, when $x = 3$ and $y = -4$

31. $3x^2 + xy - 6$, when $x = -2$ and $y = -1$

32. $2a^2 - 3ab + 2$, when $a = -1$ and $b = -2$

33. $x^3 - 2xz - 3y$, when $x = -2$, $y = -3$, and $z = 4$

34. $2x^3 + 3xyz - 2x^2$, when $x = 2$, $y = 5$, and $z = -1$

35. $a^3 - b^3$, when $a = -2$ and $b = -3$

36. $C = \frac{5}{9}(F - 32)$. Find C when $F = 5$.

37. $F = \frac{9}{5}C + 32$. Find F when $C = -10$.

38. $\ell = a + (n - 1)d$. Find ℓ, when $a = -3$, $n = 9$, and $d = -\frac{1}{2}$.

• Concept Enrichment

39. Evaluate $|x^3|$ when $x = -2$.

40. Evaluate $-|-x^3|$ when $x = -3$.

41. Evaluate $|-2x - 3y|$ when $x = -1$ and $y = 3$.

42. Evaluate $|x - y| - 3|-4x + 2y|$ when $x = -2$ and $y = -5$.

7–2 Scientific Notation

So far we have used only nonnegative integers as exponents. In this section we will define a negative exponent and use it to write numbers in expanded form.

OBJECTIVES

Upon completing this section you should be able to:

1. Write decimal numbers in expanded form using exponents.
2. Write numbers in scientific notation.

1 Expanded Form

Recall that in chapter 1 we wrote whole numbers in expanded form using nonnegative exponents as powers of 10.

Refer to section 1–3.

Example 1 Write 235 in expanded form using exponents.

Solution $235 = 2(10^2) + 3(10^1) + 5(10^0)$

SKILL CHECK 1 Write 3,671 in expanded form.

We will now define a negative exponent and use it to write decimal numbers in expanded form.

> If a is any nonnegative integer and $x \neq 0$, then $x^{-a} = \dfrac{1}{x^a}$.

Example 2 $5^{-2} = \dfrac{1}{5^2}$

Example 3 $10^{-3} = \dfrac{1}{10^3}$

SKILL CHECK 2 Write 6^{-2} using a positive exponent.

Remember in chapter 4 how we wrote a decimal in expanded form. For instance,

$$0.135 = 1\left(\frac{1}{10^1}\right) + 3\left(\frac{1}{10^2}\right) + 5\left(\frac{1}{10^3}\right).$$

Refer to section 4–1.

Using our definition of a negative exponent, we could now write the number as

$$0.135 = 1(10^{-1}) + 3(10^{-2}) + 5(10^{-3}).$$

Example 4 Write 381.67 in expanded form using exponents.

Solution $381.67 = 3(10^2) + 8(10^1) + 1(10^0) + 6(10^{-1}) + 7(10^{-2})$

SKILL CHECK 3 Write 3.524 in expanded form using exponents.

Given an expression written as a number times a power of 10, we can write it without exponents.

Example 5 Write 5.6×10^2 without exponents.

Solution $5.6 \times 10^2 = 5.6(100) = 560$

How do 5.6 and 560 compare?

Example 6 Write 2.37×10^3 without exponents.

Solution $2.37 \times 10^3 = 2.37(1{,}000) = 2{,}370$

How do 2.37 and 2,370 compare?

Notice in examples 5 and 6 that we moved the decimal point to the right the same number of places as the value of the exponent of 10.

Skill Check Answers

1. $3(10^3) + 6(10^2) + 7(10^1) + 1(10^0)$
2. $\dfrac{1}{6^2}$
3. $3(10^0) + 5(10^{-1}) + 2(10^{-2}) + 4(10^{-3})$

Example 7

$$7.85 \times 10^4 = 7.85(10{,}000) = 78{,}500$$

The decimal point is moved 4 places to the *right*.

⦿ ***SKILL CHECK 4*** Write 8.43×10^5 without exponents.

Example 8

$$7.85 \times 10^{-4} = 7.85\left(\frac{1}{10^4}\right) = \frac{7.85}{10{,}000} = 0.000785$$

The decimal point is moved 4 places to the *left*.

⦿ ***SKILL CHECK 5*** Write 3.64×10^{-3} without exponents.

⦿ ***Skill Check Answers***

4. 843,000
5. 0.00364

Exercise 7–2–1

1 Write without exponents.

1. 7^0 **2.** 18^0 **3.** 2^{-2} **4.** 2^{-3}

5. 5^{-3} **6.** 6^{-2} **7.** 10^{-5} **8.** 10^{-6}

Write in expanded form using exponents.

9. 0.8 **10.** 0.27 **11.** 0.915

12. 0.605 **13.** 7.14 **14.** 60.3

15. 128.72 **16.** 5,124.17

Write without exponents.

17. 0.4×10^1 **18.** 0.73×10^1 **19.** 4.6×10^2

20. 0.85×10^2 **21.** 3.16×10^3 **22.** 7.25×10^4

23. 61×10^{-1}

24. 34×10^{-2}

25. 605×10^{-4}

26. 103×10^{-3}

27. 7.94×10^{5}

28. 5.21×10^{7}

29. 9.17×10^{-6}

30. 4.32×10^{-9}

2 Scientific Notation

Exponents are used in many fields of science to write numbers in what is called scientific notation. If a number is either very large or very small, this method of expressing it keeps it from being cumbersome and can make computations easier.

> A number is in **scientific notation** if it is expressed as the product of a power of 10 and a number equal to or greater than 1 and less than 10.

Example 9 3.6×10^{3}; 7.25×10^{-3}; and 9.15×10^{8} are all in scientific notation.

Notice that the definition contains two parts. First we must have a number equal to or greater than 1 and less than 10, and second we must multiply by a power of 10.

Example 10 26.5×10^{3} is *not* in scientific notation since we do not have a number equal to or greater than 1 and less than 10.

26.5 is not between 1 and 10.

Example 11 5.3×4^{3} is *not* in scientific notation since we are not multiplying by a power of 10.

4^3 is not a power of 10.

Example 12 Earth is approximately 93,000,000 miles from the sun. Express this distance in scientific notation.

We know that scientific notation requires that we have 9.3 (a number equal to or greater than 1 and less than 10). Now we must determine the proper exponent for 10. The simplest way to determine this is to ask, "How many places, and in what direction, must we move the decimal point in 9.3 to get 93,000,000?" Since we must move the decimal point seven places to the right, we have 9.3×10^{7}.

⦿ ***SKILL CHECK 6*** Write 365,000,000 in scientific notation.

Example 13 Write 0.0074 in scientific notation.

First we must have 7.4. Now, how many places, and in what direction, do we move the decimal point in 7.4 to get 0.0074? Thus, $0.0074 = 7.4 \times 10^{-3}$.

To summarize: If the exponent is *positive* the decimal point is moved to the *right* that number of places. If the exponent is *negative* the decimal point is moved to the *left* that number of places.

Remember, 10 to a negative power moves the decimal point to the left.

⦿ ***SKILL CHECK 7*** Write 0.0000532 in scientific notation.

⦿ ***Skill Check Answers***

6. 3.65×10^{8}
7. 5.32×10^{-5}

Exercise 7–2–2

A State whether or not the given number is in scientific notation.

1. 3.6×10^5
2. 0.05×10^8
3. 2.78×10^{16}
4. 8.2×10^{-3}
5. 6.1×2^{10}
6. 25×10^{-3}
7. 0.645×10^4
8. 5×10^8

Write each number in scientific notation.

9. 5,000
10. 5,280
11. 728,000
12. 346,000,000
13. 0.000000235
14. 0.0000000052
15. 233,000,000,000,000
16. 0.0000000000739

Write each number without using exponents.

17. 3.201×10^5
18. 7.28×10^{-6}
19. 1.07×10^{-9}
20. 6.23×10^{23}
21. 5.02×10^{10}
22. 3.58×10^{-1}
23. 4.07×10^{-5}
24. 5.3762×10^2
25. 3.6×10^0
26. 9.9×10^1

27. The planet Mars is approximately 49,000,000 miles from Earth. Express this distance in scientific notation.

28. An enormous cloud of hydrogen gas eight million miles in diameter was discovered around the comet Bennet by NASA in 1970. Express this distance in scientific notation.

29. A light-year (the distance light travels in a year) is approximately 6.0×10^{12} miles. Express this distance without exponents.

30. The mass of the Earth is approximately 5.9×10^{24} kilograms. Express this without exponents.

31. An angstrom is a unit of length. One angstrom is approximately 1×10^{-8} cm. Express this length without exponents.

32. The diameter of the nucleus of an average atom is approximately 3.5×10^{-12} cm. Express this distance without exponents.

33. A red blood cell is approximately .001 cm in diameter. Express this length in scientific notation.

34. The Rubik's Cube puzzle has over 43,000,000,000,000,000,000 color combinations. Express this number in scientific notation.

• Concept Enrichment

35. Explain why scientific notation is a useful tool.

36. Multiply and give the answer in scientific notation: $(3.5 \times 10^4)(2.8 \times 10^8)$

7–3 Square Roots

1 Square Roots of Perfect Squares

OBJECTIVES

Upon completing this section you should be able to:

1. Find a square root of a perfect square.
2. Find an approximate value of a square root of a number using a calculator.

Using the definition of exponents, $(5)^2 = 25$. We say that 25 is the **square** of 5. We now introduce a new term in our language. If 25 is the square of 5, then 5 is said to be a *square root* of 25.

> If $x^2 = y$, then x is a **square root** of y.

Example 1 Since $2^2 = 4$, 2 is a square root of 4.

Example 2 Since $(-2)^2 = 4$, -2 is a square root of 4.

From these two examples you will note that 4 has two square roots, 2 and -2. In fact it is true that every positive number has two square roots.

Example 3 Find the square roots of 25.

Solution Since $(5)^2 = 25$ and $(-5)^2 = 25$, $+5$ and -5 are the square roots of 25.

⦿ ***SKILL CHECK 1*** Find the square roots of 9.

Whole numbers such as 4, 9, 16, and so on, whose square roots are integers, are called **perfect squares.**

Remember, integers are the whole numbers and their negatives.

Example 4 36 is a perfect square since the square roots of 36 are integers ($+6$ and -6).

⦿ ***SKILL CHECK 2*** Is 16 a perfect square?

⦿ ***Skill Check Answers***

1. $+3$ and -3
2. yes

The **principal square root** of a positive number is the positive square root.

$\sqrt{9}$ indicates the principal square root or positive square root of 9.

The symbol $\sqrt{}$ is called a **radical sign** and indicates the principal square root of a number.

Note the difference in these two problems.

It is very important that you understand the difference between these two statements.

a. Find the square roots of 25.

b. Find $\sqrt{25}$.

You may sometimes see the symbol $\pm\sqrt{}$. This means both square roots of a number are called for. For example:

$$\pm\sqrt{25} = \pm 5.$$

± 5 is the short way of writing $+5$ and -5.

For (a) the answer is $+5$ and -5 since $(+5)^2 = 25$ and $(-5)^2 = 25$.

For (b) the answer is $+5$ since the radical sign represents the principal or positive square root.

⦿ ***SKILL CHECK 3*** Find $\sqrt{4}$.

Example 5 Find the value of $\sqrt{81} - 2\sqrt{16}$.

Solution Since $\sqrt{81} = 9$ and $\sqrt{16} = 4$, we have

$$\begin{aligned}\sqrt{81} - 2\sqrt{16} &= 9 - 2(4)\\ &= 9 - 8\\ &= 1.\end{aligned}$$

⦿ ***Skill Check Answers***

3. 2
4. 4

⦿ ***SKILL CHECK 4*** Find the value of $2\sqrt{25} - \sqrt{36}$.

Exercise 7–3–1

1

1. Starting with 0, list the first 25 perfect squares.

Find the square roots of each number.

2. 4 **3.** 25 **4.** 16 **5.** 64 **6.** 49

7. 81 **8.** 100 **9.** 121 **10.** 400 **11.** 225

Find the value.

12. $\sqrt{9}$ **13.** $\sqrt{25}$ **14.** $\sqrt{144}$ **15.** $\sqrt{36}$

16. $\sqrt{1}$

17. $\sqrt{121}$

18. $\sqrt{900}$

19. $\sqrt{10{,}000}$

20. $\sqrt{289}$

21. $\sqrt{441}$

22. $3\sqrt{16}$

23. $5\sqrt{4}$

24. $-2\sqrt{25}$

25. $-4\sqrt{49}$

26. $\sqrt{25} - \sqrt{16}$

27. $\sqrt{4} - 2\sqrt{9}$

28. $3\sqrt{64} - \sqrt{36}$

29. $\sqrt{5^2 + 12^2}$

30. A formula from geometry that is used to find the length of the hypotenuse c of a right triangle is $c = \sqrt{a^2 + b^2}$, where a and b represent the lengths of the two legs of the right triangle. Find the length of the hypotenuse if $a = 3$ inches and $b = 4$ inches.

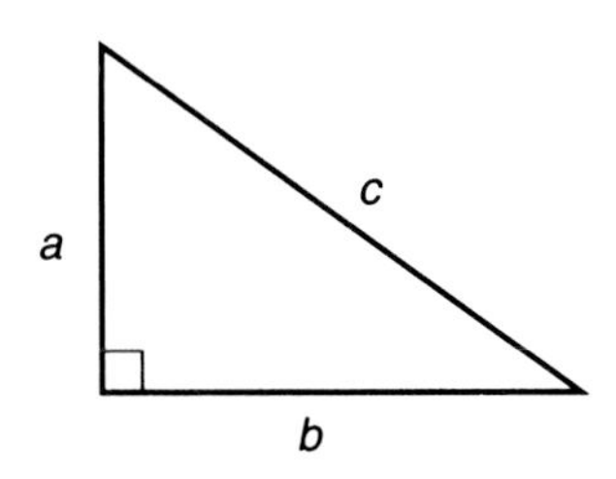

• Concept Enrichment

31. It was stated that every positive number has two square roots. Is there any number that has only one square root?

32. Of the numbers we have studied so far, why can't we find the square roots of -25?

2 Finding Approximate Values of Square Roots

Not all numbers have principal square roots that are whole numbers. In fact, some of them will be nonterminating, nonrepeating decimals. Such a number cannot be written as a rational number. A number that cannot be written as a quotient of two integers is called an **irrational number.**

We cannot find the exact value of numbers like $\sqrt{3}$ and $\sqrt{10}$.

While we cannot find the exact value of a square root of a number that is not a perfect square, we can find the value correct to several decimal places by using a calculator. To do this we use the $\boxed{\sqrt{}}$ button. For instance, to find the square root of 9 we enter $\boxed{9}$ and then press the $\boxed{\sqrt{}}$ button, obtaining the result 3.

The table on the inside back cover of this text gives the value correct to four decimal places of the square roots of the integers from 1 to 200.

Example 6 Find $\sqrt{49}$.

First enter 49. Then press the $\boxed{\sqrt{}}$ button. The result 7 will appear in the display.

We can also find the approximate square root of a number that is not a perfect square.

Recall how to round an answer.

Example 7 Find $\sqrt{108}$ rounded to four decimal places.

We enter 108 followed by pressing the $\boxed{\sqrt{}}$ button. The result in the display rounded to four decimal places is 10.3923.

⦿ ***Skill Check Answer***

5. 9.8995

⦿ ***SKILL CHECK 5*** Find $\sqrt{98}$ rounded to four decimal places.

Exercise 7–3–2

2 Use a calculator to find each value rounded to four decimal places.

1. $\sqrt{2}$ **2.** $\sqrt{5}$ **3.** $\sqrt{13}$ **4.** $\sqrt{19}$ **5.** $\sqrt{28}$

6. $\sqrt{99}$ **7.** $\sqrt{109}$ **8.** $\sqrt{161}$ **9.** $\sqrt{17.1}$ **10.** $\sqrt{38.5}$

11. $\sqrt{65.96}$ **12.** $\sqrt{31.09}$ **13.** $\sqrt{7} + \sqrt{2}$ **14.** $\sqrt{3} + \sqrt{5}$ **15.** $\sqrt{10} - \sqrt{12}$

16. $\sqrt{11} - \sqrt{30}$ **17.** $\sqrt{7} - 4\sqrt{5}$ **18.** $\sqrt{14} - 5\sqrt{8}$ **19.** $\sqrt{\frac{2}{3}}$ **20.** $\sqrt{\frac{5}{12}}$

7–4 The Pythagorean Theorem

OBJECTIVE

Upon completing this section you should be able to:

1 Use the Pythagorean theorem to find a side of a right triangle.

In chapter 5 we discussed right triangles. We mentioned that the side opposite the 90° angle is called the hypotenuse and the other two sides are called legs. An important relationship exists between the sides of a right triangle. This relationship is named for the Greek mathematician Pythagoras and is called the **Pythagorean theorem.**

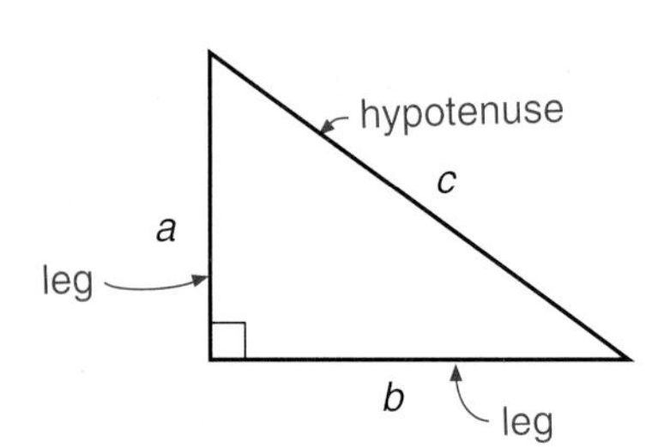

THEOREM If a and b represent the lengths of the legs of any right triangle and c represents the length of the hypotenuse, then

$$a^2 + b^2 = c^2.$$

1 Using the Theorem

Given the lengths of two sides of a right triangle, we can use the theorem to determine the length of the third side.

Example 1 Find the length of the hypotenuse of a right triangle having legs of lengths 3 inches and 4 inches.

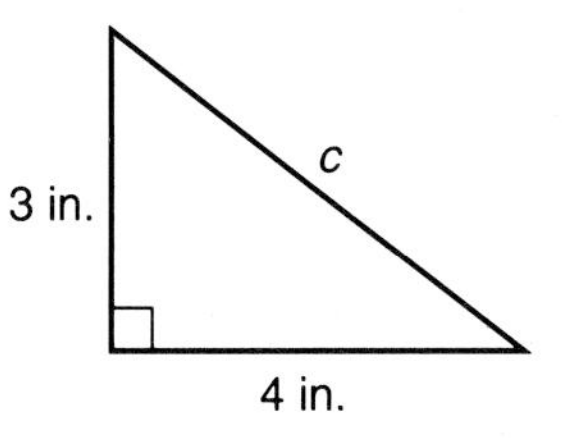

Solution Using the Pythagorean theorem, we write

$$\begin{aligned} a^2 + b^2 &= c^2 \\ 3^2 + 4^2 &= c^2 \\ 9 + 16 &= c^2 \\ 25 &= c^2 \\ c &= \sqrt{25} \\ c &= 5 \textit{ in.} \end{aligned}$$

Note we only find the principal square root since we are dealing with length.

◉ *SKILL CHECK 1* Find the length of the hypotenuse of a right triangle having legs of lengths 6 feet and 8 feet.

Example 2 Find the length of the leg of a right triangle if the other leg measures 5 inches and the hypotenuse measures 8 inches. Give the answer to two decimal places.

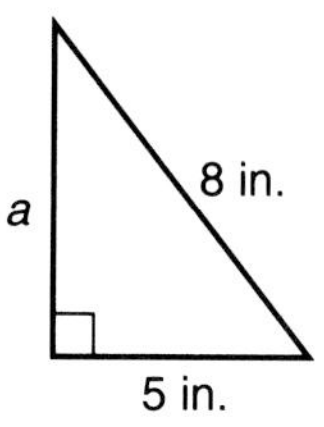

Solution Substituting in the Pythagorean theorem, we have

$$\begin{aligned} a^2 + b^2 &= c^2 \\ a^2 + 5^2 &= 8^2 \\ a^2 + 25 &= 64. \end{aligned}$$

Writing this as a related subtraction statement, we have

$$\begin{aligned} a^2 &= 64 - 25 \\ a^2 &= 39 \\ a &= \sqrt{39} \\ a &= 6.24 \textit{ in.} \end{aligned}$$

Notice that $a = \sqrt{39}$ is an exact answer where $a = 6.24$ is approximate.

◉ *SKILL CHECK 2* Find the length of a leg of a right triangle if the other leg measures 6 feet and the hypotenuse measures 9 feet. Round the answer to two decimal places.

A right triangle having two sides with the same length is an *isosceles right triangle.* The acute angles both measure 45°.

See section 5–4.

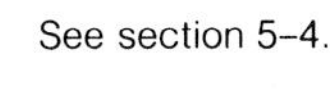

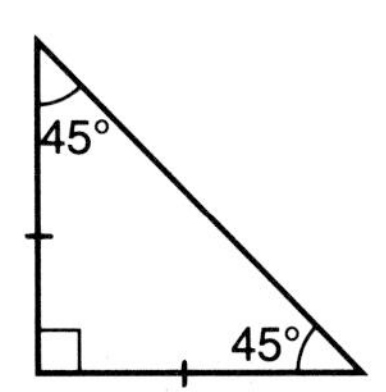

◉ *Skill Check Answers*

1. 10 ft
2. 6.71 ft

Example 3 Find the length of the hypotenuse of an isosceles right triangle if the two equal sides measure 1 inch.

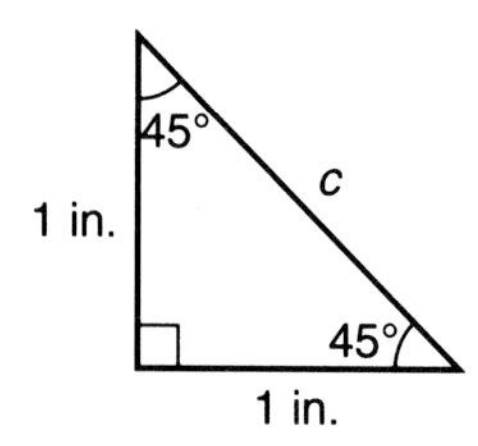

Solution Using the Pythagorean theorem, we have

$$a^2 + b^2 = c^2$$
$$(1)^2 + (1)^2 = c^2$$
$$1 + 1 = c^2$$
$$2 = c^2$$
$$\sqrt{2} = c$$

or $c = \sqrt{2}$ *in.*

Exact answer.

⦿ SKILL CHECK 3 Find the length of the hypotenuse of an isosceles right triangle if the two legs measure 3 inches each. Give an exact answer.

A right triangle whose angles measure 30°, 60°, and 90° has some special properties. One of these properties is that the length of the side opposite the 30° angle is one-half the length of the hypotenuse.

Example 4 Given a triangle whose angles measure 30°, 60°, and 90°, if the length of the hypotenuse is 2 feet, find the lengths of the other two sides (legs) of the triangle.

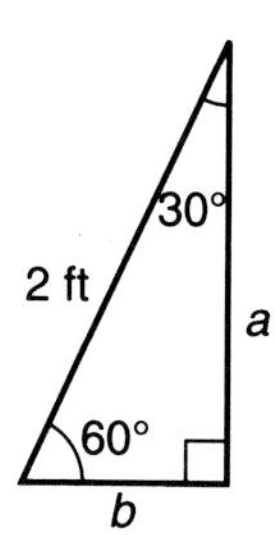

Solution As mentioned, the length of the side opposite the 30° angle is one-half that of the hypotenuse. Thus $b = 1$ ft. To find a we must use the Pythagorean theorem.

$$a^2 + b^2 = c^2$$
$$a^2 + (1)^2 = (2)^2$$
$$a^2 + 1 = 4$$
$$a^2 = 4 - 1$$
$$a^2 = 3$$

or $a = \sqrt{3}$ *ft.*

Exact answer.

⦿ SKILL CHECK 4 Given a 30°, 60°, 90° triangle, if the hypotenuse measures 8 feet, find the length of the side opposite the 60° angle. Give an exact answer.

⦿ Skill Check Answers

3. $\sqrt{18}$ inches
4. $\sqrt{48}$ ft

Exercise 7–4–1

1 Find the length of the third side of the right triangle. Give the answer to two decimal places.

1.

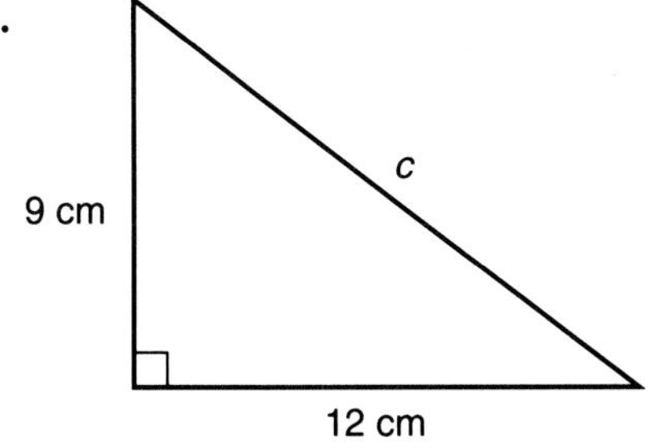

2.

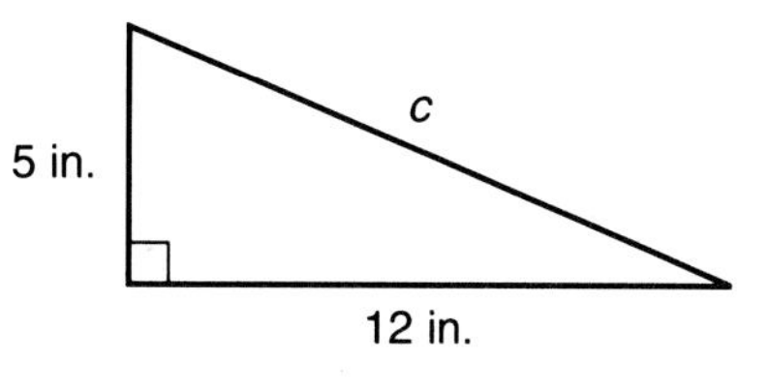

3.

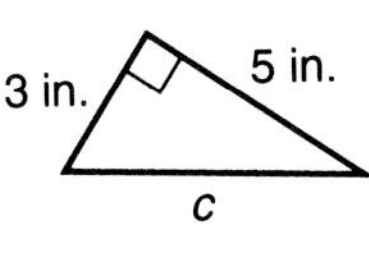

4.

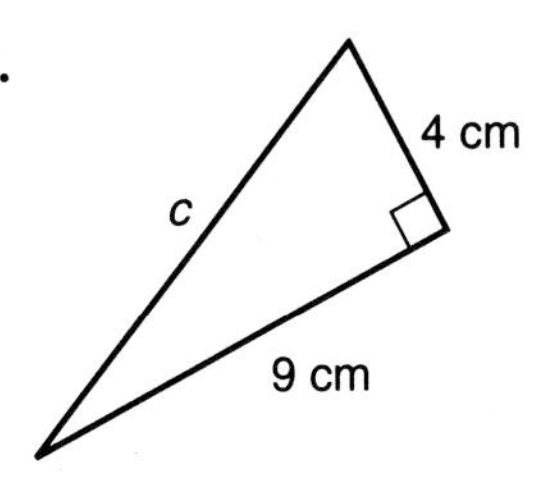

5.

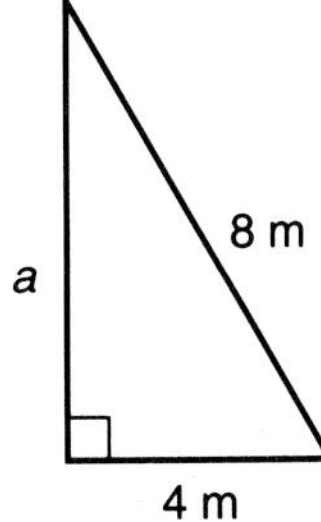

6.

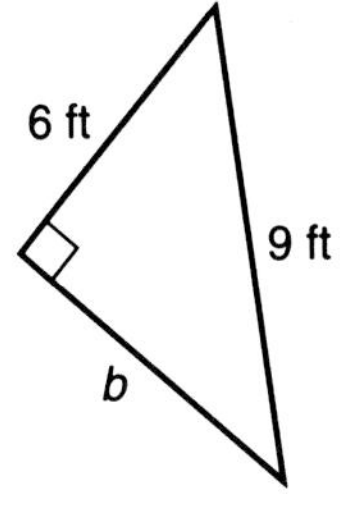

7. $a = 5$ cm, $b = 5$ cm

8. $a = 4$ in., $c = 10$ in.

9. $a = 6$ ft, $c = 12$ ft

10. $b = 11$ cm, $c = 14$ cm

Find the lengths of the sides indicated by letters. Give exact answers.

11.

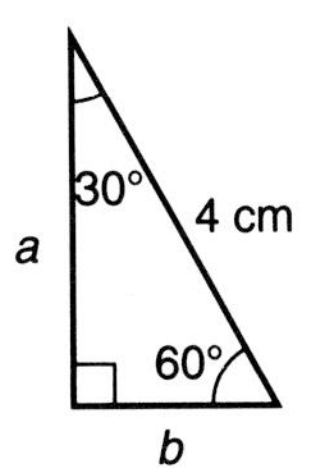

12.

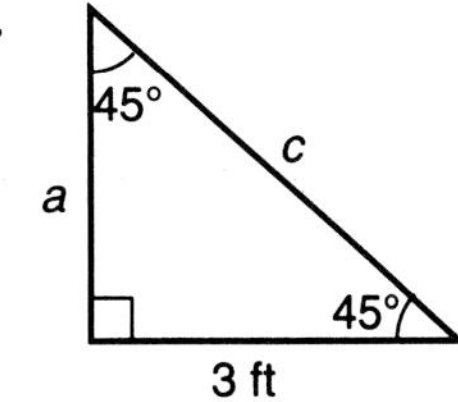

13. The length of one leg of an isosceles right triangle is 5 centimeters. Find the length of the hypotenuse. Give an exact answer.

14. The length of the hypotenuse of a 30°, 60°, 90° right triangle is 8 feet. Find the length of the other two sides. Give an exact answer.

• Concept Enrichment

Draw a diagram to help solve each problem.

15. Find the length of the diagonal of a square having a side of length 1 inch. Round the answer to two decimal places.

16. Find the length of the diagonal of a rectangle that is 10 meters long and 4 meters wide. Give the answer to two decimal places.

17. A baseball diamond is in the shape of a square with the bases at the vertices of the square. Each side is 90 feet in length. Find the distance from home plate to second base. Give the answer to two decimal places.

18. A rectangular room is 15 feet wide and 25 feet long. Find the distance from one corner to the opposite corner. Give the answer to two decimal places.

19. A vertical pole 21 feet high has a wire reaching from the top of the pole to the ground 9 feet from the pole. How long is the wire? Give the answer to two decimal places.

20. A ladder 13 feet long leans against a house. The bottom of the ladder is 7 feet from the house. Find the height above the ground where the ladder touches the house. Give the answer to two decimal places.

7–5 Real Numbers

OBJECTIVES

Upon completing this section you should be able to:

1. Identify the subsets of the set of real numbers.
2. State the properties of the real numbers.

In this chapter and the previous chapter we have extended the set of numbers with which we work to include more than just the numbers of arithmetic. At this point we wish to categorize this set of numbers and summarize their properties.

1 Subsets of the Real Numbers

The numbers first encountered in elementary arithmetic are those used in counting. They are called the **counting numbers** or **natural numbers.** This set of numbers can be written using set notation as $\{1,2,3,4, \ldots\}$. The braces indicate a set and the three dots indicate that the pattern of numbers continues indefinitely.

The set of counting numbers is extended to include zero. This set is the set of **whole numbers** and is indicated as $\{0,1,2,3, \ldots\}$. We see that the set of counting numbers is contained within the set of whole numbers. For this reason the set of counting numbers is said to be a **subset** of the set of whole numbers.

The numbers are sometimes referred to as the **elements** of the set.

The set of whole numbers is extended to include the negatives of the counting numbers, giving us the set of **integers** $\{\ldots, -3, -2, -1, 0, +1, +2, +3, \ldots\}$. We see that the set of whole numbers is a subset of the set of integers.

The set of **rational numbers** is composed of those numbers that can be expressed as a quotient of two integers. The set of integers is a subset of this set since, for example, 3 can be expressed as $\frac{3}{1}, \frac{6}{2}, \frac{-15}{-5}$, and so on. The decimal representation of a rational number will terminate or repeat.

The quotient of two numbers is sometimes called a ratio. Hence the name *rational* number.

There are some numbers however that cannot be expressed as a quotient of two integers. In section 7–3, for example, we saw that numbers such as $\sqrt{3}$ and $\sqrt{10}$ when expressed as a decimal did not repeat or terminate. Such numbers cannot be expressed as a quotient of two integers. In chapter 5, when discussing the circle, we used the number π. This number also cannot be expressed as a quotient of two integers. Such numbers are called **irrational numbers.** It is evident that the set of rational numbers is *not* a subset of the set of irrational numbers.

In some cases we used $= \frac{22}{7}$ or $\pi = 3.14$ but neither is the *exact* value of π.

The set of rational numbers and the set of irrational numbers together make up a set called the **real numbers.** The diagram below shows the relationships among the sets of numbers we have discussed.

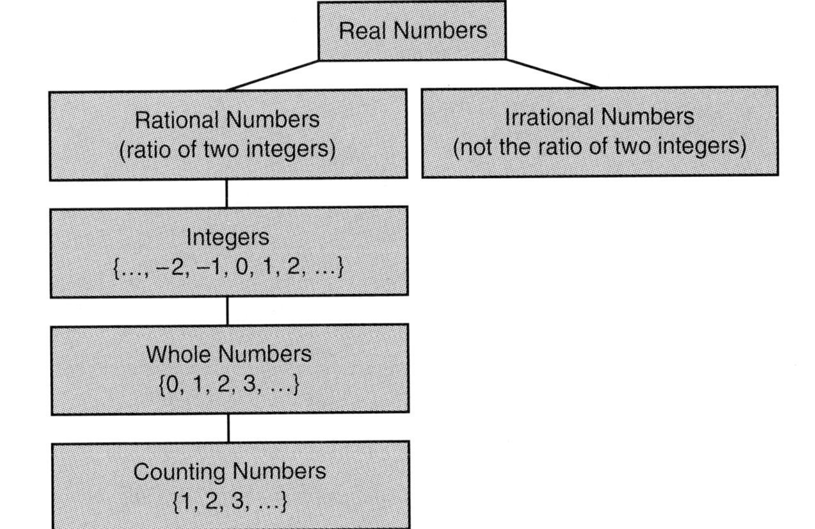

It may be helpful from time to time to refer back to this chart.

From this point on our discussions will involve the real numbers.

2 Properties of the Real Numbers

Along the way we have also mentioned several properties that numbers possess. These properties are valid for the set of real numbers and are summarized here.

Properties of the Real Numbers

1. **Closure property of addition**
   If a and b are real numbers, then $a + b$ is also a real number.
2. **Commutative property of addition**
   If a and b are real numbers, then

$$a + b = b + a.$$

3. **Associative property of addition**
If a, b, and c are real numbers, then

$$(a + b) + c = a + (b + c).$$

0 is the additive identity.

4. **Additive identity**
There is a real number (zero) such that for any real number a, $a + 0 = a$.

5. **Additive inverse** (opposite)
If a is a real number, then there is a real number $-a$ such that $a + (-a) = 0$.

This list of properties may be a useful reference in the future.

6. **Closure property of multiplication**
If a and b are real numbers, then ab is also a real number.

7. **Commutative property of multiplication**
If a and b are real numbers, then

$$ab = ba.$$

8. **Associative property of multiplication**
If a, b, and c are real numbers, then

$$(ab)c = a(bc).$$

1 is the multiplicative identity.

9. **Multiplicative identity**
There is a real number (1) such that for any real number a, $a(1) = a$.

10. **Multiplicative inverse** (reciprocal)
If a is a nonzero real number, then there is a real number $\frac{1}{a}$ such that $a\left(\frac{1}{a}\right) = 1$.

11. **Distributive property of multiplication over addition**
If a, b, and c are real numbers, then

$$a(b + c) = ab + ac.$$

Exercise 7–5–1

1 Questions 1–6 refer to the set of numbers $\{-5, -2.61, -\sqrt{3}, -\frac{1}{7}, 0, \sqrt{7}, \pi, 4.2, 6\}$.

1. List the counting numbers from the set.

2. List the whole numbers from the set.

3. List the integers from the set.

4. List the rational numbers from the set.

5. List the irrational numbers from the set.

6. List the real numbers from the set.

2 State the property of the real numbers that is illustrated.

7. $4 + 6 = 6 + 4$

8. $7 + 1$ is a real number.

9. $[(3)(-5)](-4) = (3)[(-5)(-4)]$

10. $\frac{1}{2} + (-3) = (-3) + \frac{1}{2}$

11. $(-5)(+1) = -5$

12. $2[7 + (-3)] = 2(7) + 2(-3)$

13. $[(-8) + \sqrt{6}] + 3.2 = -8 + [\sqrt{6} + 3.2]$

14. $\left(-\frac{2}{3}\right) + 0 = -\frac{2}{3}$

15. $(+3.18)(-\sqrt{5})$ is a real number.

16. $\left(+\frac{3}{7}\right) + \left(-\frac{3}{7}\right) = 0$

17. $(-5)\left(-\frac{1}{5}\right) = 1$

18. $(\sqrt{5})(2) = 2\sqrt{5}$

19. $-12 + 12 = 12 + (-12)$

20. $[36 + (-41)] + 41 = 36 + [(-41) + 41]$

• Concept Enrichment

21. Explain the distinction between the additive identity and the multiplicative identity.

22. Explain the distinction between the additive inverse of a number and the multiplicative inverse of a number.

CHAPTER 7 SUMMARY

Key Words

Section 7–1

- A **polynomial** is a sum or difference of one or more monomials.

Section 7–2

- A **negative exponent** can be used to write a decimal number in expanded form.
- **Scientific notation** is used to represent very small or very large numbers.

Section 7–3

- If $x^2 = y$, then x is a **square root** of y.
- The **principal square root** of a positive number is the positive square root.
- The symbol $\sqrt{\ }$ is called a **radical sign** and indicates the principal square root of a number.
- A **perfect square** has integers as its square roots.
- **Rounding** may be needed to approximate a square root of some numbers.

Section 7–4

- The **Pythagorean theorem** enables us to find the measure of a side of a right triangle, given the measures of the other two sides.

Section 7–5

- The **integers** are composed of the whole numbers and their negatives.
- The **rational numbers** can be expressed as a quotient of two integers.
- The **irrational numbers** cannot be expressed as a quotient of two integers.
- The **real numbers** are composed of the sets of rational numbers and the irrational numbers.
- The **properties of real numbers** enable us to perform various operations on them.

Procedures

Section 7–2

- If a is any nonnegative integer and $x \neq 0$, then $x^{-a} = \frac{1}{x^a}$.
- A number is in scientific notation if it is expressed as the product of a power of 10 and a number equal to or greater than 1 but less than 10.

Section 7–3

- We can find the principal square root (rounded to several decimal places) of any nonnegative number by using a calculator.

Section 7–4

- If a and b represent the measures of the legs of any right triangle and c represents the measure of the hypotenuse, then $a^2 + b^2 = c^2$.

CHAPTER 7 REVIEW

Simplify.

1. $(-3)(-4) - (6)(-2)$

2. $(-5)(2) + (-3)(-2)$

3. $5 + (-1)(3) - (4)(-1) - 3$

4. $7 - (-4)(-1) - (3)(2) - 4$

5. $8 + (-6)(-2) \div (-3) - 3(5)$

6. $14 - 3(8 - 11) - 14 \div 2$

7. $(-3)^5$

8. $(-4)^3$

9. $-(-1)^3$

10. $-(-2)^4$

11. $(-2)^3(-1)^3$

12. $(-2)^3(-1)^4$

13. $(3a)(-2a^3)$

14. $(-5x^2)(-2x^3)$

15. $\frac{-15xy^2}{3y}$

16. $\frac{21a^2b}{-3ab}$

17. $-5x(2x^2 - 3x + 1)$

18. $-2x(3x^2 - 5x - 3)$

19. $13x - 5[x - (3x + 4)]$

20. $5a - [-3 - 2(3a - 2)]$

21. $2[3a - 2(a + 3) + a]$

22. $10x - 3[2x + 5(x - 3)]$

23. $2x^2 - 3x[2x - 3(x - 2)]$

24. $6x^2 - 2x[3(2x - 1) - 7x]$

25. $3[x^2 - x(2x - 5)] - 2[3x - 2x(x + 4)]$

26. $2[a - b(2b - 1)] - 5[2a + 3b(b - 2)]$

Evaluate.

27. $3x^3$, when $x = -5$

28. $(2x)^3$, when $x = -2$

29. $-x^3$, when $x = -2$

30. $(-a)^3$, when $a = -2$

31. $2a^2 - 3a + 1$, when $a = -4$

32. $3x^3 - 4x - 5$, when $x = -1$

33. $5x^3 - 2xy^2 - 3y$, when $x = -3$ and $y = 4$

34. $x^2 - y^2$, when $x = -3$ and $y = -4$

35. $C = \frac{5}{9}(F - 32)$. Find C when $F = -4$.

36. $S = \frac{a}{1 - r}$. Find S when $a = -12$ and $r = \frac{1}{4}$.

Write without exponents.

37. 2^{-4}

38. 3^{-2}

39. 0.5×10^3

40. 23×10^{-5}

41. 4.12×10^4

42. 3.5×10^{-4}

Write in scientific notation.

43. 37,600,000

44. 51,300

45. 0.0000512

46. 0.00000571

47. Find the square roots of 169.

48. Find the square roots of 144.

49. Find the value of $\sqrt{196}$.

50. Find the value of $\sqrt{361}$.

Find the lengths of the sides indicated by letters. Give exact answers.

51.

3 in.
60°
c
30°
b

52.

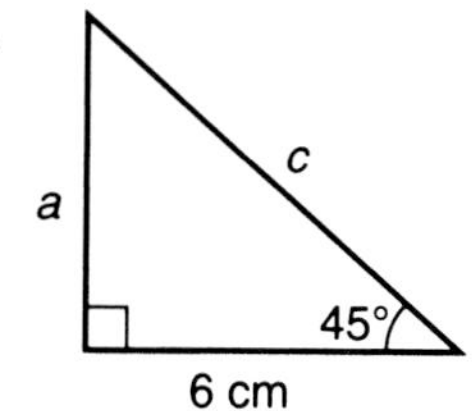

53. Find the length of the hypotenuse of a right triangle if the legs measure 3 inches and 6 inches. Round the answer to two decimal places.

54. If one leg of a right triangle measures 8 centimeters and the hypotenuse measures 10 centimeters, find the length of the third side. Round the answer to two decimal places.

State the property of the real numbers that is illustrated.

55. $(12)(-6) = (-6)(12)$

56. $5 + [(-6) + (-4)] = [5 + (-6)] + (-4)$

57. $-24 + 24 = 0$

58. $35(1) = 35$

CHAPTER 7 TEST

1. Simplify: $(5)(-2) + (-4)(-1)$

2. Simplify: $12 - 3(4 - 7) \div 3 - 2$

3. Simplify: $(-2)^5$

4. Simplify: $(-2)^3(+3)^2$

5. Simplify: $-2x^2(3x^2 - 2x - 5)$

6. Simplify: $\dfrac{-20xy^2z}{-4yz}$

7. Simplify: $5x - 3[2x - 4(x + 3)]$

8. Evaluate $2x^2 - 3xy + 4y^2$ when $x = -2$ and $y = 3$.

9. Write 24×10^{-5} without exponents.

10. Write 29,000,000 in scientific notation.

11. Write 5.13×10^4 without exponents.

12. Find the square roots of 49.

13. Find $\sqrt{225}$.

14. The length of the hypotenuse of a right triangle is 13 feet. If the length of one leg is 12 feet, find the length of the other leg.

15. Which property of the real numbers is illustrated by $a + 3 = 3 + a$?

CHAPTERS 1–7 CUMULATIVE TEST

1. Subtract: $\begin{array}{r} 401 \\ -176 \\ \hline \end{array}$

2. Find the prime factorization of 252.

3. Divide: $\frac{15}{32} \div \frac{35}{24}$

4. Replace the question mark in $\frac{3}{8}\ ?\ \frac{4}{11}$ with the inequality symbol positioned to indicate a true statement.

5. Multiply: $\left(-\frac{5}{6}\right)\left(-\frac{3}{10}\right)$

6. Two angles of a triangle measure 43° and 64°. What is the measure of the third angle?

7. What is the place value of the digit 5 in the number 452,391?

8. Reduce: $\frac{3x^2y}{15xy^2}$

9. Round 3.0537 to the nearest tenth.

10. Simplify: $13a^2 - 2a[5a - 4(3 - a)]$

11. Find: $|-12|$

12. Find the volume of a rectangular parallelepiped if the edges measure $3\frac{1}{2}$ feet by $1\frac{1}{4}$ feet by 4 feet.

13. Write 0.000356 in scientific notation.

14. Simplify: $-5 - 3 + 10 + 6 - 9$

15. Multiply: $\begin{array}{r} 16.08 \\ \times \quad 3.7 \\ \hline \end{array}$

16. Find the least common multiple of 10, 14, and 35.

17. Add: $3.24 + 68.4 + 12 + 0.59$

18. Evaluate $2x^3 - 5x^2y + 3xy^2$ when $x = -3$ and $y = -2$.

19. Simplify: $13a - (6 - 5a) - 20a + 5$

20. Divide: $5.4\overline{)183.87}$

21. Simplify: $3a + 6b - 2a + 4b$

22. Divide: $37\overline{)2{,}904}$

23. Find the length of the hypotenuse of a right triangle if the legs measure 2 feet and 5 feet. Round the answer to two decimal places.

24. The price of a stock opened in the morning at $28\frac{1}{8}$ and closed that afternoon at $31\frac{1}{2}$. What was the gain for the day?

25. A rectangular room measures 18 feet by 12 feet. If the price of baseboard molding is \$0.75 per foot, how much will it cost to purchase enough molding to go around the room?

CHAPTER

8 Pretest

Answer as many of the following problems as you can before starting this chapter. When you finish the chapter, take the test at the end and compare your scores to see how much you have learned.

1. Why isn't $2x^4y + x^{-1}y^2 - 5$ a polynomial?

2. What is the degree of the polynomial $3x^2y - 5xy + y^3$ in the variable x?

3. Classify $5x = 4x + 1$ as an identity, a contradiction, or a conditional equation.

4. Is $x = -2$ a solution of the equation $4x + 3 = 2x - 1$?

Solve for x.

5. $x + 6 = 36$

6. $\frac{2}{3}x = \frac{4}{9}$

7. $3x - 5 = -20$

8. $-4x = 64$

9. $5x - 2 = 8x + 10$

10. $\frac{1}{2}x + \frac{2}{3} = \frac{1}{2}$

11. $\frac{1}{4}\left(\frac{5}{2}x + 3\right) = \frac{1}{2}(x - 4)$

12. The area (A) of a triangle is 48 square inches and the base (b) is 12 inches long. Find the height (h). $\left(\text{Use the formula } A = \frac{1}{2}bh.\right)$

C H A P T E R

8 Solving Equations

The solution to equations is basic to the scientific progress made in modern times.

S T U D Y H E L P

Checking solutions is important at all levels of mathematics. Failure to check can cost you grade points and can cost the scientific community billions of dollars. Not checking for a computational error resulted in an out-of-focus Hubble telescope.

The solution of equations is the central theme of algebra. Skills learned manipulating the numbers and symbols of algebra allow us to find answers to problems by solving equations. In this chapter we will study some techniques for solving first-degree equations having one variable.

8–1 Polynomials

OBJECTIVES

Upon completing this section you should be able to:

1. Recognize polynomials.
2. Determine the degree of a polynomial in a particular variable.

In section 2–6 we defined a **monomial** as an expression in which the numerical coefficient and variables are related only by the operation of multiplication. We also noted that the variables must have whole number exponents.

Example 1 $5x^3y$ is a monomial since the numerical coefficient and variables are related only by multiplication and the exponents of the variables are whole numbers.

Example 2 $2a^5b^{-2}$ is not a monomial because the exponent -2 is not a whole number.

We now wish to consider expressions composed of one or more monomials.

1 Polynomials

A **polynomial** is an algebraic expression in which each term is a monomial.

Notice that this definition includes a monomial as being a polynomial.

This is a monomial.

Example 3 $3x^2y$ is a polynomial consisting of a single term.

SKILL CHECK 1 Is $5xy^3z^{-2}$ a polynomial?

Example 4 $2x^2 + 3y$ is a polynomial consisting of two terms.

Example 5 $2a^3b - 5a^2b^2 + 4ab - 5$ is a polynomial consisting of four terms.

Special names are used for some polynomials. If a polynomial has exactly two terms, it is called a **binomial**.

Two terms.

Example 6 $3x^2 + 2y$ is a binomial.

If a polynomial has three terms, it is called a **trinomial**.

Three terms.

Example 7 $2x + y^2 - z$ is a trinomial.

A monomial must have whole number exponents.

Example 8 $2x^3y - 3x^2 + xy^{-2}$ is *not* a trinomial. It is not a polynomial since the term xy^{-2} is not a monomial. It has a negative exponent.

2 Degree of a Polynomial

Another important word used in discussing polynomials is the *degree* of a polynomial.

The **degree** of a polynomial in a particular variable is the highest power of that variable in the polynomial.

4 is the highest power of x.

Example 9 $3x^4 - 12x^2 + 7x$ is a fourth-degree polynomial in x.

SKILL CHECK 2 What is the degree in x of $2x^3 - 5x^2 + 3x - 1$?

Skill Check Answers

1. no
2. 3

Example 10 $3x^2y + 2x - 7$ is a second-degree polynomial in x. It is also a first-degree polynomial in y.

2 is the highest power of x, and 1 is the higest power of y.

⊙ ***SKILL CHECK 3*** What is the degree of $4x^3y - 3xy^2 + 2x$ in y?

⊙ ***Skill Check Answer***

3. 2

Exercise 8–1–1

1 Determine if the algebraic expression is a polynomial.

1. $x^2 + 3x$

2. $5x^3y - 1$

3. $2a^3b - a^2b^2 + 3$

4. $3a^2b - 2ab^2 - 4a^{-3}b$

5. $6x^2 + 4xy - 2y$

6. $x^4y - x^5y^2 + xy - 5$

7. $2xy^{-5} + 2x^2y^5 + xy^2$

8. $3ab^4 - 5a^2b^3 + 4$

9. $4y^3 - z^3 + 4y^2z^0$

10. $3x^0y^2 + 4x^2y^2 - 3$

2

11. What is the degree in x of the polynomial $4x^3 - 3x^2 + 15x - 1$?

12. What is the degree in x of the polynomial $5x^4 + 7x^2 - 2x + 4$?

Given the polynomial $x^3y - 5x^2y^2 + 2xy^4$:

13. What is the degree of the polynomial in x?

14. What is the degree of the polynomial in y?

Given the polynomial $5x^4y + 3x^3 - 2x$:

15. What is the degree of the polynomial in x?

16. What is the degree of the polynomial in y?

Given the polynomial $2xy^2 - 5y^3 + 4$:

17. What is the degree of the polynomial in y?

18. What is the degree of the polynomial in x?

Given the polynomial $3xy - 2x^2y^2 - 6x^3y + 3x^4$:

19. What is the degree of the polynomial in y?

20. What is the degree of the polynomial in x?

8–2 Conditional Equations

OBJECTIVES

Upon completing this section you should be able to:

1. Classify an equation as an identity, a contradiction, or a conditional equation.
2. Solve simple equations mentally.

You will learn to solve many types of equations as you continue your study in algebra. First we wish to define what is meant by the word *equation* and specify the type of equation we will learn to solve.

> An **equation** in algebra is a statement in symbols that two algebraic expressions have the same value (are equal).

The equations we will be interested in at the present are those that contain polynomials.

> A **polynomial equation** is an equation involving only polynomials. The degree of the equation is the degree of the polynomial of highest degree.

In this chapter we will learn to solve first-degree polynomial equations having one variable.

1 Classifying Equations

Equations can be classified in three main types.

a. An **identity** is true for all values of the numbers and variables in it.

Example 1 $2 + 3 = 5$ is an identity.

Example 2 $2x + 3x = 5x$ is an identity since any value substituted for x will yield an equality.

For example, if we let $x = 6$ then

$$2x + 3x = 5x$$

becomes

$2(6) + 3(6)$	$5(6)$
$12 + 18$	30
30	

Try some other value for x.

Remember, when you choose a value for x, you must use that same value wherever x appears in the equation.

⦿ SKILL CHECK 1 Is $4x - x = 3x$ an identity?

b. A **contradiction** is never true for the values of the numbers and variables in it.

Example 3 $x + 1 = x + 2$ is a contradiction since no value of x will result in a true statement.

⦿ SKILL CHECK 2 Is $x - 1 = x + 1$ a contradiction?

⦿ Skill Check Answers

1. yes
2. yes

c. A **conditional equation** is true for only certain values of the variable.

Example 4 $x + 3 = 9$ is true only if $x = 6$ since no other number added to 3 would equal 9.

⦿ ***SKILL CHECK 3*** Is $x + 3 = 9$ a conditional equation?

Example 5 $3x = -3$ is true only if $x = -1$ since -1 is the only number whose product with 3 is -3.

Of the three main types the focus of our attention will be on conditional equations. Finding the values that make a conditional equation true is one of the main objectives of this text.

> A **solution** or **root** of an equation is the value of the variable or variables that make the equation a true statement.

The *solution* or *root* is said to satisfy the equation.
Solving an equation means finding the solution or root.

2 Solving Equations Mentally

Many equations can be solved mentally. Ability to solve an equation mentally will depend on your knowledge of the operations on real numbers. The better you know the facts of multiplication and addition, the more adept you will be at mentally solving equations.

Example 6 Solve mentally for x: $x + 3 = 7$

Solution To have a true statement we need a value for x that, when added to 3, will yield 7. Our knowledge of arithmetic indicates that 4 is the needed value. Therefore, the solution to the equation is $x = 4$.

What number added to 3 equals 7?

⦿ ***SKILL CHECK 4*** Solve mentally for x: $x + 5 = 7$

Example 7 Solve mentally for x: $x - 5 = 3$

Solution What number do we subtract 5 from to obtain 3? Again our experience with arithmetic tells us that $8 - 5 = 3$. Therefore, the solution is $x = 8$.

⦿ ***SKILL CHECK 5*** Solve mentally for x: $x - 1 = 6$

Example 8 Solve mentally for x: $3x = 15$

Solution What number must be multiplied by 3 to obtain 15? The solution is $x = 5$.

$3(5) = 15$

⦿ ***SKILL CHECK 6*** Solve mentally for x: $4x = 12$

Example 9 Solve mentally for x: $\frac{x}{2} = 7$

Solution What number do we divide by 2 to obtain 7? The solution is $x = 14$ since $\frac{14}{2} = 7$.

⦿ ***SKILL CHECK 7*** Solve mentally for x: $\frac{x}{3} = 7$

⦿ ***Skill Check Answers***

3. yes
4. 2
5. 7
6. 3
7. 21

6 is the only number from which we subtract 1 to obtain 5. Thus $2x$ must stand for 6.

Example 10 Solve mentally for x: $2x - 1 = 5$

Solution $6 - 1 = 5$, thus $2x = 6$. Then $x = 3$. Checking, we have

$$\begin{array}{r|l} 2x - 1 & 5 \\ 2(3) - 1 & 5 \\ 6 - 1 & \\ 5 & \end{array}$$

Regardless of how an equation is solved, it is always a good practice to check the solution by substituting the value for x in the original equation to see if an equality is indeed obtained.

⦿ SKILL CHECK 8 Solve mentally for x: $3x + 1 = 10$

Many students think that when they have found the solution to an equation, the problem is finished. *Not so!* The final step should *always* be to check the solution.

Example 11 A student solved the equation $5x - 3 = 4x + 2$ and found an answer of $x = 6$. Was this right or wrong?

Solution Does $x = 6$ satisfy the equation $5x - 3 = 4x + 2$? To check we substitute 6 for x in the equation to see if we obtain a true statement.

$$\begin{array}{r|l} 5x - 3 & 4x + 2 \\ 5(6) - 3 & 4(6) + 2 \\ 30 - 3 & 24 + 2 \\ 27 & 26 \end{array}$$

This is not a true statement, so the answer $x = 6$ is wrong.

Another student solved the same equation and found $x = 5$.

$$\begin{array}{r|l} 5x - 3 & 4x + 2 \\ 5(5) - 3 & 4(5) + 2 \\ 25 - 3 & 20 + 2 \\ 22 & 22 \end{array}$$

This is a true statement, so $x = 5$ is correct.

⦿ SKILL CHECK 9 Is $x = 2$ a solution to the equation $3x + 2 = 5x - 2$?

⦿ *Skill Check Answers*

8. 3
9. yes

Exercise 8–2–1

1 Classify each equation as an identity, a contradiction, or a conditional equation.

1. $3x + x = 4x$

2. $2x + 1 = 5$

3. $x - 2 = 11$

4. $6x = 4x + 2x$

5. $5x + 1 = 5x$

6. $2x - 1 = 2x + 1$

7. $20x - 8x = 7x + 5x$

8. $x + 5 = 2x - 1$

9. $4x - 3 = 17$

10. $3x + 2x = 4x + 9$

2 Solve mentally and check by substitution.

11. $x + 4 = 7$

12. $2x = 6$

13. $x - 3 = 8$

14. $x + 1 = 10$

15. $3x = -12$

16. $2x = -16$

17. $\frac{x}{2} = 5$

18. $\frac{x}{4} = 4$

19. $5x = 30$

20. $x + 9 = 11$

21. $5x = 3$

22. $2x + 1 = 11$

23. $9x = 3$

24. $12 + x = 25$

25. $x + 5 = 0$

26. $3x - 5 = 22$

27. $2x - 1 = 13$

28. $x + 3 = 0$

29. $\frac{x}{3} + 5 = 7$

30. $\frac{x}{2} - 4 = 6$

• Concept Enrichment

31. Is -2 a solution of the equation $7x - 2 = 3x + 22$?

32. Find the solution(s) to the equation $x^2 = 9$.

33. How many solutions do you think there are to any first-degree conditional equation in one variable?

34. How many solutions do you think there are to any second-degree conditional equation in one variable? Then what about $x^2 = 0$?

8–3 The Addition Property of Equality

OBJECTIVES

Upon completing this section you should be able to:

1. Use the addition property of equality to solve equations.
2. Solve some basic applied problems using the addition property.

In the previous section you solved equations mentally and checked the solutions for correctness. Some equations are not easily solved mentally and in this and subsequent sections of this chapter we will develop an orderly procedure for solving any first-degree polynomial equation in one variable.

Fundamental to the techniques to be developed in this and following sections is the idea of equivalent equations.

> Two equations are **equivalent** if they have the same solution or solutions.

Verify that this is true by substituting $x = 2$ in both equations.

Example 1 $3x = 6$ and $2x + 1 = 5$ are equivalent equations because in both cases $x = 2$ is the solution.

Techniques for solving equations will involve processes for changing an equation to an equivalent equation. If a complicated equation such as $2x - 4 + 3x = 7x + 2 - 4x$ can be changed to a simple equation $x = 3$, and the equation $x = 3$ is equivalent to the original equation, then we have solved the equation.

The procedure for changing a given equation to an equivalent equation involves the properties of equality. We have used some properties of equality many times in the previous chapters without formally stating them.

These properties are the basis for working with equations.

> **Reflexive property** of equality: $a = a$.
> **Symmetric property** of equality: If $a = b$, then $b = a$.
> **Transitive property** of equality: If $a = b$ and $b = c$, then $a = c$.

The reflexive property is the rather obvious statement that a quantity is equal to itself.

The symmetric property allows us to switch sides of an equation. If we have the statement $3 = x$, we probably would rather write it as $x = 3$ since we read from left to right.

The transitive property is sometimes stated as "quantities equal to the same thing are equal to each other."

We have used the transitive property many times. For instance if we state "$x = 5 + 3$ so $x = 8$," we have used the transitive property. A complete statement would be "$x = 5 + 3$ and $5 + 3 = 8$, so $x = 8$."

1 The Addition Property

Another important property of equality is the *addition property*.

Since $5 = 5$, we can write $5 + 3 = 5 + 3$ or $8 = 8$.

> **Addition property** of equality: If $a = b$, then $a + c = b + c$.

We will now restate the addition property as it applies to changing a given equation to an equivalent equation.

The quantity to be added may be positive or negative.

> If the same quantity is added to both sides of an equation, the resulting equation is equivalent to the original equation.

Example 2 Solve for x if $x + 7 = 12$.

Solution Even though this equation can easily be solved mentally, we wish to use the addition property to show a technique. Your thinking should be in this manner: "Since my goal is to solve for x, I must have x by itself on one side of the equation. I now have $x + 7$ and must find a way to get x alone in an equation equivalent to the given equation." We know that if we add (-7) to 7 the result is zero and also that $x + 0$ is x. So we proceed as follows:

$$x + 7 = 12$$
$$x + 7 + (-7) = 12 + (-7) \quad \text{(addition rule)}$$
$$x + 0 = 12 - 7$$
$$x = 5$$

We add (-7) to both sides of the equation.

Check: $x + 7 = 12$

$$\begin{array}{c|c} 5 + 7 & 12 \\ 12 & \end{array}$$

⦿ ***SKILL CHECK 1*** Solve for x: $x + 2 = 6$

Remember that "adding the negative" is the definition of subtraction. So if it is easier, subtract the same quantity from both sides.

Example 3 Solve for x: $x - 8 = -4$

Solution

$$x - 8 = -4$$
$$x - 8 + 8 = -4 + 8 \quad \text{(adding 8 to both sides)}$$
$$x + 0 = 4$$
$$x = 4$$

Note that $x + 0$ may be written simply as x since zero added to any quantity equals the quantity itself.

Check: $x - 8 = -4$

$$\begin{array}{c|c} 4 - 8 & -4 \\ -4 & \end{array}$$

⦿ ***SKILL CHECK 2*** Solve for x: $x - 3 = -5$

Example 4 Solve for x: $x + \frac{1}{2} = \frac{2}{5}$

Solution

$$x + \frac{1}{2} = \frac{2}{5}$$
$$x + \frac{1}{2} - \frac{1}{2} = \frac{2}{5} - \frac{1}{2}$$
$$x = \frac{4}{10} - \frac{5}{10}$$
$$x = -\frac{1}{10}$$

Subtract $\frac{1}{2}$ from both sides. (That is, add $-\frac{1}{2}$ to both sides.)

Check: $x + \frac{1}{2} = \frac{2}{5}$

$$\begin{array}{c|c} -\frac{1}{10} + \frac{1}{2} & \frac{2}{5} \\ -\frac{1}{10} + \frac{5}{10} & \\ \frac{4}{10} & \\ \frac{2}{5} & \end{array}$$

⦿ ***Skill Check Answers***

1. 4
2. -2

⦿ ***SKILL CHECK 3*** Solve for x: $x + \frac{2}{3} = \frac{1}{6}$

Example 5 Solve for x: $3x = 2x - 3$

Solution Note that here the variable occurs on both sides of the equation. Our goal is to arrive at an equation of the form $x =$ (some number), so we elect to subtract $2x$ from both sides [or add $(-2x)$ to both sides].

Recall how to add and subtract like terms.

$$3x = 2x - 3$$
$$3x - 2x = 2x - 2x - 3$$
$$x = -3$$

Remember that checking your solution is an important step in solving equations.

Check: $3x = 2x - 2$

$3(-3)$	$2(-3) - 3$
-9	$-6 - 3$
-9	-9

⦿ ***SKILL CHECK 4*** Solve for x: $5x = 4x - 1$

2 Applications

Example 6 The selling price S of an item is obtained by using the formula $S = C + M$, where C represents the cost to the merchant and M represents the margin (profit and other expenses). Find the margin of an item that cost $12.50 and sold for $21.95.

$S = \$21.95$
$C = \$12.50$

Solution Substituting in the formula, we have

$$S = C + M$$
$$21.95 = 12.50 + M$$

Subtract 12.50 from both sides.

$$21.95 - 12.50 = M$$
$$9.45 = M$$

Symmetric property.

$$M = \$9.45.$$

Check: $S = C + M$

21.95	$12.50 + 9.45$
	21.95

⦿ ***Skill Check Answers***

3. $-\frac{1}{2}$
4. -1
5. $21.75

⦿ ***SKILL CHECK 5*** Using the formula from example 6, find the cost if the selling price is $32.50 and the margin is $10.75.

Exercise 8–3–1

1 Solve for x and check.

1. $x + 3 = 5$

2. $x + 4 = 7$

3. $x + 7 = 2$

4. $x + 4 = 11$

5. $x - 7 = 2$

6. $x - 4 = 6$

7. $x - 1 = -3$

8. $x - 3 = -2$

9. $x + 4 = -5$

10. $x + 6 = -1$

11. $x + 2.4 = 1.6$

12. $x + 3.4 = 2.7$

13. $x - 3.7 = 0.5$

14. $x - 1.3 = 2.9$

15. $x - \frac{1}{2} = \frac{3}{5}$

16. $x - \frac{2}{3} = \frac{1}{2}$

17. $x + \frac{1}{3} = \frac{2}{5}$

18. $x + \frac{3}{5} = \frac{1}{4}$

19. $7x = 6x + 2$

20. $3x = 2x + 1$

21. $x + 5 = 5$

22. $x + 3 = -3$

23. $2x - 4 = 3x$

24. $5 + 3x = 4x$

25. $x - 3.6 = 2.7$

26. $x - 11.5 = 4.6$

27. $x + 12 = 10$

28. $x + 9 = 5$

29. $x + \frac{3}{4} = 4\frac{2}{5}$

30. $x + 4 = \frac{1}{2}$

31. $8 + x = 2x$

32. $3 + 3x = 4x$

33. $5x + 2 = 4x + 10$

34. $2x + 3 = x + 1$

2

35. Two sides of a triangle are 10.1 meters and 13.5 meters. If the perimeter of the triangle is 47 meters, find the length of the third side. (Use $P = a + b + c$.)

36. What is the cost of an item that sold for $12.95 if the margin was $3.05? (Use $S = C + M$.)

37. Using the formula $M = S - C$, determine the selling price of an article that cost \$38.75 and sold with a margin of \$5.65.

38. Find the margin M on the sale of an item if the profit P was \$10.35 and the overhead O was \$3.54. (Use $P = M - O$.)

• Concept Enrichment

39. The total earnings (E) of a waitress are equal to the sum of the tips (t) and the product of the hourly rate (r) and the number of hours (h) worked ($E = rh + t$). If Jane worked eight hours at \$1.50 per hour and her total earnings were \$50.75, what did she make in tips?

40. The distance (s) in meters above the ground at a given time (t) of an object dropped from a height (h) is given by the formula $s = h - 4.9t^2$. Find the height that an object was dropped from if it strikes the ground in three seconds. (Hint: Use $s = 0$.)

41. What property allows us to write $5 = x$ as $x = 5$?

42. Why don't we need a subtraction property of equality?

8–4 The Multiplication Property of Equality

OBJECTIVES

Upon completing this section you should be able to:

1 Use the multiplication property of equality to solve equations.

2 Solve some basic applied problems using the multiplication property.

We need one more property of equality to complete the procedure for solving any first-degree polynomial equation in one variable.

1 Multiplication Property

Multiplication property of equality: If $a = b$, then for any number c, $ac = bc$.

We will restate this property in terms of equivalent equations.

If each side of an equation is multiplied by the same nonzero number, the resulting equation is equivalent to the original equation.

Notice that this property states we must multiply each side of an equation by a nonzero number. Multiplying each side by zero yields $0 = 0$, which, while true, is useless.

Example 1 Solve for x: $\frac{2}{3}x = 18$

Solution We want to obtain $1x$ alone on one side of the equation. The coefficient of x is $\frac{2}{3}$. To obtain a coefficient of 1 we must multiply $\frac{2}{3}$ by $\frac{3}{2}$, so we multiply both sides of the equation by $\frac{3}{2}$.

Recall that a number multiplied by its reciprocal (multiplicative inverse) is 1.

$$\frac{2}{3}x = 18$$
$$\frac{3}{2}\left(\frac{2}{3}x\right) = \frac{3}{2}(18)$$
$$x = 27$$

Check: $\frac{2}{3}x = 18$

$$\frac{2}{3}(27) \;\Big|\; 18$$
$$18$$

⦿ ***SKILL CHECK 1*** Solve for x: $\frac{3}{4}x = 12$

Since division is defined as multiplying by the reciprocal, we can also divide each side of an equation by the same nonzero number.

Example 2 Solve for x: $5x = 20$

Solution We could multiply each side of the equation by $\frac{1}{5}$ but it might be easier just to divide each side by 5.

5 is the reciprocal of $\frac{1}{5}$.

$$5x = 20$$
$$\frac{5x}{5} = \frac{20}{5}$$
$$x = 4$$

Check: $5x = 20$

$$5(4) \;\Big|\; 20$$
$$20$$

⦿ ***SKILL CHECK 2*** Solve for x: $3x = 15$

Example 3 Solve for x: $-3x = 12$

Solution We can divide each side by -3.

Or we could multiply each side by $-\frac{1}{3}$.

$$\frac{-3x}{-3} = \frac{12}{-3}$$
$$x = -4$$

Check: $-3x = 12$

$$-3(-4) \;\Big|\; 12$$
$$12$$

⦿ ***SKILL CHECK 3*** Solve for x: $-2x = 10$

⦿ ***Skill Check Answers***

1. 16
2. 5
3. −5

We can multiply *or* divide each side of an equation by the same nonzero number. In fact, we may want to multiply and divide in separate steps.

Example 4 Solve for x if $\frac{2}{3}x = \frac{3}{5}$.

Solution One method of solving this equation is to multiply both sides by $\frac{3}{2}$.

We multiply by the reciprocal of the coefficient of x.

$$\frac{3}{2}\left(\frac{2}{3}x\right) = \frac{3}{2}\left(\frac{3}{5}\right)$$
$$x = \frac{9}{10}$$

Another method of solving this problem involves multiplying and dividing in separate steps. This method may at times make the computation simpler. In this method we wish to first eliminate all fractions in the equation $\frac{2}{3}x = \frac{3}{5}$.

If we multiply both sides of the equation by 3, we would eliminate the fraction $\frac{2}{3}$ but not the fraction $\frac{3}{5}$. If instead we multiply by 5, we would eliminate $\frac{3}{5}$ but not $\frac{2}{3}$. To eliminate both $\frac{2}{3}$ and $\frac{3}{5}$ we must multiply by a number divisible by both 3 and 5. It is best to use the smallest of such numbers (the least common multiple). The LCM of 3 and 5 is 15. Recall that the LCM is also called the least common denominator (LCD) of the two fractions. Multiplying both sides by 15 yields

Multiplying by the LCD will always eliminate all fractions in an equation.

$$(15)\left(\frac{2}{3}x\right) = (15)\left(\frac{3}{5}\right)$$
$$\frac{15}{1}\left(\frac{2}{3}x\right) = \frac{15}{1}\left(\frac{3}{5}\right)$$
$$10x = 9$$
$$\frac{10x}{10} = \frac{9}{10}$$
$$x = \frac{9}{10}.$$

Would you get the same solution if you multiplied both sides by 30 or 45 or 60?

We see that either method yields the same solution.

Check: $\frac{2}{3}x = \frac{3}{5}$

$$\left(\frac{2}{3}\right)\left(\frac{9}{10}\right) \;\Big|\; \frac{3}{5}$$
$$\frac{3}{5}$$

⦿ ***Skill Check Answer***

4. $\frac{5}{9}$

⦿ ***SKILL CHECK 4*** Solve for x: $\frac{3}{5}x = \frac{1}{3}$

2 Applications

The multiplication or division property of equality is also useful in solving many types of applied problems.

Example 5 The formula for finding the area A of a triangle is $A = \frac{1}{2}bh$, where b represents the base of the triangle and h represents the altitude to the base. Find the altitude if the base measures 13 cm and the area is 39 cm².

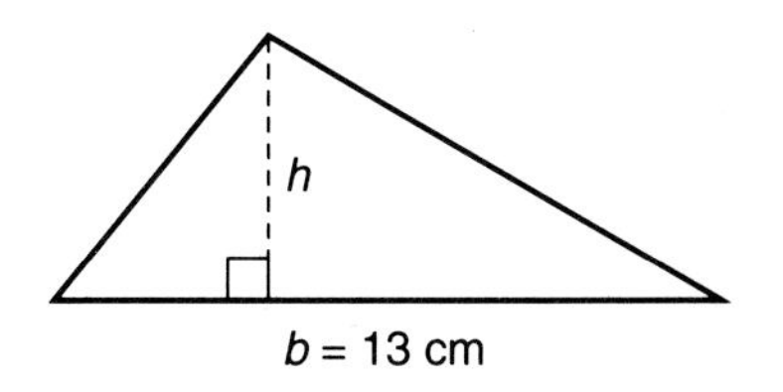

Solution To solve a problem involving a formula we first use the substitution principle.

$$A = \frac{1}{2}bh$$

$$39 \text{ cm}^2 = \frac{1}{2}(13 \text{ cm})h$$

$$78 \text{ cm}^2 = (13 \text{ cm})h \qquad \text{Multiply both sides by 2.}$$

$$\frac{78 \text{ cm}^2}{13 \text{ cm}} = h \qquad \text{Divide both sides by 13 cm.}$$

$$h = 6 \text{ cm}$$

Check: $A = \frac{1}{2}bh$

$$39 \text{ cm}^2 \;\Big|\; \frac{1}{2}(13 \text{ cm})(6 \text{ cm})$$

$$\frac{1}{2}(78 \text{ cm}^2)$$

$$39 \text{ cm}^2$$

◉ ***SKILL CHECK 5*** Find the length of the base of a triangle if the altitude is 9 in. and the area is 63 in.².

◉ ***Skill Check Answer***

5. 14 in.

Exercise 8–4–1

1 Solve for x.

1. $\frac{1}{2}x = 6$

2. $\frac{1}{3}x = 12$

3. $\frac{2}{5}x = 10$

4. $\frac{3}{5}x = 9$

5. $\frac{2}{3}x = -8$

6. $\frac{5}{8}x = -14$

7. $\frac{3}{5}x = -15$

8. $\frac{2}{3}x = -6$

9. $-\frac{3}{4}x = 8$

10. $-\frac{7}{8}x = 42$

11. $2x = 24$

12. $3x = 27$

13. $-4x = 20$

14. $-5x = 15$

15. $3x = 1$

16. $4x = 2$

17. $4x = -2$

18. $7x = -21$

19. $-x = 14$

20. $-x = -23$

21. $\frac{2}{3}x = \frac{5}{8}$

22. $\frac{3}{5}x = \frac{5}{6}$

23. $\frac{5}{8}x = -\frac{1}{4}$

24. $-\frac{2}{3}x = \frac{1}{6}$

25. $\frac{5}{6}x = \frac{2}{3}$

26. $\frac{2}{5}x = \frac{4}{15}$

27. $-\frac{3}{8}x = \frac{1}{2}$

28. $\frac{7}{3}x = -\frac{5}{8}$

29. $\frac{5}{9}x = 3\frac{1}{3}$

30. $\frac{3}{4}x = 4\frac{1}{2}$

31. The formula for the area of a rectangle is $A = \ell w$, where A is area, ℓ is length, and w is width. If the area of a rectangle is 391 square meters and the length is 23 meters, find the width.

$A = 391$ m^2 w

23 m

32. The formula for the perimeter (P) of a square is given by $P = 4s$, where s represents the length of one side. Find the length of a side of a square whose perimeter is 52 inches.

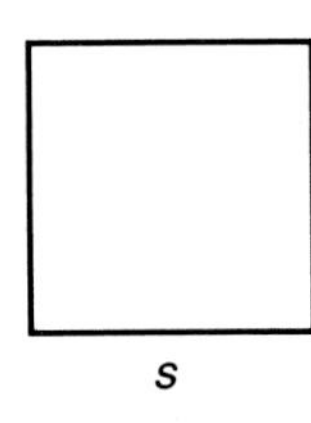

s

33. The formula for the area of a triangle is $A = \frac{1}{2}bh$, where A is area, b is the base, and h is the altitude. If the area of the triangle is 279 cm² and the altitude is 18 cm, find the length of the base.

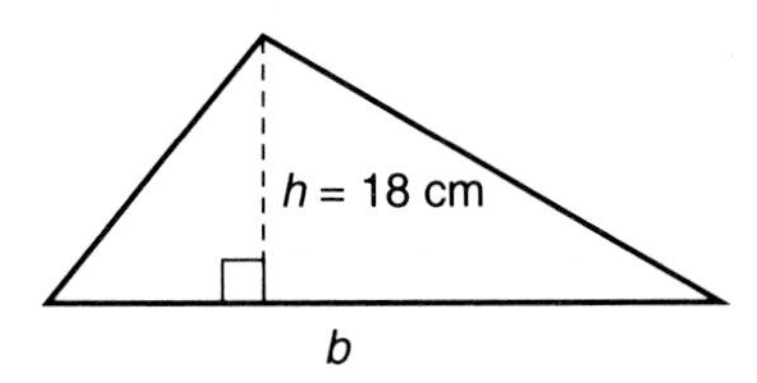

34. Find the altitude h of a triangle if the area A is 12 cm² and the base b is 3 cm. $\left(\text{Use } A = \frac{1}{2}bh.\right)$

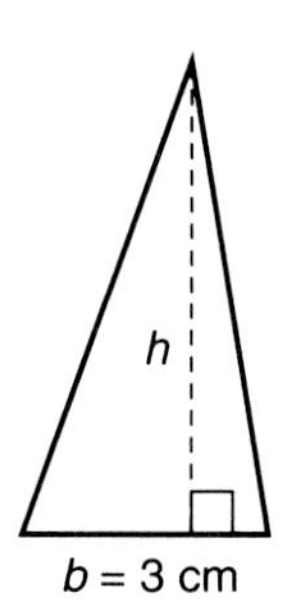

35. The formula $d = rt$ gives the distance d traveled at a constant rate r in the time t. If a man drives 347 kilometers in five hours, what was his average rate?

36. If the perimeter of a rectangle is 39.8 cm and the width is 7.5 cm, what is the length? (Use $P = 2\ell + 2w$.)

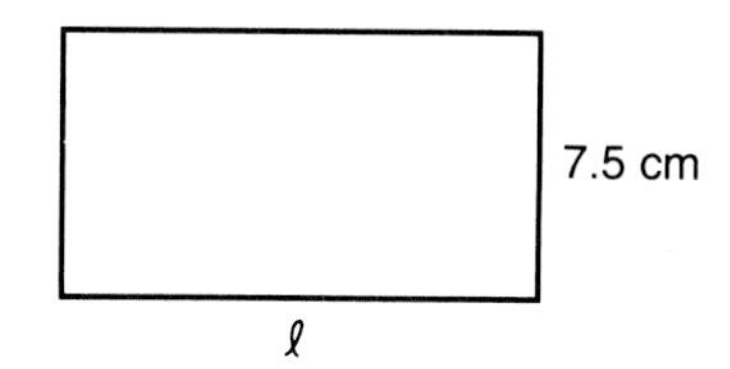

37. $\frac{9}{5}C = F - 32$ is a relationship between Fahrenheit (F) and Celsius (C) temperatures. Find the Celsius temperature if the temperature is 86° F.

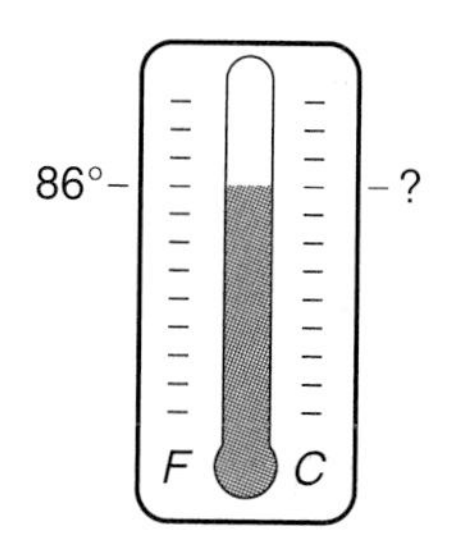

38. A relationship between Fahrenheit (F) and Celsius (C) temperature is given by $5F = 9C + 160$. Find the Fahrenheit temperature if the Celsius temperature is −20° C.

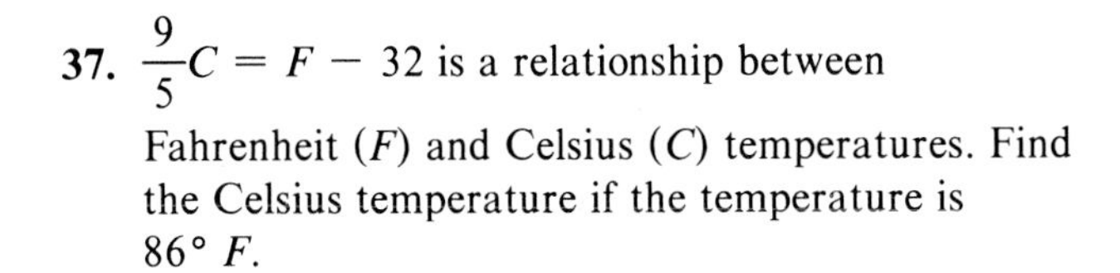

• Concept Enrichment

39. The distance (s) in feet that a falling object travels in t seconds is given by $s = \frac{1}{2}gt^2$, where g is the acceleration due to gravity. Find g if an object fell 144 feet in 3 seconds.

40. The volume of a rectangular solid is given by the formula $V = \ell wh$, where ℓ is length, w is width, and h is height. If a room is $16\frac{1}{2}$ feet long and 15 feet wide, how high must it be to have a volume of 2,970 cubic feet?

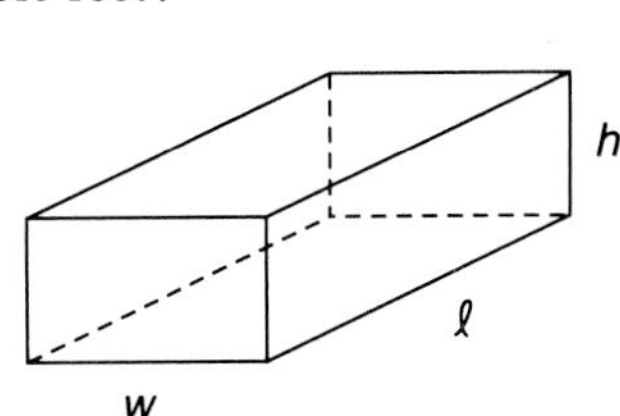

41. Show two ways of solving $-5x = 35$.

42. Why don't we need a separate division property of equality?

8–5 Using More Than One Property to Solve Equations

OBJECTIVES

Upon completing this section you should be able to:

1. Use more than one property to solve equations.
2. Apply the orderly procedure established in this section to systematically solve equations.

1 Using Both Properties

In solving some equations it may be necessary to use both the multiplication (division) property and addition (subtraction) property.

Example 1 Solve for x: $2x + 3 = 21$

Solution To solve this equation we will need to use more than a single property. We first wish to obtain an equivalent equation having the term involving the variable on one side and everything else on the other side. To accomplish this we need to add -3 to both sides (or subtract 3 from each side).

$$2x + 3 - 3 = 21 - 3$$
$$2x = 18$$

We can then multiply each side by $\frac{1}{2}$ (divide each side by 2), obtaining

Notice that we have used both the addition and multiplication properties in solving this equation.

$$\frac{2x}{2} = \frac{18}{2}$$
$$x = 9.$$

Check:
$$2x + 3 = 21$$
$$2(9) + 3 \quad | \quad 21$$
$$18 + 3$$
$$21$$

◉ ***SKILL CHECK 1*** Solve for x: $2x + 5 = 17$

Example 2 Solve for x: $3x - 4 = -19$

Solution We first add 4 to both sides of the equation.

Addition property.

$$3x - 4 = -19$$
$$3x - 4 + 4 = -19 + 4$$
$$3x = -15$$

We next divide each side by 3.

Multiplication (division) property.

$$\frac{3x}{3} = \frac{-15}{3}$$
$$x = -5$$

Check:
$$3x - 4 = -19$$
$$3(-5) - 4 \quad | \quad -19$$
$$-15 - 4$$
$$-19$$

◉ ***Skill Check Answer***

1. 6

⦿ ***SKILL CHECK 2*** Solve for x: $5x - 3 = -18$

Example 3 Solve for x: $5x = 14 - 2x$

Solution Here our goal of obtaining x alone on one side would suggest we eliminate $-2x$ on the right by adding $2x$ to both sides.

$$5x = 14 - 2x$$
$$5x + 2x = 14 - 2x + 2x$$

Addition property.

$$7x = 14$$
$$\frac{7x}{7} = \frac{14}{7}$$

Divide each side by 7.

$$x = 2$$

Check: $5x = 14 - 2x$

$5x$	$14 - 2x$
$5(2)$	$14 - 2(2)$
10	$14 - 4$
	10

Note that we check by always substituting the solution in the *original* equation.

⦿ ***SKILL CHECK 3*** Solve for x: $3x = 12 - x$

Example 4 Solve for x: $2x + 6 = 3x - 9$

Solution Here we have a slightly more involved task. Again, we wish to have all terms involving the variable on one side of the equation. We begin by adding $-3x$ to each side (subtract $3x$ from each side).

$$2x + 6 = 3x - 9$$
$$2x - 3x + 6 = 3x - 3x - 9$$
$$-x + 6 = -9$$

Note that we could also add $(-2x)$ to both sides of the equation obtaining

$$6 = x - 9.$$

Then adding $(+9)$ to both sides we have

$$15 = x$$

or

$$x = 15.$$

We next add -6 to each side (subtract 6 from each side).

$$-x + 6 - 6 = -9 - 6$$
$$-x = -15$$

Since we are solving for x and not $-x$, we must multiply each side by -1.

$$(-1)(-x) = (-1)(-15)$$
$$x = 15$$

Check: $2x + 6 = 3x - 9$

$2x + 6$	$3x - 9$
$2(15) + 6$	$3(15) - 9$
$30 + 6$	$45 - 9$
36	36

⦿ ***SKILL CHECK 4*** Solve for x: $3x + 7 = 4x - 5$

Some equations contain parentheses. In solving such equations the first step should be to eliminate the parentheses. This will enable us to use the properties more easily.

Example 5 Solve for x: $2(x + 5) = 4 - x$

Solution We first use the distributive property to eliminate the parentheses.

$$2(x + 5) = 4 - x$$
$$2x + 10 = 4 - x$$

Make sure you multiply each term inside the parentheses by 2.

Adding x to each side, we obtain

$$2x + x + 10 = 4 - x + x$$
$$3x + 10 = 4.$$

⦿ ***Skill Check Answers***

2. -3
3. 3
4. 12

Adding -10 to each side gives

$$3x + 10 - 10 = 4 - 10$$
$$3x = -6.$$

Finally, dividing each side by 3 yields

$$x = -2.$$

Check:
$$\begin{array}{r|l} 2(x+5) & = 4 - x \\ 2(-2+5) & 4-(-2) \\ 2(3) & 4+2 \\ 6 & 6 \end{array}$$

SKILL CHECK 5 Solve for x: $5(x + 3) = 3 - x$

Example 6 Solve for x: $\frac{5}{8}x + 3 = 5\frac{1}{2}$

Solution This equation involves fractions, so the first thing we will do is eliminate them by multiplying each side by the least common denominator 8.

Remember that multiplying by the LCD will eliminate all fractions.

$$\frac{5}{8}x + 3 = 5\frac{1}{2}$$
$$(8)\left(\frac{5}{8}x + 3\right) = (8)\left(5\frac{1}{2}\right)$$

We use the distributive property and multiply each term inside the parentheses by 8. In effect, this means we must multiply every term of the equation by the same quantity.

Before multiplying, change any mixed numbers to improper fractions. In this example change $5\frac{1}{2}$ to $\frac{11}{2}$.

$$(8)\left(\frac{5}{8}x\right) + 8(3) = 8\left(\frac{11}{2}\right)$$
$$5x + 24 = 44$$

Be careful that *each* term on both sides of the equation is multiplied by the same number.

$$5x + 24 = 44$$

Subtract 24 from each side.

$$5x + 24 - 24 = 44 - 24$$
$$5x = 20$$

Divide each side by 5.

$$\frac{5x}{5} = \frac{20}{5}$$
$$x = 4$$

Check:
$$\begin{array}{r|l} \frac{5}{8}x + 3 & = 5\frac{1}{2} \\ \frac{5}{8}(4) + 3 & 5\frac{1}{2} \\ \frac{5}{2} + 3 & \\ \frac{11}{2} & \\ 5\frac{1}{2} & \end{array}$$

Skill Check Answers
5. -2
6. -8

SKILL CHECK 6 Solve for x: $\frac{5}{6}x + 11 = 4\frac{1}{3}$

Example 7 Solve for x: $\frac{2}{3}x = \frac{1}{8}(5 - 8x)$

Solution First eliminate the parentheses.

$$\frac{2}{3}x = \frac{5}{8} - x$$

Multiply each term inside the parentheses by $\frac{1}{8}$.

Next we see that the least common multiple of 3 and 8 is 24, so we multiply each side of the equation by 24.

$$24\left(\frac{2}{3}x\right) = 24\left(\frac{5}{8}\right) - 24(x)$$
$$16x = 15 - 24x$$
$$40x = 15$$
$$x = \frac{15}{40}$$
$$x = \frac{3}{8}$$

Remember that *each* term must be multiplied by 24.

Answers should be given in simplest form.

Check: $\frac{2}{3}x = \frac{1}{8}(5 - 8x)$

$$\frac{2}{3}\left(\frac{3}{8}\right) \;\Big|\; \frac{1}{8}\left[5 - 8\left(\frac{3}{8}\right)\right]$$
$$\frac{1}{4} \;\Big|\; \frac{1}{8}(5 - 3)$$
$$\Big|\; \frac{1}{4}$$

Remember to always check the answer by substituting it in the *original* equation.

⊙ ***SKILL CHECK 7*** Solve for x: $\frac{4}{5}x = \frac{1}{2}(3 - 2x)$

2 An Orderly Procedure for Solving Equations

We see that some equations require the application of more than one property to arrive at a solution. At this point we will introduce an orderly procedure for solving the type of equations studied in this chapter. This procedure is not the only approach to solving equations, but experience has indicated that following this order provides a smoother more mistake-free method.

An Orderly Procedure for Solving Equations

1. Eliminate parentheses by applying the distributive property.
2. Eliminate fractions by multiplying each term of the equation by the LCD of all fractions in the equation.
3. Add or subtract like terms on each side of the equation.
4. Use the addition property to obtain the terms involving the variable on one side of the equation and all other terms on the other side. Again add or subtract like terms.
5. Divide each side of the equation by the coefficient of the variable.
6. Check the solution by substituting it in the original equation.

⊙ ***Skill Check Answer***

7. $\frac{5}{6}$

Example 8 Solve for x: $2x + 11 - \frac{2}{3}x = \frac{1}{2}(x + 2)$

Solution We will use the 6 steps in the procedure for solving equations.

Step 1. Eliminate parentheses.

$$2x + 11 - \frac{2}{3}x = \frac{1}{2}x + 1$$

Step 2. Multiply by 6 (LCD) to eliminate fractions.

$$6(2x) + 6(11) - 6\left(\frac{2}{3}x\right) = 6\left(\frac{1}{2}x\right) + 6(1)$$

$$12x + 66 - 4x = 3x + 6$$

Step 3. Combine like terms.

$$8x + 66 = 3x + 6$$

Step 4. Add $-3x$ and -66 to each side.

$$5x = -60$$

Step 5. Divide each side by 5.

$$x = -12$$

Step 6. Check.

Check: $2x + 11 - \frac{2}{3}x = \frac{1}{2}x + 1$

$2(-12) + 11 - \frac{2}{3}(-12)$	$\frac{1}{2}(-12) + 1$
$-24 + 11 + 8$	$-6 + 1$
-5	-5

⊙ ***SKILL CHECK 8*** Solve for x: $x + 2 - \frac{3}{4}x = \frac{1}{3}(x + 9)$

Some applications may involve equations that require the use of both the addition and multiplication properties in their solution.

Example 9 A relationship between Fahrenheit (F) and Celsius (C) temperature is given by the formula $F = \frac{9}{5}C + 32$. Find the Celsius temperature if the Fahrenheit temperature is 14°.

Solution We will substitute 14° for F in the formula and then solve for C.

$$F = \frac{9}{5}C + 32$$

$$14 = \frac{9}{5}C + 32$$

Multiply each term by the LCD 5.

$$5(14) = 5\left(\frac{9}{5}C\right) + 5(32)$$

$$70 = 9C + 160$$

Subtract 160 from each side.

$$70 - 160 = 9C$$

$$-90 = 9C$$

Divide each side by 9.

$$-10 = C$$

$$C = -10°$$

Check: $F = \frac{9}{5}C + 32$

14	$\frac{9}{5}(-10) + 32$
	$-18 + 32$
	14

⊙ ***SKILL CHECK 9*** Find the Celsius temperature if the Fahrenheit temperature is 5°.

⊙ ***Skill Check Answers***

8. -12
9. $-15°$

Exercise 8–5–1

1 Solve for x.

1. $3x + 1 = 10$

2. $5x + 7 = 13$

3. $14x - 5 = 9$

4. $8x - 3 = 1$

5. $2x + 9 = 1$

6. $3x + 10 = 4$

7. $4x - 5 = -29$

8. $2x - 7 = -21$

9. $2x + 9 = 3$

10. $4x + 13 = 5$

11. $3x = 2(5 - x)$

12. $2x = 5 - 4x$

13. $3x = 8x + 5$

14. $6x = x - 21$

15. $4x = x - 5$

16. $2x = 4x + 5$

17. $5x + 4 = 2x + 25$

18. $8x + 5 = 2x + 5$

19. $4(3x + 1) = x - 3$

20. $3x - 4 = 8 - x$

21. $8 - 2x = 3 - 3x$

22. $3x - 2 = 10 - 3x$

23. $\frac{1}{3}x + \frac{7}{6} = \frac{1}{2}$

24. $\frac{1}{5}(4x + 25) = \frac{1}{5}$

25. $\frac{7}{8}x + 4 = 2\frac{3}{4}$

26. $\frac{1}{4}x - \frac{2}{3} = 1\frac{1}{2}$

27. $\frac{1}{2}(x + 24) = \frac{3}{4}x$

28. $2x - \frac{3}{5} = \frac{2}{3}x$

29. $2x - \frac{2}{3} = \frac{3}{5}x + 2$

30. $4x - \frac{3}{8} = \frac{3}{4}(x + 4)$

31. $\frac{5}{8}x - \frac{1}{2} = \frac{3}{5}x + 3$

32. $\frac{1}{3}x + \frac{3}{8} = 2x + \frac{3}{4}$

33. $\frac{4}{5}x - 10 = \frac{2}{3}x + 2 - \frac{1}{5}x$

34. $\frac{4}{5}x + 6 - \frac{1}{3}x = \frac{2}{5}x + 9$

2

35. $x - \frac{1}{5}x + 1 = \frac{1}{3}(x - 5)$

36. $\frac{1}{2}\left(\frac{4}{3}x + x - 1\right) = \frac{1}{3}(x - 15)$

37. $\frac{2}{3}x = \frac{3}{5} + \frac{1}{2}x - 11$

38. $x + \frac{6}{7} + \frac{2}{3}x = \frac{1}{5}(3x + 20)$

39. $2\left(3x - \frac{1}{3}\right) = \frac{5}{8}x + 4 + \frac{2}{3}x$

40. $\frac{1}{5}x + \frac{1}{2} - \frac{2}{3}x = \frac{1}{3}x + 4$

41. Using the formula $F = \frac{9}{5}C + 32$, find C when $F = 77°$.

42. Using the formula $F = \frac{9}{5}C + 32$, find the Celsius temperature at which water boils at sea level. This occurs when the Fahrenheit temperature is 212°.

• Concept Enrichment

43. If the perimeter of a rectangle is $7\frac{2}{3}$ feet and the length measures $2\frac{1}{2}$ feet, find the measure of the width. (Use the formula $P = 2\ell + 2w$.)

44. The perimeter of a rectangle is 36.4 centimeters. If the width measures 5.7 centimeters, what is the measure of the length?

CHAPTER 8 SUMMARY

Key Words

Section 8–1

- A **monomial** is an expression in which the numerical coefficient and variables are related only by the operation of multiplication and the exponents of the variables are whole numbers.
- A **polynomial** is the indicated sum or difference of one or more monomials.
- A **binomial** is a polynomial having exactly two terms.
- A **trinomial** is a polynomial having exactly three terms.
- The **degree** of a polynomial in a variable is the highest power of that variable in the polynomial.

Section 8–2

- An **equation** in algebra is a statement in symbols that two algebraic expressions have the same value.
- A **polynomial equation** is an equation involving only polynomials.
- An **identity** is true for all values of the numbers and variables in it.
- A **contradiction** is never true.
- A **conditional equation** is true for only certain values of the variable.
- A **solution** or **root** of an equation is the value of the variable or variables that make the equation a true statement.

Section 8–3

- **Equivalent equations** have the same solution or solutions.
- The **reflexive property** of equality states $a = a$.
- The **symmetric property** of equality states that if $a = b$, then $b = a$.
- The **transitive property** of equality states that if $a = b$ and $b = c$, then $a = c$.
- The **addition property** of equality states that if $a = b$, then $a + c = b + c$.

Section 8–4

- The **multiplication property** of equality states that if $a = b$, then for any number c, $ac = bc$.

Procedures

Section 8–3

- If the same quantity is added to (subtracted from) both sides of an equation, the resulting equation is equivalent to the original equation.

Section 8–4

- If each side of an equation is multiplied (divided) by the same nonzero number, the resulting equation is equivalent to the original equation.

Section 8–5

- An orderly procedure for solving an equation is given in the following steps.
    1. Eliminate parentheses by applying the distributive property.
    2. Eliminate fractions by multiplying each term of the equation by the LCD of all fractions in the equation.
    3. Add or subtract like terms on each side of the equation.
    4. Use the addition property to obtain the terms involving the variable on one side of the equation and all other terms on the other side. Again add or subtract like terms.
    5. Divide each side of the equation by the coefficient of the variable.
    6. Check the solution by substituting it in the original equation.

CHAPTER 8 REVIEW

1. Explain why $2x^2y - 3xy^3 + y^4 + 5$ is a polynomial.

2. Explain why $3a^2b^3 - 2ab^4 - a^{-1}b^5$ is not a polynomial.

3. What is the degree in x of the polynomial $x^3y^5 + 2x^2y^6 - 4y - 2$?

4. What is the degree in y of the polynomial $3x^2y - 2x^5 + x^3y^4 - 6$?

Classify each statement as an identity, a contradiction, or a conditional equation.

5. $2x = x + 3$

6. $5x - x = 4x$

7. $6x - 2x = 3x + x$

8. $7x - 3 = 7x$

9. $6x = 3x + 3$

10. $2x + 3 - x = x + 3$

Solve for x.

11. $x + 2 = 5$

12. $x - 4 = 3$

13. $7x = -56$

14. $x + 25 = 6$

15. $x + 36 = 12$

16. $2x = -104$

17. $x + 13 = 52$

18. $x - 7 = 68$

19. $-3x = 24$

20. $3x = x + 8$

21. $\frac{2}{3}x = 4$

22. $-5x = 40$

23. $7x = 5x + 9$

24. $\frac{1}{5}x = 20$

25. $-x = 9$

26. $-x = 3$

27. $-7x = 2x + 27$

28. $5x + 3 = 2x$

29. $\frac{1}{3} + 2x = \frac{1}{2}(x + 8)$

30. $6x = 78$

31. $4x + 5 = 2x - 17$

32. $\frac{2}{5}x = \frac{4}{7}$

33. $x - \frac{2}{5} = \frac{1}{7}$

34. $4x + \frac{3}{4} = \frac{1}{2}(x + 6)$

35. $\frac{5}{6}x - 3 = \frac{1}{2}$

36. $14x = 168$

37. $\frac{2}{3}x - 18 = \frac{1}{2}x + 5$

38. $5x + 2 = x - 19$

39. $3x + 4 = 19$

40. $\frac{3}{8}x = \frac{5}{6}$

41. $\frac{4}{7}x - 5 = \frac{1}{2}x$

42. $7x - \frac{1}{2} = 4\left(\frac{1}{5}x + 1\right)$

43. $-11x = 166$

44. $4x - 1 = 3x - 5$

45. $\frac{2}{3} + 5x = 3\left(\frac{1}{2} + x\right)$

46. $8x + 5 = x - 37$

47. $3x + 5 = 2x + 5$

48. $\frac{5}{6}x - \frac{2}{3} = \frac{1}{2}x + 1$

49. $3\frac{1}{2}x = 4\frac{1}{3}$

50. $x - 6 = 14 - 3x$

51. $\frac{1}{2}(3x - 4) = \frac{1}{5} + \frac{1}{3}x - 9\frac{1}{5}$

52. $\frac{1}{2}(x - 1) = \frac{2}{3}\left(1 - \frac{3}{4}x\right)$

53. $\frac{1}{4}\left(3 + \frac{8}{3}x\right) = \frac{1}{2}(x + 4)$

54. $x - \frac{7}{8} = 3\left(\frac{1}{5}x + \frac{1}{4} - \frac{1}{6}x\right)$

55. The area (A) of a rectangle is 225 square feet. Find the length (ℓ) if the width (w) is 12.5 feet. Use the formula $A = \ell w$.

56. The volume (V) of a rectangular solid is 60 cubic meters. Find the length (ℓ) of the solid if the width (w) is 5 meters and the height (h) is $\frac{2}{3}$ meter. Use the formula $V = \ell wh$.

57. The wages (w) made by a worker are equal to the product of the hourly rate (r) and the number of hours worked (t). The formula is given by $w = rt$. Find the hourly rate of a worker earning $217.50 in $37\frac{1}{2}$ hours.

58. A relationship between Fahrenheit (F) and Celsius (C) temperatures is given by $9C = 5F - 160$. Find the Fahrenheit temperature if the Celsius temperature is $-5°$.

59. Find the width (w) of a rectangle if the perimeter (P) is 400 cm and the length (ℓ) is 115.8 cm. Use the formula $P = 2\ell + 2w$.

60. The perimeter (P) of a certain rectangle is $15\frac{2}{3}$ feet. Find the width (w) of the rectangle if the measure of the length (ℓ) is $5\frac{1}{2}$ feet. Use the formula $P = 2\ell + 2w$.

CHAPTER 8 TEST

1. Explain why $4x^3y^2 - 3xy + x^{-2}y^5 - 1$ is not a polynomial.

2. What is the degree in y of the polynomial $x^5y^2 + x^3y^4 - 3xy^2 + 3$?

3. Classify $5x - 2 = 3x - 2 + 2x$ as an identity, a contradiction, or a conditional equation.

4. Is $x = 4$ a solution of the equation $4x - 3 = x + 1$?

Solve for x.

5. $x + 12 = 3$

6. $4x - 9 = 27$

7. $6x = 78$

8. $\frac{3}{4}x = \frac{1}{8}$

9. $3x + 13 = 1 - x$

10. $\frac{1}{2}x - 1 = \frac{1}{3}x + 4$

11. $\frac{1}{2}\left(\frac{3}{4}x + 1 - 2x\right) = \frac{2}{3}x + \frac{3}{4}$

12. The length (ℓ) of a rectangle is $10\frac{1}{2}$ feet. If the perimeter (P) of the rectangle is $35\frac{2}{3}$ feet, find the measure of the width (w). Use the formula $P = 2\ell + 2w$.

CHAPTERS 1–8 CUMULATIVE TEST

1. Subtract: $\begin{array}{r} 10{,}813 \\ -\ \ 954 \\ \hline \end{array}$

2. Multiply: $\begin{array}{r} 1{,}043 \\ \times\ 145 \\ \hline \end{array}$

3. Reduce: $\dfrac{105x^2y^3}{126xy^5}$

4. Divide: $\dfrac{3a^2}{25bc} \div \dfrac{21ac}{45b^3}$

5. Solve for x: $x + 15 = 7$

6. Simplify: $(-2)(6)(-5)(-1)$

7. Simplify: $7 - 15 + 4 - (-6) - 3$

8. Multiply: $\begin{array}{r} 9.17 \\ \times 23.4 \\ \hline \end{array}$

9. One acute angle of a right triangle measures 54°. What is the measure of the other acute angle?

10. Divide: $16.3\overline{)91.932}$

11. What is the place value of the digit 4 in the number 241,637?

12. Find the least common multiple of 18 and 60.

13. Round 31.999 to the nearest hundredth.

14. Write the number 0.0000183 in scientific notation.

15. Solve for x: $\dfrac{2}{3}x = 18$

16. Simplify: $8xy + 4x^2 - 12xy - 3x^2$

17. Find the prime factorization of 1,960.

18. The measure of the hypotenuse of a right triangle is 10 meters. The measure of one leg is 6 meters. Find the measure of the other leg.

19. Simplify: $x^2y + 5xy + 4x^2y - 3x^2y - 3xy$

20. Add: $5\frac{1}{3} + 3\frac{5}{6}$

21. Simplify: $x(2x - 3) - 4x[x - 2(1 - x)]$

22. Solve for x: $\frac{4}{5}x - \frac{1}{2} = \frac{3}{4}x + 2$

23. Find the circumference of a circle if the radius measures $3\frac{1}{2}$ inches. Use $\pi = \frac{22}{7}$.

24. The area of a triangle is 28 square feet and the altitude is 7 feet. Find the length of the base.

25. What is the cost for tile to cover a rectangular kitchen floor that measures 15 feet by 12 feet if the tile is priced at $4.50 per square foot?

CHAPTER

9 Pretest

Answer as many of the following problems as you can before starting this chapter. When you finish the chapter, take the test at the end and compare your scores to see how much you have learned.

1. Write the ratio 24:84 as a fraction in reduced form.

2. Are the ratios 16:36 and 24:56 equal?

3. Is $\frac{6}{16} = \frac{15}{40}$ a proportion?

4. Is $\frac{16}{24} = \frac{12}{18}$ a proportion?

5. Solve for x: $\frac{x}{16} = \frac{3}{12}$

6. Solve for x: $\frac{6}{x} = \frac{8}{12}$

7. Solve for x: $\frac{3}{4} = \frac{x}{14}$

8. If the ratio of passing grades to failing grades is 7 to 2, how many students failed if 28 passed?

9. If an automobile uses $8\frac{1}{4}$ gallons of gasoline to travel 231 miles, how many miles can it travel on 12 gallons?

10. If 2 teaspoons of an instant coffee are used for three cups of coffee, how many are needed for 18 cups?

11. The speed limit on some interstate highways is 65 miles per hour. What is the speed limit in kilometers per hour? Round the answer to the nearest tenth. (1 mile = 1.609 kilometers)

12. Mr. Smith's lawn mower has a 7-liter gasoline tank. How many gallons will the tank hold? Round the answer to the nearest hundredth. (1 gallon = 3.785 liters)

13. a. Express $\frac{3}{8}$ as a percent.

b. Express 0.002 as a percent.

14. The school service club has 40 members. A proposal to change the meeting time was defeated by a vote of 23 to 17. What percent of the members voted for the change?

15. Mr. Jones must pay 0.5% interest each month on the balance of his home loan. If the balance is now $17,850, how much interest will he have to pay this month?

CHAPTER

9 Ratio, Proportion, and Percent

If you go shopping, watch television, read a newspaper, or look at a retail catalog, you cannot escape the topics covered in this chapter.

STUDY HELP

Estimation is an excellent tool in checking a solution. Equally important is the "reasonableness" of your answer. Asking yourself, "Is this answer of the magnitude it should be?" may many times start you searching for an error.

Ratios and proportions arise continually in our daily living and we may have used them without knowing them by name. In this chapter we will define ratios and show how they are used in proportions. The techniques established in solving equations will be utilized to find solutions to proportions. We will also examine various problems that use ratios and proportions in their solutions. Percent will be introduced as a ratio and problems involving percent will be solved using equations.

9–1 Ratios

OBJECTIVES

Upon completing this section you should be able to:

1. Express a ratio as a fraction.
2. Determine if two ratios are equal.

In this section we wish to define a ratio and determine when two ratios are equal.

1 Writing a Ratio as a Fraction

> A **ratio** is the quotient of two numbers, usually written as a fraction.

The ratio of a to b can be written as $a{:}b$ or as $\frac{a}{b}$, but the fraction form is more common. Note the value of b cannot be zero as division by zero is not allowed.

Example 1 Write the ratio of 8 to 12 as a common fraction.

The number after the word "to" is always the denominator.

The answer is $\frac{8}{12}$.

Example 2 Write the ratio of 2 to 3 as a common fraction.

What number is after the word "to"?

The answer is $\frac{2}{3}$.

Reduce the fraction $\frac{8}{12}$.

Are the ratios in examples 1 and 2 equal? The answer is yes since $\frac{8}{12} = \frac{2}{3}$.

⦿ SKILL CHECK 1 Write the ratio 3 to 5 as a common fraction.

2 Equal Ratios

> Two ratios are **equal** if they are the same when expressed as fractions in reduced form.

Example 3 Is the ratio of 15 to 18 the same as the ratio of 10 to 12?

Solution

$\frac{15}{18} = \frac{5}{6}$ in reduced form. (Divide numerator and denominator by 3.)

$\frac{10}{12} = \frac{5}{6}$ in reduced form. (Divide numerator and denominator by 2.)

Therefore 15 to 18 is the same as 10 to 12.

⦿ SKILL CHECK 2 Is the ratio of 6 to 9 the same as the ratio of 8 to 12?

⦿ *Skill Check Answers*

1. $\frac{3}{5}$
2. yes

Example 4 Write the ratio of $\frac{1}{2}$ to $\frac{3}{8}$ as a common fraction.

Solution

$$\frac{\frac{1}{2}}{\frac{3}{8}} = \frac{1}{2} \div \frac{3}{8} = \left(\frac{1}{2}\right)\left(\frac{8}{3}\right) = \frac{4}{3}$$

Recall the procedure for simplifying a complex fraction.

⊙ ***SKILL CHECK 3*** Write the ratio of $\frac{1}{3}$ to $\frac{5}{6}$ as a common fraction.

Example 5 Write the ratio of 0.04 to 0.12 as a common fraction.

Solution

$$\frac{0.04}{0.12} = \frac{4}{12} = \frac{1}{3}$$

Multiply numerator and denominator by 100.

⊙ ***SKILL CHECK 4*** Write the ratio of 0.03 to 0.15 as a common fraction.

> The ratio of a quantity to a unit is referred to as a **rate**.

An automobile traveling at a *rate* of 55 miles per hour expresses the ratio of miles (distance) to 1 hour (a unit of time).

Example 6 If Jane rode her bicycle in a race and traveled 60 miles in 3 hours, express the ratio of miles to hours.

Solution

$$60 \text{ to } 3 \text{ or } \frac{60}{3}$$

Notice that this ratio is $\frac{\text{miles}}{\text{hours}}$ and can be read miles per hour when the denominator is 1. Since $\frac{60}{3} = \frac{20}{1} = 20$, we say that Jane rode at the average speed or rate of 20 miles per hour (abbreviated mph).

The fraction bar is read "per."

Example 7 Batting averages in baseball are computed as the ratio of the number of hits to the number of times at bat. This ratio is then expressed as a decimal to the nearest thousandth. If Johnny B. had 14 hits in 38 times at bat, express his batting average as a fraction and also as a decimal to the nearest thousandth.

Don't be confused when you hear sportscasters give batting averages. They usually give them without the decimal point.

Solution

$$\frac{\text{hits}}{\text{at bat}} = \frac{14}{38} = \frac{7}{19}, \text{ as a fraction.}$$

$$7 \div 19 = 0.368 \text{ to the nearest thousandth.}$$

Divide $19\overline{)7.0000}$

⊙ ***SKILL CHECK 5*** If a baseball player had 10 hits in 34 times at bat, express the batting average as both a common fraction and a decimal rounded to the nearest thousandth.

⊙ ***Skill Check Answers***

3. $\frac{2}{5}$ 4. $\frac{1}{5}$ 5. $\frac{5}{17}$, 0.294

Exercise 9–1–1

1 Write the ratios as common fractions in reduced form.

1. 5 to 7 **2.** 2 to 5 **3.** 8 to 9 **4.** 1 to 4 **5.** 6 to 15

6. 10 to 55 **7.** 3:17 **8.** 12:30 **9.** 16:48 **10.** 72:90

11. $\frac{3}{4}$ to $\frac{5}{8}$ **12.** $\frac{1}{3}:\frac{3}{5}$ **13.** 0.03 to 0.27 **14.** 0.12:3.6 **15.** $2\frac{1}{2}$ to $3\frac{1}{8}$

16. $0.5:3\frac{1}{4}$ **17.** $1.4:\frac{3}{5}$

2 Determine if the pairs of ratios are equal.

18. $\frac{1}{2}, \frac{5}{10}$ **19.** $\frac{2}{3}, \frac{8}{12}$ **20.** $\frac{18}{30}, \frac{3}{5}$ **21.** $\frac{18}{24}, \frac{3}{4}$ **22.** $\frac{7}{21}, \frac{2}{3}$

23. $\frac{3}{7}, \frac{36}{84}$ **24.** $\frac{32}{64}, \frac{7}{14}$ **25.** $\frac{12}{16}, \frac{16}{24}$ **26.** $\frac{42}{98}, \frac{48}{84}$ **27.** $\frac{24}{32}, \frac{36}{48}$

• Concept Enrichment

28. It rained 2 inches in 3 hours. Express the ratio of inches to hours.

29. A car traveled 120 miles in 3 hours. Express the ratio of miles to hours in reduced form.

30. An 8-ounce can of vegetables costs 77 cents. Express the ratio of cents to ounces and find the quotient to the nearest hundredth of a cent.

31. Baseball player Hank Aaron had 3,771 hits from 12,364 times at bat. Find his batting average by expressing the ratio of hits to the number of times at bat and give the quotient as a decimal rounded to three decimal places.

32. Edith McGuire, of the United States, ran the 200-meter dash in 23 seconds. Express the ratio of meters to seconds and give the quotient as a decimal rounded to one decimal place.

9–2 Proportions

OBJECTIVE

Upon completing this section you should be able to:

1. Determine if a given statement is a proportion.

Now that we have defined ratio, we wish to discuss another concept called a proportion.

1 Determining a Proportion

For four numbers a, b, c, and d, if the ratio $\frac{a}{b}$ is equal to the ratio $\frac{c}{d}$, then the statement $\frac{a}{b} = \frac{c}{d}$ is a **proportion**.

Example 1 $\frac{4}{6} = \frac{2}{3}$ is a proportion.

SKILL CHECK 1 Is $\frac{5}{10} = \frac{1}{2}$ a proportion?

Example 2 $\frac{3}{4} = \frac{75}{100}$ is a proportion.

SKILL CHECK 2 Is $\frac{1}{4} = \frac{25}{100}$ a proportion?

Example 3 $\frac{5}{10} = \frac{10}{20}$ is a proportion.

Reduce both fractions.

SKILL CHECK 3 Is $\frac{4}{12} = \frac{5}{15}$ a proportion?

In section 9–1 we found that two ratios are equal if they are the same when expressed as fractions in reduced form. Another method, called **cross multiplication,** is derived as follows.

Given the proportion $\frac{a}{b} = \frac{c}{d}$, we use the method of solving fractional equations and multiply both sides by the LCD, which is (bd).

$$(bd)\left(\frac{a}{b}\right) = (bd)\left(\frac{c}{d}\right)$$

This gives

$$ad = bc.$$

Skill Check Answers

1. yes
2. yes
3. yes

This useful fact is stated: if $\frac{a}{b} = \frac{c}{d}$, then $ad = bc$.

In the proportion

$$\frac{a}{b} \times \frac{c}{d}$$

we sometimes refer to ad and bc as the **cross products.** We can state this in words as the following rule.

WARNING

Note that this rule applies only when a single fraction occurs on each side of the equation.

> Two ratios are equal if and only if their cross products are equal.

You could also reduce each of these fractions.

Example 4 Is $\frac{8}{12} = \frac{6}{9}$ a proportion?

Solution We find the cross products $(8)(9) = 72$ and $(12)(6) = 72$. Since the cross products are equal, the two ratios are equal and therefore the statement is a proportion.

⦿ ***SKILL CHECK 4*** Is $\frac{3}{9} = \frac{5}{15}$ a proportion?

Example 5 Is $\frac{7}{15} = \frac{9}{19}$ a proportion?

Solution $(7)(19) = 133$ and $(15)(9) = 135$. Since the cross products are not equal, the ratios are not equal and the statement is not a proportion.

⦿ ***SKILL CHECK 5*** Is $\frac{6}{9} = \frac{5}{7}$ a proportion?

Example 6 If Kathy cycled 2 miles in 8 minutes and Bobby cycled 5 miles in 20 minutes, were they traveling at the same rate?

WARNING

Make sure that *both* ratios are miles to minutes.

Solution We need to determine if $\frac{2}{8} = \frac{5}{20}$ is true. Since $(2)(20) = 40$ and $(8)(5) = 40$, we see the statement is true. Therefore, they were traveling at the same rate.

⦿ ***SKILL CHECK 6*** If Kathy rode 4 miles in 20 minutes and Bobby rode 3 miles in 15 minutes, did they travel at the same rate?

⦿ ***Skill Check Answers***

4. yes 5. no 6. yes

Exercise 9–2–1

1 Determine if each is a proportion.

1. $\frac{1}{2} = \frac{3}{6}$

2. $\frac{2}{3} = \frac{6}{9}$

3. $\frac{3}{5} = \frac{6}{10}$

4. $\frac{3}{4} = \frac{12}{16}$

5. $\frac{12}{15} = \frac{4}{5}$

6. $\frac{6}{14} = \frac{3}{7}$

7. $\frac{2}{5} = \frac{6}{10}$

8. $\frac{8}{9} = \frac{4}{3}$

9. $\frac{6}{12} = \frac{5}{10}$

10. $\frac{3}{6} = \frac{5}{8}$

11. $\frac{4}{12} = \frac{8}{16}$

12. $\frac{14}{16} = \frac{21}{24}$

13. $\frac{12}{28} = \frac{6}{14}$

14. $\frac{5}{9} = \frac{35}{63}$

15. $\frac{25}{49} = \frac{5}{7}$

16. $\frac{14}{21} = \frac{12}{18}$

17. $\frac{15}{20} = \frac{16}{24}$

18. $\frac{36}{39} = \frac{24}{26}$

19. $\frac{36}{42} = \frac{24}{28}$

20. $\frac{24}{42} = \frac{47}{74}$

21. Do the ratios 3:8 and 12:24 form a proportion?

22. Do the ratios 5:9 and 40:72 form a proportion?

23. Do the ratios 18:21 and 30:35 form a proportion?

24. Do the ratios 42:72 and 63:99 form a proportion?

• Concept Enrichment

25. In a certain class there were 24 women and 30 men. In a second class there were 20 women and 25 men. Do the ratios of women to men in the two classes form a proportion?

26. In a certain sample 21 people favored a sales tax increase while 28 were opposed to it. In another sample 36 were in favor and 48 were opposed. Do the ratios in the two samples form a proportion?

27. Jim cycled 3 miles in 15 minutes and George cycled 7 miles in 35 minutes. Did they travel at the same rate?

28. One car traveled 132 miles in 3 hours. A second car traveled 88 miles in two hours. Did they travel at the same rate?

29. A 6-ounce jar of mustard sells for 90 cents and a 4-ounce jar sells for 60 cents. Are they priced at the same rate per ounce?

30. A 12-ounce can of beans sells for 98 cents and an 8-ounce can sells for 67 cents. Are they priced at the same rate per ounce?

9–3 Problems Involving Proportions

OBJECTIVES

Upon completing this section you should be able to:

1. Solve a proportion.
2. Use proportions to solve various applied problems.

1 Solving Proportions

We can find a missing number in a proportion by treating the proportion as a fractional equation and using the skills learned in chapter 8 to solve the equation.

Example 1 Find x so that $\frac{x}{8} = \frac{5}{4}$.

Solution Multiplying both sides of the equation by the LCD gives

The checking of the answers here will be left to you.

$$(8)\left(\frac{x}{8}\right) = (8)\left(\frac{5}{4}\right)$$
$$x = 10.$$

SKILL CHECK 1 Solve for x: $\frac{x}{8} = \frac{3}{4}$

Example 2 Find x so that $\frac{2}{6} = \frac{x}{15}$.

Solution The LCD is 30.

$$(30)\left(\frac{2}{6}\right) = (30)\left(\frac{x}{15}\right)$$
$$10 = 2x$$
$$5 = x$$

or

$$x = 5$$

SKILL CHECK 2 Solve for x: $\frac{7}{9} = \frac{x}{18}$.

We could use the method of cross multiplication, discussed in the previous section, to solve a proportion.

Example 3 Find x such that $\frac{5}{x} = \frac{2}{30}$.

Solution Cross multiplying, we obtain

Remember, in a proportion the cross products are equal.

$$2x = (5)(30)$$
$$2x = 150$$
$$x = 75$$

SKILL CHECK 3 Solve for x: $\frac{6}{x} = \frac{2}{5}$

2 Applications

Example 4 If Sally earned \$37.00 in 5 hours, how much would she earn (at the same hourly rate) in 7 hours?

Solution Using the ratio of $\frac{\text{earnings}}{\text{hours}}$, we set up the proportion

If one ratio is $\frac{\text{earnings}}{\text{hours}}$, then the other ratio must also be $\frac{\text{earnings}}{\text{hours}}$.

$$\frac{37}{5} = \frac{x}{7}.$$

Cross multiplying gives

$$5x = 259$$
$$x = 51.80.$$

So Sally would earn \$51.80 in 7 hours.

Skill Check Answers

1. 6
2. 14
3. 15

⦿ SKILL CHECK 4 How much will Sally earn in 16 hours?

Example 5 Mapmakers use ratios so that maps will be accurate. If a certain road map states that 1 inch = 30 miles, what is the distance (in miles) between two cities if they measure $2\frac{1}{4}$ inches apart on the map?

When ratios are used, the map is said to be *to scale*.

Solution We let x represent the distance (in miles) that we wish to find. If we use the ratio of $\frac{\text{inches}}{\text{miles}}$, we can say that the ratio of 1 inch to 30 miles is equal to the ratio of $2\frac{1}{4}$ inches to x miles.

$$\frac{1}{30} = \frac{2\frac{1}{4}}{x}$$

$$(1)(x) = (30)\left(\frac{9}{4}\right)$$

$$x = \frac{135}{2} = 67\frac{1}{2} \text{ miles}$$

$\frac{\text{inches}}{\text{miles}} = \frac{\text{inches}}{\text{miles}}$

⦿ SKILL CHECK 5 What is the distance in miles between two cities that measure $3\frac{3}{4}$ inches apart if 1 inch = 30 miles?

In chapter 5 we discussed similar triangles. In similar polygons the corresponding angles are congruent and the measures of the corresponding sides are proportional. For instance, suppose that ΔABC and $\Delta A'B'C'$ are similar.

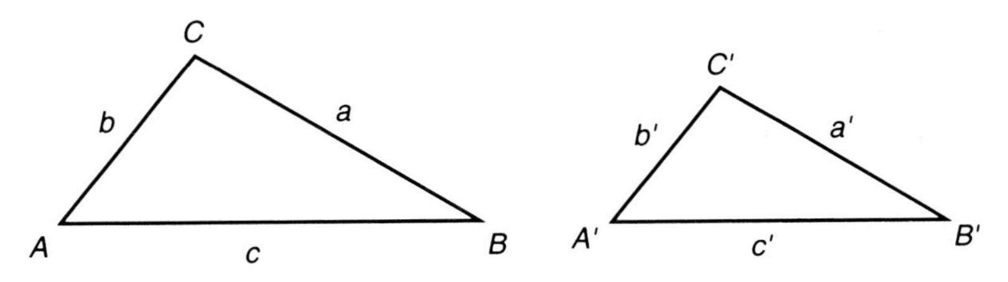

If $\angle A = \angle A'$, $\angle B = \angle B'$, and $\angle C = \angle C'$, then the measures of the corresponding sides are proportional. In other words, the ratios $\frac{a}{a'}$, $\frac{b}{b'}$, and $\frac{c}{c'}$ are equal.

Example 6 In ΔABC and $\Delta A'B'C'$, $\angle A = \angle A'$, $\angle B = \angle B'$, $\angle C = \angle C'$, $a = 12$ inches, $c = 15$ inches, and $c' = 7$ inches. Find a'.

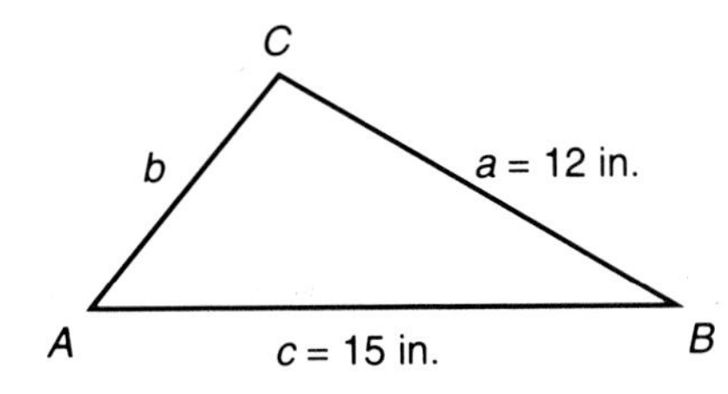

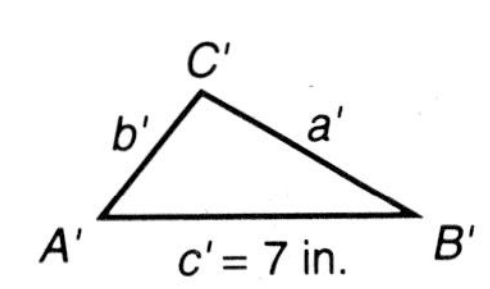

⦿ *Skill Check Answers*

4. $118.40
5. $112\frac{1}{2}$ miles

$\frac{\text{small } \Delta}{\text{large } \Delta} = \frac{\text{small } \Delta}{\text{large } \Delta}$

Solution Since the angles of ΔABC are congruent to the corresponding angles of $\Delta A'B'C'$, the triangles are similar. Then the measures of the corresponding sides are proportional and we can write

$$\frac{a'}{a} = \frac{c'}{c}$$
$$\frac{a'}{12} = \frac{7}{15}$$
$$15a' = 12(7)$$
$$15a' = 84$$
$$a' = \frac{84}{15}$$
$$a' = 5\frac{3}{5} \text{ inches.}$$

⦿ ***Skill Check Answer***

6. $4\frac{2}{3}$ in.

⦿ ***SKILL CHECK 6*** If $b = 10$ inches, find b'.

Exercise 9–3–1

1 Solve for the missing number x.

1. $\frac{1}{2} = \frac{x}{8}$ **2.** $\frac{x}{4} = \frac{9}{12}$ **3.** $\frac{1}{5} = \frac{4}{x}$ **4.** $\frac{2}{3} = \frac{x}{12}$ **5.** $\frac{5}{2} = \frac{x}{8}$

6. $\frac{4}{7} = \frac{x}{28}$ **7.** $\frac{x}{72} = \frac{5}{12}$ **8.** $\frac{3}{4} = \frac{27}{x}$ **9.** $\frac{3}{x} = \frac{21}{28}$ **10.** $\frac{4}{5} = \frac{24}{x}$

11. $\frac{x}{15} = \frac{3}{5}$ **12.** $\frac{16}{72} = \frac{x}{9}$ **13.** $\frac{4}{x} = \frac{3}{15}$ **14.** $\frac{x}{12} = \frac{12}{18}$ **15.** $\frac{x}{10} = \frac{6}{4}$

16. $\frac{x}{2} = \frac{3}{5}$ **17.** $\frac{6}{x} = \frac{5}{9}$ **18.** $\frac{35}{28} = \frac{15}{x}$

19. What number has the same ratio to 15 as 2 has to 3?

20. What number has the same ratio to 49 as 2 has to 7?

2 Use proportions to solve.

21. If Ted earns \$35 in 5 hours, how much would he earn in 8 hours?

22. If Brenda earns \$360 in 8 days, how much does she earn in 5 days?

23. If 3 cans of spray paint are needed to paint 2 lawn chairs, how many cans are needed to paint 12 chairs?

24. A 6-foot man has a 5-foot shadow. A tree has a 20-foot shadow. How tall is the tree?

25. The distance between two cities on a road map is 5 inches. If 1 inch represents 50 miles, what is the distance in miles between the two cities?

26. If 2 pounds of grass seed cover 300 square feet, how many pounds are needed for 1,875 square feet?

27. If 3 pounds of hamburger cost \$3.87, what is the cost of 5 pounds?

28. The ratio of teeth on gear A to those on gear B is 5 to 7. If gear B has 35 teeth, how many teeth does gear A have?

29. If two cities on a map are 5 inches apart and the actual distance between them is 275 miles, what is the actual distance in miles between two cities that are 8 inches apart?

30. The sales tax on an item costing \$75 is \$5.25. What is the sales tax on an item costing \$114?

31. In ΔABC and $\Delta A'B'C'$, $\angle A = \angle A'$, $\angle B = \angle B'$, and $\angle C = \angle C'$. If $a = 6$ feet, $a' = 4$ feet, and $b = 5$ feet, find b'.

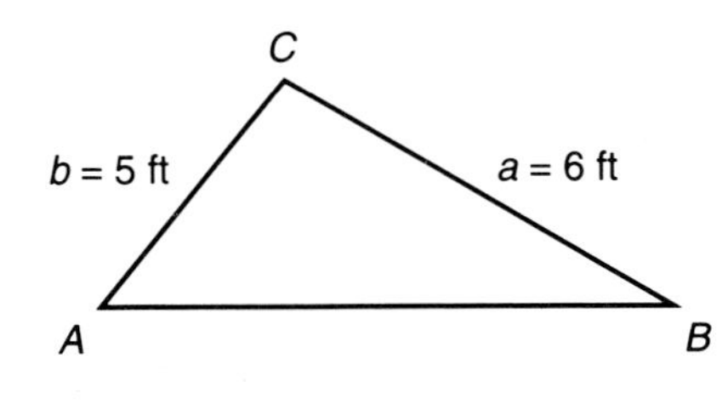

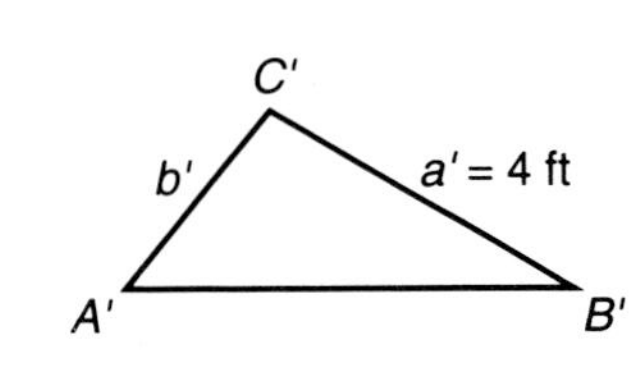

32. In ΔABC and $\Delta A'B'C'$, $\angle A = \angle A'$, $\angle B = \angle B'$, and $\angle C = \angle C'$. If $b = 1.4$ cm, $b' = 3.2$ cm, and $c = 3.5$ cm, find c'.

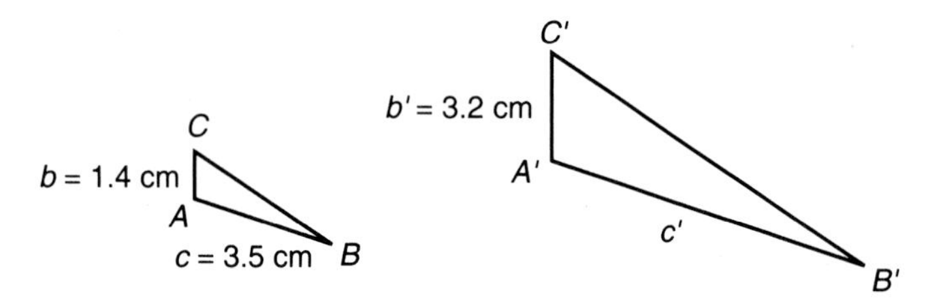

33. Find the lengths of sides a and b.

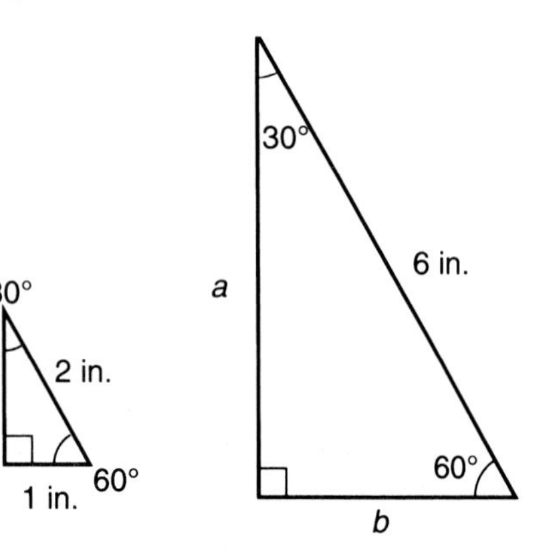

34. Find the lengths of sides a and c.

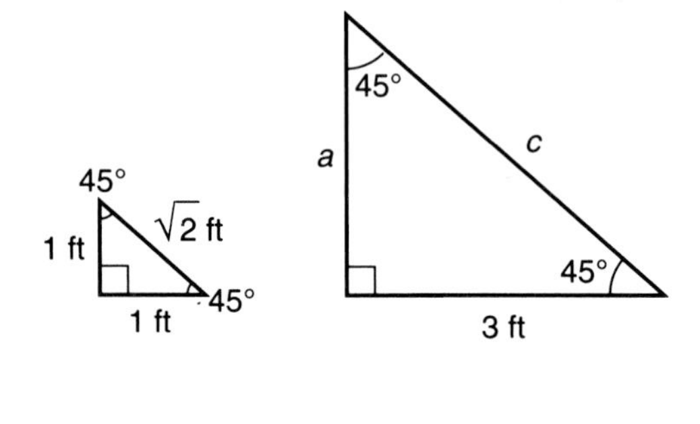

35. Find the lengths of the sides indicated by letters.

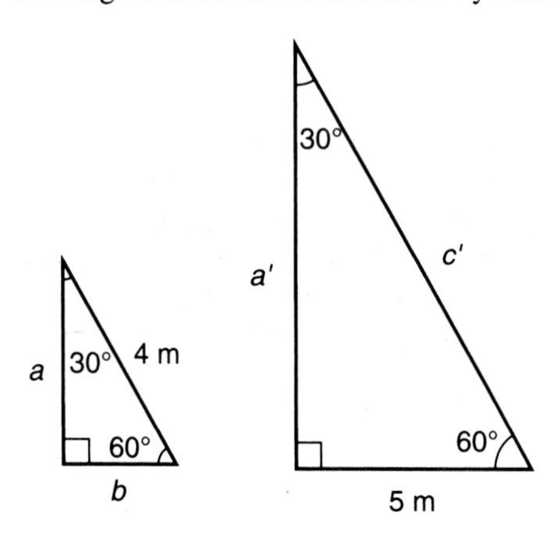

36. Two polygons are similar and the ratio of the measures of the corresponding sides is 5 : 9. If the measure of one side of the smaller polygon is 8 centimeters, find the measure of the corresponding side of the larger polygon.

37. If the property tax on a home valued at \$74,000 is \$814, what is the tax on a home valued at \$55,000?

38. On a road map two cities are 4 inches apart and the actual distance between them is 280 miles. What is the actual distance in miles between two cities that are $2\frac{3}{4}$ inches apart on the map?

39. A company that manufactures light bulbs found that from a sample of 1,250 light bulbs 3 were defective. How many could they expect to be defective out of 25,000 light bulbs?

40. An investment of $12,500 earned $1,687.50 interest in a year. How much interest would an investment of $40,000 have earned?

41. If an automobile uses $5\frac{1}{2}$ gallons of gasoline to travel 176 miles, how many gallons are needed to travel 254 miles? Round the answer to the nearest tenth.

42. A woman's investments of stocks to bonds were in the ratio of 5 to 7. If she had $18,000 invested in stocks, how much did she have invested in bonds?

43. Hockey player Phil Esposito played 1,241 games in 17 seasons. How many games did he average playing in 3 seasons?

44. If the legal restrictions for room capacity require 3 cubic meters of air space per person, how many people can legally occupy a room containing 144 cubic meters?

45. A wildlife management team, conducting a study on a particular lake, caught, tagged, and released 75 black bass. Several weeks later they caught 125 black bass and observed that 3 of them were tagged. What is the total number of black bass they could expect to be present in the lake?

46. The "golden ratio" is approximately 1 to 1.618. This ratio is believed by some to be the width-to-length ratio of a rectangle most pleasing to the eye. Assuming this to be true, what width would a book cover be if the length is 10.5 inches and it is to be most pleasing to the eye? Round the answer to one decimal place.

9–4 Metric Measurement

OBJECTIVES

Upon completing this section you should be able to:

1. Convert between the American and metric systems of measurement using proportions.
2. Convert units within the metric system.

Metric measurement is becoming more common in the United States. Converting measurements from American to metric or metric to American makes use of ratio and proportion. The following tables give the equivalent values of various metric measurements.

LENGTH

1 kilometer (km) = 1,000 meters (m)
1 hectometer (hm) = 100 meters
1 dekameter (dam) = 10 meters
1 decimeter (dm) = 0.1 meter
1 centimeter (cm) = 0.01 meter
1 millimeter (mm) = 0.001 meter

The most commonly used units of metric length are the kilometer, meter, centimeter, and millimeter.

MASS

1 kilogram (kg) = 1,000 grams (g)
1 milligram (mg) = 0.001 gram

Technically these measures refer to **mass** and there is a distinction between mass and weight. A given object has a different weight on earth than it does on the moon, but its mass does not change. However, in everyday usage these measures are generally accepted as referring to **weight**. For instance, a package of coffee is labeled as net weight = 13 ounces or 368 grams.

VOLUME

1 kiloliter (kl) = 1,000 liters (l)
1 milliliter (ml) = 0.001 liter

1 Converting Between American and Metric

The most common relationships between the American and metric systems of measurement are listed here.

The *meter* was originally defined as one ten-millionth of the distance from the equator to the North Pole on a line running through Paris, France. Since then, it has been redefined based on wavelengths of krypton 86.

1 inch = 2.54 centimeters	1 centimeter = 0.3937 inch
1 yard = 0.9144 meter	1 meter = 1.094 yards
1 mile = 1.609 kilometers	1 kilometer = 0.6214 mile
1 ounce = 28.35 grams	1 gram = 0.0353 ounce
1 pound = 0.4536 kilogram	1 kilogram = 2.205 pounds
1 quart = 0.9464 liter	1 liter = 1.057 quarts
1 gallon = 3.785 liters	

The measurements given have been rounded and are therefore approximate.

With these conversion factors we can use proportions to convert from one system to the other.

Example 1 An article weighs 30 kilograms. How many pounds does it weigh? Round the answer to two decimal places.

Solution We let x = the number of pounds. From the table we see that 1 pound = 0.4536 kilogram. Using this information, we set up a proportion.

Notice that we must be careful in setting up the proportion to have like ratios on both sides of the equal sign. In this case we will have

$$\frac{\text{pounds}}{\text{kilograms}} = \frac{\text{pounds}}{\text{kilograms}}$$

$$\frac{x}{30} = \frac{1}{0.4536}$$

$$x = \frac{30}{0.4536}$$

$$x = 66.14 \text{ pounds.}$$

⦿ ***SKILL CHECK 1*** If an item weighs 50 kilograms, what is its weight in pounds rounded to the nearest pound?

Example 2 A sign at Carnstown, Florida, states "Miami 89 miles." How many kilometers is it from Carnstown to Miami? Round the answer to one decimal place.

In the United States some road signs give distance in both miles and kilometers.

Solution Since 1 mile = 1.609 kilometers, we have

$$\frac{1}{1.609} = \frac{89}{x}$$

$$x = (89)(1.609)$$

$$x = 143.2 \text{ kilometers.}$$

$$\frac{\text{miles}}{\text{kilometers}} = \frac{\text{miles}}{\text{kilometers}}$$

⦿ ***SKILL CHECK 2*** How many kilometers are there in 100 miles? Round the answer to the nearest kilometer.

Example 3 It took 43 liters to fill a car's gas tank. How many gallons was this? Round the answer to two decimal places.

Solution

$$\frac{1}{3.785} = \frac{x}{43}$$

$$\frac{43}{3.785} = x$$

$$x = 11.36 \text{ gallons}$$

$$\frac{\text{gallons}}{\text{liters}} = \frac{\text{gallons}}{\text{liters}}$$

⦿ ***SKILL CHECK 3*** How many gallons are there in 50 liters? Round the answer to one decimal place.

⦿ ***Skill Check Answers***

1. 110 lb
2. 161 km
3. 13.2 gal

Exercise 9-4-1

1 Convert. Round the answer to one decimal place.

1. 30 centimeters to inches

2. 36 inches to centimeters

3. 40 yards to meters

4. 10 meters to yards

5. 350 centimeters to inches

6. 2 pounds to kilograms

7. 2 quarts to liters

8. 500 centimeters to inches

9. 5 yards to meters

10. 75 meters to yards

11. 30 miles to kilometers

12. 75 centimeters to inches

13. 310 grams to ounces

14. 4 ounces to grams

15. 2,500 kilometers to miles

16. 5 pounds to kilograms

17. 100 meters to yards

18. 100 yards to meters

19. 1.75 liters to quarts

20. 125 kilograms to pounds

21. 950 kilometers to miles

22. 16 ounces to grams

• Concept Enrichment

23. The cooling system of a certain automobile contains 13.5 liters. How many quarts does it contain?

24. The speed limit in Canada is 100 kilometers per hour. What is the speed limit in miles per hour?

25. Find your weight in kilograms.

26. Find your height in centimeters.

27. A station sells gasoline by the liter. If you bought 45 liters, how many gallons did you buy?

28. A fuel tank in a Buick Century has a capacity of 15.5 gallons. How many liters will it hold?

2 Converting Within the Metric System

We can also use proportions to convert from one unit to another within the metric system itself.

Example 4 Change 574 centimeters to meters.

Solution Let x = the number of meters. From the table of metric length we see that 1 cm = 0.01 meter. Setting up a proportion of

$$\frac{\text{centimeters}}{\text{meters}} = \frac{\text{centimeters}}{\text{meters}},$$

we have

$$\frac{574}{x} = \frac{1}{0.01}$$
$$x = (0.01)(574)$$
$$x = 5.74 \text{ meters.}$$

Obtained by cross multiplying.

SKILL CHECK 4 Change 864 centimeters to meters.

Example 5 Change 1.75 liters to milliliters.

Solution Since 1 ml = 0.001 liter, we can set up the proportion

$$\frac{1.75}{x} = \frac{0.001}{1}$$
$$0.001x = 1.75$$
$$x = \frac{1.75}{0.001}$$
$$x = 1{,}750 \text{ milliliters.}$$

$\frac{\text{milliliters}}{\text{liters}} = \frac{\text{milliliters}}{\text{liters}}$

We must divide by 0.001.

Notice from the statement 1 ml = 0.001ℓ that if we multiply each side by 1,000 we have 1ℓ = 1,000 ml. Setting up a proportion using this, we can write

$$\frac{1.75}{x} = \frac{1}{1{,}000}$$
$$x = 1{,}000(1.75)$$
$$x = 1{,}750 \text{ ml.}$$

Here we multiply by 1,000.

SKILL CHECK 5 Change 0.25 liters to milliliters.

Skill Check Answers

4. 8.64 m
5. 250 ml

Example 6 A roll of tape contains 1,500 centimeters. How many millimeters are there on 3 rolls?

Solution We first must convert 1,500 centimeters to millimeters. From the table of metric length we see that 1 cm = 0.01 m or multiplying each side by 100 we have 1 m = 100 cm. We also see that 1 mm = 0.001 m or multiplying each side by 1,000 we have 1 m = 1,000 mm. Since 1 m = 100 cm and also 1 m = 1,000 mm, we have 100 cm = 1,000 mm. Dividing each side by 100, we obtain 1 cm = 10 mm. Then we may write the proportion

$$\frac{\text{centimeters}}{\text{millimeters}} = \frac{\text{centimeters}}{\text{millimeters}}$$

$$\frac{1{,}500}{x} = \frac{1}{10}$$

$$x = 1{,}500(10)$$

$$x = 15{,}000 \text{ millimeters.}$$

This is the amount on 1 roll so on 3 rolls we would have

$$3(15{,}000 \text{ mm}) = 45{,}000 \text{ millimeters.}$$

⊙ ***Skill Check Answer***

6. 20,000 mm

⊙ ***SKILL CHECK 6*** How many millimeters are there in 2,000 centimeters?

Exercise 9–4–2

2 Convert.

1. 218 centimeters to meters

2. 38 centimeters to meters

3. 4.6 meters to centimeters

4. 0.16 meters to centimeters

5. 1,840 millimeters to meters

6. 95 millimeters to meters

7. 35 millimeters to centimeters

8. 8 millimeters to centimeters

9. 4,280 meters to kilometers

10. 5,062 meters to kilometers

11. 16.5 grams to milligrams

12. 3.6 grams to milligrams

13. 5,920 grams to kilograms

14. 820 grams to kilograms

15. 6 liters to milliliters

16. 19 liters to milliliters

17. 30.5 milliliters to liters

18. 128.6 milliliters to liters

19. A roll of wallpaper is 518 centimeters long. How many meters are on 6 rolls?

20. A ball of twine has 19.4 meters of string on it. A second ball has 684 centimeters of string. How many millimeters of string are on both balls?

21. Each tablet contains 75 milligrams of a certain medication. How many grams of the medication are contained in a bottle of 50 tablets?

22. A certain prescription of 100 pills contains a total of 12.5 grams of a drug. How many milligrams of the drug are in each pill?

23. A bottle contains 750 milliliters. How many liters are in a case of 12 bottles?

24. A bottle contains 1,750 milliliters. How many liters are in a case of 6 bottles?

• Concept Enrichment

25. Can you state a rule in terms of simply moving the decimal point for changing centimeters to meters?

26. Can you state a rule in terms of moving the decimal point for changing liters to milliliters?

9–5 Percent

OBJECTIVES

Upon completing this section you should be able to:

1. Express a percent as a common fraction.
2. Express a percent as a decimal number.
3. Express a decimal number as a percent.
4. Express a common fraction as a percent.

The symbol %, read "percent," is encountered in many areas of our daily life. We hear statements such as: "The prime rate rose 1%," "14% of the class made A's," "57% of the people surveyed disagreed with the proposal," and so on.

Percent means "per hundred" and indicates a ratio of a number to 100.

Example 1 75% means $\frac{75}{100}$.

Example 2 150% means $\frac{150}{100}$.

1 Expressing a Percent as a Common Fraction

The meaning of percent therefore allows us to express any percent as a common fraction.

Example 3 Express 24% as a common fraction in reduced form.

"Per" means divide.

Solution

$$24\% = \frac{24}{100} = \frac{6}{25}$$

⊙ SKILL CHECK 1 Express 58% as a common fraction in reduced form.

Example 4 Express $33\frac{1}{3}\%$ as a common fraction in reduced form.

Solution

$$33\frac{1}{3}\% = \frac{33\frac{1}{3}}{100} = \frac{100}{3} \cdot \frac{1}{100} = \frac{1}{3}$$

⊙ SKILL CHECK 2 Express $66\frac{2}{3}\%$ as a common fraction in reduced form.

The % symbol may have originated as a short way of writing $\frac{1}{100}$.

Example 5 Express 12.5% as a common fraction in reduced form.

Solution

$$12.5\% = \frac{12.5}{100} = \frac{125}{1{,}000} = \frac{1}{8}$$

⊙ SKILL CHECK 3 Express 6.25% as a common fraction in reduced form.

2 Expressing a Percent as a Decimal

See section 4–5.

In our study of decimals we learned that dividing a number by 100 repositions the decimal point two places to the left. Thus for instance,

$$75\% = \frac{75}{100} = 0.75$$

In other words, we note that 75% = 0.75. When working problems involving percent, we usually express a percent in decimal form. Hence the following rule.

> To express a percent in decimal form reposition the decimal point two places to the *left* and write the resulting number without a percent sign.

⊙ *Skill Check Answers*

1. $\frac{29}{50}$
2. $\frac{2}{3}$
3. $\frac{1}{16}$

Example 6 Express 37% as a decimal number.

Solution
$$37\% = 0.37$$

The decimal is moved two places to the left.

⦿ ***SKILL CHECK 4*** Express 63% as a decimal number.

Example 7 Express 273% as a decimal number.

Solution
$$273\% = 2.73$$

Two places to the left.

⦿ ***SKILL CHECK 5*** Express 104% as a decimal number.

Example 8 Express 0.5% as a decimal number.

Solution
$$0.5\% = 0.005$$

⦿ ***SKILL CHECK 6*** Express 0.03% as a decimal number.

3 Expressing a Decimal as a Percent

In some situations we may wish to change a decimal number to a percent. If we express the decimal number as a ratio of a number to 100, the numerator of this ratio will be the desired percent. To accomplish this we multiply the decimal number by $\frac{100}{100}$. In other words, we multiply the decimal number by 100 and place it over 100. For instance,

$\frac{100}{100} = 1$. Multiplying by 1 does not change the value of a number.

$$0.23 = (0.23)\left(\frac{100}{100}\right) = \frac{(0.23)(100)}{100} = \frac{23}{100}$$

In our study of decimals we learned that multiplying a number by 100 repositions the decimal point two places to the right. $0.23(100) = 23$. Now since we have expressed 0.23 as $\frac{23}{100}$, we have $0.23 = 23\%$. This illustrates the following rule.

See section 4-4.

> To express a decimal number as a percent reposition the decimal point two places to the *right* and write the resulting number with a percent sign.

Example 9 Express 0.61 as a percent.

Solution
$$0.61 = 61\%$$

The decimal is moved two places to the right.

⦿ ***SKILL CHECK 7*** Express 0.21 as a percent.

Example 10 Express 1.6 as a percent.

Solution
$$1.6 = 160\%$$

Two places to the right.

⦿ ***SKILL CHECK 8*** Express 5.4 as a percent.

Example 11 Express 0.003 as a percent.

Solution
$$0.003 = 0.3\%$$

⦿ ***SKILL CHECK 9*** Express 0.014 as a percent.

4 Expressing a Common Fraction as a Percent

We sometimes need to express a common fraction as a percent. Since we already learned to change a common fraction to a decimal number and a decimal number to a percent, we have the following rule.

⦿ ***Skill Check Answers***

4. 0.63
5. 1.04
6. 0.0003
7. 21%
8. 540%
9. 1.4%

To express a common fraction as a percent first change the fraction to a decimal number and then express the decimal number as a percent.

Example 12 Express $\frac{7}{8}$ as a percent correct to one decimal place.

Solution Dividing 8 into 7, we obtain

$$\frac{7}{8} = 0.875 = 87.5\%$$

⦿ ***SKILL CHECK 10*** Express $\frac{5}{8}$ as a percent correct to one decimal place.

Example 13 Express $\frac{1}{3}$ as a percent correct to one decimal place.

WARNING

Don't confuse decimal percents with decimal numbers.

Solution $\frac{1}{3} = 0.333 = 33.3\%$ to one decimal place.

⦿ ***SKILL CHECK 11*** Express $\frac{5}{6}$ as a percent correct to one decimal place.

We can also express a fraction as an exact percent if we use the concept of proportion.

Example 14 Express $\frac{7}{8}$ as an exact percent.

Recall $x\% = \frac{x}{100}$.

Solution Since a percent x can be written as the fraction $\frac{x}{100}$, we can write the proportion

$$\frac{x}{100} = \frac{7}{8}$$

Cross multiplying gives

$$8x = 700$$
$$x = \frac{700}{8} = 87\frac{1}{2}\%$$

We see that $\frac{7}{8} = \frac{87\frac{1}{2}}{100} = 87\frac{1}{2}\%$

⦿ ***SKILL CHECK 12*** Express $\frac{5}{8}$ as an exact percent.

Example 15 Express $\frac{1}{3}$ as an exact percent.

Solution If we let x represent the percent, then we write

$$\frac{x}{100} = \frac{1}{3}$$
$$3x = 100$$
$$x = \frac{100}{3} = 33\frac{1}{3}\%$$

⦿ ***Skill Check Answers***

10. 62.5%
11. 83.3%
12. $62\frac{1}{2}\%$

⊙ **SKILL CHECK 13** Express $\frac{5}{6}$ as an exact percent.

⊙ ***Skill Check Answer***

13. $83\frac{1}{3}\%$

Exercise 9–5–1

1 Express each percent as a common fraction in reduced form.

1. 50% **2.** 20% **3.** 65%

4. 150% **5.** $24\frac{1}{2}\%$ **6.** $31\frac{1}{5}\%$

7. $83\frac{1}{3}\%$ **8.** 1.75% **9.** 87.5%

10. 112.8%

2 Express each percent as a decimal number.

11. 15% **12.** 30% **13.** 3%

14. 1.8% **15.** 15.6% **16.** 0.19%

17. 238% **18.** 100% **19.** 0.09%

20. 18.64%

3 Express each decimal number as a percent.

21. 0.92 **22.** 0.5 **23.** 3.2

24. 0.06

25. 0.403

26. 8

27. 0.072

28. 2.03

29. 0.004

30. 1.018

4 Express each fraction as a percent. If not exact, round the answer to one decimal place.

31. $\frac{1}{2}$

32. $\frac{1}{5}$

33. $\frac{3}{4}$

34. $\frac{9}{10}$

35. $\frac{7}{100}$

36. $\frac{1}{6}$

37. $\frac{3}{8}$

38. $\frac{4}{7}$

39. $\frac{7}{4}$

40. $\frac{10}{3}$

Express each fraction as an exact percent.

41. $\frac{1}{7}$

42. $\frac{1}{9}$

43. $\frac{2}{3}$

44. $\frac{1}{6}$

45. $\frac{5}{6}$

46. $\frac{3}{8}$

47. $\frac{5}{8}$

48. $\frac{5}{7}$

49. $\frac{7}{3}$

50. $\frac{10}{3}$

• Concept Enrichment

51. What does percent mean?

52. Explain the distinction between decimal percents and decimal numbers.

9-6 Problems Involving Percents

1 Solving Percent Problems

OBJECTIVE

Upon completing this section you should be able to:

1 Solve various problems involving percents.

Problems involving percent can be solved using proportions since a percent indicates a ratio of a number to 100.

Example 1 Find 25% of 860.

Solution 25% can be expressed as the ratio $\frac{25}{100}$. We will let x represent a number that has the same ratio to 860 as 25 has to 100. This gives us the equation

$$\frac{x}{860} = \frac{25}{100}$$
$$100x = (860)(25)$$
$$x = 215$$

Thus 215 is 25% of 860.

◉ *SKILL CHECK 1* Find 30% of 210.

◉ *Skill Check Answer*

1. 63

Example 2 53 is 20% of what number?

$20\% = \frac{20}{100}$

Solution Here we are told that 53 is in the same ratio to some number as 20 is to 100. If x represents the unknown number, we have

$$\frac{20}{100} = \frac{53}{x}$$
$$20x = (100)(53)$$
$$x = 265$$

Thus 53 is 20% of 265.

⦿ ***SKILL CHECK 2*** 128 is 80% of what number?

Example 3 17 is what percent of 20?

Solution In this problem we must find an unknown number that has the same ratio to 100 as 17 has to 20. If we let x represent the unknown number, then we can write

x represents a percent.

$$\frac{x}{100} = \frac{17}{20}$$
$$20x = (100)(17)$$
$$x = 85$$

So 17 is 85% of 20.

⦿ ***SKILL CHECK 3*** 26 is what percent of 40?

Sometimes a quantity may be greater than 100% of a given number.

Example 4 15 is what percent of 12?

Solution We must find a number that has the same ratio to 100 as 15 has to 12. If we let x represent the unknown number, we have

Notice in this proportion that 15 is greater than 12. Thus we would expect x to be greater than 100.

$$\frac{x}{100} = \frac{15}{12}$$
$$12x = (100)(15)$$
$$x = 125$$

Thus 15 is 125% of 12.

⦿ ***SKILL CHECK 4*** 36 is what percent of 25?

Example 5 In a math class of 35 students, 7 students earned a grade of A. What percent of the class earned an A?

Solution We must find a number that has the same ratio to 100 as 7 has to 35.

$$\frac{x}{100} = \frac{7}{35}$$
$$35x = (100)(7)$$
$$x = 20$$

So 20% of the class earned an A.

⦿ ***SKILL CHECK 5*** If 12 students in a class of 50 received an A, what percent does this represent?

⦿ ***Skill Check Answers***

2. 160
3. 65%
4. 144%
5. 24%

Example 6 Bill has saved $1,285 toward buying an outboard motor for his boat. He claims this is 80% of his goal. What is his goal?

Solution The ratio of 1,285 to some number must be the same as the ratio of 80 to 100. If x represents the unknown number, then

We should reason here that the goal will be greater than $1,285.

$$\frac{80}{100} = \frac{1{,}285}{x}$$
$$80x = (100)(1{,}285)$$
$$x = 1{,}606.25$$

Bill's goal is $1,606.25.

◉ ***SKILL CHECK 6*** $425 is 85% of the purchase price of an item. What is the purchase price?

Example 7 The population of a certain town increased from 36,400 in 1980 to 42,800 in 1990. What is the percent of increase in the original population rounded to the nearest tenth of a percent?

Solution We must first find the increase in population.

$$\text{Increase} = 42{,}800 - 36{,}400 = 6{,}400$$

Our task now is to find a number that has the same ratio to 100 as 6,400 has to 36,400. If we let x represent the unknown number, then

$$\frac{x}{100} = \frac{6{,}400}{36{,}400}$$
$$36{,}400x = (100)(6{,}400)$$
$$x = 17.6 \text{ (rounded to one decimal place)}$$

WARNING

Percent increase or decrease is always based on the *original* number, which in this case is 36,400.

Therefore the percent of increase in population from 1980 to 1990 is 17.6% to the nearest tenth of a percent.

◉ ***SKILL CHECK 7*** A person who weighed 132 pounds last month weighs 138 pounds this month. What is the percent of increase over the weight last month? Round the answer to the nearest tenth of a percent.

◉ ***Skill Check Answers***

6. $500
7. 4.5%

Exercise 9–6–1

1

1. Find 20% of 80.

2. Find 35% of 70.

3. What is 16% of 60?

4. Find 28% of 150.

5. 66% of 230 is what number?

6. Find 83% of 420.

7. 6 is 60% of what number?

8. 8 is 20% of what number?

9. 93 is 75% of what number?

10. 18 is 4% of what number?

11. 36 is 16% of what number?

12. 162 is 9% of what number?

13. 3 is what percent of 4?

14. 7 is what percent of 15?

15. What percent of 24 is 5?

16. What percent of 25 is 8?

17. 25 is what percent of 28?

18. What percent of 55 is 42?

19. Frank borrowed $5,000 at 14% annual interest and was to pay only interest the first year. How much should he pay the first year?

20. A person makes $320 during a certain week. Twenty percent of the wages is withheld for federal taxes. How much is withheld?

21. There were 45 members present at a club meeting. If that number represents 60% of the total membership, what is the total membership?

22. A community group has collected donations amounting to $120,000. If this represents 80% of its goal, what is its goal?

23. Ray has money invested at 14% annual interest. If his interest for the year was $3,360, how much did he have invested?

24. A credit card company charges interest at the rate of 1.5% per month on the balance. What was the amount of the balance during a month if the interest charge was $5.40?

25. An exploration party had traveled $78\frac{3}{10}$ miles up the Amazon River. They were told by their guide that this represents only 2% of the total length of the river. How long is the Amazon River?

26. During a snowstorm Buffalo, New York, received 7.09 inches of snow. If this represents 8% of the total annual average for the city, what is the total yearly snowfall? Round the answer to the nearest tenth of an inch.

27. A calculator regularly priced at $39.00 was offered with a $5.85 rebate from the manufacturer. What percent was the rebate of the regular price?

28. A discount of $10.50 was given on a pair of shoes regularly priced at $59.50. What percent of the regular price was the discount? Round the answer to the nearest tenth of a percent.

29. In a math class consisting of 55 students 24 were females. What percent of the class was female? Round the answer to the nearest tenth of a percent.

30. In one season the New York Yankees played a total of 162 games and won 103 of them. What percent of the games did they win? Round the answer to the nearest tenth of a percent.

31. A certain model of an automobile that cost $9,800 last year is priced at $10,350 this year. What is the percent of increase in the original price rounded to the nearest tenth of a percent?

32. By installing insulation a homeowner was able to reduce the normal $120 per month utility bill to $96. What is the percent of decrease of the original utility bill?

CHAPTER 9 SUMMARY

Key Words

Section 9–1

- A **ratio** is the indicated quotient of two numbers.
- A **rate** is the ratio of a quantity to a unit.

Section 9–2

- A **proportion** is a statement that two ratios are equal.
- In a proportion the **cross products** are equal.

Section 9–4

- **Metric measure** is based on powers of ten.

Section 9–5

- **Percent** means per hundred and indicates a ratio of a number to 100.

Procedures

Section 9–1

- To determine if two ratios are equal express them as common fractions in reduced form.

Section 9–2

- To determine if two ratios form a proportion, use the method of cross multiplication.

Section 9–3

- To solve a problem involving a proportion treat the proportion as a fractional equation.

Section 9–4

- To change from one unit of measure to another form a proportion and find its solution.

Section 9–5

- To express a percent in decimal form reposition the decimal point two places to the left and write the resulting number without a percent sign.
- To express a decimal number as a percent reposition the decimal point two places to the right and write the resulting number with a percent sign.
- To express a common fraction as a percent first change the fraction to a decimal number and then express the decimal number as a percent.

Section 9–6

- Many types of problems involving percents are solved by forming an equation based on a proportion.

CHAPTER 9 REVIEW

Write the ratio as a fraction in reduced form.

1. 3 to 8 **2.** 4 to 6 **3.** 12:27 **4.** 18:42

Determine if the pairs of ratios are equal.

5. $\frac{3}{5}, \frac{15}{25}$ **6.** $\frac{12}{18}, \frac{18}{27}$ **7.** $\frac{12}{36}, \frac{21}{54}$ **8.** $\frac{36}{78}, \frac{24}{52}$

Determine if each is a proportion.

9. $\frac{1}{2} = \frac{5}{10}$ **10.** $\frac{3}{8} = \frac{12}{34}$ **11.** $\frac{6}{9} = \frac{4}{6}$ **12.** $\frac{10}{14} = \frac{30}{42}$

13. $\frac{27}{72} = \frac{15}{45}$ **14.** $\frac{10}{24} = \frac{30}{72}$ **15.** $\frac{21}{35} = \frac{27}{45}$ **16.** $\frac{32}{52} = \frac{48}{78}$

Solve for x.

17. $\frac{1}{4} = \frac{5}{x}$ **18.** $\frac{8}{x} = \frac{2}{5}$ **19.** $\frac{7}{3} = \frac{x}{6}$ **20.** $\frac{4}{x} = \frac{6}{18}$

21. $\frac{12}{x} = \frac{15}{4}$ **22.** $\frac{15}{4} = \frac{x}{6}$ **23.** $\frac{x}{3} = \frac{6}{9}$ **24.** $\frac{x}{27} = \frac{54}{36}$

Convert. Round the answer to one decimal place.

25. 7.5 yd to meters **26.** 18 qt to liters **27.** 215 cm to inches

28. 45 g to ounces

29. 20 km to miles

30. 135 lb to kilograms

31. 416.2 cm to meters

32. 12.3 liters to milliliters

Express each percent as a decimal number.

33. 37%

34. 9%

35. 13.8%

36. 206%

37. 0.7%

38. 12.04%

Express each decimal as a percent.

39. 0.55

40. 0.3

41. 0.06

42. 5.2

43. 0.016

44. 4.198

Express each fraction as a percent. If not exact, round the answer to the nearest tenth of a percent.

45. $\frac{2}{5}$

46. $\frac{3}{16}$

47. $\frac{7}{10}$

48. $\frac{21}{40}$

49. $\frac{5}{2}$

50. $\frac{12}{5}$

51. Find 40% of 90.

52. Find 35% of 60.

53. Find 56% of 240.

54. Find 120% of 40.

Find each to the nearest tenth of a percent.

55. 3 is what percent of 15?

56. What percent of 66 is 12?

57. 56 is what percent of 320?

58. What percent of 900 is 150?

Find the number. Round to two decimal places.

59. 3 is 60% of what number?

60. 14 is 83% of what number?

61. 128 is 11% of what number?

62. 17.4 is 55% of what number?

63. What number has the same ratio to 24 as 5 has to 8?

64. A car traveled 228 miles in 6 hours. Express the ratio of miles to hours in reduced form.

65. One car traveled 114 miles in 3 hours and a second car traveled 190 miles in 5 hours. Did they travel at the same average rate?

66. The distance between two cities on a road map is 3 inches. If the actual distance is 135 miles, what is the actual distance between two cities that are 5 inches apart on the map?

67. Gear A has 30 teeth and gear B has 35. If this same ratio is to be maintained and gear A is replaced by a gear with 12 teeth, how many teeth must be on the gear replacing gear B?

68. A car travels 187.6 miles on 6.7 gallons of gasoline. How many gallons are needed to travel 266 miles?

69. What is the capacity in quarts of a 1.75-liter bottle? Round the answer to two decimal places.

70. A swimming event in the Olympic Games is the 1,500-meter freestyle. What is the distance in yards? Round the answer to one decimal place.

71. In a survey of 345 people 28% were under age 20. How many were under age 20? Round the answer to the nearest whole number.

72. Find the interest paid on $28,000 for one year if the annual rate is 13.5%.

73. A dress regularly priced at $88.50 was marked down by $19.00. What percent was the discount of the regular price? Round the answer to the nearest tenth of a percent.

74. During the 1985–86 season the Boston Celtics played 82 games and won 67 of them. What percent of the games did they win? Round the answer to the nearest tenth of a percent.

75. Baseball player Ty Cobb had 4,191 hits in his career, giving him a batting average of 0.367, which means he got a hit 36.7% of his times at bat. Find the number of times he was at bat. Round the answer to the nearest whole number.

76. A woman invested some money at an annual interest rate of 12.5%. If she received $1,687.50 interest during the year, find how much she had invested.

CHAPTER 9 TEST

1. Write the ratio 75:90 as a fraction in reduced form.

2. Are the ratios 12:28 and 15:35 equal?

3. Is $\frac{6}{15} = \frac{14}{35}$ a proportion?

4. Is $\frac{12}{36} = \frac{6}{9}$ a proportion?

5. Solve for x: $\frac{x}{6} = \frac{15}{9}$

6. Solve for x: $\frac{10}{x} = \frac{4}{14}$

7. Solve for x: $\frac{6}{5} = \frac{x}{4}$

8. The ratio of boys to girls in grade four of a certain school is 6 to 5. If there are 15 girls in the class, how many boys are in the class?

9. If an automobile uses 3 gallons of gasoline to travel 69 miles, how many gallons are necessary to travel 161 miles?

10. In a certain population 3 out of 5 people have brown eyes. If 75 people are selected at random from this population, how many could be expected to have brown eyes?

11. If the speed limit on a certain road is 55 miles per hour, what is the speed limit in kilometers per hour? Round the answer to the nearest tenth. (1 kilometer = 0.6214 mile)

12. It took 14 gallons of gasoline to fill a tank. How many liters would have filled the tank? Round the answer to the nearest whole number. (1 gallon = 3.785 liters)

13. **a.** Express $\frac{5}{8}$ as a percent.

b. Express 1.3 as a percent.

14. In a survey 560 people were asked to give a "yes" or "no" opinion on a question. 231 answered "yes." What percent answered "yes"?

15. On a loan of $800 Bill paid $76 interest for one year. What was the yearly rate of interest?

CHAPTERS 1–9 CUMULATIVE TEST

1. Write the number 1,200,300 in words.

2. Divide: $48\overline{)106{,}512}$

3. Find the least common multiple of 6, 14, and 21.

4. Multiply: $\begin{array}{r} 18.35 \\ \times\ 21.8 \\ \hline \end{array}$

5. Reduce: $\dfrac{6a^2b}{21abc}$

6. Simplify: $-4 + (-3) - (-7) - 5 + 1$

7. Find: $|-(+16)|$

8. Find the area of a triangle if the base measures 12.50 centimeters and the altitude is 9.00 centimeters.

9. Subtract: $2\dfrac{5}{12} - 1\dfrac{3}{8}$

10. Simplify: $2a^2b + 5ab - a^2b - 2ab$

11. Evaluate $2a^2b + 3bc - ab^2c$ when $a = -1$, $b = 3$, and $c = -2$.

12. Express 4.615 as a percent.

13. Find the hypotenuse of a right triangle if the legs measure 3 inches and 5 inches. Round the answer to two decimal places.

14. Replace the question mark in $\dfrac{5}{13}\ ?\ \dfrac{2}{5}$ with the inequality symbol positioned to indicate a true statement.

15. Round 15.70185 to the nearest thousandth.

16. State the number of terms in the expression $5a - 4b + 5c - 1$.

17. Divide: $11.3\overline{)244.193}$

18. Solve for x: $x - 3 = -5$

19. Convert 830 meters to kilometers.

20. Find the prime factorization of 819.

21. Solve for x: $-\frac{x}{3} = 15$

22. Simplify: $\frac{(-3)(-10)}{(-5)(6)}$

23. Find the measure of an angle if its supplement measures 28°.

24. Solve for x: $2x - 4 = 3\left(\frac{1}{2}x + 1\right)$

25. Add: $\left(-\frac{3}{5}\right) + \left(+\frac{4}{7}\right)$

26. Subtract: $261.3 - 78.62$

27. Write 53,000,000 in scientific notation.

28. A person works $7\frac{1}{3}$ hours per day for six days. What is the total number of hours worked?

29. A rectangular garden is 25.5 feet long and 20.0 feet wide. How much will it cost to fence the garden if the price of fencing is $1.95 per foot?

30. If 15 pounds of lawn fertilizer are needed to cover 5,000 square feet, how many pounds are needed to cover 1,500 square feet?

C H A P T E R

10 Pretest

Answer as many of the following problems as you can before starting this chapter. When you finish the chapter, take the test at the end and compare your scores to see how much you have learned.

Given the following set of numbers, answer questions 1–4.
{15, 6, 3, 15, 12, 3, 15, 12, 9, 7}

1. Find the mean.

2. Find the median.

3. Find the mode(s).

4. Find the range.

5. To receive a letter grade of B in a course a student must have an average of 80 on five tests. If a student has scores of 78, 82, 90, and 84 on the first four tests, what is the lowest score the student can receive on the fifth test to receive a B?

The following chart indicates the percentage of adults who regularly watch certain types of television shows. Use this information to answer questions 6–10.

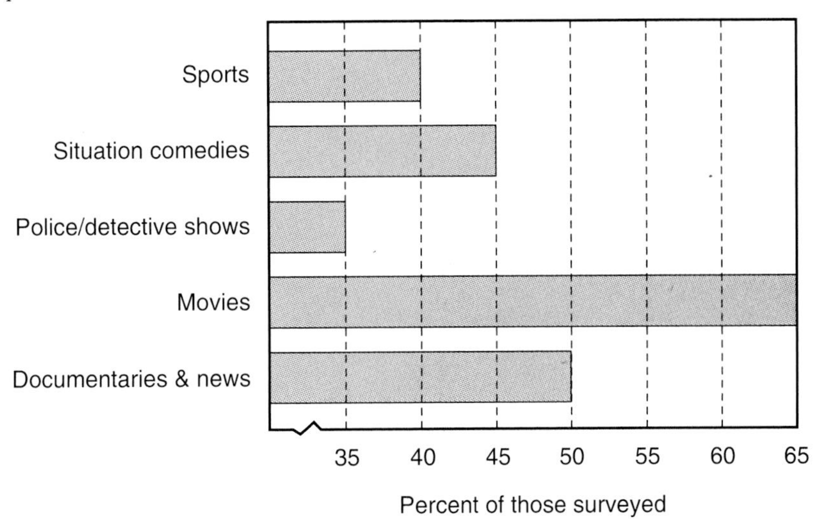

6. What percent of the surveyed adults regularly watch situation comedies?

7. What percent of adults watch news and documentaries on a regular basis?

8. Which type of show is regularly watched by 40% of the adults surveyed?

9. Which type of show is regularly watched most often?

10. Which type of show is regularly watched least often?

The following graph shows the results of a survey conducted in a midwestern town to determine how people feel about banning firearms. Use this information to answer questions 11–15.

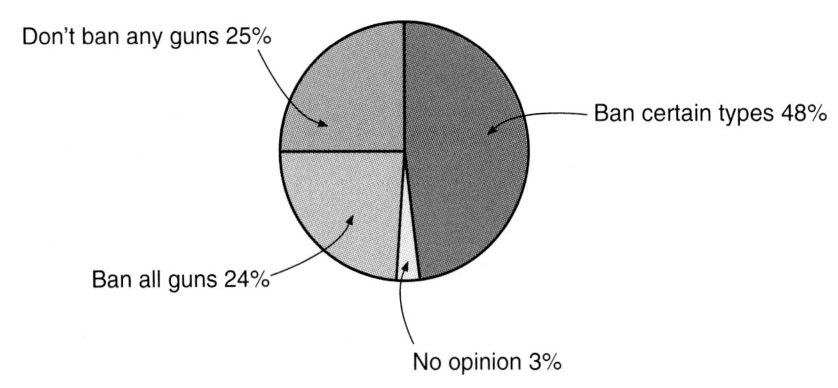

11. What percent feel that all guns should be banned?

12. What percent feel that no guns should be banned?

13. If a surveyed person thought that only machine guns should be banned, which group is he or she included in?

14. What percent feel that all or at least some guns should be banned?

15. If 2,000 people were surveyed, how many had no opinion?

CHAPTER

10 Interpreting Data from Statistics and Graphs

Ideas can often be expressed more effectively in pictures than in words. This is as true in some areas of mathematics as in other fields of study.

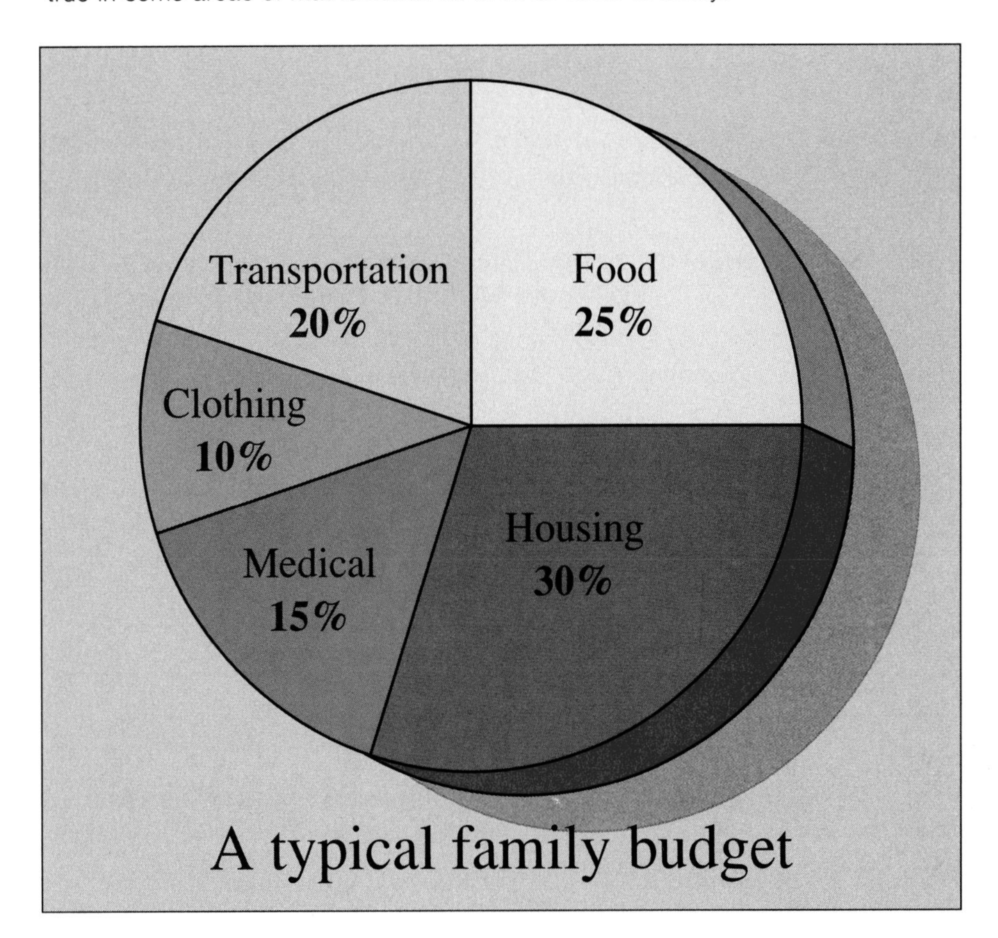

A typical family budget

STUDY HELP

Don't fail to utilize any supplementary materials that are available to you. This may include such things as audiotapes, videotapes, student solutions manual, additional problem sets, tutorial help, and other materials available in a math lab. Consult your instructor regarding the availability of these items.

Data is information used as the basis for decision making. **Statistics** is the mathematics of the collection, organization, and interpretation of numerical data. In this chapter we will learn how to obtain statistical information from a set of data and properly interpret various types of graphs.

10–1 Statistics

OBJECTIVES

Upon completing this section you should be able to:

1. Find the mean of a set of data.
2. Find the median of a set of data.
3. Find the mode of a set of data.
4. Find the range of a set of data.

A number that in some way describes a set of data is a **statistic.** One way to examine a collection of numerical data is to find a number that describes the *central tendency* of the set. That is, a number that in some way is representative of the set.

1 Mean

In section 4–5 we discussed **arithmetic mean** or **mean.** To find the mean of a set of scores (numbers) we divide the sum of the scores by the number of scores in the set.

$$\text{Mean} = \frac{\text{sum of scores}}{\text{number of scores}}$$

The arithmetic mean is sometimes referred to as the **arithmetic average** or simply **average.**

Example 1 The weights of seven men are found to be 160, 178, 190, 210, 170, 203, and 198 pounds. Find the average weight (mean) to the nearest pound.

Solution We must add the weights and divide by 7.

Notice that no one actually weighs 187 lbs.

```
 160           187
 178       7)1309
 190          7
 210          60
 170          56
 203           49
 198           49
1,309           0
```

The average weight or mean is 187 pounds.

SKILL CHECK 1 Find the mean of the set of numbers $\{120, 153, 108, 239, 195, 211\}$.

Example 2 Grade point average is a concern of almost every college student. The grade point average (GPA) is determined by dividing the total grade points by the total credit hours.

$$\text{GPA} = \frac{\text{total grade points}}{\text{total credit hours}}$$

Susan took five three-credit-hour courses with the following results: one A, two B's, one C, and one D. Her school awards grade points as follows:

A—4 grade points per credit hour
B—3 grade points per credit hour
C—2 grade points per credit hour
D—1 grade point per credit hour
F—0 grade points per credit hour

What is her grade point average for the courses?

Skill Check Answer

1. 171

Solution We arrange the information as follows.

Grade	Credit hours	Grade points	
A	3	12	3 credit hr × 4
B	6	18	6 credit hr × 3
C	3	6	3 credit hr × 2
D	3	3	3 credit hr × 1
	15	39	

She has a total of 39 grade points for 15 credit hours, so we divide 39 by 15.

$$\frac{39}{15} = 2.6$$

She has a grade point average of 2.6.

What would the grade point average be if she received all C's?

◉ *SKILL CHECK 2* The next semester Susan took five three-credit-hour courses and earned three A's and two C's. Find her grade point average for that semester.

Sometimes setting up an equation is useful in solving certain questions involving averages.

Example 3 To obtain a grade of B in a particular mathematics class a student must have an average of 80 on five tests. On the first four tests a student received grades of 84, 73, 89, and 91. What is the lowest score the student can receive on the fifth test to receive a grade of B in the course?

Solution The sum of the five tests must be at least $5(80) = 400$ to earn a B. If we let x represent the score on the fifth test, then the sum of the five tests is $84 + 73 + 89 + 91 + x$ and since their sum must be 400, we can write

$$84 + 73 + 89 + 91 + x = 400.$$

Solving, we obtain

$$\begin{aligned} 337 + x &= 400 \\ x &= 400 - 337 \\ x &= 63. \end{aligned}$$

Thus if the student obtains a score of 63 in the fifth test, a grade of B will be received.

Of course the student should not try for just a 63 but should try to get the best score possible.

◉ *SKILL CHECK 3* If in example 3 the student received grades of 91, 68, 87, and 92, what is the lowest score the student can receive on the fifth test to earn a grade of B?

2 Median

A second measure of central tendency is the *median* of a set of numbers.

> After arranging a set of scores in order from smallest to greatest, the **median** is the middle score of the set if there is an odd number of scores. If the set contains an even number of scores, the median is the mean of the two middle scores.

◉ *Skill Check Answers*

2. 3.2
3. 62

Example 4 Find the median of the following set of numbers: {20, 33, 7, 19, 51, 35, 31}.

Solution First arrange the numbers in order from smallest to greatest.

What is the mean of this set of numbers?

7, 19, 20, 31, 33, 35, 51

(31 ← middle number)

There is an odd number of scores so the median is 31.

⊙ ***SKILL CHECK 4*** Find the median of the set of numbers {39, 25, 7, 42, 51, 10, 46}.

Example 5 Ten students received the following scores on a math test: 65, 81, 70, 100, 93, 80, 83, 90, 78, 84. Find the median score.

Solution Again, arrange the numbers from smallest to greatest.

Note there are two middle scores, 81 and 83.

65, 70, 78, 80, 81, 83, 84, 90, 93, 100

$$\frac{81 + 83}{2} = \frac{164}{2} = 82$$

The median score is 82.

⊙ ***SKILL CHECK 5*** Find the median of the set of numbers {19, 41, 12, 38, 2, 37, 16, 50}.

3 Mode

Another statistic that is often used is the *mode.*

> The **mode(s)** of a set of scores is the score or scores that occur most often in the set.

Notice that a set of numbers has only one mean and only one median but it can have more than one mode. If each number occurs in a set only once, then the set is said to have no mode or it can be regarded that each number is a mode of the set.

Example 6 Find the mode or modes of the following set of numbers: {38, 3, 5, 7, 3, 41, 26, 9}.

Solution We first arrange the numbers from smallest to greatest.

This set has only one mode.

3, 3, 5, 7, 9, 26, 38, 41

(3, 3 ← mode)

The mode is 3 since it is the number that occurs most often.

⊙ ***SKILL CHECK 6*** Find the mode(s) of the set of numbers {21, 6, 14, 31, 28, 14, 12, 52}.

⊙ ***Skill Check Answers***

4. 39
5. 28
6. 14

Example 7 Find the mode or modes of the following set of numbers: {32, 41, 38, 32, 56, 41, 38, 55, 41, 38, 39}.

Solution Arranging the numbers in order, we have

$$32, 32, \underbrace{38, 38, 38}, 39, \underbrace{41, 41, 41}, 55, 56.$$

38 and 41 are both modes because each number occurs three times.

There are two modes.

⊙ ***SKILL CHECK 7*** Find the mode(s) of the set of numbers {19, 23, 27, 19, 29, 27, 16, 23, 27, 15, 23}.

4 Range

The final statistic we will mention is *range*.

> The **range** of a set of numbers is the difference between the greatest and the least numbers in the set.

Example 8 The president of a certain company has an annual salary of \$87,000 and the lowest paid employee earns \$12,000. What is the range of salaries in the company?

Solution The greatest salary is \$87,000 and the least is \$12,000. Subtracting, we obtain

$$\$87{,}000 - \$12{,}000 = \$75{,}000.$$

The range is thus \$75,000.

⊙ ***SKILL CHECK 8*** Find the range of the set of numbers {21, 63, 14, 39, 11, 25}.

⊙ ***Skill Check Answers***

7. 23 and 27
8. 52

Exercise 10–1–1

For each set of numbers find the

1 a. mean,

2 b. median,

3 c. mode or modes,

4 d. range.

1. {5, 5, 9, 10, 11}

2. {3, 10, 10, 13, 16}

3. {7, 3, 8, 7, 9}

4. {10, 13, 8, 13, 6}

5. {12, 6, 5, 7, 4, 5}

6. {4, 8, 3, 7, 9, 4}

7. {7, 5, 3, 18, 7, 3, 11, 10}

8. {7, 7, 4, 9, 3, 15, 7, 4}

9. {11, 7, 6, 7, 12, 11, 7, 15, 8, 11}

10. {14, 6, 15, 10, 14, 8, 10, 15, 9, 13}

11. The high temperatures for five consecutive days were 74°, 69°, 73°, 80°, and 74° F.
   a. Find the average high temperature for the five days.
   b. Find the median.
   c. Find the mode(s).
   d. Find the range.

12. The weights of eight people are 135 lb, 180 lb, 145 lb, 116 lb, 175 lb, 124 lb, 145 lb, and 180 lb.
   a. Find the average weight per person rounded to one decimal place.
   b. Find the median.
   c. Find the mode(s).
   d. Find the range.

13. The low temperatures for eight consecutive nights were 10°, −7°, 0°, −10°, 4°, −10°, −4°, −7° C.
   a. Find the average low temperature for the eight nights.
   b. Find the median.
   c. Find the mode(s).
   d. Find the range.

14. The low temperatures for six consecutive nights were −12°, 0°, 5°, 0°, −6°, −11°.
   a. Find the average low temperature for the six nights.
   b. Find the median.
   c. Find the mode(s).
   d. Find the range.

15. In a certain semester a student received the following grades:

Course	Grade	Credit hours
History	C	3
Algebra	B	3
Biology	A	4
English	A	3
Drama	B	2
Physical Ed.	C	1

If a grade of A receives 4 grade points per credit hour, B receives 3 grade points, and C receives 2 grade points, determine the student's grade point average for the semester rounded to one decimal place.

16. In a certain quarter a student received the following grades:

Course	Grade	Credit hours
Social Science	A	3
Algebra	D	3
Art	B	2
Drafting	F	2
Journalism	A	3
Physical Ed.	D	1

If the grade points are awarded as follows,
A—4 points, B—3 points, C—2 points, D—1 point, and F—0 points, find the student's grade point average for the quarter.

17. To obtain a grade of B in a particular course a student must have an average of 80 on six tests. If a student has scores of 84, 69, 93, 81, and 74 on the first five tests, what is the minimum score needed on the sixth test to guarantee a B in the course?

18. To obtain a grade of A in a course a student must have an average of 90 on five tests. A student receives scores of 98, 79, 91, and 88 on the first four tests. What is the minimum score needed on the fifth test to guarantee a grade of A?

• Concept Enrichment

19. Which statistic(s), (mean, median, mode, range), will always refer to a score in the set of scores?

20. Which statistic(s), (mean, median, mode, range), can have more than one value?

10–2 Pictographs and Tables

OBJECTIVES

Upon completing this section you should be able to:

1. Read and interpret information from a pictograph.
2. Read and interpret information from a chart or table.

Various types of graphs and tables are used to show numerical data in an organized way. The purpose of using a graph or table is that it is usually easier to compare amounts as we interpret information. In this section we will discuss pictographs (picture graphs) and tables.

1 Pictographs

A **pictograph** uses rows of pictures or symbols to represent numbers. A pictograph must have a key that gives a value to each symbol.

The following pictograph shows the amount of money in passbook accounts of a certain bank at the end of each month for the first six months of a year.

Note that each symbol represents $200,000.

Month	Passbook deposits
January	$ $ $ $
February	$ $ $ $ $
March	$ $ $ $ $ $
April	$ $ $
May	$ $ $ $ $
June	$ $ $ $ $ $ $
	$ = $200,000

We will use this information in examples 1–3.

Example 1 How much money was on deposit at the end of January?

Solution There are four symbols shown for January and the key indicates that each of them represents $200,000. Thus 4($200,000) = $800,000 was on deposit at the end of that month.

⦿ ***SKILL CHECK 1*** How much money was on deposit at the end of March?

Example 2 In which month was the least amount of money on deposit?

Solution Here we see one advantage of a pictograph. At a glance we can see that April had the least amount on deposit (only three symbols).

How much money was on deposit at the end of April?

⦿ ***SKILL CHECK 2*** In which month was the greatest amount of money on deposit?

⦿ ***Skill Check Answers***

1. $1,200,000
2. June

Example 3 How much money was on deposit at the end of May?

Solution Here we see a disadvantage of a pictograph since a partial symbol is given and we may not be certain what the exact value of it is. We estimate that the partial symbol is about one-half a full symbol. So the value for May is

$$4\frac{1}{2}(\$200,000) = \frac{9}{2}(\$200,000)$$
$$= \$900,000.$$

If 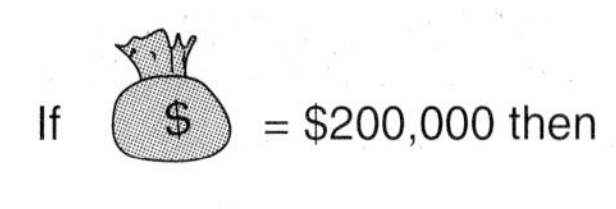 = $200,000 then

= $100,000

◉ ***SKILL CHECK 3*** How much money was on deposit at the end of February?

2 Tables and Charts

Tables or **charts** are sometimes useful in showing numerical data. The following table contains the same data as shown in the previous pictograph.

Month	Passbook Deposits
January	$ 800,000
February	$ 950,000
March	$1,200,000
April	$ 600,000
May	$ 900,000
June	$1,400,000

Notice that in a table the amounts can be as precise as we wish.

The visual representation of the data in a pictograph has the advantage of showing variations at a glance. A table can be used to give more accurate data.

The following portion of a federal income tax table will be used in examples 4 and 5.

If your taxable income is—		And you are—			
At least	But less than	Single	Married filing jointly •	Married filing sepa-rately	Head of a house-hold
		Your tax is—			
19,000					
19,000	19,050	2,854	2,854	3,218	2,854
19,050	19,100	2,861	2,861	3,232	2,861
19,100	19,150	2,869	2,869	3,246	2,869
19,150	19,200	2,876	2,876	3,260	2,876
19,200	19,250	2,884	2,884	3,274	2,884
19,250	19,300	2,891	2,891	3,288	2,891
19,300	19,350	2,899	2,899	3,302	2,899
19,350	19,400	2,906	2,906	3,316	2,906
19,400	19,450	2,914	2,914	3,330	2,914
19,450	19,500	2,925	2,921	3,344	2,921
19,500	19,550	2,939	2,929	3,358	2,929
19,550	19,600	2,953	2,936	3,372	2,936
19,600	19,650	2,967	2,944	3,386	2,944
19,650	19,700	2,981	2,951	3,400	2,951
19,700	19,750	2,995	2,959	3,414	2,959
19,750	19,800	3,009	2,966	3,428	2,966
19,800	19,850	3,023	2,974	3,442	2,974
19,850	19,900	3,037	2,981	3,456	2,981
19,900	19,950	3,051	2,989	3,470	2,989
19,950	20,000	3,065	2,996	3,484	2,996

This table is for incomes between $19,000 and $20,000.

◉ ***Skill Check Answer***

3. $950,000

At least	But less than	Single	Married filing jointly •	Married filing sepa-rately	Head of a house-hold
		Your tax is—			
19,000					
19,000	19,050	2,854	2,854	3,218	2,854
19,050	19,100	2,861	2,861	3,232	2,861
19,100	19,150	2,869	2,869	3,246	2,869
19,150	19,200	2,876	2,876	3,260	2,876
19,200	19,250	2,884	2,884	3,274	2,884
19,250	19,300	2,891	2,891	3,288	2,891
19,300	19,350	2,899	2,899	3,302	2,899
19,350	19,400	2,906	2,906	3,316	2,906
19,400	19,450	2,914	2,914	3,330	2,914
19,450	19,500	2,925	2,921	3,344	2,921
19,500	19,550	2,939	2,929	3,358	2,929
19,550	19,600	2,953	2,936	3,372	2,936
19,600	19,650	2,967	2,944	3,386	2,944
19,650	19,700	2,981	2,951	3,400	2,951
19,700	19,750	2,995	2,959	3,414	2,959
19,750	19,800	3,009	2,966	3,428	2,966
19,800	19,850	3,023	2,974	3,442	2,974
19,850	19,900	3,037	2,981	3,456	2,981
19,900	19,950	3,051	2,989	3,470	2,989
19,950	20,000	3,065	2,996	3,484	2,996

Example 4 If a single person has a taxable income of \$19,363, how much is that person's tax?

Solution Looking at the table, we find that \$19,363 is between \$19,350 and \$19,400. We look across in the corresponding column marked "single" and find the tax is \$2,906. This amount represents the tax due for any taxable income in this range.

⦿ ***SKILL CHECK 4*** If a single person has a taxable income of \$19,790, how much is that person's tax?

At least	But less than	Single	Married filing jointly •	Married filing sepa-rately	Head of a house-hold
		Your tax is—			
19,000					
19,000	19,050	2,854	2,854	3,218	2,854
19,050	19,100	2,861	2,861	3,232	2,861
19,100	19,150	2,869	2,869	3,246	2,869
19,150	19,200	2,876	2,876	3,260	2,876
19,200	19,250	2,884	2,884	3,274	2,884
19,250	19,300	2,891	2,891	3,288	2,891
19,300	19,350	2,899	2,899	3,302	2,899
19,350	19,400	2,906	2,906	3,316	2,906
19,400	19,450	2,914	2,914	3,330	2,914
19,450	19,500	2,925	2,921	3,344	2,921
19,500	19,550	2,939	2,929	3,358	2,929
19,550	19,600	2,953	2,936	3,372	2,936
19,600	19,650	2,967	2,944	3,386	2,944
19,650	19,700	2,981	2,951	3,400	2,951
19,700	19,750	2,995	2,959	3,414	2,959
19,750	19,800	3,009	2,966	3,428	2,966
19,800	19,850	3,023	2,974	3,442	2,974
19,850	19,900	3,037	2,981	3,456	2,981
19,900	19,950	3,051	2,989	3,470	2,989
19,950	20,000	3,065	2,996	3,484	2,996

Example 5 If the taxable income is \$19,870 for a married person filing jointly and \$3,468 in taxes has already been withheld, how much of a refund should be received?

Solution We see that \$19,870 is between \$19,850 and \$19,900. The tax shown under the "married filing jointly" column is \$2,981. Since \$3,468 has already been paid, the refund should be

$$\$3,468 - \$2,981 = \$487.$$

⦿ ***SKILL CHECK 5*** If in example 5 only \$2,540 had been withheld, how much does the person still owe?

⦿ ***Skill Check Answers***

4. \$3,009
5. \$441

Exercise 10–2–1

1 This pictograph shows the enrollment at a college for five consecutive years. Refer to the pictograph to answer questions 1–10.

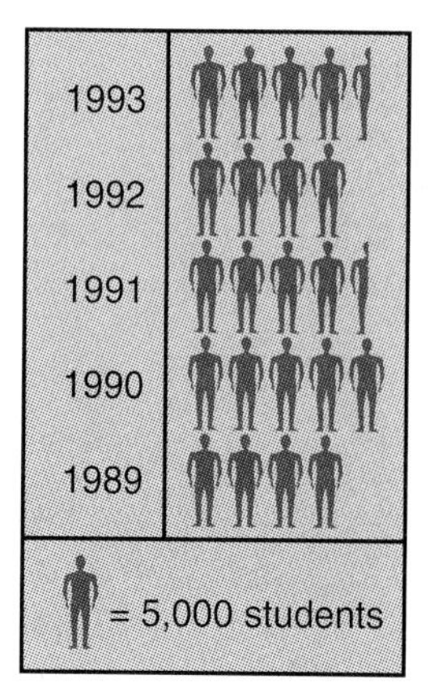

1. What was the enrollment in 1989?

2. What was the enrollment in 1991?

3. In what year was the enrollment the greatest?

4. Between what two consecutive years was there the greatest increase in enrollment?

5. In which two consecutive years was there a decrease in enrollment?

6. What was the increase in enrollment from 1992 to 1993?

7. What was the total drop in enrollment from 1990 to 1992?

8. What was the percent drop in enrollment from 1990 to 1992?

9. What was the percent change in enrollment from 1992 to 1993?

10. What was the percent change in enrollment from 1991 to 1992? Round the answer to the nearest tenth of a percent.

This pictograph shows the number of football touchdown passes thrown in a single season by four of the leading passers. Refer to the pictograph to answer questions 11–20.

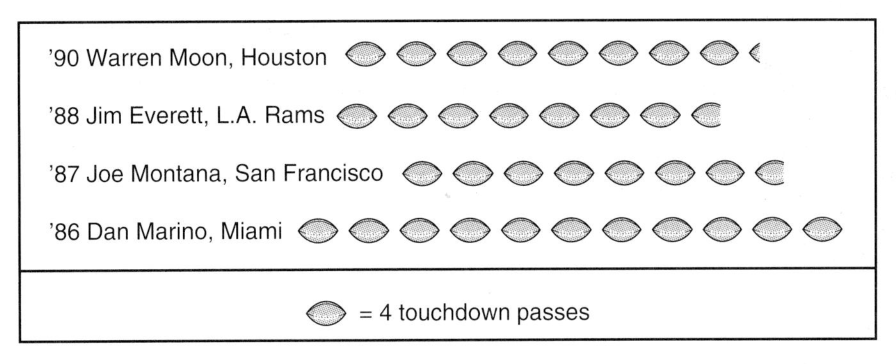

11. How many touchdown passes are represented by one football?

12. How many touchdown passes were thrown by Warren Moon in 1990?

13. How many touchdown passes were thrown by Joe Montana in 1987?

14. Which player threw the most touchdown passes in a single season?

15. Which player threw the second most number of touchdown passes in a single season?

16. In which year were the most touchdown passes thrown by a single passer?

17. How many more touchdown passes did Dan Marino throw in 1986 than did Joe Montana in 1987?

18. What passer has the same record of touchdown passes as Joe Montana?

19. Express the number of touchdown passes thrown by Joe Montana in 1987 as a percentage of those thrown by Dan Marino in 1986. Round the answer to the nearest percent.

20. Express the number of touchdown passes thrown by Warren Moon in 1990 as a percentage of those thrown by Dan Marino in 1986.

This table indicates the federal tax due on various taxable incomes. Use the table to answer questions 21–28.

If your taxable income is— At least	But less than	And you are— Single	Married filing jointly •	Married filing separately	Head of a household
		Your tax is—			
14,000					
14,000	14,050	2,104	2,104	2,104	2,104
14,050	14,100	2,111	2,111	2,111	2,111
14,100	14,150	2,119	2,119	2,119	2,119
14,150	14,200	2,126	2,126	2,126	2,126
14,200	14,250	2,134	2,134	2,134	2,134
14,250	14,300	2,141	2,141	2,141	2,141
14,300	14,350	2,149	2,149	2,149	2,149
14,350	14,400	2,156	2,156	2,156	2,156
14,400	14,450	2,164	2,164	2,164	2,164
14,450	14,500	2,171	2,171	2,171	2,171
14,500	14,550	2,179	2,179	2,179	2,179
14,550	14,600	2,186	2,186	2,186	2,186
14,600	14,650	2,194	2,194	2,194	2,194
14,650	14,700	2,201	2,201	2,201	2,201
14,700	14,750	2,209	2,209	2,209	2,209
14,750	14,800	2,216	2,216	2,216	2,216
14,800	14,850	2,224	2,224	2,224	2,224
14,850	14,900	2,231	2,231	2,231	2,231
14,900	14,950	2,239	2,239	2,239	2,239
14,950	15,000	2,246	2,246	2,246	2,246
15,000					
15,000	15,050	2,254	2,254	2,254	2,254
15,050	15,100	2,261	2,261	2,261	2,261
15,100	15,150	2,269	2,269	2,269	2,269
15,150	15,200	2,276	2,2[illegible]	[illegible]276	2,276
15,200	15,250	[illegible]	[illegible]	[illegible]	2,284
15,250	15,3[illegible]	[illegible]	[illegible]	[illegible]	[illegible]91
15,300	[illegible]	[illegible]	[illegible]	[illegible]	[illegible]
15[illegible]	[illegible]	[illegible]	[illegible]	[illegible]	[illegible]

If your taxable income is— At least	But less than	And you are— Single	Married filing jointly •	Married filing separately	Head of a household
		Your tax is—			
17,000					
17,000	17,050	2,554	2,554	2,658	2,554
17,050	17,100	2,561	2,561	2,672	2,561
17,100	17,150	2,569	2,569	2,686	2,569
17,150	17,200	2,576	2,576	2,700	2,576
17,200	17,250	2,584	2,584	2,714	2,584
17,250	17,300	2,591	2,591	2,728	2,591
17,300	17,350	2,599	2,599	2,742	2,599
17,350	17,400	2,606	2,606	2,756	2,606
17,400	17,450	2,614	2,614	2,770	2,614
17,450	17,500	2,621	2,621	2,784	2,621
17,500	17,550	2,629	2,629	2,798	2,629
17,550	17,600	2,636	2,636	2,812	2,636
17,600	17,650	2,644	2,644	2,826	2,644
17,650	17,700	2,651	2,651	2,840	2,651
17,700	17,750	2,659	2,659	2,854	2,659
17,750	17,800	2,666	2,666	2,868	2,666
17,800	17,850	2,674	2,674	2,882	2,674
17,850	17,900	2,681	2,681	2,896	2,681
17,900	17,950	2,689	2,689	2,910	2,689
17,950	18,000	2,696	2,696	2,924	2,696
18,000					
18,000	18,050	2,704	2,704	2,938	2,704
18,050	18,100	2,711	2,711	2,952	2,711
18,100	18,150	2,719	2,719	2,966	2,719
18,150	18,200	2,726	2,726	2,980	[illegible]
18,200	18,250	2,734	2,734	[illegible]	[illegible]
18,250	18,300	2,741	2,74[illegible]	[illegible]	[illegible]
18,300	18,350	2,749	[illegible]	[illegible]	[illegible]
[illegible]50	18,400	2,75[illegible]	[illegible]	[illegible]	[illegible]
[illegible]450	[illegible]	[illegible]	[illegible]	[illegible]	[illegible]

If your taxable income is— At least	But less than	And you are— Single	Married filing jointly •	Married filing separately	Head of a household
		Your tax is—			
20,000					
20,000	20,050	3,079	3,004	3,498	3,004
20,050	20,100	3,093	3,011	3,512	3,011
20,100	20,150	3,107	3,019	3,526	3,019
20,150	20,200	3,121	3,026	3,540	3,026
20,200	20,250	3,135	3,034	3,554	3,034
20,250	20,300	3,149	3,041	3,568	3,041
20,300	20,350	3,163	3,049	3,582	3,049
20,350	20,400	3,177	3,056	3,596	3,056
20,400	20,450	3,191	3,064	3,610	3,064
20,450	20,500	3,205	3,071	3,624	3,071
20,500	20,550	3,219	3,079	3,638	3,079
20,550	20,600	3,233	3,086	3,652	3,086
20,600	20,650	3,247	3,094	3,666	3,094
20,650	20,700	3,261	3,101	3,680	3,101
20,700	20,750	3,275	3,109	3,694	3,109
20,750	20,800	3,289	3,116	3,708	3,116
20,800	20,850	3,303	3,124	3,722	3,124
20,850	20,900	3,317	3,131	3,736	3,131
20,900	20,950	3,331	3,139	3,750	3,139
20,950	21,000	3,345	3,146	3,764	3,146
21,000					
21,000	21,050	3,359	3,154	3,778	3,154
21,050	21,100	3,373	3,161	3,792	3,161
21,100	21,150	3,387	3,169	3,806	3,169
[illegible]	[illegible]0 21,200	3,401	3,176	3,820	3,176
[illegible]	[illegible]	[illegible]415	3,184	3,834	3,184
[illegible]	[illegible]	[illegible]191	3,848	3,191	
[illegible]	[illegible]	[illegible]	[illegible]862	3,199	
[illegible]	[illegible]	[illegible]	[illegible]	[illegible]206	

21. What is the tax due for a single person having a taxable income of $17,682?

22. What is the tax due for a single person having a taxable income of $14,501?

23. What is the tax due for a married person filing jointly having a taxable income of $20,934?

24. What is the tax due for a married person filing separately having a taxable income of $17,089?

25. A person filing as head of a household has a taxable income of $17,553. What is the tax due on that amount?

26. A married person has a taxable income of $20,756. How many dollars less in tax would be paid if the person files jointly instead of separately?

27. A single person with a taxable income of $14,865 had taxes amounting to $2,943 withheld from paychecks during the year. How much of a refund should the person receive?

28. A person qualifying as head of a household had $2,950 in taxes withheld during the year. If the taxable income was $20,014, how much in taxes is still owed?

Use this chart for income tax rates to answer questions 29–32.

Schedule X—Use if your filing status is Single

If the amount on Form 1040, line 37, is: Over—	But not over—	Enter on Form 1040, line 38	of the amount over—
$0	$19,450	.......15%	$0
19,450	47,050	$2,917.50 + 28%	19,450
47,050	97,620	10,645.50 + 33%	47,050
97,620	....	Use Worksheet below to figure your tax.	

Schedule Z—Use if your filing status is Head of household

If the amount on Form 1040, line 37, is: Over—	But not over—	Enter on Form 1040, line 38	of the amount over—
$0	$26,050	.......15%	$0
26,050	67,200	$3,907.50 + 28%	26,050
67,200	134,930	15,429.50 + 33%	67,200
134,930	....	Use Worksheet below to figure your tax.	

Schedule Y-1—Use if your filing status is Married filing jointly or Qualifying widow(er)

If the amount on Form 1040, line 37, is: Over—	But not over—	Enter on Form 1040, line 38	of the amount over—
$0	$32,450	.......15%	$0
32,450	78,400	$4,867.50 + 28%	32,450
78,400	162,770	17,733.50 + 33%	78,400
162,770	....	Use Worksheet below to figure your tax.	

Schedule Y-2—Use if your filing status is Married filing separately

If the amount on Form 1040, line 37, is: Over—	But not over—	Enter on Form 1040, line 38	of the amount over—
$0	$16,225	.......15%	$0
16,225	39,200	$2,433.75 + 28%	16,225
39,200	123,570	8,866.75 + 33%	39,200
123,570	....	Use Worksheet below to figure your tax.	

29. A person filing a single status has a taxable income on line 37 of Form 1040 of $51,904. From the chart determine the tax to be written on line 38.

30. If the amount shown on line 37 of Form 1040 is $52,510, what is the tax on line 38 for a married person filing separately?

31. If a person files as head of a household, what is the tax to be entered on line 38 if the amount on line 37 is $51,904?

32. A married person filing jointly enters $75,240 on line 37. What is the tax amount to be entered on line 38?

Calculator Problems

The following table shows the population of the United States for census years 1950–1990. Use this table to answer questions 33–38.

Year	Population
1990	249,632,692
1980	226,545,805
1970	203,302,031
1960	179,323,175
1950	150,697,361

33. How many more people were counted in the United States in 1990 than in 1950?

34. How many more people were counted in the United States in 1990 than in 1970?

35. What is the percentage increase in population from 1980 to 1990? Express the answer to the nearest tenth of a percent.

36. What is the percentage increase in population from 1950 to 1990? Express the answer to the nearest tenth of a percent.

37. Between which two consecutive censuses did the percentage of population increase the most?

38. Between which two consecutive censuses did the percentage of population increase the least?

10–3 Bar Graphs and Line Graphs

OBJECTIVES

Upon completing this section you should be able to:

1. Read and interpret bar graphs.
2. Read and interpret line graphs.

Bar graphs and **line graphs** are used to show comparisons of data, as are pictographs. However, bar graphs and line graphs are more accurate than pictographs because they use two scales instead of a key.

1 Bar Graphs

The following bar graph shows the number of students enrolled in mathematics courses at Studious College for the years 1988 through 1992.

Notice how easy it is to compare the size of the enrollments for the various years.

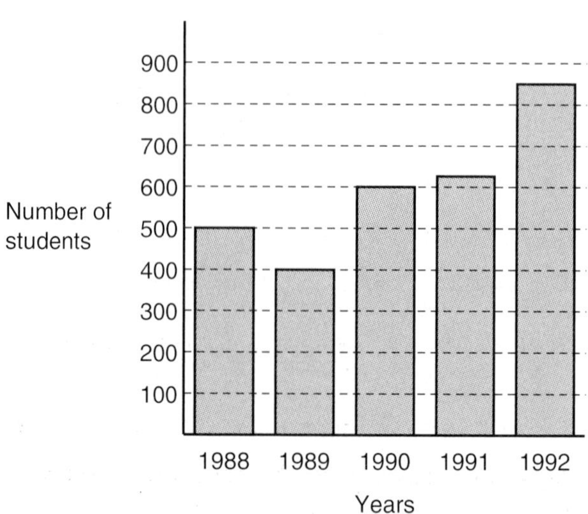

Notice the two scales, years and number of students. We will use this information in examples 1–3.

Example 1 About how many students were enrolled in mathematics courses in 1990?

Solution From the scale at the left we note that the bar for 1990 is at about 600. Thus, this represents the enrollment for that year.

⦿ ***SKILL CHECK 1*** About how many students were enrolled in mathematics courses in 1988?

Example 2 For which two years was the enrollment about the same?

Solution At a glance we see that 1990 and 1991 had approximately the same enrollment.

⦿ ***SKILL CHECK 2*** In what year was the enrollment the lowest?

Example 3 What was the range of enrollment during the five years?

Solution The range is the difference between the highest enrollment (1992) and the lowest enrollment (1989). Thus the range is $850 - 400 = 450$.

Recall the definition of range from section 10-1.

⦿ ***SKILL CHECK 3*** What was the median enrollment during the five years?

2 Line Graphs

In a line graph the data are represented by points instead of a bar or symbol. The points are connected by line segments to show upward and downward trends. Line graphs, like bar graphs, must have two scales.

The following line graph shows the same information as the previous bar graph—the number of students enrolled in mathematics courses at Studious College for the years 1988–1992.

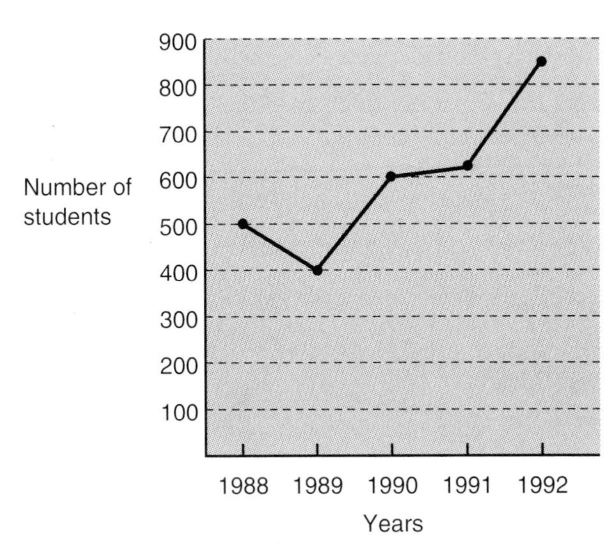

We use this information in examples 4 and 5.

Example 4 Between what two years did the enrollment decline?

Solution At a glance we see the enrollment declined from 1988 to 1989.

Notice how easy it is to see upward and downward trends with a line graph.

⦿ ***SKILL CHECK 4*** Between what two years did the enrollment remain about the same?

Example 5 Between what two years did the enrollment increase the most?

Solution Again, at a glance we see that the sharpest increase was from 1991 to 1992.

⦿ ***Skill Check Answers***

1. 500
2. 1989
3. 600
4. 1990–1991

◉ SKILL CHECK 5 About how much did the enrollment increase between 1989 and 1990?

A bar graph is used to show comparison of data but it does not have to be over a period of time. It can be used to show populations of different cities, the number of games won by different teams, the heights of different individuals, and so on.

A line graph is used to show a change over a period of time. It can be used to show the high temperatures of a particular location for a series of months, the Dow-Jones Industrial Average for a period of weeks, the sales of a store for several years, and so on.

Bar graphs and line graphs can be used to compare two or more sets of data.

The following line graph shows the miles per gallon of fuel used for two automobiles traveling at various speeds.

Note we must be told what the dashed and solid lines represent.

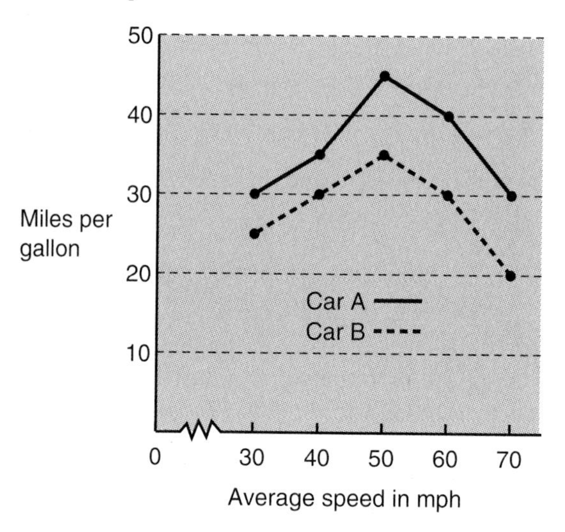

This information is used in examples 6–8.

Example 6 At what speed did each automobile obtain the highest miles per gallon?

Solution Both car A and car B obtained the highest miles per gallon at an average speed of 50 mph.

◉ SKILL CHECK 6 At what speed did car B obtain the lowest miles per gallon?

Example 7 Which car is more fuel efficient?

Which line represents car A?

Solution Car A gets higher miles per gallon at all speeds shown.

Example 8 How does an average speed above 50 mph affect the miles per gallon of each car?

Notice again how easy it is to see trends.

Solution The miles per gallon of fuel drops drastically at speeds above 50 mph for each car.

◉ Skill Check Answers

5. 200
6. 70 mph

Exercise 10–3–1

1 The following bar graph shows the percentage of the total population projected to be over age 65 in the year 2025 for several selected countries.

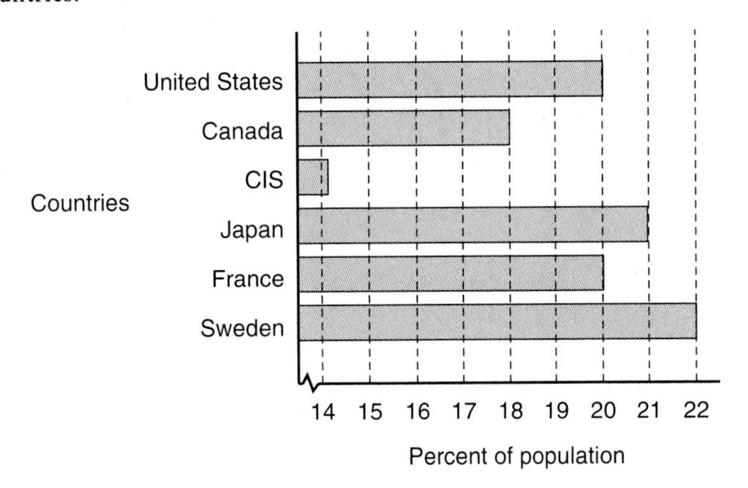

Use this information to answer questions 1–10.

1. What percent of the population of the United States will be over 65 years of age in 2025?

2. What percent of the population of the Commonwealth of Independent States (CIS) will be over 65 years of age in 2025?

3. What two countries will have about the same percentage of population over age 65?

4. What percentage of the population of Canada will be over age 65?

5. Which country will have the greatest percentage over age 65?

6. Which country will have the lowest percentage over age 65?

7. How many percentage points higher will the United States be over Canada in its population over age 65?

8. France will be how many percentage points higher than the CIS in its population over age 65?

9. What percent of the population of Japan will be 65 or younger?

10. What percent of the population of Sweden will be 65 or younger?

The following graph shows the areas of several states.

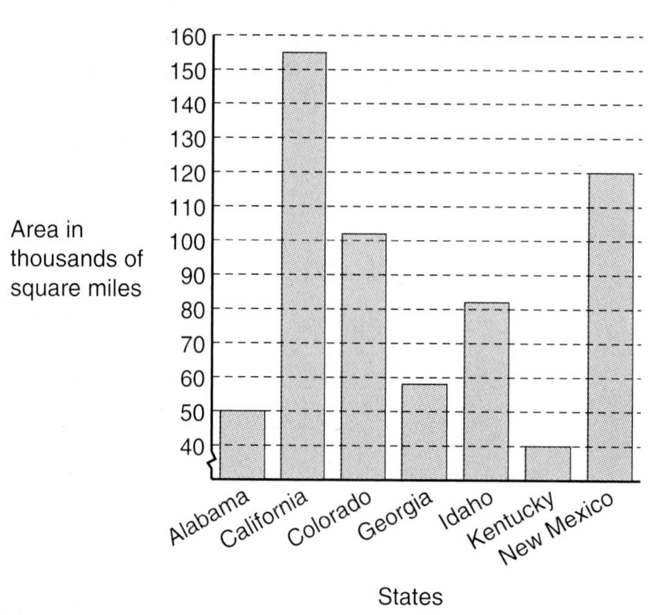

Use this graph to answer questions 11–20.

11. What is the area of California to the nearest 10,000 square miles?

12. What is the area of Colorado to the nearest 10,000 square miles?

13. Which state has an area of approximately 80,000 square miles?

14. Which state has an area of approximately 120,000 square miles?

15. Which state has the greatest area of those shown?

16. Which state has the least area of those shown?

17. How much larger is Georgia than Alabama?

18. How much larger is Colorado than Kentucky?

19. What percentage in area is Kentucky compared to California?

20. How much larger is New Mexico than Idaho expressed as a percentage?

The following line graph shows the enrollment at a college for ten consecutive years.

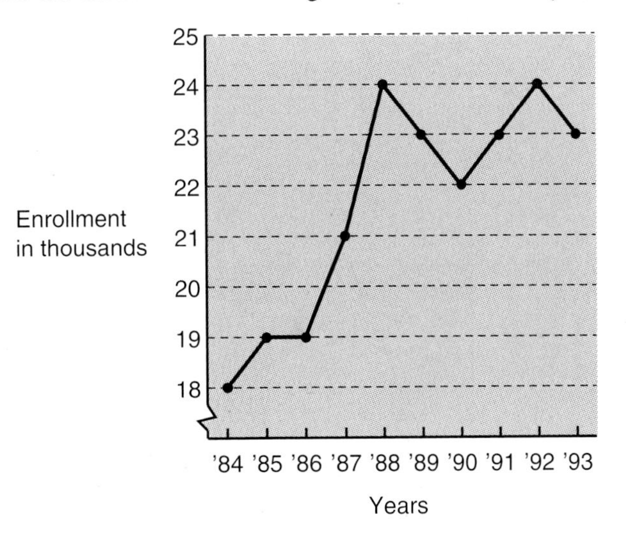

Use this information to answer questions 21–30.

21. What was the enrollment in 1990?

22. What was the enrollment in 1985?

23. Which year(s) had the greatest enrollment?

24. Which year had the lowest enrollment?

25. Which year between 1988 and 1993 had the lowest enrollment?

26. Which year between 1990 and 1993 had the greatest enrollment?

27. How many more students were enrolled in 1988 than in 1986?

28. What was the range in enrollment for the ten years?

29. What was the percentage increase in enrollment from 1987 to 1988? Round the answer to the nearest whole number.

30. What was the percentage decrease in enrollment from 1988 to 1990? Round the answer to the nearest whole number.

The following graph shows the gross sales of two different stores owned by the same company for the first six months of a year.

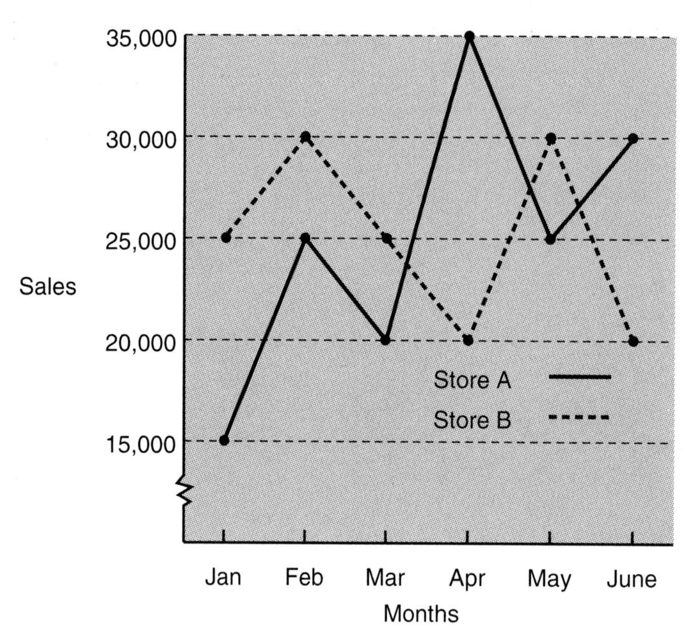

Use this information to answer questions 31–40.

31. What were the sales of store A in January?

32. What were the sales of store B in January?

33. Which store had the greatest gross sales and in which month did it occur?

34. In which two months did store B have its highest sales?

35. In which two months did store A outsell store B?

36. Which store had greater sales in March?

37. What was the total sales for stores A and B in May?

38. How much greater were the sales of store A than store B in April?

39. What was the range of sales for store A during this period?

40. What was the range of sales for store B over the six months?

10–4 Circle Graphs

1 Reading and Interpreting Circle Graphs

OBJECTIVE

Upon completing this section you should be able to:

1 Read and interpret circle graphs.

A **circle graph** (sometimes called a pie graph) shows relationships of data by the relative size of sectors of a circle. The entire circle represents 100% and each sector represents a percent of the whole. We recall from chapter 5 that a complete revolution is 360 degrees. Thus the number of degrees in the central angle of each sector is a percent of 360°.

The following circle graph shows the various components of a local landfill.

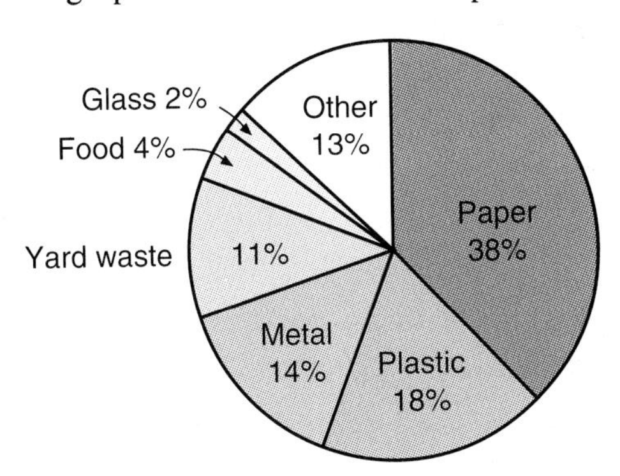

Notice that the sum of the percentages shown is 100%.

Examples 1–4 refer to this graph.

Example 1 Which component accounts for the greatest percentage of the landfill?

Solution The graph indicates that the largest sector is paper (38%). Thus paper accounts for the greatest percentage of the landfill.

◉ ***SKILL CHECK 1*** Which component accounts for the least percentage of the landfill?

Example 2 What percent of the landfill is composed of metal or plastic?

Solution Metal comprises 14% and plastic 18% of the landfill. The part of the landfill that contains metal or plastic is the sum of the two percentages.

$$14\% + 18\% = 32\%$$

This indicates that about one-third of the landfill is composed of metal and plastic.

◉ ***SKILL CHECK 2*** What percent is composed of yard waste or food?

Example 3 What group would include old automobile and truck tires?

Solution The only category shown that would include tires is "other."

◉ ***SKILL CHECK 3*** What group would include soft drink cans?

Example 4 Glass, metal, plastic, and paper together make up what percent of the landfill?

Solution Adding the percentages of the four groups gives

$$2\% + 14\% + 18\% + 38\% = 72\%.$$

◉ ***SKILL CHECK 4*** What percent of the landfill contains no paper?

◉ ***Skill Check Answers***

1. glass
2. 15%
3. metal
4. 62%

A nonprofit organization provided the following information regarding how donations are allocated.

What is the total percent for all categories?

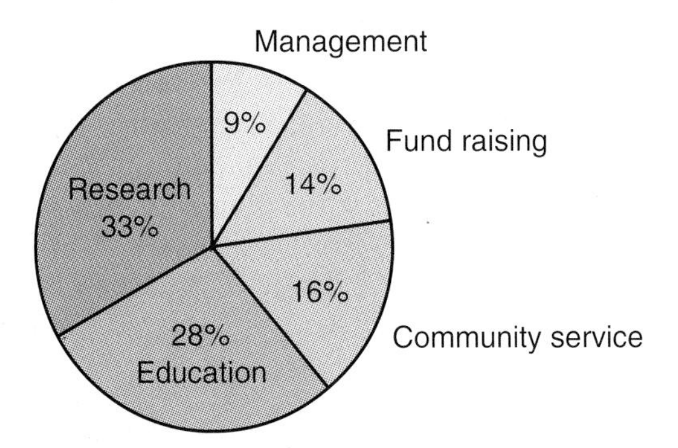

We use this information in examples 5–8.

Example 5 What percent of donations is used for research?

Solution We look at the sector labeled "research" and see that 33% is used for that category.

Which is the largest category?

SKILL CHECK 5 What percent of donations is used for education?

Example 6 What percent is spent for education and community service combined?

Solution We add the percentages of the two categories. Education uses 28% and community service uses 16%. Thus $28\% + 16\% = 44\%$.

SKILL CHECK 6 What percent is spent for fund raising and management?

Example 7 If a person donates \$30.00 to this organization, how much actually goes to research, education, and community service?

Solution The total percent for research, education, and community service is

$$33\% + 28\% + 16\% = 77\%.$$

So 77% of \$30.00 is

Recall: $77\% = 0.77$

$$(0.77)(\$30.00) = \$23.10$$

to be used for these expenses.

SKILL CHECK 7 If a person donates \$40, how much goes to fund raising and management?

Example 8 What is the measure, to the nearest degree, of the central angle of the sector representing community service?

Solution We must find 16% of 360°.

$16\% = 0.16$

$$0.16(360°) = 57.6°$$

The measure to the nearest degree is 58°.

Example 9 If you are making a circle graph showing how a family's income is spent and one item is "rent: 29%," what is the measure, to the nearest degree, of the central angle of the sector representing rent?

Solution We must find 29% of 360°.

$$0.29(360°) = 104.4°$$

Thus the measure to the nearest degree is 104°.

Skill Check Answers

5. 28%
6. 23%
7. \$9.20

⊙ **SKILL CHECK 8** What is the measure of the central angle of a sector representing 45% of a circle?

⊙ ***Skill Check Answer***

8. 162°

Exercise 10–4–1

1 The following circle graph shows a person's classes of investments as a percentage of total money invested.

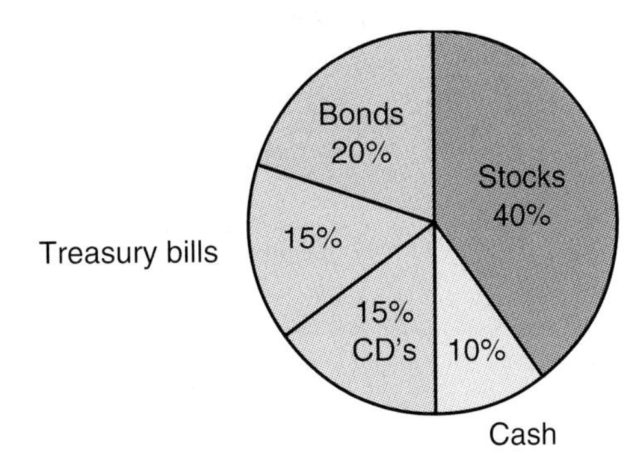

Use this information to answer questions 1–10.

1. What percent is invested in bonds?

2. What percent is invested in CD's?

3. In which class is the person invested the heaviest?

4. In which two classes is the person invested equally?

5. What percent is invested in stocks and bonds combined?

6. What percent is invested in Treasury Bills and CD's combined?

7. If all the cash was used to buy CD's, what percent would then be in CD's?

8. If one-half of the stocks were converted to cash, what percent would then be in cash?

9. If $60,000 is invested in stocks, how much is invested in bonds?

10. If $60,000 is invested in stocks, how much is invested in Treasury Bills?

A toy manufacturer estimates its toy sales to be shared by different types of stores as indicated by the following circle graph.

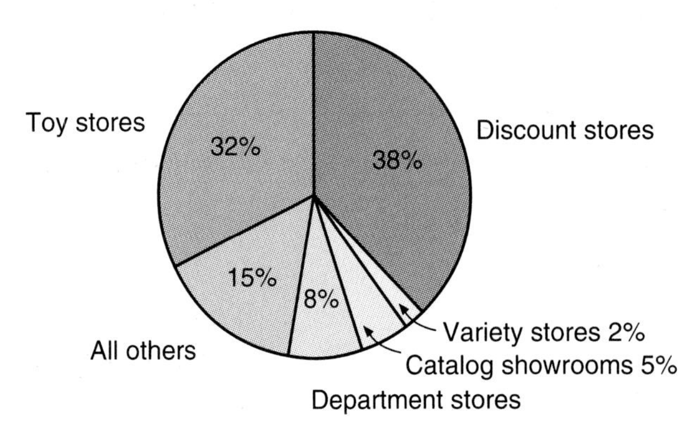

Use this graph to answer questions 11–20.

11. What percent of the toy sales are made by department stores?

12. What percent of the toy sales are made by toy stores?

13. Which type of store sells the most toys?

14. Which type of store sells the least toys?

15. Which two types of stores account for 70% of the toy sales?

16. What percent of toy sales are made by catalog showroom and department stores combined?

17. What is the measure of the central angle of the sector representing catalog showrooms?

18. What is the measure of the central angle of the sector representing toy stores? Round the answer to the nearest whole number.

19. If department store toy sales for the manufacturer amount to 11.2 million dollars, what is the amount of its toy sales by variety stores?

20. If department store toy sales for the manufacturer amount to 11.2 million dollars, what are its total sales for toys by all stores combined?

The following graph shows the budget of a single working person as a percentage of income (after taxes).

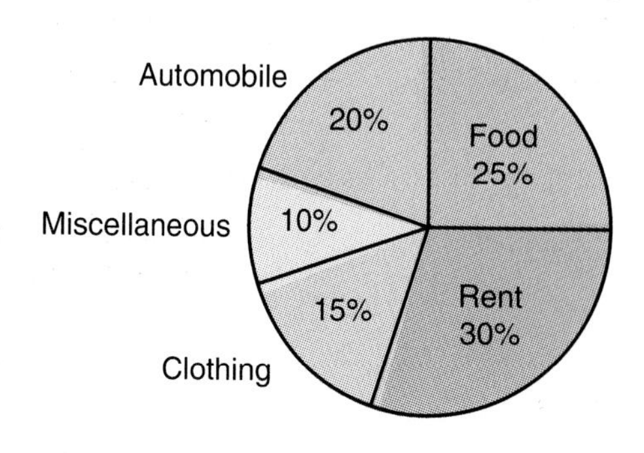

Use this information to answer questions 21–32.

21. What percent of income is budgeted for clothing?

22. What percent of income is budgeted for rent?

23. Which item is the greatest expense?

24. Which item is the least expense?

25. What fractional part of the income is represented by automobile expenses?

26. What fractional part of the income is represented by food expenses?

27. What percent of the income is budgeted for food and clothing combined?

28. What percent of the income is budgeted for rent and food combined?

29. What is the measure of the central angle of the sector representing rent?

30. What is the measure of the central angle of the sector representing food?

31. If the person's income (after taxes) for this month is $1,500, how much is budgeted for rent?

32. If the person's income (after taxes) for this month is $1,500, how much is budgeted for clothing?

CHAPTER 10 SUMMARY

Key Words

Section 10–1

- **Data** is information used as the basis for decision making.
- **Statistics** is the mathematics of the collection, organization, and interpretation of numerical data.
- A **statistic** is a number that describes a set of data.
- The **arithmetic mean** or **arithmetic average** and the **median** are measures of central tendency.
- The **mode(s)** of a set of scores is the score or scores that occur most often in the set.
- The **range** of a set of numbers is the difference between the greatest and the least numbers in the set.

Section 10–2

- A **pictogragh** uses rows of symbols to represent numbers. Each symbol represents a particular value.
- **Tables** and **charts** are also used to show numerical data. They can give more accurate data than a pictograph.

Section 10–3

- **Bar graphs** are used to show comparisons of data. They contain two scales.
- **Line graphs** are used to show comparisons of data over a period of time.

Section 10–4

- **Circle graphs** use sectors of a circle to show relationships of data as percentages of the whole.

Procedures

Section 10–1

- To find the mean of a set of scores divide the sum of the scores by the number of scores in the set.

$$\text{Mean} = \frac{\text{sum of scores}}{\text{number of scores}}$$

- To find the grade point average (GPA) divide the total grade points by the total credit hours.

$$\text{GPA} = \frac{\text{total grade points}}{\text{total credit hours}}$$

- To find the median of an ordered set of scores determine the middle score if there is an odd number of scores. If there is an even number of scores, use the mean of the two middle scores.
- To find the mode(s) look for the score or scores that occur most often in the set.
- To find the range subtract the least number from the greatest number in the set.

Section 10–4

- To find the measure of a sector of a circle multiply 360° by the percentage represented by that sector.

CHAPTER 10 REVIEW

For each of the following sets of numbers, find the (a) mean, (b) median, (c) mode(s), and (d) range.

1. {21, 18, 15, 36, 18}

2. {11, 9, 16, 10, 11}

3. {13, 10, 12, 20, 13, 10, 16, 6}

4. {3, 9, 5, 3, 24, 3, 9, 7}

5. A student received the following grades for a semester.

Course	Grade	Credit Hours
English	B	3
Humanities	A	4
Drama	C	2
Chemistry	C	3
Calculus	B	5
Physical Ed.	D	1

If the grade points are awarded as follows, A: 4 points, B: 3 points, C: 2 points, and D: 1 point; find the student's grade point average for the semester rounded to one decimal place.

6. To obtain a grade of A in a mathematics class a student must have an average of 90 on six tests. If a student obtained scores of 98, 92, 87, 79, and 88 on the first five tests, what is the lowest score that can be obtained on the sixth test to assure a grade of A?

The following pictograph shows the number of telephones in service for three separate towns.

Town A	
Town B	
Town C	
	= 10,000 telephones

Use this pictograph to answer questions 7–16.

7. Which town has the greatest number of telephones?

8. Which town has the fewest number of telephones?

9. How many telephones are in service in town A?

10. How many telephones are in service in town C?

11. How many more telephones are in service in town A than in town C?

12. How many more telephones are in service in town B than in town C?

13. What is the total number of telephones in service in all three towns?

14. What is the total number of telephones in service in towns A and C?

15. What percent of the total telephones in service in all three towns are in use in town A? Round the answer to the nearest percent.

16. What percent of the total telephones in service in all three towns are in use in town B? Round the answer to the nearest percent.

The following table shows the total calories and calories from fat in several foods.

Food	Total Calories	Fat Calories
Prime rib (3 oz)	346	270
Round steak (3 oz)	162	45
Fried chicken (1/2 breast)	218	81
Broiled chicken (1/2 breast)	142	27
Microwave popcorn (1 cup)	60	36
Air-popped popcorn (1 cup)	30	—
Potato chips (1 oz)	114	72
Unsalted pretzels (1 oz)	110	—
Butter pecan ice cream (1/2 cup)	160	90
Butter almond ice milk (1/2 cup)	100	18

Use this table to answer questions 17–26.

17. How many total calories are in 3 ounces of prime rib?

18. How many total calories are in 1/2 broiled chicken breast?

19. How many calories come from fat in 1 cup of microwave popcorn?

20. How many calories come from fat in 1/2 cup of butter pecan ice cream?

21. What percent of the total calories in 3 ounces of prime rib comes from fat?

22. What percent of the total calories in 3 ounces of round steak comes from fat?

23. What percent of the total calories in 1/2 cup of butter pecan ice cream comes from fat? Round the answer to the nearest whole number.

24. What percent of the total calories in 1/2 cup of butter almond ice milk comes from fat?

25. Which two foods have no calories derived from fat?

26. Which has the lesser percentage of calories derived from fat, 1/2 broiled chicken breast or 3 ounces of round steak?

The following bar graph shows enrollment at some selected colleges.

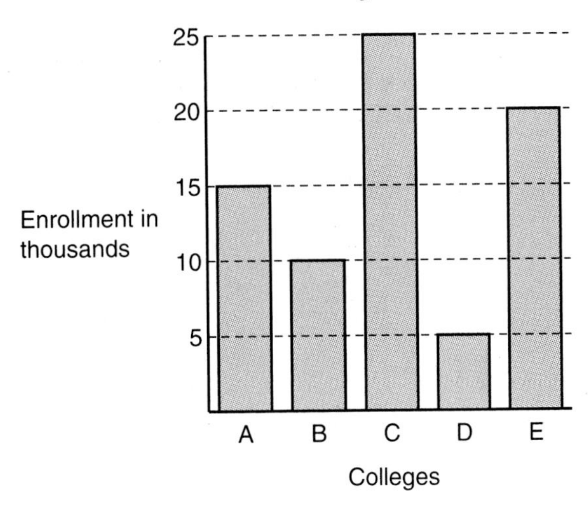

Use this information to answer questions 27–36.

27. What is the enrollment in college A?

28. What is the enrollment in college D?

29. What college has an enrollment of 10,000 students?

30. What college has an enrollment of 25,000 students?

31. Which college has the greatest enrollment?

32. Which college has the least enrollment?

33. Which colleges have an enrollment greater than 10,000 students?

34. Which colleges have an enrollment less than 20,000 students?

35. How many more students are enrolled in college A than college B?

36. How many more students are enrolled in college C than college B?

The following line graph shows the number of VCR's manufactured by a company over a period of 8 years.

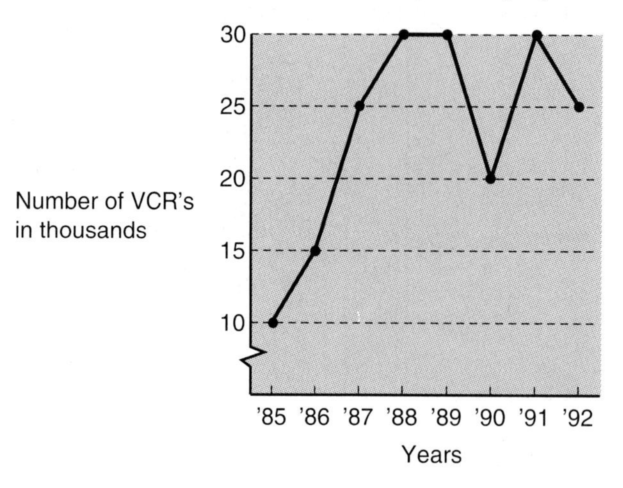

Use this information to answer questions 37–46.

37. How many VCR's were produced in 1987?

38. How many VCR's were produced in 1990?

39. In what years were the greatest number of VCR's produced?

40. In what year were the fewest VCR's produced?

41. Between which two consecutive years did production drop the most?

42. Between which two consecutive years did production increase the greatest?

43. What was the total number of VCR's produced in 1986, 1987, and 1988?

44. What was the total number of VCR's produced in the eight-year period?

45. What was the percentage increase in production from 1987 to 1988?

46. What was the percentage decrease in production from 1991 to 1992? Round the answer to the nearest whole number.

The following circle graph shows the estimated uses for batteries sold by a particular retail store. It indicates the different types of battery-operated devices as a percentage of purchases.

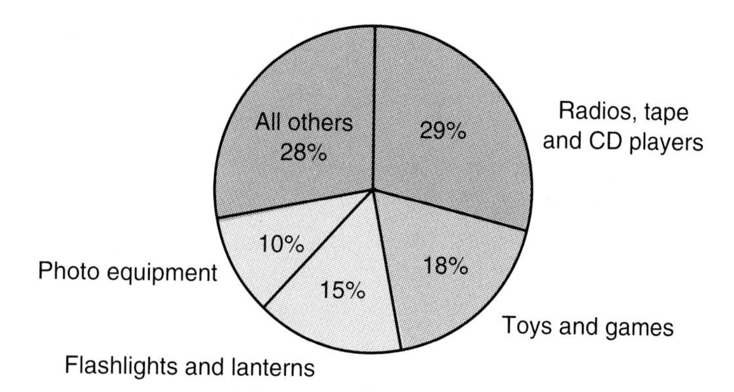

Use this information to answer questions 47–58.

47. What percent of battery-operated devices purchased are toys and games?

48. What percent of battery-operated devices purchased are in the group of radios, tape and CD players?

49. What percent of purchases include photo equipment, flashlights, and lanterns?

50. What percent of purchases include toys, games, flashlights, and lanterns?

51. Which group accounts for the greatest sales?

52. Which group accounts for the least sales?

53. Which group would include battery-operated clocks?

54. Which group would include a battery-operated wristwatch?

55. What is the measure of the central angle of the sector representing toys and games?

56. What is the measure of the central angle of the sector representing flashlights and lanterns?

57. If the store sold 8,500 battery-operated devices, how many flashlights/lanterns did it sell?

58. An average household that purchased all its battery-operated appliances from this store has 12 battery-operated devices. How many flashlights/lanterns does it have? Round the answer to the nearest whole number.

CHAPTER 10 TEST

Given the following set of numbers, answer questions 1–4.
{10, 8, 4, 14, 6, 8, 3, 15, 4}

1. Find the mean.
2. Find the median.
3. Find the mode(s).
4. Find the range.
5. A student obtained the following scores in the first four tests: 90, 95, 88, and 79. If an average score of 90 is needed to obtain a grade of A, what is the lowest score the student can receive in the fifth test to obtain a grade of A?

The following line graph shows monthly exports of US manufactured goods for a six-month period.

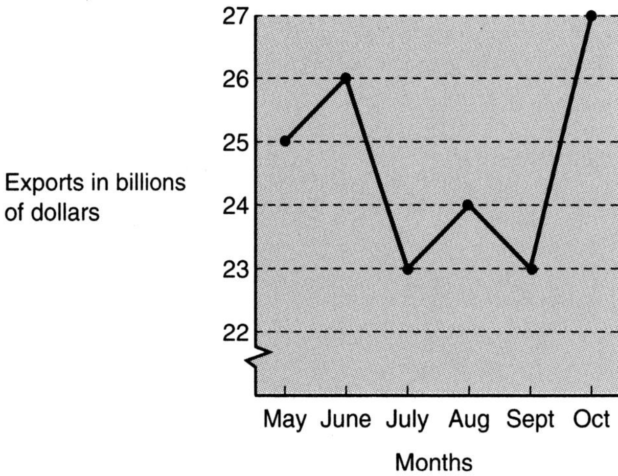

Use this information to answer questions 6–10.

6. What month had the most exports?
7. What was the amount of exports in August?
8. Between which two consecutive months was the increase in exports the greatest?
9. What was the total value of exports for May, June, and July?
10. What was the percentage increase in exports from September to October? Round the answer to the nearest whole number.

The following circle graph shows the percent of various final grades in a class.

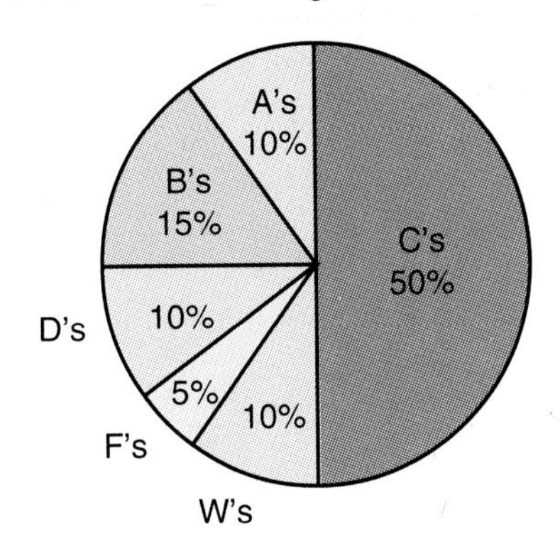

Use this information to answer questions 11–15.

11. What percent of the class received an A?

12. What percent of the class received an A or B?

13. What percent of the class received a passing grade (D or better)?

14. If 4 students received an A, how many received a B?

15. What is the measure of the central angle of the sector representing those who received a B?

CHAPTERS 1–10 CUMULATIVE TEST

1. Subtract: $\begin{array}{r} 5{,}412 \\ -1{,}569 \\ \hline \end{array}$

2. Divide: $34\overline{)110{,}216}$

3. Multiply: $\left(\frac{3a}{5b}\right)\left(\frac{25b^3}{9a^2}\right)$

4. Multiply: $\begin{array}{r} 3.65 \\ \times\ 2.6 \\ \hline \end{array}$

5. Find: $|+32|$

6. Evaluate $x^2y - 3xy + 2y^2z$ when $x = -5$, $y = 4$, and $z = -1$.

7. Solve for x: $x + 4 = -1$

8. Express $\frac{3}{5}$ as a percent.

9. Find the least common multiple of 10, 14, and 15.

10. Write the number 2,005,021 in words.

11. Add: $3\frac{3}{8} + 11\frac{5}{6}$

12. Divide and round the answer to the nearest hundredth: $8.3\overline{)54.63}$

13. Simplify: $-6 + 5 - (-2) + 1 - 10$

14. Convert 15.5 centimeters to meters.

15. Find the arithmetic mean for the set of numbers $\{12, 19, 23, 25, 27, 41, 63\}$.

16. Find the prime factorization of 396.

17. Replace the question mark in $\frac{4}{7} ? \frac{3}{5}$ with the inequality symbol positioned to indicate a true statement.

18. Two angles of a triangle measure 21° and 130°. Find the measure of the third angle.

19. Write 0.00000104 in scientific notation.

20. Solve for x: $\frac{x}{2} = -10$

21. Divide: $-21 \div \left(\frac{3}{7}\right)$

22. Change 0.74 to a common fraction in reduced form.

23. Round 61,553 to the nearest hundred.

24. Solve for x: $\frac{1}{2}(x - 3) = \frac{2}{3}x + 1$

25. The length of one leg of an isosceles right triangle is 3 inches. Find the length of the hypotenuse. Round the answer to two decimal places.

26. Write the ratio 42 : 56 as a common fraction in reduced form.

27. Find the volume of a pyramid if its altitude is 10 inches and its base is a square having a side that measures 6 inches.

28. What will it cost to tile a rectangular floor that measures 12 feet by 8 feet if the price of the tile is $2.90 per square foot?

29. The salary of a worker increased from $16,950 last year to $18,306 this year. What was the percent of increase from last year to this year?

30. The graph shows the monthly sales of salesperson A and salesperson B for the first six months of a year.

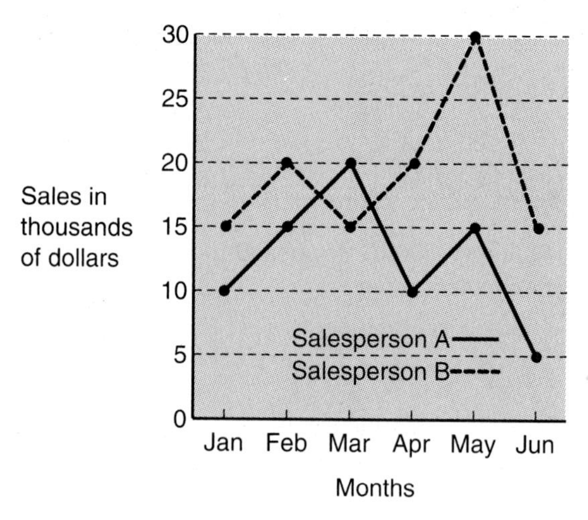

In which month(s) did salesperson A sell more than salesperson B?

CHAPTER

11 Pretest

Before beginning this chapter, answer as many of the following questions as you can. When you finish the chapter, take the test at the end and compare the scores to see how much you have learned.

1. Write an algebraic expression for twice a number x, decreased by nine.

2. Write an algebraic expression for one-third of a number y, increased by one.

3. Write an algebraic expression for 17.3% of a number x.

4. Write an algebraic expression for the value in cents of n nickels.

5. One number is eight times another number. If q represents the smaller number, write an algebraic expression in terms of q for the larger number.

6. The width of a rectangle is 4 inches shorter than the length. Express the width and length in terms of x.

7. A model A graphing calculator is priced $25 higher than model B. If p represents the price of model A, write an algebraic expression in terms of p to represent the price of model B.

8. The length of rope A is three feet longer than twice the length of rope B. The sum of the two lengths is 27 feet. Find the length of each rope.

9. The measure of the second side of a triangle is twice the first, and the measure of the third side is two inches more than the first. If the perimeter of the triangle is 26 inches, find the measure of each side.

10. A beginning karate class has twice as many students as the advanced class. The intermediate class has three more students than the advanced class. How many students are in the advanced class if the total enrollment for the three classes is 43?

CHAPTER

11 Analyzing Word Problems

Strategies for solving word problems are utilized in fields such as architecture, science, business, agriculture, and finance.

STUDY HELP

Reading a mathematics book carefully cannot be overemphasized. Paying special attention to explanations and working through each example are necessary if you are going to learn and retain the concepts presented. Careful reading becomes even more important when problems are stated in words.

Throughout this text you have solved many word problems and learned important techniques such as reading the problem carefully and being sure to answer the question asked. Some of the problems involved using formulas and others required you to form an equation from the given information. Equations and the procedure for solving them are basic to solving word problems. The task then becomes forming the correct equation from the given information. In this chapter we will study techniques for outlining and analyzing a word problem to form an equation that will lead to the solution of the problem.

11–1 Translating Phrases from Words to Algebra

OBJECTIVE

Upon completing this section you should be able to:

1. Change a word phrase to an algebraic expression.

The primary task when attempting to solve a word problem is one of translation. The problem is written in one language and must be translated into another—the language of algebra. This translation process must be precise if we are to be successful in solving the problem.

Example 1 Statement: A number x increased by seven is twelve.
Algebraic translation: $x + 7 = 12$

Do these two statements have the same meaning? Does the algebraic equation make the same statement as the English sentence? If so, we have correctly translated from one language to another and can easily solve for the missing number.

1 Changing Word Phrases to Algebraic Expressions

Before we begin to outline completed sentences or solve problems we need to review the meaning of certain **phrases.**

Example 2 Write an algebraic expression for: A certain number n, decreased by four.

Solution Since n represents the "certain number" and "decreased by" means subtraction, then the algebraic expression is $n - 4$.

Note that a phrase will yield only an algebraic expression and *not* a complete equation. It is impossible to solve for the unknown without a complete equation.

SKILL CHECK 1 Write an algebraic expression for: A certain number n, increased by seven.

Example 3 Write an algebraic expression for: Five times a certain number p.

Solution Since p represents the "certain number," the expression is $5p$.

There are key words that give clues to the operations to be used.
Addition—words such as "increased by," "sum," "more than," "greater than," "total."
Subtraction—words such as "decreased by," "less than," "difference," "diminished by."
Multiplication—words such as "times," "of," "product," "twice."
Division—words such as "quotient," "divided by."

SKILL CHECK 2 Write an algebraic expression for: Five divided by a certain number p.

Example 4 Write an algebraic expression for: Five more than a certain number.

Solution No letter is specified to represent the variable so, in this case, we must select a letter to represent "a certain number." If we select x to represent the variable and recognize that "more than" means addition, the expression is $x + 5$.

SKILL CHECK 3 Write an algebraic expression for: Seven less than a certain number.

Example 5 Write an algebraic expression for: Seven more than twice a certain number.

Solution Again, let x represent the unknown number. Then twice the number would translate as $2x$ and seven more than $2x$ would be $2x + 7$.

SKILL CHECK 4 Write an algebraic expression for: Seven less than three times a certain number.

Skill Check Answers

1. $n + 7$
2. $\frac{5}{p}$
3. $x - 7$
4. $3x - 7$

The letter x is often used to represent an unknown number (variable). However, any letter can be used since it has no significance other than as a placeholder.

Example 6 Write an algebraic expression for: 5% of a given number.

Solution If we let n represent the given number, then 5% of n can be written $0.05n$ since the word "of" implies multiplication.

Recall: 5% = 0.05

⊙ SKILL CHECK 5 Write an algebraic expression for: 15% of a given number.

Example 7 Write an algebraic expression for: The value in cents of d dollars.

Solution The value of a dollar is 100 cents. Therefore to indicate the value of d dollars we multiply d by 100, obtaining the expression $100d$.

⊙ SKILL CHECK 6 Write an algebraic expression for: The value in ounces of p pounds.

⊙ Skill Check Answers

5. $0.15x$
6. $16p$

Exercise 11–1–1

1 Write an algebraic expression for each phrase. If no letter is specified, use x to represent the variable.

1. A number x, increased by five
2. A number y, decreased by three
3. Five more than a given number a
4. Eight less than a given number b
5. A given number N, less nine
6. A given number A less than nine
7. Ten greater than a given number q
8. The sum of x and 4
9. Twice a certain number r
10. Eight times a certain number n
11. Twice a certain number, increased by two
12. Three times a certain number, increased by six
13. Six times a certain number, decreased by four
14. Nine times a certain number, decreased by one
15. One-fifth a certain number
16. Half a certain number
17. Three less than a number
18. Four less than twice a number
19. The product of a and 8
20. The product of x and y
21. Half the product of x and 5

xy

22. Three times the product of a and b

23. Half the sum of b and 7

24. Twice the sum of x and 3

25. Three times the difference of x and 9

26. 25% of a number

27. 15% of a number

28. 7% of a number

29. 10% of a number

30. 16.8% of a number

31. 9.5% of a number

32. The value in cents of n nickels

33. The value in cents of d dimes

34. The value in cents of q quarters

35. The total value in cents of d dimes and q quarters

36. The number of days in w weeks

37. The number of months in y years

38. The number of weeks in y years

39. The number of weeks in d days

40. The number of years in m months

• Concept Enrichment

41. In each of the preceding problems a variable was used to represent a number. Why can't we find an exact value for the number?

42. Why is an accurate translation from the language of words to the language of algebra so important?

11–2 Representing Unknowns as Algebraic Expressions

1 Expressing Relationships between Unknowns

OBJECTIVE

Upon completing this section you should be able to:

1 Express a relationship between two or more unknowns in a given statement by using one unknown.

When more than one unknown number is involved in a problem, we try to **outline** the problem by using a letter, such as x, to represent one of the numbers. We then use the relationships given in the problem to express the other unknowns using the same letter.

It may not always be possible to relate the unknowns in a problem using only one unknown. You will be able to do so, however, with all problems in this chapter.

Example 1 The Sears Tower in Chicago is eight stories taller than the Empire State Building in New York. Write algebraic expressions for the height of each building, using one unknown.

Solution From the information given we do not know how many stories either building has. We will choose one of the buildings and represent the number of stories by x.

Let x = number of stories in the Empire State Building

The information states that the Sears Tower is eight stories taller than the Empire State Building. We therefore can write

$x + 8$ = number of stories in the Sears Tower.

We could instead choose x to represent the number of stories in the Sears Tower. Then $x - 8$ would represent the number of stories in the Empire State Building.

SKILL CHECK 1 One person is four inches taller than another. If we let x represent the height of the shorter person, write an expression for the height of the taller person in terms of x.

Example 2 The length of a rectangle is three meters more than the width. Write expressions for the length and width using one unknown.

Solution Both the length and width are unknown quantities so we must write expressions that give one of these quantities in terms of the other. First we will let x represent the width.

Let x = width of the rectangle.

Now since we are told that the length is three more than the width, we can write

$x + 3$ = length of the rectangle.

We can draw a sketch and label the sides as follows:

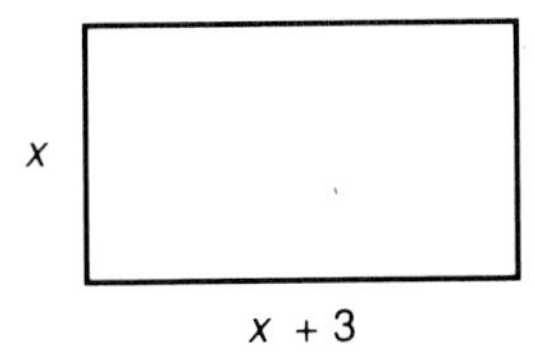

It is sometimes helpful to use a diagram to see the relationships between unknowns.

Suppose we had decided to let x represent the length of the rectangle. We would write

let x = length of the rectangle.

Now we would have to recognize that the statement "the length is three more than the width" is the same as saying "the width is three less than the length." Then

$x - 3$ = width of the rectangle.

Skill Check Answer

1. $x + 4$

We would label the rectangle as follows:

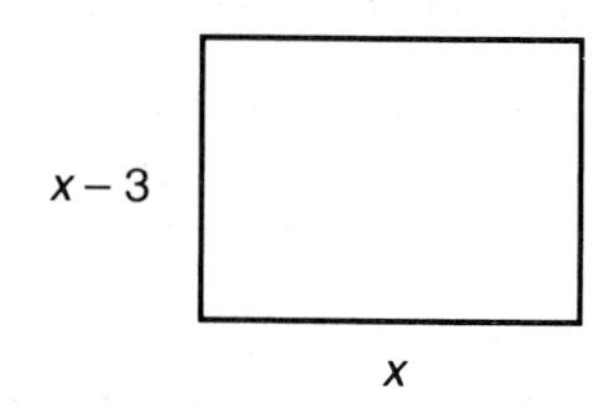

⊙ *SKILL CHECK 2* One person is five years older than another. If we let a represent the age of the older person, write an expression for the age of the younger person in terms of a.

We see that there is more than one way to identify the unknowns. Therefore it is extremely important to designate which unknown is represented by each algebraic expression.

Example 3 The width of a rectangle is one-fourth the length. Write expressions for the length and width using one unknown.

Solution At first glance we might think that a fraction will be necessary here. If we were to let x represent the length, we would have a fraction representing the width. This certainly would not be incorrect, but we can avoid the use of fractions by careful selection.

If we let x = length then $\frac{1}{4}x$ = width.

If we note that "the width is one-fourth the length" is the same as "the length is four times the width," we may let

$$x = \text{width}$$

then

$$4x = \text{length.}$$

Is the width one-fourth the length?

x

$4x$

Example 4 The sum of two numbers is 20. Write expressions for both numbers using one unknown.

Solution Let N = first number.
Then $20 - N$ = second number.

⊙ *SKILL CHECK 3* In the preceding example let N = second number and write an expression for the first number.

Example 5 Express algebraically the relationship of the unknowns in this sentence: A certain number is four more than a second number and is three less than a third number.

Solution To express these three numbers algebraically first decide which will be represented by the letter we choose.

For instance, if x represents the first number, we have the following:

$$x = \text{first number}$$
$$x - 4 = \text{second number}$$
$$x + 3 = \text{third number.}$$

Which number to represent by x is arbitrary, but reread the relationships several times to see if one option is preferable over another.

Ask yourself, "Is the first number four more than the second?" and "Is the first number three less than the third?" If the answers are "yes," you have correctly outlined the sentence.

⊙ *Skill Check Answers*

2. $a - 5$
3. $20 - N$

Now suppose in the same example we decide to allow x to represent the second number. We then have

$$x + 4 = \text{first number}$$
$$x = \text{second number}$$
$$x + 7 = \text{third number.}$$

Ask the same questions again, "Is the first number four more than the second?" and "Is the first number three less than the third?" This outline is also correct.

◉ *SKILL CHECK 4* Referring to example 5, represent the three numbers if x represents the third number.

Example 6 Write algebraic expressions for four consecutive integers. (Examples of consecutive integers are 1, 2, 3, 4, and so on.)

Solution Each consecutive integer is one greater than the preceding integer. If we select n to represent the first integer, then the four consecutive integers are represented as

$$n = \text{first integer}$$
$$n + 1 = \text{second integer}$$
$$n + 2 = \text{third integer}$$
$$n + 3 = \text{fourth integer.}$$

Note that each integer is one greater than the preceding one.

◉ *Skill Check Answer*

4. $x - 3 = \text{first number}$
$x - 7 = \text{second number}$
$x = \text{third number}$

Exercise 11–2–1

1 Express the unknowns in terms of x.

1. A certain number is eight more than a second number. Write expressions for the numbers.
2. A certain number is five less than a second number. Write expressions for the numbers.
3. The Sears Tower in Chicago is 104 feet taller than the World Trade Center in New York. Write expressions for the height of each building.
4. The height of the Empire State Building in New York is 204 feet less than that of the Sears Tower in Chicago. Write expressions for the height of each building.
5. During the second week of production of a new car, General Motors produced 612 units more than it did during the first week. Write expressions for the number of cars produced each week.
6. A student purchased a biology book and a math book. The biology book cost five dollars more than the math book. Write expressions for the cost of each book.
7. A 7:00 A.M. math class has eight fewer students than a 9:00 A.M. class. Write expressions for the number of students in each class.
8. The capacity of computer A is 32 kilobytes more than computer B. Write expressions for the capacity of each computer.

9. The population of San Francisco is twice that of Long Beach. Write expressions for the population of each city.

10. The population of El Paso is half that of Boston. Write expressions for the population of each city.

11. The length of a rectangle is five meters more than the width. Write expressions for the length and width.

$L = 5 + x$
$W = x$

12. The width of a rectangle is one-third the length. Write expressions for the length and width.

13. The length of a rectangle is three centimeters more than twice the width. Write expressions for the length and width.

$W = x$
$L = 3 + 2x$

14. The width of a rectangle is three inches more than half the length. Write expressions for the length and width.

15. The sum of two numbers is 40. Write expressions for the numbers.

16. The total income of a married couple is $50,000. Write expressions for the income of each person.

17. The difference of two numbers is 18. Write expressions for the two numbers.

18. Computer A is more expensive than computer B. If the difference in their cost is $250, write expressions for the cost of each.

19. An individual makes two investments that total $10,000. Write expressions to represent the two investments.

20. A person has $8,000 less invested in stocks than in bonds. Write expressions for the amount invested in each.

21. The price of this year's Chevrolet Cavalier is 5% higher than last year's. Write expressions for last year's price and this year's price.

22. This year's enrollment at a college is 7% less than last year's. Write expressions for the enrollment for last year and this year.

23. One number is four more than the second and nine less than the third. Write expressions for the three numbers.

24. The price of a certain model of a Texas Instruments calculator is six dollars more than a Sharp calculator and three dollars less than one made by Hewlett-Packard. Write expressions for the three prices.

25. One number is half the second and three times the third. Write expressions for the three numbers.

26. On a particular flight the cost of a coach ticket is half the cost of a first-class ticket and thirty dollars more than an economy ticket. Write expressions for the cost of the three tickets.

27. Write expressions for three consecutive even integers. (Examples of consecutive even integers are 2, 4, 6, 8, 10, and so on.)

28. Write expressions for three consecutive odd integers. (Examples of consecutive odd integers are 1, 3, 5, 7, 9, and so on.)

29. The cost of a house today is three times its cost in 1970 and twice its cost in 1976. Write expressions for the cost in 1970, 1976, and now.

30. The capacity of computer A is twice that of computer B and 32 kilobytes less than computer C. Write expressions for the capacity of the three computers.

31. Jane has five dollars more than Bob and thirteen dollars less than Jim. Write expressions for the numbers of dollars each person has.

32. The load capacity of a Ford pick-up truck is eight cubic feet more than a Nissan and nine cubic feet more than a Toyota. Write expressions for the load capacity of the three trucks.

33. A Buick Century obtains four miles per gallon more than a Chevrolet Caprice and eight miles per gallon less than a Toyota Corolla. Write expressions for the mileage for the three cars.

34. The state of Ohio has half the population of California and twice the population of Indiana. Write expressions for the population of the three states.

35. A certain amount is invested in an account for one year at 14% interest. Write expressions for the original amount, the interest earned, and the total amount in the account at the end of the year.

36. The length of one side of a triangle is twice the length of the second side and four less than the length of the third side. Write expressions for the lengths of the three sides.

37. A meter stick is cut into two pieces. Write expressions for the length of each piece in centimeters. (Recall that 1 m = 100 cm.)

38. The atomic weight of sulfur is twice that of oxygen and eight times that of helium. Write expressions for the atomic weight of each element.

39. A person has twice as many dimes as nickels and two more quarters than dimes. Write expressions for the number of each kind of coin the person has.

40. Write an expression for the total value, in cents, of the coins discussed in question 39.

- **Concept Enrichment**

41. If a person's income is 7% higher this year than last year, can we express the unknowns as

$$x = \text{last year's income}$$

and $$1.07x = \text{this year's income?}$$

42. If the price of computer A is 5% more than computer B, is it also true that the price of computer B is 5% less than computer A? Explain.

11–3 A Strategy for Solving Word Problems

OBJECTIVE

Upon completing this section you should be able to:

1 Solve a word problem by applying a step-by-step strategy.

In the preceding exercises we outlined relationships between unknowns within a statement. If we are to solve a problem and find the unknown numbers, there must be within the problem a sentence that yields an equation. In this section we will outline some of the same problems from section 11–2. We will then form an equation from the given information and solve for all unknowns asked for in the problem.

1 A Step-by-Step Strategy

The strategy we will use follows these six steps. Notice in each example how these steps aid us in arriving at a correct solution.

A Step-by-Step Strategy for Solving Word Problems

Step 1 Determine what is to be found (including how many unknowns).
Step 2 Use the given information to write algebraic expressions for each of the unknowns.
Step 3 Write an equation that relates the unknowns to each other.
Step 4 Solve the equation.
Step 5 Make sure you have answered the question.
Step 6 Check your answers to make sure they agree with the original problem.

The first step in this process is to read and re-read the problem until we determine what is being asked for. Then we must look at the given information and outline the problem so as to arrive at an equation relating the unknown quantities.

Example 1 The Sears Tower in Chicago is eight stories taller than the Empire State Building in New York. If the total number of stories in both buildings is 212, find the number of stories in each building.

Solution Notice that we are asked to find the number of stories in each building. Thus there are two unknowns to find.

This is Step 1.

Looking further we see a relationship between the unknowns that leads to the outline

let $x =$ the number of stories in the Empire State Building

then $x + 8 =$ the number of stories in the Sears Tower.

Step 2.

In example 1 of the previous section we had enough information for the above outline. This time, however, we have the additional statement that the total number of stories is 212. In words we say

Number of stories in the Empire State Building	+	Number of stories in the Sears Tower	=	Total number of stories

The algebraic translation is

$$x + (x + 8) = 212.$$

Step 3.

Now solving this equation we first eliminate parentheses, obtaining

$$\begin{aligned} x + x + 8 &= 212 \\ 2x + 8 &= 212 \\ 2x &= 204 \\ x &= 102. \end{aligned}$$

Step 4.

Notice that we have only found the number of stories in the Empire State Building. The question asked us to find the number of stories in *each* building. Thus to find the number of stories in the Sears Tower we must substitute 102 for x in the expression $x + 8$, obtaining

$$x + 8 = (102) + 8 = 110.$$

The answers are

number of stories in the Empire State Building = 102
number of stories in the Sears Tower = 110.

Step 5.

Check: 110 is eight more than 102, and the sum of 110 and 102 is 212.

Step 6.

Example 2 The length of a rectangle is three meters more than the width. Find the length and width if the perimeter of the rectangle is 26 meters.

Solution We are asked to find the dimensions of the rectangle. From the information given we outline as follows:

Let x = width.
Then $x + 3$ = length.

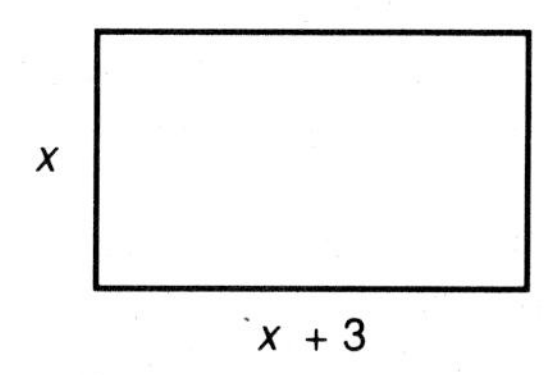

Since the perimeter is equal to the sum of twice the length and twice the width, or $P = 2\ell + 2w$, we can write the equation

$$2(x) + 2(x + 3) = 26.$$

Solving, we obtain

$$\begin{aligned} 2x + 2x + 6 &= 26 \\ 4x + 6 &= 26 \\ 4x &= 20 \\ x &= 5. \end{aligned}$$

Eliminate parentheses.

Also $x + 3 = 8.$

Thus the width is 5 meters and the length is 8 meters.

Get in the habit of always summarizing your answers.

Check: The length (8) is three more than the width (5), and the perimeter is $2(8) + 2(5) = 16 + 10 = 26$.

Checking your answers is one of the most important parts of the solution.

⦿ SKILL CHECK 1 Work example 2 letting x = length and $x - 3$ = width.

Example 3 The width of a rectangle is one-fourth the length. If the perimeter is 200 centimeters, find the length.

Solution Here we are asked to find the length of a rectangle and are given a relationship between the length and width. We outline as follows:

This particular way of stating the unknowns avoids fractions.

$$\text{Let } x = \text{width.}$$
$$\text{Then } 4x = \text{length.}$$

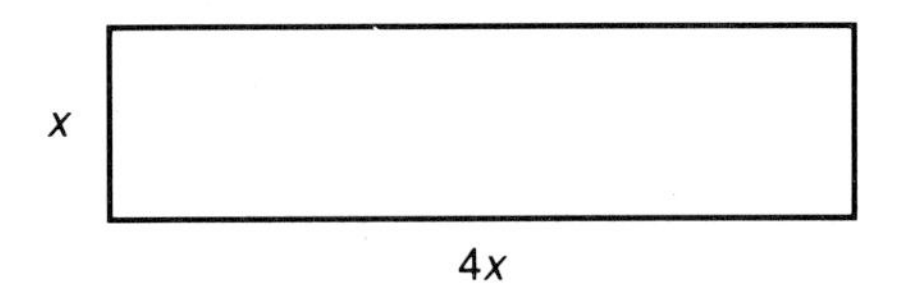

Again, the perimeter is twice the width plus twice the length. Therefore

$$\begin{aligned} 2(x) + 2(4x) &= 200 \\ 2x + 8x &= 200 \\ 10x &= 200 \\ x &= 20. \end{aligned}$$

What does x represent?

The problem asks for the length only. Thus,

$$4x = 4(20) = 80.$$

The length is 80 cm.

Check: The width (20) is one-fourth the length (80). Also, the perimeter is $2(20) + 2(80) = 40 + 160 = 200$.

⦿ SKILL CHECK 2 Work example 3 again using x = length and $\frac{1}{4}x$ = width.

Example 4 The sum of two numbers is 20. Their difference is 4. Find the numbers.

Solution To find the two numbers we can let

$$n = \text{first number.}$$
$$\text{Then } 20 - n = \text{second number.}$$

Since the difference of the two numbers is 4, we have the statement

$$(\text{first number}) - (\text{second number}) = 4.$$

The algebraic translation is

$$n - (20 - n) = 4.$$

We could also write the equation $(20 - n) - n = 4$ since we don't know which number is larger. Solve the above equation.

Solving, we obtain

$$\begin{aligned} 2n &= 24 \\ n &= 12. \\ \text{Also } 20 - n &= 8. \end{aligned}$$

The two numbers are 8 and 12.

Check: The sum of 8 and 12 is 20. The difference of 8 and 12 is 4.

⦿ Skill Check Answers

1. width = 5 m
   length = 8 m
2. 80 cm

Example 5 A certain number is four more than a second number and three less than a third number. Find the numbers if their sum is 23.

Solution In this problem we are asked to find the numbers and are given a statement about their sum.

If we let x represent the first number, we have

$$x = \text{first number}$$
$$x - 4 = \text{second number}$$
$$x + 3 = \text{third number.}$$

Again, we could just as well let x represent the second or third number. The important thing to remember is to label each algebraic expression with the number that it represents.

The statement "their sum is 23" gives the equation

$$x + (x - 4) + (x + 3) = 23.$$

If we now solve the equation, we obtain

$$3x - 1 = 23$$
$$3x = 24$$
$$x = 8.$$

Leaving the answer as $x = 8$ would not be a solution to the problem. We are asked to "find the numbers." The answer must be

$$\text{first number} = x = 8$$
$$\text{second number} = x - 4 = 4$$
$$\text{third number} = x + 3 = 11.$$

WARNING

Do not just solve the equation for x and think you have solved the problem. Check to see what x represents and reread the problem to see what it is asking for.

We check the problem by noting that the first number (8) is four more than the second (4) and is three less than the third (11) and that their sum is 23.

◉ ***SKILL CHECK 3*** Work example 5 using x to represent the second number.

◉ ***Skill Check Answer***

3. First number = 8
Second number = 4
Third number = 11

Exercise 11–3–1

1 Solve each problem using the six steps discussed in this section.

1. A certain number is eight more than a second number. The sum of the two numbers is 22. Find the numbers.

2. A certain number is five less than a second number. The sum of the two numbers is 33. Find the numbers.

3. The Sears Tower in Chicago is 104 feet taller than the World Trade Center in New York. If the sum of their heights is 2,804 feet, find the height of each building.

4. The height of the Empire State Building in New York is 204 feet less than the height of the Sears Tower in Chicago. If the sum of their heights is 2,704 feet, find the height of the Empire State Building.

5. During the second week of production of a new car, General Motors produced 612 units more than it did during the first week. If the total production for the two weeks was 18,416, find the production for each week.

6. A student purchased a biology book and a math book. The biology book cost five dollars more than the math book and the total cost of both books was \$37.00. Find the cost of the biology book.

7. A 7:00 A.M. math class has eight fewer students than a 9:00 A.M. class. Find the number of students in each class if the total number of students in both classes is 84.

8. The capacity of computer A is 32 kilobytes more than the capacity of computer B. The total capacity of both computers is 96 kilobytes. Find the capacity of each computer.

9. The population of San Francisco is twice that of Long Beach. If the total population of both cities is 1,074,000, find the population of San Francisco.

10. The population of El Paso is half that of Boston. Find the population of each city if the total population for both cities is 960,000.

11. The length of a rectangle is five meters more than the width. Find the length and width if the perimeter is 46 meters.

12. The width of a rectangle is one-third the length. The perimeter is 384 feet. Find the length and width.

13. The length of a rectangle is three centimeters more than twice the width. If the perimeter is 192 centimeters, find the length.

14. The width of a rectangle is three inches more than half the length. If the perimeter is 108 inches, find the width.

15. The sum of two numbers is 40. Their difference is 12. Find the numbers.

16. The total income of a married couple is \$50,000. The wife earns \$5,800 more than the husband. Find the earnings of each.

17. The difference of two numbers is 18. Their sum is 82. Find the numbers.

18. Computer A is \$250 more expensive than computer B. If the total cost of both computers is \$11,090, find the cost of each.

19. An individual makes two investments that total \$10,000. If the first investment is \$3,256 more than the second, find the amount of each investment.

20. A person has \$8,000 less invested in stocks than in bonds. If the total invested in both is \$86,500, find the amount invested in stocks.

21. A certain number is 37 more than a second number. If the sum of the two numbers is 15, find the numbers.

22. A certain number is 25 less than a second number. Find the numbers if their sum is -13.

23. Two angles are supplementary (the sum of their measures is 180°). One of the angles is twice the measure of the other. Find the measures of the two angles.

24. Find the measures of two supplementary angles if one angle is 30° less than twice the other.

25. Two angles are complementary (the sum of their measures is 90°). The measure of one of the angles is four times that of the other. Find the measures of the two angles.

26. Find the measures of two complementary angles if the measure of one is 14° greater than three times the other.

27. One number is four more than a second number and nine less than a third number. Find the three numbers if their sum is 50.

28. The price of a certain model of a Texas Instruments calculator is six dollars more than a Sharp calculator and three dollars less than one made by Hewlett-Packard. If the total price of the three calculators is \$46.50, find the price of each.

29. On a particular flight the cost of a coach ticket is half the cost of a first-class ticket and thirty dollars more than an economy ticket. If the total cost of the three tickets is $406, find the cost of each.

30. Jane has five dollars more than Bob and thirteen dollars less than Jim. If they have a total of $197, find the amount each person has.

31. A Buick Regal obtains four miles per gallon more gas mileage than a Chevrolet Caprice and eight miles per gallon less than a Toyota Corolla. If the total gas mileage for the three cars is 70 mpg, find the mileage of each.

32. The state of Ohio has half the population of California and twice the population of Indiana. If the total population of the three states is 38,500,000, find the population of each state.

33. One side of a triangle is twice the length of the second side and four centimeters less than that of the third side. If the perimeter of the triangle is 84 centimeters, find the length of each side.

34. The atomic weight of sulfur is twice that of oxygen and eight times that of helium. If the total atomic weights of the three elements is 52, find the atomic weight of each.

CHAPTER 11 SUMMARY

Key Words

Section 11–1

- **Word phrases** need to be translated into algebraic expressions when solving word problems.

Section 11–2

- An **outline** expresses all unknowns in a problem in terms of one unknown.

Section 11–3

- A **strategy** for solving word problems involves following a series of steps.

Procedures

Section 11–3

- To solve a word problem follow these steps:
  - Step 1 Determine what is to be found (including how many unknowns).
  - Step 2 Use the given information to write algebraic expressions for each of the unknowns.
  - Step 3 Write an equation that relates the unknowns to each other.
  - Step 4 Solve the equation.
  - Step 5 Make sure you have answered the question.
  - Step 6 Check your answers to make sure they agree with the original problem.

CHAPTER 11 REVIEW

Write an algebraic expression for each phrase. If no letter is specified, use x to represent the variable.

1. Write an algebraic expression for twice a number n, decreased by seven.

2. Write an algebraic expression for half a number c, increased by three.

3. Write an algebraic expression for the sum of a given number r and 5.

4. Write an algebraic expression for the value in cents of Q quarters.

5. Write an algebraic expression for the number of meters in y centimeters.

6. A certain number is seven more than three times another number. Write expressions for the numbers.

7. The sum of two numbers is 84. Write expressions for the numbers.

8. Write an expression for the interest when a certain amount is invested for a year at 16%.

9. The sum of three numbers is 181. The first number is twice the third. Write expressions for the three numbers.

10. Write an expression for the total amount received at the end of one year if x dollars are invested at 17.3% interest.

11. If x represents one number and $x + 5$ represents another number, write a mathematical statement that indicates their sum is 23.

12. If one number is represented by x and a second number is represented by $3x - 1$, write a mathematical statement that indicates their sum is 19.

13. Write an equation to show that a number represented by $3x$ less a number represented by $x - 7$ is equal to 13.

14. Write an equation to show that the sum of two numbers, represented by x and $x - 6$, less a third number represented by $3x$, is equal to -11.

15. Express algebraically the fact that a certain number added to 9% of the number is 1.26.

16. One item costs $1.25 more than another item. Write an equation to show that the sum of the costs of the two items is $5.00.

17. A certain number, less a second number that is five more than three times the first number, is equal to -7. Find the numbers.

18. The length of one side of a triangle is twice that of the second and four less than that of the third. If the perimeter is 64, find the three sides.

19. A meter stick is cut into two pieces so that one piece is 6 centimeters longer than the other. Find the lengths of the two pieces.

20. A man has twice as many dimes as nickels, and two more quarters than dimes. The total number of coins is 22. How many of each kind of coin does he have?

21. Roger is seven years older than Dave. The sum of their ages is 63. Find their ages.

22. Ellen is four years older than Kathy, and Barbara is two years younger than twice Kathy's age. How old is each of the women if the sum of their ages is 58?

23. A 40.0-meter length of rope is cut into three pieces. The second piece is three times as long as the first, and the third piece is five meters shorter than twice the length of the second. Find the length of each piece.

24. The sum of three numbers is 123. The second number is five less than twice the first. The third number is seven more than the sum of the first two numbers. Find the numbers.

CHAPTER 11 TEST

1. Write an algebraic expression for three times a number x, increased by six.

$3x+6$

2. Write an algebraic expression for one-half a number y, decreased by seven.

$\frac{1}{2}y-7$

3. Write an algebraic expression for 9.4% of a number n.

$9.4\%n = 0.094n$

4. Write an algebraic expression for the number of weeks in d days.

5. One number is eight less than another. If r represents the smaller number, write an algebraic expression in terms of r for the larger number.

6. Juan is 3 years older than Peter. If k represents Juan's age, express Peter's age in terms of k.

7. The number of students enrolled in an algebra class is five less than twice the number enrolled in a chemistry class. If the number of students enrolled in the chemistry class is represented by x, write an algebraic expression to represent the enrollment of the algebra class.

8. The length of a rectangle is five feet longer than twice the width. If the perimeter of the rectangle is 28 feet, find the length.

9. A compact disc player costs $250 more than a VCR unit. If the total cost of both the CD and VCR is $890, find the cost of each.

10. A piece of rope that is 63 meters long is cut into three pieces such that the second piece is twice as long as the first, and the third piece is three meters longer than the second. Find the length of each piece.

CHAPTERS 1–11 CUMULATIVE TEST

1. What is the place value of the digit 2 in the number 3,241,056?

2. Find the difference of 10,143 and 4,168.

3. Find the prime factorization of 252.

4. Replace the question mark in $\frac{3}{5}$? $\frac{7}{12}$ with the inequality symbol positioned to indicate a true statement.

5. Subtract: $5\frac{1}{2} - 3\frac{5}{6}$

6. Find the volume of a sphere having a radius of $1\frac{1}{2}$ meters.

7. Solve for x: $x - 8 = 10$

8. Find: $|-28|$

9. Simplify:
$16a^2b + 12ab^2 - 7a^2b + 3ab - 5ab^2 + 4ab$

10. Simplify: $\frac{3a^3}{5b} \div \frac{9a}{10b^2}$

11. Add: $\frac{3}{16} + \frac{5}{12}$

12. Evaluate: $\frac{(-20)(-3)(-1)}{(-2)(-5)}$

13. Subtract: $32 - 18.031$

14. The altitude of a triangle measures 8 feet and the area is 68 square feet. Find the length of the base.

15. Solve for x: $-3x = 21$

16. Round 23,846 to the nearest thousand.

17. Find the product: $(4a^2)(2a)(a^4)$

18. Round 129.9999 to the nearest thousandth.

19. Write the ratio 90:225 as a common fraction in reduced form.

20. Multiply: (21.05)(1.6)

21. Simplify: $8 - (-2) - 15 + 4 - 9$

22. Write 35,700,000 in scientific notation.

23. Simplify: $15a^2 - 3ab - 4a^2 + ab$

24. Multiply: $\begin{array}{r} 263 \\ \times\ 76 \\ \hline \end{array}$

25. Evaluate $3xy^2 + 5xy - 2x^2y$ if $x = -3$ and $y = -2$.

26. Divide: $9.3\overline{)22.413}$

27. Simplify:
$2x(3x^2 - x + 1) - 3x(2x^2 - 4x + 7)$

28. Solve for x: $\frac{1}{3}(2x - 1) = x + \frac{1}{2}(x + 3)$

29. Find the mode(s) for the set of numbers: $\{6, 8, 8, 8, 11, 12, 14, 25, 25\}$.

30. Express $\frac{3}{8}$ as a percent.

31. Find the least common multiple of 15, 12, and 35.

32. Write an algebraic expression to represent one-half a number n, decreased by three.

33. Find the area of a triangle if the base measures 18.5 centimeters and the altitude is 12.4 centimeters.

34. The price of a certain model of automobile is 5.5% greater this year than last year. If x represents last year's price, represent this year's price in terms of x.

35. If one leg of a right triangle measures 4.00 feet and the hypotenuse measures 8.00 feet, find the length of the third side rounded to the nearest hundredth.

36. Two people divided some money in the ratio of 5 to 8. If the person with the smaller amount received $12.00, how much did the person with the larger share receive?

37. The circle graph shows a budget for a particular family. If the family has an income of $2,500 for a month, how much is budgeted for food?

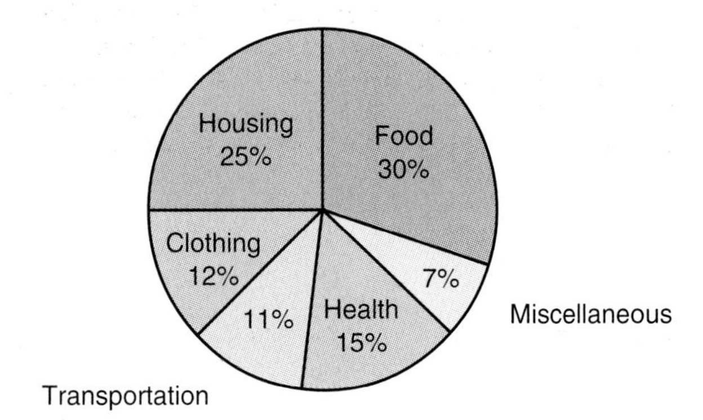

38. June has a balance of $344.00 due on a credit card. If the credit card company charges 1.5% interest per month on the balance, how much interest will June have to pay this month?

39. What will it cost to fence a rectangular field that measures 100 feet by 250 feet if the price of fencing is $18.00 for a four-feet section?

40. A student purchased a chemistry book and a math book for a total of $75. Find the price of the chemistry book if it cost $11 more than the math book.

Answer Appendix

Chapter 1 Pretest

The number in brackets after each answer refers to the section of the chapter that discusses that type of problem.
1. Multiplication [1–1] **2.** Division and addition [1–1] **3.** Ten thousands [1–2] **4.** Thirty-eight thousand, twenty-six [1–2] **5.** $24 > 8$ [1–2] **6.** $2(10^3) + 0(10^2) + 1(10^1) + 0(10^0)$ [1–3] **7.** 1,002 [1–4] **8.** 1,116 [1–4] **9.** 6,907 [1–4] **10.** 198 [1–5] **11.** 108 [1–5] **12.** 5,009 [1–5] **13.** 2 [1–6] **14.** x [1–6] **15.** $9a$ [1–7] **16.** $7x$ [1–7] **17.** $ab + 6a^2b$ [1–7] **18.** 620 [1–8] **19.** 15,000 [1–8] **20.** $118,900 [1–2] **21.** 751 miles [1–4] **22.** $355 [1–5]

1–1–1

1. Addition **3.** Subtraction **5.** Multiplication **7.** Multiplication **9.** Division **11.** Multiplication and addition **13.** Division and subtraction **15.** $2 + 6$ **17.** $5 - y$ **19.** $8a$ **21.** $4 \div x$ or $\frac{4}{x}$ **23.** $a + b$ **25.** $m - n$ **27.** $(8)(5)$ **29.** $9 \div 3$ or $\frac{9}{3}$ **31.** Numbers being added or subtracted **33.** Addition, subtraction, multiplication, and division

1–2–1

1. Ten thousands **3.** Hundreds **5.** Millions **7.** Ten millions **9.** Three hundred twenty-one **11.** Two thousand, one hundred twenty-nine **13.** Ten thousand, one hundred twenty-four **15.** Three hundred twenty thousand, four **17.** Two million, five hundred three thousand, one hundred **19.** Fifteen million, two hundred thousand, six hundred twenty-one **21.** Seven hundred sixty-three million, seven hundred sixty-three thousand, seven hundred sixty-three **23.** Twenty-two billion, five hundred nineteen million, fifty-four thousand, one hundred eleven **25.** 326 **27.** 1,801 **29.** 26,042 **31.** 206,201 **33.** 68,045,308 **35.** 511,412,100 **37.** Twenty-six thousand, seven hundred ninety-eight dollars **39.** Thirty-one thousand, forty-five miles **41.** Four thousand, five hundred fifty-two **43.** 2,596 **45.** $52,904 **47.** 250,000 **49.** Ten

1–2–2

1. $5 > 3$ **3.** $4 > 1$ **5.** $2 < 5$ **7.** $1 < 2$ **9.** $12 < 27$ **11.** $24 < 43$ **13.** $15 > 0$ **15.** $200 < 400$ **17.** $2{,}125 > 2{,}115$ **19.** $1{,}972{,}823 < 1{,}973{,}000$ **21.** five

1–3–1

1. $(2)(2)(2)$ **3.** $(5)(5)(5)(5)$ **5.** $(x)(x)(x)(x)(x)$ **7.** 5^3 **9.** 1 **11.** $2(10^2) + 5(10^1) + 1(10^0)$ **13.** $6(10^2) + 1(10^1) + 9(10^0)$ **15.** $2(10^3) + 4(10^2) + 5(10^1) + 8(10^0)$ **17.** $3(10^1) + 0(10^0)$ **19.** $5(10^2) + 0(10^1) + 0(10^0)$ **21.** $3(10^3) + 0(10^2) + 4(10^1) + 1(10^0)$ **23.** $6(10^3) + 0(10^2) + 0(10^1) + 0(10^0)$ **25.** $2(10^4) + 3(10^3) + 4(10^2) + 2(10^1) + 6(10^0)$ **27.** $4(10^3) + 5(10^2) + 0(10^1) + 2(10^0)$ **29.** $8(10^4) + 1(10^3) + 2(10^2) + 9(10^1) + 4(10^0)$ **31.** $1(10^4) + 5(10^3) + 8(10^2) + 0(10^1) + 4(10^0)$ **33.** $2(10^4) + 3(10^3) + 0(10^2) + 3(10^1) + 8(10^0)$ **35.** $1(10^5) + 7(10^4) + 2(10^3) + 2(10^2) + 5(10^1) + 4(10^0)$ **37.** $3(10^5) + 5(10^4) + 0(10^3) + 4(10^2) + 0(10^1) + 0(10^0)$ **39.** $5(10^6) + 4(10^5) + 6(10^4) + 1(10^3) + 0(10^2) + 0(10^1) + 6(10^0)$ **41.** To express a number in a compact form

1-4-1

1. Additive identity **3.** Commutative property of addition **5.** Associative property of addition **7.** Associative property of addition **9.** Commutative property of addition **11.** 20 **13.** 23 **15.** 30 **17.** 31 **19.** 86 **21.** 70 **23.** 115 **25.** 112 **27.** 121 **29.** 128 **31.** 779 **33.** 981 **35.** 931 **37.** 1,052 **39.** 1,066 **41.** 1,360 **43.** 1,618 **45.** 2,031 **47.** 11,278 **49.** 7,714 **51.** 19,106 **53.** 4,344 **55.** 10,283 **57.** 5,322 **59.** 102,371 **61.** $40 **63.** 83 pounds **65.** $88 **67.** 1,029 miles **69.** 537 electoral votes

1-5-1

1. 3 **3.** 3 **5.** 11 **7.** 10 **9.** 22 **11.** 21 **13.** 11 **15.** 321 **17.** 232 **19.** 520 **21.** 7 **23.** 32 **25.** 101 **27.** 243 **29.** 300 **31.** No

1-5-2

1. 15 **3.** 18 **5.** 18 **7.** 17 **9.** 468 **11.** 188 **13.** 176 **15.** 164 **17.** 88 **19.** 118 **21.** 28 **23.** 28 **25.** 166 **27.** 159 **29.** 38

1-5-3

1. 22 **3.** 156 **5.** 487 **7.** 378 **9.** 1,616 **11.** 4,056 **13.** 375 **15.** 219 **17.** 109 **19.** 3,268 **21.** 198 **23.** 232 **25.** 4,009 **27.** 11,809 **29.** 1,208 **31.** 248 **33.** 98 **35.** 436 **37.** 806 **39.** 23 days **41.** 287 pages **43.** 26 degrees **45.** 1,078 ft **47.** 397 ft **49.** $8

1-6-1

1. 2 **3.** 1 **5.** 2 **7.** 1 **9.** 3 **11.** 3 **13.** 4 **15.** 4 **17.** 1 **19.** 3
21. 3 is the coefficient, x is the base, 2 is the exponent **23.** 12 is the coefficient, x is the base, 1 is the exponent **25.** Terms are parts being added or subtracted and factors are parts being multiplied.

1-7-1

1. Yes **3.** Yes **5.** No **7.** No **9.** Yes

1-7-2

1. $11x$ **3.** $4x$ **5.** $4a$ **7.** $8x^2$ **9.** $6a^3$ **11.** $16xy$ **13.** xy **15.** $4a + 6b$ **17.** $9a^2b + 11ab^2$ **19.** $22abc + ab$ **21.** $ab + ac$ **23.** $3a^2 + 8a$ **25.** $8x^3y^2$ **27.** $7x^2y - 5xy^2 - 9x^2y^2$ **29.** $M + 8m$ **31.** Distributive property

1-8-1

1. 60 **3.** 30 **5.** 520 **7.** 290 **9.** 4,210 **11.** 400 **13.** 200 **15.** 2,400 **17.** 1,100 **19.** 7,000 **21.** 58,000 **23.** 26,000 **25.** 35,000 **27.** 40,000 **29.** 19,000 **31.** 140,000 **33.** 460,000 **35.** 260,000 **37.** 340,000 **39.** 600,000 **41.** 158,460 **43.** 158,500 **45.** 200,000 **47.** 300,000 **49.** 296,000 **51.** 5,555,560 **53.** 5,555,600 **55.** 5,560,000 **57.** 10,000 **59.** 10,000

Chapter 1 Review

The number in brackets after each answer refers to the section of the chapter that discusses that type of problem.
1. Division [1-1] **3.** Multiplication and subtraction [1-1] **5.** Addition and division [1-1] **7.** Tens [1-2] **9.** Ten thousands [1-2] **11.** Ten thousand, three hundred twenty-five [1-2] **13.** Seven hundred twenty-four million, eight hundred three thousand, two hundred forty-one [1-2] **15.** 523,201 [1-2] **17.** $8 > 0$ [1-2] **19.** $3,099 < 3,103$ [1-2]
21. $8(10^2) + 2(10^1) + 7(10^0)$ [1-3] **23.** $4(10^4) + 0(10^3) + 3(10^2) + 9(10^1) + 6(10^0)$ [1-3] **25.** 28 [1-4] **27.** 97 [1-4] **29.** 123 [1-4] **31.** 110 [1-4] **33.** 605 [1-4] **35.** 1,000 [1-4] **37.** 1,282 [1-4] **39.** 8,417 [1-4] **41.** 10,719 [1-4]

43. 233 [1–5] **45.** 259 [1–5] **47.** 197 [1–5] **49.** 3,178 [1–5] **51.** 22,179 [1–5] **53.** 2 [1–6] **55.** 1 [1–6] **57.** 3 [1–6] **59.** $9a - 6b$ [1–7] **61.** $5x^2 + 9x - 4y$ [1–7] **63.** $a^2b + 4a^2$ [1–7] **65.** 350 [1–8] **67.** 1,740 [1–8] **69.** 21,300 [1–8] **71.** 46,000 [1–8] **73.** 460,000 [1–8] **75.** Three thousand, nine hundred fifteen [1–2] **77.** 1,815 [1–2] **79.** $373 [1–4]

Chapter 1 Test

The number in brackets after each answer refers to the section of the chapter that discusses that type of problem.
1. Division [1–1] **2.** Multiplication and subtraction [1–1] **3.** Thousands [1–2] **4.** Fifty-two thousand, six hundred nineteen [1–2] **5.** $18 < 23$ [1–2] **6.** $3(10^3) + 0(10^2) + 0(10^1) + 5(10^0)$ [1–3] **7.** 1,003 [1–4] **8.** 1,347 [1–4] **9.** 5,364 [1–4] **10.** 295 [1–5] **11.** 409 [1–5] **12.** 4,007 [1–5] **13.** 3 [1–6] **14.** 1 [1–6] **15.** $4a$ [1–7] **16.** $8x$ [1–7] **17.** $7x^2y + 6xy$ [1–7] **18.** 380 [1–8] **19.** 12,000 [1–8] **20.** 387,906 [1–2] **21.** 20,306 ft [1–4] **22.** $8,838 [1–5]

Chapter 2 Pretest

The number in brackets after each answer refers to the section of the chapter that discusses that type of problem.
1. 432 [2–1] **2.** 722,545 [2–1] **3.** 257,000 [2–1] **4.** 8 R2 [2–2] **5.** 64 [2–2] **6.** 47 R16 [2–2] **7.** 3 [2–3] **8.** 11 [2–3] **9.** 3 [2–3] **10.** 10 [2–3] **11.** 12 [2–4] **12.** 36 [2–4] **13.** 227 [2–4] **14.** x^7 [2–5] **15.** $8a^3b^4$ [2–6] **16.** $(2^2)(3^2)(5)$ [2–7] **17.** 21 [2–8] **18.** $42x^2y$ [2–8] **19.** 50 [2–4] **20.** $256 [2–2] **21.** 744 hours [2–1] **22.** $117 [2–1]

2–1–1

1. Multiplicative identity **3.** Commutative property of multiplication **5.** Associative property of multiplication **7.** Commutative property of multiplication **9.** Multiplicative identity **11.** 175 **13.** 292 **15.** 153 **17.** 492 **19.** 2,125 **21.** 714 **23.** 532 **25.** 1,728 **27.** 3,744 **29.** 13,464 **31.** 32,982 **33.** 390 **35.** 1,540 **37.** 28,768 **39.** 143,992 **41.** 33,800 **43.** 41,600 **45.** 72,400 **47.** 381,300 **49.** 641,982 **51.** 1,406,000 **53.** 1,434,823 **55.** 1,275,425 **57.** 3,033,577 **59.** 2,512,000 **61.** 32,050 **63.** 1,750 **65.** 1,700 **67.** 3,000 **69.** 30,000 **71.** 806,940 **73.** 1,970 **75.** $90 **77.** 630 miles **79.** $772 **81.** $11,808; $2,813

2–1–2

1. 120 **3.** 2,110 **5.** 3,200 **7.** 20,500 **9.** 132,000 **11.** 160,000 **13.** 50,300,000 **15.** 2,091,000 **17.** 2,000 **19.** 6,100,000

2–2–1

1. 4 **3.** 5 **5.** 7 **7.** 7 **9.** 7 **11.** 9 R3 **13.** 8 R1 **15.** 9 R1 **17.** 8 R1 **19.** 9 R2

2–2–2

1. 12 **3.** 23 **5.** 15 **7.** 13 **9.** 15 R4 **11.** 35 R5 **13.** 254 R1 **15.** 64 R2 **17.** 420 R2 **19.** 302 R4 **21.** 3,107 **23.** 5,051 R2 **25.** $38 **27.** 58 **29.** $56 **31.** 6 R2 **33.** 26 mpg **35.** 4 R4

2–2–3

1. 36 **3.** 63 **5.** 42 R14 **7.** 304 R22 **9.** 340 **11.** 34 **13.** 563 **15.** 115 R6 **17.** 403 R10 **19.** 2,203 **21.** 53 **23.** 83 R91 **25.** 285 **27.** 3,004 **29.** 2,005 R20 **31.** $3,012 **33.** 56 cases, 6 cans **35.** $173 **37.** 78 gallons **39.** 21 gross **41.** 24 mpg **43.** 12 R26

2–3–1

1. 1 **3.** 7 **5.** 10 **7.** 2 **9.** 21 **11.** 3 **13.** 38 **15.** 24 **17.** 15 **19.** 42 **21.** 24 **23.** 98 **25.** 1 **27.** 3 **29.** 42

-3-2

1. 15 **3.** 0 **5.** 2 **7.** 16 **9.** 40 **11.** 18 **13.** 36 **15.** 14 **17.** 14 **19.** 74 **21.** 17 **23.** 23 **25.** 47 **27.** 14 **29.** 22 **31.** Subtraction was performed before multiplication and division

2-3-3

1. 6 **3.** 8 **5.** 4 **7.** 64 **9.** 9 **11.** 9 **13.** 11 **15.** 31 **17.** 5 **19.** 40 **21.** 28 **23.** 28 **25.** 8 **27.** 46 **29.** 5

2-4-1

1. 16 **3.** 75 **5.** 56 **7.** 414 **9.** 261 **11.** 8 **13.** 42 **15.** 288 **17.** 44 **19.** 73 **21.** 648 **23.** 35 **25.** 1,402 **27.** 2,500 **29.** 66 **31.** 38 **33.** 32 **35.** 41 **37.** 3,840 **39.** 125 **41.** 50 is divisible by 10.

2-5-1

1. x^7 **3.** a^6 **5.** y^{11} **7.** x^{14} **9.** w^{12} **11.** a^5 **13.** x^8 **15.** a^3b^4 **17.** x^6y **19.** a^5b^3 **21.** x^4y^4 **23.** a^5b **25.** $x^{13}y^3$ **27.** a^8b^5 **29.** a^6bc^6 **31.** We must add the exponents, not multiply them.

2-6-1

1. Yes **3.** Yes **5.** No **7.** Yes **9.** No

2-6-2

1. $10x^5$ **3.** $12x^8$ **5.** $21x^5$ **7.** $66a^{10}$ **9.** $32a^2b^3$ **11.** $35x^2y^2$ **13.** $6x^3y^4$ **15.** $30a^5bc$ **17.** $30x^{11}$ **19.** $42a^4b^4$ **21.** $48x^3y^6$ **23.** $165a^3b^3c^4$ **25.** $120x^3y^9z^3$ **27.** $60x^2y^2z$ **29.** $60a^2b^3c^5d^2$ **31.** The expression involves division by x.

2-7-1

1. 1,2,4 **3.** 1,13 **5.** 1,2,5,10 **7.** 1,3,9,27 **9.** 1,2,3,6,7,14,21,42 **11.** Yes **13.** No **15.** Yes **17.** Yes **19.** No **21.** No. It does not have exactly two distinct factors. It only has one.

2-7-2

1. $(3)(7)$ **3.** $(2^2)(3)$ **5.** 2^4 **7.** $(2^2)(3)(5)$ **9.** $(2^3)(7)$ **11.** $(3^2)(5)$ **13.** $(2)(5)(7)$ **15.** $(7)(11)$ **17.** $(2^2)(7^2)$ **19.** $(2)(3)(5)(13)$ **21.** $(3)(5)(7)(11)$ **23.** $(11)(17)$ **25.** $(2)(11)(17)$ **27.** $(5)(17)(19)$ **29.** $(2^3)(11)(31)$ **31.** $(2)(3^2)(7)(41)$ **33.** $(2^2)(3^2)(7)(11)$ **35.** 2^7 **37.** $(3^2)(5)(11)(13)$ **39.** $(2)(3^2)(17)(43)$ **41.** $(3^2)(5^2)(7)$ **43.** $(2^3)(5)(7^2)$ **45.** $(3)(5^2)(7)(13)$ **47.** $(7)(11^2)$ **49.** $(3^2)(11)(17)(23)$ **51.** Expressing the number as a product of primes

2-8-1

1. 4 **3.** 2 **5.** 1 **7.** 42 **9.** 18 **11.** 1 **13.** 21 **15.** 18 **17.** 9 **19.** x^2 **21.** a^2b **23.** $21x^4$ **25.** $3xy^2$ **27.** $9x^3$ **29.** 4 **31.** 12 **33.** $42a$ **35.** 1 **37.** $36x^3yz$ **39.** 63

2-8-2

1. 6 **3.** 12 **5.** 12 **7.** 90 **9.** $72x^3$ **11.** $36a^3b^3$ **13.** 72 **15.** $120x^2y$ **17.** 12 **19.** 24 **21.** 120 **23.** $252a^3b^2$ **25.** $270x^5y^7$ **27.** 864 **29.** 7,140 **31.** $144a^3bc$ **33.** 90 **35.** $72a^2b^2c$ **37.** 360 **39.** $108x^2y^2$ **41.** A multiple of a number is the product of that number and a whole number. The LCM of a set of numbers is the smallest number that is a multiple of each number in the set.

Chapter 2 Review

The number in brackets after each answer refers to the section of the chapter that discusses that type of problem.
1. 498 [2-1] **3.** 782 [2-1] **5.** 11,050 [2-1] **7.** 2,280 [2-1] **9.** 65,178 [2-1] **11.** 54,800 [2-1] **13.** 360,460 [2-1] **15.** 620,464 [2-1] **17.** 27 [2-2] **19.** 391 [2-2] **21.** 42 R3 [2-2] **23.** 56 [2-2] **25.** 25 R6 [2-2] **27.** 2,005 R20 [2-2] **29.** 3,005 [2-2] **31.** 640 [2-2] **33.** 4 [2-3] **35.** 15 [2-3] **37.** 11 [2-3] **39.** 49 [2-3] **41.** 26 [2-3] **43.** 20 [2-3] **45.** 40 [2-4] **47.** 54 [2-4] **49.** 116 [2-4] **51.** 60 [2-4] **53.** a^{13} [2-5] **55.** x^5y^4 [2-5] **57.** $15x^6$ [2-6] **59.** $16a^2b^3$ [2-6] **61.** $(2^2)(3)(11)$ [2-7] **63.** $(2^2)(3^2)(5)(7)$ [2-7] **65.** 2 [2-8] **67.** $4xy$ [2-8] **69.** 6 [2-8] **71.** 20 [2-8] **73.** $24a^2b^3$ [2-8] **75.** 180 [2-8] **77.** $72x^2y^2z$ [2-8] **79.** $14,940 [2-1] **81.** 13 [2-2]

Chapter 2 Test

The number in brackets after each answer refers to the section of the chapter that discusses that type of problem.
1. 768 [2-1] **2.** 501,410 [2-1] **3.** 32,100,000 [2-1] **4.** 9 R1 [2-2] **5.** 58 [2-2] **6.** 45 R18 [2-2] **7.** 5 [2-3] **8.** 9 [2-3] **9.** 16 [2-3] **10.** 8 [2-3] **11.** 54 [2-4] **12.** 216 [2-4] **13.** 84 [2-4] **14.** a^7b^2 [2-5] **15.** $12a^4b^5$ [2-6] **16.** $(2^3)(3)(17)$ [2-7] **17.** $14xy^2$ [2-8] **18.** 60 [2-8] **19.** 84 [2-4] **20.** $29 [2-2] **21.** 192 slices [2-1] **22.** $830 [2-1]

Chapters 1–2 Cumulative Test

The number in brackets after each answer refers to the chapter and section that discusses that type of problem.
1. Multiplication [1-1] **2.** Ten thousands [1-2] **3.** 32,823 [2-1] **4.** 204 [2-2] **5.** Thirty thousand sixteen [1-2] **6.** 9 [2-3] **7.** 16 [2-4] **8.** $30 > 28$ [1-2] **9.** 439 [1-4] **10.** 4 [1-6] **11.** 27 [2-4] **12.** $(2^3)(3)(7)$ [2-7] **13.** 2,567 [1-5] **14.** $2x + 9y$ [1-7] **15.** $14x^3y^6$ [2-6] **16.** 510,000 [1-8] **17.** $6ab^3$ [2-8] **18.** 30 [2-8] **19.** 28,403 [1-2] **20.** 48 months [2-2]

Chapter 3 Pretest

The number in brackets following each answer refers to the section of the chapter that discusses that type of problem.
1. $\frac{2a}{7b^3}$ [3-1] **2.** $\frac{119}{12}$ [3-5] **3.** 12 [3-4] **4.** 45 [3-4] **5.** $\frac{2}{15}$ [3-2] **6.** $\frac{18a}{35b}$ [3-3] **7.** $\frac{9}{11}$ [3-4] **8.** $\frac{1}{4}$ [3-4] **9.** 9 [3-5] **10.** $\frac{32x + 21y}{56x^2y^2}$ [3-4] **11.** $\frac{3}{22}$ [3-3] **12.** $\frac{12b - 5a}{20ab}$ [3-4] **13.** $\frac{17}{9}$ [3-4] **14.** $\frac{3}{22}$ [3-2] **15.** $3\frac{5}{12}$ [3-5] **16.** $\frac{10}{21}$ [3-3] **17.** $\frac{4a}{9c}$ [3-2] **18.** $12\frac{5}{36}$ [3-5] **19.** $\frac{9}{16}$ [3-5] **20.** $\frac{7}{9} > \frac{8}{11}$ [3-4] **21.** $475 [3-2] **22.** 32 loads [3-3] **23.** $2\frac{1}{15}$ hours [3-5]

3-1-1

1. $\frac{2}{3}$ **3.** $\frac{3}{4}$ **5.** $\frac{2}{5}$ **7.** $\frac{3}{5}$ **9.** x^2 **11.** $\frac{1}{a^2}$ **13.** $\frac{3}{7}$ **15.** $\frac{2}{3}$ **17.** $\frac{1}{2x}$ **19.** $\frac{a}{4}$ **21.** $\frac{1}{3}$ **23.** $\frac{1}{9}$ **25.** $\frac{1}{4}$ **27.** $\frac{1}{2}$ **29.** $\frac{3}{4a}$ **31.** $\frac{2y}{3x}$ **33.** $\frac{6}{7}$ **35.** $\frac{3}{8}$ **37.** $\frac{10x^2}{27y^2}$ **39.** $\frac{4}{11}$ **41.** $\frac{28a^2}{39b}$ **43.** $\frac{4xy}{13}$ **45.** $\frac{7}{12}$ **47.** $\frac{4a^2}{9b^2}$ **49.** $\frac{2}{3}$ **51.** $\frac{3}{14}$ **53.** $\frac{63x^3}{68y^4}$ **55.** $\frac{3y^5}{4x}$ **57.** $\frac{6}{11}$ **59.** $\frac{7}{12}$ **61.** If the numerator is less than the denominator, it is a proper fraction. If not, it is an improper fraction.

3-2-1

1. $\frac{3}{10}$ **3.** $\frac{1}{8}$ **5.** $\frac{10}{21}$ **7.** $\frac{9}{20}$ **9.** $\frac{a^3}{b^4}$ **11.** $\frac{x^4}{y^2}$ **13.** $\frac{2}{7}$ **15.** $\frac{5}{11}$ **17.** xy^2 **19.** $\frac{b}{a}$ **21.** $\frac{2}{5}$ **23.** $\frac{10}{3}$ **25.** $\frac{2}{x}$ **27.** $\frac{3a}{5}$ **29.** $\frac{3}{40}$ **31.** $\frac{3}{10}$ **33.** $\frac{b^2}{2a}$ **35.** $\frac{4}{25}$ **37.** $\frac{y}{18a^2}$ **39.** $\frac{6}{11}$ **41.** $\frac{a^2}{8b}$ **43.** $\frac{27}{40}$ **45.** $\frac{4}{15}$ **47.** $\frac{3}{10a}$ **49.** 1 **51.** 4

. $3ab^2$ **55.** $9a^2b^2$ **57.** 5 **59.** 1 **61.** 60 **63.** 9 **65.** $\frac{3}{7}$ **67.** 1 **69.** 13 **71.** 18 ounces **73.** $\frac{1}{2}$ liter

75. $\frac{1}{3}$ cup sugar, $\frac{1}{8}$ tsp. vanilla, $\frac{1}{4}$ cup flour **77.** \$75 **79.** \$90

3-3-1

1. $\frac{8}{9}$ **3.** $\frac{3}{4}$ **5.** 20 **7.** $\frac{1}{20}$ **9.** $\frac{2}{3}$ **11.** 1 **13.** $\frac{2}{3}$ **15.** $\frac{3}{2}$ **17.** x **19.** $\frac{2b}{3a}$ **21.** $\frac{25}{18}$ **23.** 1 **25.** $\frac{35x}{36}$ **27.** $\frac{3ab}{2}$

29. $\frac{5}{7}$ **31.** 10 **33.** $\frac{6}{7}$ **35.** $\frac{48ay}{49x}$ **37.** $\frac{25a}{2}$ **39.** $\frac{2}{25}$ **41.** 2 **43.** $\frac{xy}{112}$ **45.** 18 **47.** $\frac{9}{20}$ **49.** $\frac{aby}{4x}$ **51.** $\frac{1}{10}$

53. $\frac{2}{3}$ **55.** $\frac{3bx^2y}{10a}$ **57.** $\frac{5ad}{24bc^2}$ **59.** $\frac{8}{15}$ **61.** 9 **63.** 6 **65.** $\frac{10}{3}$ **67.** $\frac{7a^3}{15b^3}$ **69.** $\frac{7}{9}$ **71.** 100 students **73.** 15 bows

75. 6 furlongs **77.** 32 pails **79.** 16 loads **81.** 1

3-4-1

1. $\frac{3}{5}$ **3.** $\frac{2}{5}$ **5.** $\frac{6}{x}$ **7.** $\frac{a+b}{7}$ **9.** $\frac{a-2b}{9}$ **11.** $\frac{1}{2}$ **13.** $2x$ **15.** $\frac{13}{a}$ **17.** $\frac{x}{3}$ **19.** $\frac{2a-b}{xy}$ **21.** $\frac{9}{x}$ **23.** $\frac{1}{2}$

25. $\frac{8}{9a}$ **27.** 3 **29.** $\frac{13}{20ab}$ **31.** $\frac{1}{x}$ **33.** $\frac{1}{4a}$ **35.** $\frac{2a}{b}$ **37.** $\frac{16}{19}$ **39.** $\frac{3a}{8bc}$

3-4-2

1. 6 **3.** 6 **5.** ab **7.** x^2y **9.** 48 **11.** 84 **13.** $10x^2y$ **15.** 18 **17.** 72 **19.** xyz **21.** $10xy^2$ **23.** $4a^2b$
25. $60a^2b^3$ **27.** $12x^2y$ **29.** $144x^2y^2$

3-4-3

1. 3 **3.** 4 **5.** x **7.** 16 **9.** $3ab^2$ **11.** $12ab$ **13.** 18 **15.** 12 **17.** $65a$ **19.** $48axy^3$

3-4-4

1. $\frac{1}{3} < \frac{2}{5}$ **3.** $\frac{3}{4} > \frac{5}{8}$ **5.** $\frac{2}{4} < \frac{6}{8}$ **7.** $\frac{3}{9} = \frac{5}{15}$ **9.** $\frac{5}{6} < \frac{6}{7}$ **11.** $\frac{5}{16} < \frac{3}{8}$ **13.** $\frac{8}{20} = \frac{6}{15}$ **15.** $\frac{3}{8} > \frac{4}{11}$

3-4-5

1. $\frac{7}{10}$ **3.** $\frac{8}{15}$ **5.** $\frac{2y+3x}{xy}$ **7.** $\frac{7}{24}$ **9.** $\frac{1}{36}$ **11.** $\frac{9}{10}$ **13.** $\frac{2a-1}{a^2}$ **15.** $\frac{5}{36}$ **17.** $\frac{ay+b}{xy}$ **19.** $\frac{1}{36}$ **21.** $\frac{6}{5}$

23. $\frac{5x-3a}{ax^2}$ **25.** $\frac{29}{40}$ **27.** $\frac{25}{48}$ **29.** $\frac{5ax+8}{30x^2}$ **31.** $\frac{10b-5}{12ab}$ **33.** $\frac{3}{8}$ **35.** $\frac{4y^3-5x}{x^2y^4}$ **37.** $\frac{2x^2+11y}{6x^2y^2}$ **39.** $\frac{31}{48}$

41. $\frac{16ax-15}{36x^2}$ **43.** $\frac{1}{6}$ **45.** $\frac{7}{8}$ **47.** $\frac{29}{20}$ **49.** $\frac{bc+3ac+4ab}{abc}$ **51.** $\frac{11}{54}$ **53.** $\frac{63b+40a}{144a^2b^2}$ **55.** $\frac{7ay^4-4ax}{8x^2y^5}$ **57.** $\frac{1}{3}$

59. $\frac{35}{18}$ **61.** $\frac{40ab+18a+21}{96a^2b}$ **63.** $\frac{31}{90}$ **65.** $\frac{68xz-65ay}{70x^2y^2z^2}$ **67.** $\frac{101}{210}$ **69.** $\frac{5}{24}$ **71.** $\frac{17}{24}$ cup **73.** $\frac{1}{16}$ pound

75. $\frac{3}{10}$ mile **77.** $\frac{37}{24}$ acres **79.** $\frac{15}{8}$ tons

3-5-1

1. $\frac{9}{2}$ **3.** $\frac{11}{3}$ **5.** $\frac{53}{8}$ **7.** $\frac{47}{6}$ **9.** $\frac{49}{4}$ **11.** $2\frac{1}{2}$ **13.** $4\frac{1}{2}$ **15.** $2\frac{4}{7}$ **17.** $6\frac{4}{5}$ **19.** $13\frac{1}{4}$

3-5-2

1. $10\frac{5}{6}$ **3.** $7\frac{7}{8}$ **5.** $1\frac{19}{22}$ **7.** 45 **9.** $2\frac{2}{3}$ **11.** 30 **13.** $13\frac{5}{7}$ **15.** $5\frac{45}{56}$ **17.** $3\frac{5}{9}$ **19.** $16\frac{4}{5}$ **21.** $1\frac{2}{5}$

23. $5\frac{1}{15}$ **25.** $32\frac{1}{2}$ hours **27.** $34\frac{2}{15}$ mpg **29.** $1\frac{1}{3}$ pounds

3-5-3

1. $7\frac{2}{3}$ **3.** $5\frac{3}{4}$ **5.** $6\frac{11}{15}$ **7.** $12\frac{1}{8}$ **9.** $17\frac{7}{12}$ **11.** $8\frac{23}{24}$ **13.** $10\frac{2}{63}$ **15.** $17\frac{29}{60}$ **17.** $15\frac{5}{12}$ cups **19.** $20\frac{7}{12}$ hours

3-5-4

1. $2\frac{1}{2}$ **3.** $2\frac{2}{3}$ **5.** $4\frac{1}{2}$ **7.** $3\frac{1}{8}$ **9.** $1\frac{1}{12}$ **11.** $10\frac{22}{35}$ **13.** $1\frac{1}{3}$ **15.** $8\frac{7}{12}$ **17.** $22\frac{7}{12}$ ounces **19.** $7\frac{3}{4}$ yards

21. $14\frac{5}{12}$ hours

Chapter 3 Review

The number in brackets after each answer refers to the section of the chapter that discusses that type of problem.

1. $\frac{3}{7}$ [3-1] **3.** $\frac{3}{8a^2}$ [3-1] **5.** $\frac{6}{7}$ [3-1] **7.** $\frac{2y^2}{7x}$ [3-1] **9.** 15 [3-4] **11.** $66ab$ [3-4] **13.** $\frac{69}{7}$ [3-5] **15.** $7\frac{5}{8}$ [3-5]

17. $\frac{5}{18}$ [3-2] **19.** $\frac{2}{3}$ [3-3] **21.** $\frac{11}{12}$ [3-4] **23.** $\frac{1}{4}$ [3-4] **25.** $\frac{12}{13a}$ [3-4] **27.** $\frac{2b^3}{7a}$ [3-2] **29.** $\frac{1}{24x}$ [3-3] **31.** $\frac{13}{20}$ [3-2]

33. $\frac{8}{5}$ [3-4] **35.** $\frac{15b - 14a}{21ab}$ [3-4] **37.** $\frac{4x + 5a^2y}{6a^2b}$ [3-4] **39.** $\frac{20b^2}{a}$ [3-2] **41.** $\frac{14}{3}$ [3-3] **43.** $2\frac{11}{48}$ [3-5]

45. $\frac{15ay^2 + 14x}{36x^2y^3}$ [3-4] **47.** $\frac{13}{2ab}$ [3-2] **49.** $1\frac{1}{2}$ [3-5] **51.** $6\frac{1}{2}$ [3-5] **53.** $\frac{28z - 15x^2y}{36xy^2z}$ [3-4] **55.** 10 [3-5]

57. 9 [3-3] **59.** $\frac{81}{80}$ [3-4] **61.** $6\frac{13}{24}$ [3-5] **63.** $4\frac{4}{5}$ [3-5] **65.** $\frac{25}{27a^2}$ [3-3] **67.** $\frac{31}{20}$ [3-4] **69.** $\frac{19}{24}$ [3-5]

71. $5\frac{5}{12}$ [3-5] **73.** $\frac{2}{27y^2}$ [3-2] **75.** $\frac{51}{8}$ [3-4] **77.** $9\frac{29}{72}$ [3-5] **79.** 2 [3-5] **81.** $6\frac{41}{42}$ [3-5] **83.** $12\frac{1}{2}$ [3-5]

85. $3\frac{3}{5}$ [3-5] **87.** $\frac{3}{4} > \frac{8}{11}$ [3-4] **89.** $\frac{3}{13} < \frac{1}{4}$ [3-4] **91.** \$150 [3-2] **93.** 128 seconds [3-3] **95.** $43\frac{1}{3}$ hours [3-5]

97. $15\frac{5}{12}$ yards [3-5] **99.** $\frac{3}{5}$ second [3-5]

Chapter 3 Test

The number in brackets after each answer refers to the section of the chapter that discusses that type of problem.

1. $\frac{3x}{5y^2}$ [3-1] **2.** $\frac{58}{5}$ [3-5] **3.** 15 [3-4] **4.** 24 [3-4] **5.** $\frac{2}{21}$ [3-2] **6.** $\frac{33y}{50x}$ [3-3] **7.** $\frac{11}{13}$ [3-4] **8.** $\frac{1}{3}$ [3-4]

9. 12 [3-5] **10.** $\frac{18c + 20ab}{45a^2bc}$ [3-4] **11.** $\frac{3x}{32}$ [3-3] **12.** $\frac{35y - 24x}{42xy}$ [3-4] **13.** $\frac{71}{36}$ [3-4] **14.** $\frac{2}{15}$ [3-2] **15.** $2\frac{13}{24}$ [3-5]

16. $\frac{4}{15}$ [3–3] **17.** $\frac{3}{10xy}$ [3–2] **18.** $13\frac{5}{8}$ [3–5] **19.** $\frac{5}{6}$ [3–5] **20.** $\frac{5}{9} < \frac{7}{12}$ [3–4] **21.** 9 tons [3–2]

22. 27 doses [3–3] **23.** $11\frac{7}{12}$ oz [3–5]

Chapters 1–3 Cumulative Test

The number in brackets after each answer refers to the chapter and section that discusses that type of problem.

1. Tens [1–2] **2.** $x + 5y$ [1–7] **3.** $\frac{4}{7}$ [3–1] **4.** 2,114 [1–4] **5.** 50,000 [1–8] **6.** $\frac{2}{21}$ [3–3] **7.** 198 R23 [1–8]

8. 1,198 [1–5] **9.** $\frac{11}{18}$ [3–4] **10.** $(2^3)(5^2)(11)$ [2–7] **11.** 5 [1–6] **12.** $\frac{26}{3}$ [3–5] **13.** $201 < 300$ [1–2] **14.** $\frac{3ab}{2}$ [3–2]

15. 120 [2–8] **16.** 524,001 [2–1] **17.** $15a^2b^3$ [2–6] **18.** 30 [2–4] **19.** $4\frac{1}{15}$ [3–5] **20.** $1,696 [2–1]

Chapter 4 Pretest

The number in brackets after each answer refers to the section of the chapter that discusses that type of problem.

1. $(3)(10^1) + (8)(10^0) + (7)\left(\frac{1}{10^1}\right) + (5)\left(\frac{1}{10^2}\right) + (2)\left(\frac{1}{10^3}\right)$ [4–1] **2.** Four hundred six and ninety-eight hundredths [4–1]

3. 5.69 [4–2] **4.** $\frac{3}{40}$ [4–7] **5.** 6.375 [4–7] **6.** 23.75 [4–3] **7.** 14.787 [4–3] **8.** 3.6 [4–5] **9.** 11.34 [4–6]

10. $1.58a^3b^3$ [4–4] **11.** 29.31 [4–3] **12.** $14.285x$ [4–3] **13.** 12.227 [4–5] **14.** 11.46 [4–6] **15.** 145.306 [4–4]
16. 502.07 [4–3] **17.** 23.549 [4–3] **18.** 5.06 [4–5] **19.** 2.28068 [4–4] **20.** 0.007 [4–6] **21.** $266.71 [4–3]
22. $155.70 [4–4]

4–1–1

1. $(3)\left(\frac{1}{10^1}\right)$ **3.** $(0)\left(\frac{1}{10^1}\right) + (4)\left(\frac{1}{10^2}\right)$ **5.** $(3)\left(\frac{1}{10^1}\right) + (0)\left(\frac{1}{10^2}\right) + (9)\left(\frac{1}{10^3}\right)$

7. $(1)\left(\frac{1}{10^1}\right) + (0)\left(\frac{1}{10^2}\right) + (5)\left(\frac{1}{10^3}\right) + (3)\left(\frac{1}{10^4}\right)$ **9.** $(3)(10^1) + (8)(10^0) + (1)\left(\frac{1}{10^1}\right) + (2)\left(\frac{1}{10^2}\right) + (4)\left(\frac{1}{10^3}\right)$

11. Six tenths **13.** One hundred twenty-nine thousandths **15.** Eight hundred two thousandths **17.** Twenty seven and three hundredths **19.** Sixteen and four ten thousandths **21.** 0.4 **23.** 0.361 **25.** 125.204 **27.** 100.007 **29.** One hundred five and seventy-three hundredths dollars

4–2–1

1. 0.5 **3.** 0 **5.** 3.416 **7.** 3.42 **9.** 12.35 **11.** 11 **13.** 1,310 **15.** 9.0537 **17.** 0 **19.** 2.0 **21.** 0.1010 **23.** 0 **25.** 5,200 **27.** 0.25001 **29.** 5.0 **31.** 10.00 **33.** 5.60 indicates we have rounded to the nearest hundredth. **35.** $41.61

4–3–1

1. 6.895 **3.** 19.355 **5.** 7.888 **7.** 21.183 **9.** 70.9114 **11.** 10.07 **13.** 20.04 **15.** 28.03 **17.** 44.5615 **19.** 27.017 **21.** 103.3195 **23.** 12.13 **25.** 14.83 **27.** 5.079 **29.** 6.838 **31.** 3.738 **33.** 5.997 **35.** 2.91 **37.** 25.48 **39.** 5.76 **41.** 2.296 **43.** $1.26a$ **45.** $1.84y$ **47.** $3.24a + 3.74b$ **49.** $72.00 **51.** $15.05 **53.** 32.4 gallons **55.** 915.5 miles **57.** $103.20 **59.** 22.5 **61.** $13.22 **63.** $262.66

4–4–1

1. 1.28 **3.** 0.32 **5.** 0.924 **7.** 1.692 **9.** 13.0192 **11.** 47.43 **13.** 0.04464 **15.** 31.46 **17.** 13,217 **19.** 0.6813 **21.** $0.7x^2$ **23.** $17.2a^5$ **25.** $0.82x^4y$ **27.** $0.9a^3b^2$ **29.** $0.07x^3y^4$ **31.** 8.917 **33.** 7.0392 **35.** 19.2168 **37.** 0.0732 **39.** 222.794 **41.** 24.7303 **43.** $3.30 **45.** $17.90 **47.** $2,634.96 **49.** $10.41 **51.** 158.88 sq ft **53.** It is somewhere near 50. **55.** $10.41

4–4–2

1. 35 **3.** 161.3 **5.** 241.7 **7.** 30,120 **9.** 4,210 **11.** 82.59 **13.** 31.9 **15.** 4.1 **17.** 213,510 **19.** 0.14

4–5–1

1. 0.7 **3.** 3.5 **5.** 7.04 **7.** 0.017 **9.** 16.04 **11.** 4.013 **13.** 1.483 **15.** 0.31265 **17.** 4.83 **19.** 0.66 **21.** 4.96 **23.** 3.42 **25.** 0.01 **27.** 3.474 **29.** 3.077 **31.** 2.854 **33.** 0.030 **35.** Quotient **37.** An infinite number

4–5–2

1. 1.83 **3.** 4.126 **5.** 0.136 **7.** 0.1294 **9.** 0.0831 **11.** 0.0293 **13.** 0.0001 **15.** 0.01003 **17.** 0.03496 **19.** 0.00204

4–5–3

1. 83.8 **3.** 21.3 points per game **5.** $6.21 **7.** $18,871.67 **9.** 27.0 years old **11.** 295.8 miles

4–6–1

1. 3.5 **3.** 3.1 **5.** 25.1 **7.** 10.6 **9.** 2.14 **11.** 3.7 **13.** 8.22 **15.** 4.07 **17.** 0.014 **19.** 0.005 **21.** 31.5 **23.** 1,620 **25.** 33.4 **27.** 287.1 **29.** 10.8 **31.** 23.2 **33.** 212.8 **35.** 2.7 **37.** 11.43 **39.** 7.05 **41.** 21.06 **43.** 4.27 **45.** $0.69 per pound **47.** 23.8 miles per gallon **49.** 1.95 meters per second **51.** Fundamental principle of fractions

4–7–1

1. $\frac{7}{10}$ **3.** $\frac{2}{5}$ **5.** $\frac{17}{50}$ **7.** $2\frac{9}{10}$ **9.** $\frac{31}{250}$ **11.** $\frac{119}{500}$ **13.** $\frac{52}{125}$ **15.** $7\frac{13}{125}$ **17.** $11\frac{31}{40}$ **19.** $\frac{72}{625}$

4–7–2

1. 0.5 **3.** 0.8 **5.** 0.125 **7.** $0.1\overline{6}$ **9.** $0.\overline{18}$ **11.** $0.\overline{428571}$ **13.** $3.\overline{3}$ **15.** 2.75 **17.** 0.214 **19.** 5.810 **21.** 10.909 **23.** 3.379 **25.** The only factor of 8 is 2. **27.** 0.667

4–7–3

1. 0.29 **3.** 5.6 **5.** 16.35 **7.** 1.76 **9.** 4.15 **11.** 1.39 **13.** 4.0306 **15.** 2.536 **17.** 3.92 **19.** 19.1

Chapter 4 Review

The number in brackets after each answer refers to the section of the chapter that discusses that type of problem.

1. $(2)\left(\frac{1}{10^1}\right) + (4)\left(\frac{1}{10^2}\right)$ [4–1] **3.** $(7)(10^0) + (3)\left(\frac{1}{10^1}\right) + (0)\left(\frac{1}{10^2}\right) + (7)\left(\frac{1}{10^3}\right) + (4)\left(\frac{1}{10^4}\right)$ [4–1]
5. Forty-nine hundredths [4–1] **7.** Fifteen and seventy-five thousandths [4–1] **9.** 0.403 [4–1] **11.** 0.81 [4–2]
13. 6.397 [4–2] **15.** 1.00 [4–2] **17.** 0.5 [4–2] **19.** 10 [4–2] **21.** $\frac{83}{100}$ [4–7] **23.** $\frac{36}{125}$ [4–7] **25.** $17\frac{12}{25}$ [4–7]
27. 0.958 [4–7] **29.** 5.778 [4–7] **31.** 0.0628 [4–4] **33.** 0.18 [4–5] **35.** 59.30 [4–3] **37.** 83.714 [4–6]
39. 27.66 [4–3] **41.** 15.84 [4–3] **43.** 19.206 [4–3] **45.** 2.223 [4–4] **47.** 0.0004 [4–4] **49.** 3.4 [4–6]
51. 23.189 [4–3] **53.** $0.57a^2b^3$ [4–4] **55.** 3,247 [4–4] **57.** 460.4 [4–3] **59.** 37.77 [4–3] **61.** 0.008 [4–6]
63. 2.329 [4–5] **65.** 79.39188 [4–4] **67.** 61.828 [4–3] **69.** 21.004 [4–3] **71.** 29.159 [4–6] **73.** 513.541 [4–3]
75. 19.51248 [4–4] **77.** 26.475 [4–5] **79.** 144.99 [4–3] **81.** $247.637xy$ [4–3] **83.** 17.798 [4–3] **85.** 11.04356 [4–4]
87. 3.081 [4–5] **89.** 3.074 [4–6] **91.** $12.52 [4–3] **93.** $247.59 [4–3] **95.** $7.09 [4–3] **97.** $17.03 [4–4]
99. $279.92 [4–4] **101.** 3.46 inches per month [4–5] **103.** 1.35 inches per hour [4–5] **105.** $2.09 per pound [4–6]

Chapter 4 Test

The number in brackets after each answer refers to the section of the chapter that discusses that type of problem.

1. $(5)(10^1) + (2)(10^0) + (6)\left(\frac{1}{10^1}\right) + (4)\left(\frac{1}{10^2}\right) + (9)\left(\frac{1}{10^3}\right)$ [4–1] **2.** Two hundred ninety-eight and thirty-seven hundredths [4–1] **3.** 8.7 [4–2] **4.** $\frac{14}{125}$ [4–7] **5.** 13.714 [4–7] **6.** 42.42 [4–3] **7.** 58.768 [4–3] **8.** 7.4 [4–5] **9.** 15.23 [4–6] **10.** $5.024x^2y^3$ [4–4] **11.** 35.661 [4–3] **12.** $17.392a$ [4–3] **13.** 21.919 [4–5] **14.** 8.340 [4–6] **15.** 1,344.84 [4–4] **16.** 386.211 [4–3] **17.** 23.869 [4–3] **18.** 7.08 [4–5] **19.** 1.37347 [4–4] **20.** 0.009 [4–6] **21.** $5.03 [4–3] **22.** $10.60 [4–4]

Chapters 1–4 Cumulative Test

The number in brackets after each answer refers to the chapter and section that discusses that type of problem.

1. 8,310 [1–8] **2.** 5,202 [1–4] **3.** 548,488 [2–1] **4.** $14x^2y$ [2–8] **5.** 48 [2–8] **6.** $\frac{14a}{15b^2c}$ [3–1] **7.** 24 [3–4] **8.** $\frac{5ax}{12y^2}$ [3–3] **9.** $\frac{9a - 5b}{12a^2b^2}$ [3–4] **10.** 15 [3–5] **11.** $\frac{33}{125}$ [4–7] **12.** 10.626 [4–4] **13.** 4.37 [4–6] **14.** $15x^5y^6$ [2–6] **15.** 28 [2–3] **16.** $29.95 [4–5] **17.** 3 [1–6] **18.** 115 [2–4] **19.** 51 [2–4] **20.** $9x^2y + 4y^2$ [1–7]

Chapter 5 Pretest

The number in brackets after each answer refers to the section of the chapter that discusses that type of problem.

1. 43° [5–1] **2.** 56° [5–1] **3.** 5 [5–2] **4.** 9 [5–2] **5.** $33\frac{1}{2}$ in. [5–3] **6.** 77.7 cm² [5–3] **7.** $\angle B = 70°$, $\angle C = 40°$ [5–4] **8.** 56 in.² [5–4] **9.** 28.26 ft [5–5] **10.** $\frac{11}{14}$ m² [5–5] **11.** 18.7 ft³ [5–6] **12.** $33\frac{11}{16}$ in.³ [5–6] **13.** $525 [5–7] **14.** 8,100 gal [5–7]

5–1–1

1. A line segment is part of a line and has two endpoints. **3.** No **5.** 65° **7.** 16° **9.** 132° **11.** 55° **13.** 45° **15.** 45° **17.** 60° **19.** 120° **21.** 60° **23.** $\angle 1$ and $\angle 3$, $\angle 4$ and $\angle 6$ **25.** $\angle 1$ and $\angle 6$ **27.** $\frac{15}{16}$ in. **29.** $1\frac{11}{16}$ in. **31.** 15 mm **33.** 22 mm **35.** 34° **37.** 115° **39.** 250° **41.** Since $\angle b + \angle c = 180°$, we can write $\angle b = 180° - \angle c$. Also since $\angle d + \angle c = 180°$, we can write $\angle d = 180° - \angle c$. Therefore $\angle b = \angle d$.

5–2–1

1. Closed **3.** Not closed **5.** Closed **7.** Not a polygon **9.** Not a polygon **11.** Polygon **13.** 5 diagonals

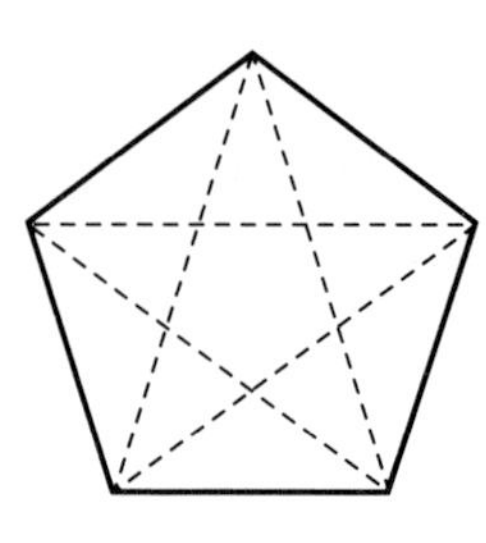

15. 5 diagonals **17.** 14 diagonals **19.** 35 diagonals **21.** Convex **23.** Concave **25.** Concave **27.** Quadrilateral **29.** Pentagon **31.** Octagon **33.** 9 **35.** 0 **37.** **39.** All 6 sides are equal in length. **41.** No

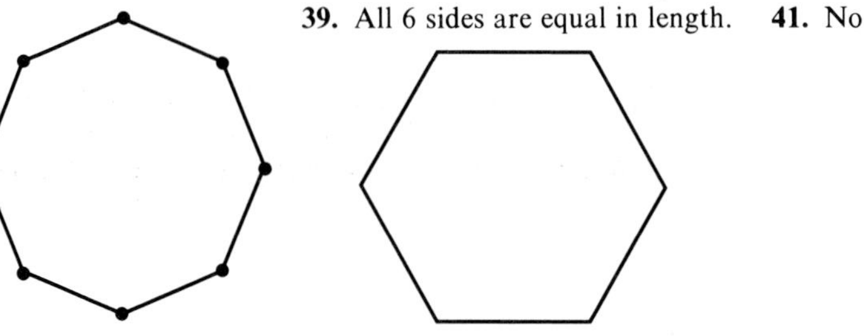

5–3–1

1. Parallelogram **3.** Yes **5.** 24.1 ft **7.** $24\frac{1}{3}$ ft **9.** 20 ft **11.** 14 cm **13.** 39.4 in. **15.** $36\frac{1}{2}$ ft **17.** $P = 4s$
19. 32 ft **21.** 21.6 cm **23.** 410 ft

5–3–2

1. $A = s^2$ **3.** 5.3 ft² **5.** 8.5 ft² **7.** $17\frac{1}{2}$ ft² **9.** 18 cm² **11.** 48 in.² **13.** 185.0 cm² **15.** $16\frac{1}{2}$ m² **17.** $20\frac{1}{4}$ yd²
19. 8 m² **21.** Perimeter is the distance around and area is the measure of the interior.

5–4–1

1. $\angle C = 90°$, right triangle **3.** $\angle B = 80°$, acute triangle **5.** $\angle A = 99°$, obtuse triangle **7.** Isosceles **9.** Scalene
11. Equilateral **13.** $\angle B = 52°$ **15.** $\angle A = \angle B = 65°$ **17.** $\angle B = 53°$, $\angle C = 74°$ **19.** 45°, 45°, 90° **21.** Yes
23. 720° **25.** 144° **27.** The sum of the measures of the interior angles of a triangle is 180°. One angle of a right triangle measures 90°. The sum of the measures of the other two angles must be 90°. Thus they are complementary.

5–4–2

1. 18 cm **3.** 42.9 in. **5.** 15 in.² **7.** 11.2 cm² **9.** 12.2 in.² **11.** 24 ft **13.** 35 in.²

5–5–1

1. 8 in. **3.** 90° **5.** 33° **7.** 22 in. **9.** 44 cm **11.** 19.5 ft **13.** 9.4 m **15.** $12\frac{4}{7}$ in.² **17.** 77 in.² **19.** 226.9 m²
21. 46.5 in.² **23.** 71.0 cm² **25.** $C = \pi r$

5–6–1

1. 105 ft³ **3.** $82\frac{1}{2}$ ft³ **5.** 27 cm³ **7.** 396 m³ **9.** $905\frac{1}{7}$ cm³ **11.** 42.9 in.³ **13.** 90 in.³ **15.** 66 cm³ **17.** 3,375 cm³
19. 179.5 in.³ **21.** 3,000 ft³ **23.** 792 in.³

5–7–1

1. 180 pieces **3.** 2,210 ft **5.** $13.20 **7.** $660.00 **9.** $107.52 **11.** $1,588.50 **13.** $271.32 **15.** 9,474 ft²
17. 16 sheets **19.** $16.02

Chapter 5 Review

The number in brackets after each answer refers to the section of the chapter that discusses that type of problem.
1. 66° [5–1] **3.** 98° [5–1] **5.** 73° [5–1] **7.** 107° [5–1] **9.** 73° [5–1] **11.** [5–2]

13. 8 [5–2] **15.** 4 [5–2] **17.** 27 [5–2] **19.** 20 [5–2] **21.** 600 ft² [5–3] **23.** 64 ft [5–1] **25.** 40° [5–4]
27. $\angle A = 30°$, $\angle B = 30°$ [5–4] **29.** 37.68 ft [5–5] **31.** 64 in.³ [5–6] **33.** 85 ft² [5–4] **35.** 1,100 cm³ [5–6]
37. 64 ft [5–3] **39.** 13,266.5 ft² [5–5] **41.** 66 cm³ [5–6] **43.** 324 ft² [5–7] **45.** 94 ft² [5–7] **47.** $675.00 [5–7]

Chapter 5 Test

The number in brackets after each answer refers to the section of the chapter that discusses that type of problem.

1. 144° [5–1] **2.** 142° [5–1] **3.** 10 [5–2] **4.** 20 [5–2] **5.** 21.2 m [5–3] **6.** 33 ft² [5–3] **7.** $\angle A = 56°, \angle B = 56°$ [5–4] **8.** 56 cm² [5–4] **9.** $7\frac{6}{7}$ ft [5–5] **10.** 314 in.² [5–5] **11.** $68\frac{3}{4}$ cm³ [5–6] **12.** $14\frac{1}{7}$ m³ [5–6] **13.** $962.50 [5–7] **14.** 301.44 ft² [5–7]

Chapters 1–5 Cumulative Test

The number in brackets after each answer refers to the chapter and section that discusses that type of problem.

1. 3,503 [1–4] **2.** 15 [2–4] **3.** $\frac{3x}{5y^2}$ [3–1] **4.** 18.935 [4–2] **5.** 19.74 [4–4] **6.** $xy + 7x^2$ [1–7] **7.** $2a^2 + 2a$ [1–7] **8.** 42 [2–8] **9.** $\frac{17}{50}$ [4–7] **10.** $7\frac{13}{24}$ [3–5] **11.** 102° [5–1] **12.** $(3^3)(11)$ [2–7] **13.** 125 cm³ [5–6] **14.** $\frac{4}{9}$ [3–2] **15.** 1,229 R27 [2–2] **16.** 15 in.² [5–4] **17.** 108 [1–5] **18.** $\frac{13}{45}$ [3–4] **19.** 4 [1–6] **20.** 17,014 [2–1] **21.** 32.6 [4–6] **22.** $38\frac{1}{2}$ ft² [5–5] **23.** $\frac{8}{27}$ [3–3] **24.** 142.5 ft² [5–3] **25.** $1.26 [4–4]

Chapter 6 Pretest

The number in brackets after each answer refers to the section of the chapter that discusses that type of problem.

1. 5.9 [6–1] **2.** $-\frac{3}{7}$ [6–1] **3.** −13 [6–1] **4.** $(-6) + (+9) = +3$ [6–2]

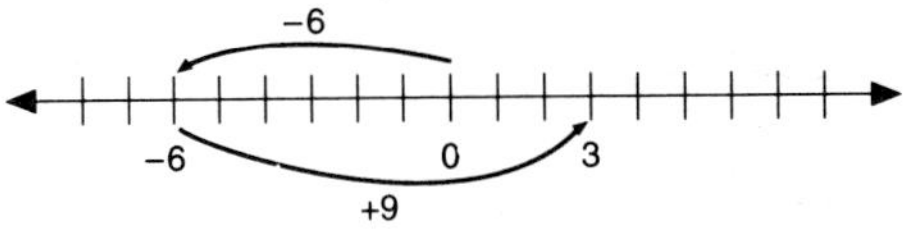

-6
-6 0 3
+9

5. −13 [6–3] **6.** −22 [6–3] **7.** −29 [6–4] **8.** 14 [6–4] **9.** 9 [6–4] **10.** 29 [6–4] **11.** $7x^2y$ [6–5] **12.** $13 + 2x$ [6–5] **13.** 8 [6–4] **14.** −70 [6–6] **15.** 48 [6–6] **16.** −36 [6–6] **17.** 4 [6–6] **18.** $-\frac{3}{2}$ [6–6] **19.** −4 [6–6] **20.** 3 [6–6]

6–1–1

1. −7 **3.** +3 **5.** $6 < 10$

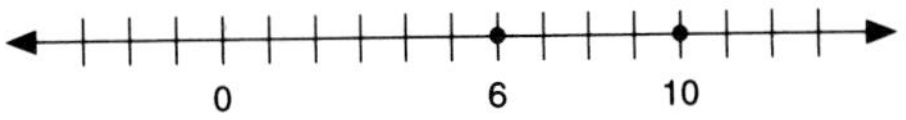

0 6 10

7. $-3 < 3$

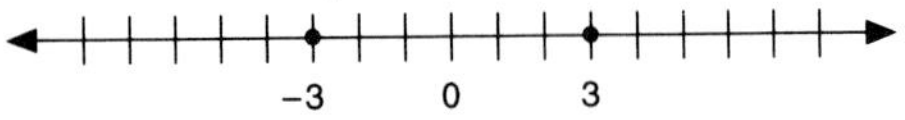

-3 0 3

9. $4 > 1.5$

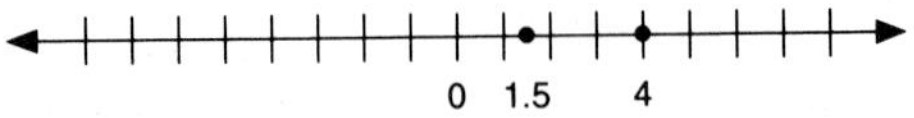

0 1.5 4

11. $-2 > -\frac{5}{2}$

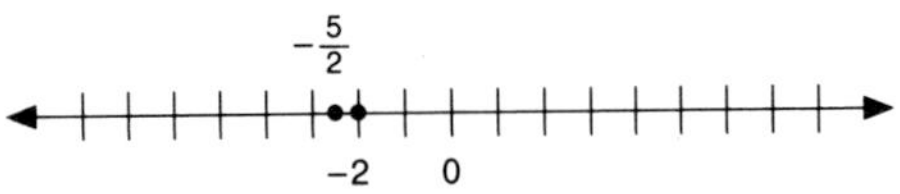

$-\frac{5}{2}$
-2 0

13. $0 < 7$

0 7

15. −20 **17.** −5 **19.** −6

6–1–2

1. −5 **3.** +6 **5.** −14 **7.** $-\frac{1}{2}$ **9.** 0 **11.** $-5\frac{3}{4}$ **13.** −4 **15.** +15 **17.** −4.25 **19.** $-\frac{1}{2}$ **21.** +9 **23.** +4 **25.** 0 **27.** $\frac{2}{3}$ **29.** 3.16 **31.** −5 **33.** 0 **35.** The absolute value of 0 is not positive.

6-2-1

1. +7 **3.** −2 **5.** +2 **7.** +\$7 **9.** −\$24 **11.** −4 **13.** +4 **15.** −8° **17.** +53° **19.** +1

6-2-2

1.

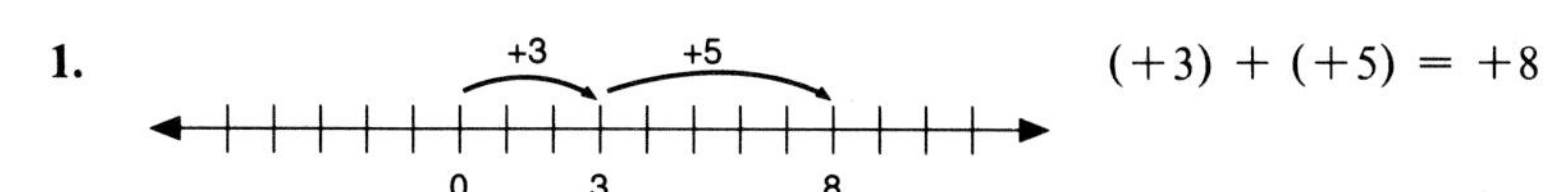

$(+3) + (+5) = +8$

3. $(-4) + (+10) = +6$

5. $(-2) + (+7) = +5$

7.

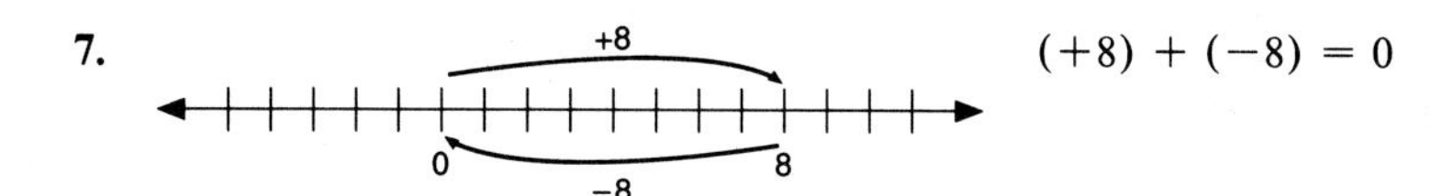

$(+8) + (-8) = 0$

9. $(+9) + (-15) = -6$

11.

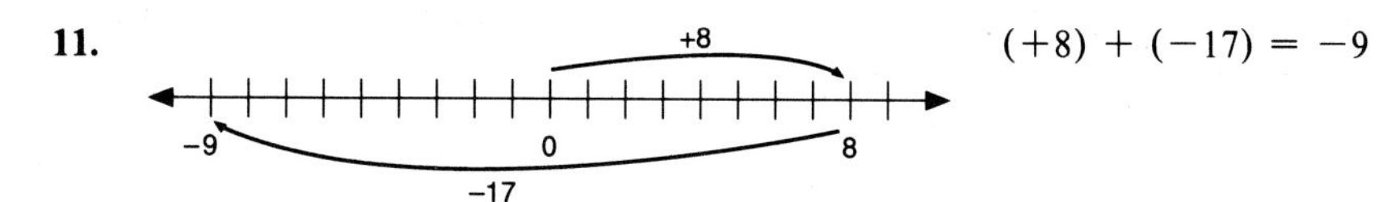

$(+8) + (-17) = -9$

13.

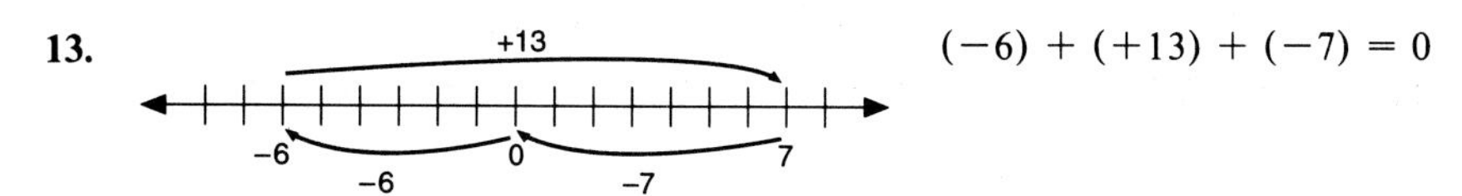

$(-6) + (+13) + (-7) = 0$

15.

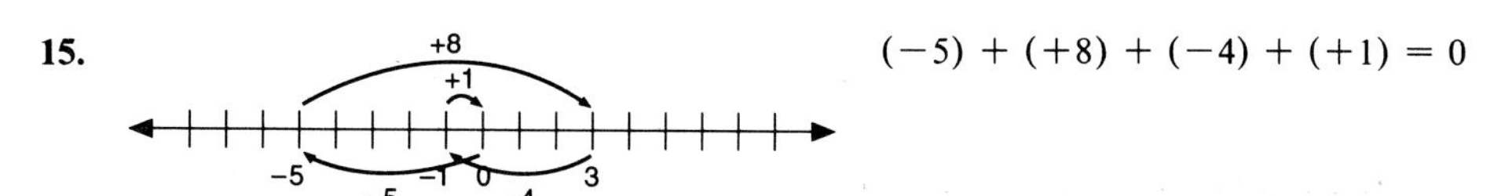

$(-5) + (+8) + (-4) + (+1) = 0$

17.

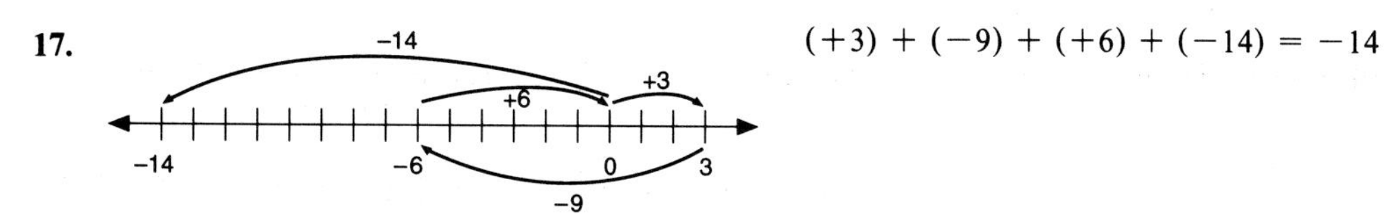

$(+3) + (-9) + (+6) + (-14) = -14$

19.

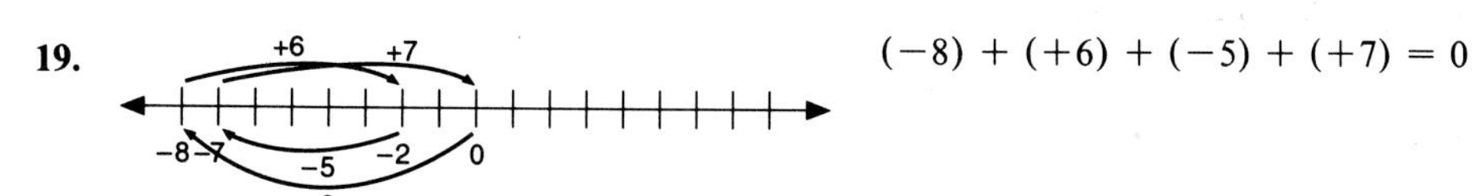

$(-8) + (+6) + (-5) + (+7) = 0$

21.

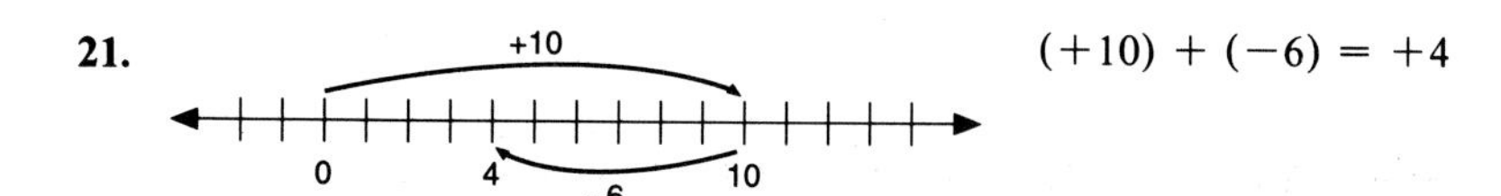

$(+10) + (-6) = +4$

23.

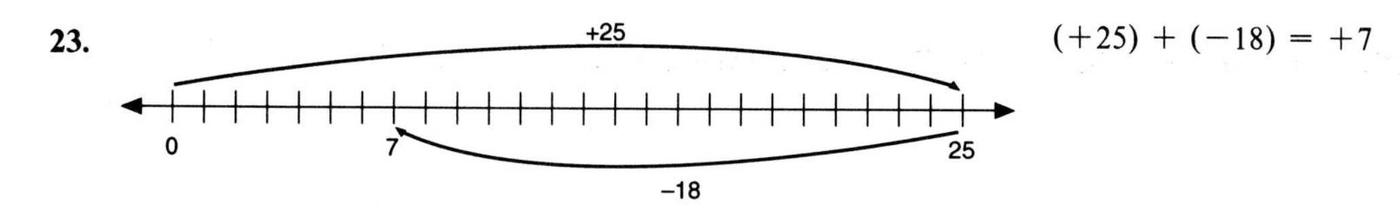

$(+25) + (-18) = +7$

25. The numbers show the direction of movement on the number line.

27. −2 −3 −5 −8 −6 −5 +2 −3 0 $(-5) + (+2) + (-3) + (-2) = -8$

6-3-1

1. $+12$ **3.** $+17$ **5.** -23 **7.** -31 **9.** $+25$ **11.** -28 **13.** $+5$ **15.** -6 **17.** -6 **19.** 0 **21.** -4 **23.** -3
25. $-\frac{2}{7}$ **27.** -8.6 **29.** $-\frac{11}{12}$ **31.** $+2$ **33.** -10 **35.** -1 **37.** -11 **39.** -18 **41.** $(+2) + (-6) = -4$
43. $(+50) + (+15) + (-9) = +56$ **45.** Commutative property

6-4-1

1. $+2$ **3.** -3 **5.** $+15$ **7.** -23 **9.** -4 **11.** $-\frac{1}{12}$ **13.** $+3.47$ **15.** 0 **17.** -10 **19.** -13 **21.** -4 **23.** $+5$
25. That number with the opposite sign **27.** Add negative three to eight

6-4-2

1. 6 **3.** -5 **5.** 3 **7.** -6 **9.** -5 **11.** 3 **13.** -9 **15.** -27 **17.** 6 **19.** -32 **21.** -3.1 **23.** -18
25. $-\frac{2}{7}$ **27.** -1.4 **29.** $-\frac{1}{8}$

6-4-3

1. 6 **3.** 1 **5.** 3 **7.** 1 **9.** 9 **11.** -3 **13.** 14 **15.** -9 **17.** 17 **19.** 40

6-4-4

1. 3 **3.** -6 **5.** 2 **7.** 3 **9.** 12 **11.** 4 **13.** -14 **15.** -19 **17.** 7 **19.** -6 **21.** 5 **23.** -6 **25.** 4 **27.** 22
29. 6

6-5-1

1. $13x$ **3.** $-3x$ **5.** $-4xy$ **7.** $17x^3 + 3x^2$ **9.** $-7ab^2c - 5abc$ **11.** 0 **13.** $-2ab$ **15.** $-15a^2b + 5ab^2$
17. $ab + 3ac$ **19.** 0 **21.** $3x - 1$ **23.** 1 **25.** $2x - 3$ **27.** $10x - 9$ **29.** $5x + 4$ **31.** $10a - b$ **33.** $a + 3b$
35. $6x + 2$ **37.** $4x + 2y - 5$ **39.** Distributive property

6-6-1

1. -40 **3.** -40 **5.** -21 **7.** 0 **9.** -182 **11.** -4 **13.** $-\frac{3}{5}$ **15.** -7.5 **17.** -11.73 **19.** $-\frac{1}{4}$ **21.** -2.1

23. -40 **25.** -48 **27.** -60 **29.** -35 **31.** 0 **33.** -6 **35.** $-9\frac{1}{2}$ **37.** -165.6 **39.** 0

6-6-2

1. $5x + 20$ **3.** $-6x - 8$ **5.** $42x - 28$ **7.** $6x^2 + 12x + 6$ **9.** $-2x^2 - 6x - 2$ **11.** $-x - 15$ **13.** $12x + 2$
15. -3 **17.** $7x - 5y$ **19.** $3x - 6$ **21.** $10a - 10b$ **23.** $-3w - 89$ **25.** $13x - 18$ **27.** $3a - 15b - 18$

6-6-3

1. 6 **3.** 8 **5.** 21 **7.** 45 **9.** 15 **11.** $\frac{1}{5}$ **13.** -40 **15.** -21 **17.** 6.0 **19.** 60 **21.** -12 **23.** 60 **25.** -1 **27.** 0 **29.** 150 **31.** 30 **33.** 168 **35.** -300 **37.** -80 **39.** -37 **41.** Positive

6-6-4

1. 3 **3.** -3 **5.** 5 **7.** -5 **9.** -4 **11.** 7 **13.** 4 **15.** -27 **17.** -7 **19.** 4 **21.** $-\frac{14}{15}$ **23.** -2 **25.** $\frac{7}{2}$

27. $-\frac{1}{21}$ **29.** $-\frac{5}{2}$ **31.** -2 **33.** 12 **35.** -2 **37.** 2 **39.** -3

Chapter 6 Review

The number in brackets after each answer refers to the section of the chapter that discusses that type of problem.

1. $5 > 2$ [6–1]

0 2 5

3. $-5 < -2$ [6–1]

−5 −2 0

5. -10 [6–1] **7.** $+\frac{3}{8}$ [6–1] **9.** $+4$ [6–1] **11.** -3 [6–1]

13.

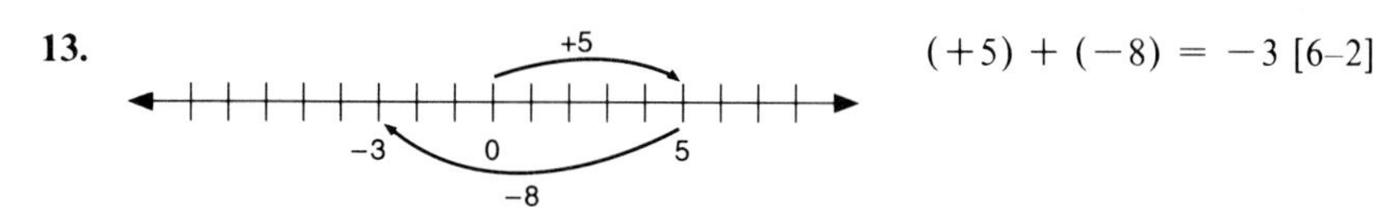

$(+5) + (-8) = -3$ [6–2]

15. $(+2) + (-5) + (+1) = -2$ [6–2]

+1 +2
−3 −2 0 2
−5

17. -2 [6–3] **19.** $+\frac{1}{3}$ [6–3] **21.** -7 [6–3] **23.** -1 [6–4] **25.** 10 [6–4]

27. -2 [6–4] **29.** 22 [6–4] **31.** -48 [6–6] **33.** $\frac{3}{14}$ [6–6] **35.** 280 [6–6] **37.** 12 [6–6] **39.** $-\frac{2}{3}$ [6–6] **41.** 3 [6–6] **43.** -60 [6–6] **45.** -6 [6–6] **47.** 0 [6–5] **49.** $x - 12$ [6–5] **51.** $9x + 3$ [6–5] **53.** $17 - 6x$ [6–6] **55.** $14 - 7a$ [6–6]

Chapter 6 Test

The number in brackets after each answer refers to the section of the chapter that discusses that type of problem.

1. $+\frac{5}{8}$ [6–1] **2.** -6.9 [6–1] **3.** -28 [6–1]

4.

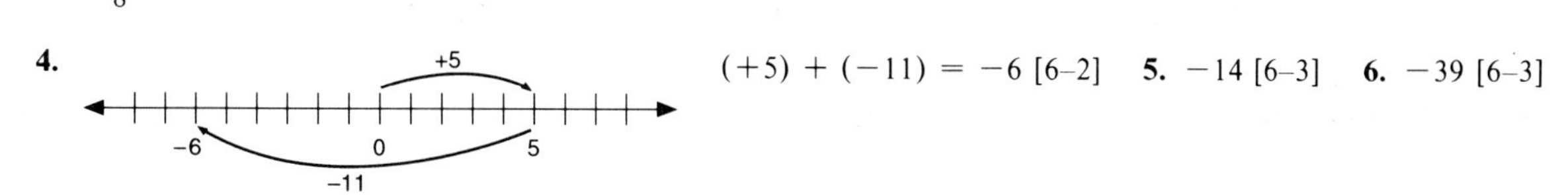

$(+5) + (-11) = -6$ [6–2] **5.** -14 [6–3] **6.** -39 [6–3]

7. -20 [6–4] **8.** 4 [6–4] **9.** -7 [6–4] **10.** 36 [6–4] **11.** $4ab + 3a^2$ [6–5] **12.** $10 + 2a$ [6–5] **13.** -9 [6–4] **14.** 55 [6–6] **15.** -96 [6–6] **16.** -192 [6–6] **17.** 13 [6–6] **18.** $-\frac{1}{10}$ [6–6] **19.** -12 [6–6] **20.** 8 [6–6]

Chapters 1-6 Cumulative Test

The number in brackets after each answer refers to the chapter and section that discusses that type of problem.

1. Hundred thousands [1-2] **2.** 613,210 [2-1] **3.** $\frac{7}{4}$ [3-4] **4.** 21.4 [4-6] **5.** -3 [6-1] **6.** 0.6804 [4-4]

7. $2a + b$ [1-7] **8.** 60 [2-8] **9.** $\frac{8}{49}$ [3-5] **10.** 18 [2-3] **11.** 7.8 ft^2 [5-4] **12.** 6.81 [4-2] **13.** 27° [5-1]

14. $18 < 24$ [1-2] **15.** 121,000 [1-8] **16.** $\frac{x}{y}$ [3-2] **17.** $\frac{63}{1,000}$ [4-7] **18.** -2 [6-3] **19.** $-2x + 9$ [6-5]

20. 53.80 m [5-3] **21.** -4 [6-4] **22.** $(2^2)(3)(5^2)$ [2-7] **23.** -5 [6-6] **24.** $1\frac{3}{4}$ liters [3-5] **25.** $866.25 [5-7]

Chapter 7 Pretest

The number in brackets after each answer refers to the section of the chapter that discusses that type of problem.
1. -1 [7-1] **2.** 11 [7-1] **3.** -125 [7-1] **4.** -108 [7-1] **5.** $-3x^3 + 6x^2 - 12x$ [7-1] **6.** $-4a$ [7-1]
7. $32x - 50$ [7-1] **8.** -54 [7-1] **9.** 216,000 [7-2] **10.** 3.19×10^{-6} [7-2] **11.** 0.0000302 [7-2]
12. $+11$ and -11 [7-3] **13.** 13 [7-3] **14.** 6 in. [7-4] **15.** Associative property of multiplication [7-5]

7-1-1

1. -5 **3.** -7 **5.** 12 **7.** -24 **9.** 36 **11.** 14 **13.** 22 **15.** 5 **17.** -32 **19.** -4 **21.** 72 **23.** $2a^2$

25. $-10x^3$ **27.** $-8x^4$ **29.** $-2x^2 + 5x$ **31.** $-9a^3 + 12a^2 + 3a$ **33.** $-6a^4 + 6a^3 - 8a^2$ **35.** $3y$ **37.** $-\frac{a}{3}$

39. $-5a^2b$ **41.** $\frac{-7y}{5x}$ **43.** $5a^2 - 6a$ **45.** $x^3 - 14x^2 + 15x$ **47.** Any even number **49.** Distributive property

7-1-2

1. $10x - 3$ **3.** $-38x + 65y$ **5.** $4x + 9$ **7.** $21a - 42$ **9.** $6x^2 + 2x$ **11.** $-9x^2 - 16x$ **13.** $-x^2 + 4xy + 6y$
15. $-19x - 20y^2 + 122y$ **17.** -2 **19.** -16 **21.** 25 **23.** 36 **25.** 25 **27.** -24 **29.** 14 **31.** 8 **33.** 17 **35.** 19
37. 14 **39.** 8 **41.** 7

7-2-1

1. 1 **3.** $\frac{1}{4}$ **5.** $\frac{1}{125}$ **7.** $\frac{1}{100,000}$ **9.** 8×10^{-1} **11.** $9(10^{-1}) + 1(10^{-2}) + 5(10^{-3})$ **13.** $7(10^0) + 1(10^{-1}) + 4(10^{-2})$
15. $1(10^2) + 2(10^1) + 8(10^0) + 7(10^{-1}) + 2(10^{-2})$ **17.** 4 **19.** 460 **21.** 3,160 **23.** 6.1 **25.** 0.0605 **27.** 794,000
29. 0.00000917

7-2-2

1. Yes **3.** Yes **5.** No **7.** No **9.** 5.0×10^3 **11.** 7.28×10^5 **13.** 2.35×10^{-7} **15.** 2.33×10^{14} **17.** 320,100
19. 0.00000000107 **21.** 50,200,000,000 **23.** 0.0000407 **25.** 3.6 **27.** 4.9×10^7 miles **29.** 6,000,000,000,000 miles
31. 0.00000001 cm **33.** 1.0×10^{-3} cm **35.** It provides a means of writing very large or very small numbers in a compact form.

7-3-1

1. 0, 1, 4, 9, 16, 25, 36, 49, 64, 81, 100, 121, 144, 169, 196, 225, 256, 289, 324, 361, 400, 441, 484, 529, 576 **3.** ± 5 **5.** ± 8
7. ± 9 **9.** ± 11 **11.** ± 15 **13.** 5 **15.** 6 **17.** 11 **19.** 100 **21.** 21 **23.** 10 **25.** -28 **27.** -4 **29.** 13 **31.** 0

7-3-2

1. 1.4142 **3.** 3.6056 **5.** 5.2915 **7.** 10.4403 **9.** 4.1352 **11.** 8.1216 **13.** 4.0600 **15.** −0.3018 **17.** −6.2985 **19.** 0.8165

7-4-1

1. 15 cm **3.** 5.83 in. **5.** 6.93 m **7.** 7.07 cm **9.** 10.39 ft **11.** $a = \sqrt{12}$ cm, $b = 2$ cm **13.** $\sqrt{50}$ cm **15.** 1.14 in. **17.** 127.28 ft **19.** 22.85 ft

7-5-1

1. 6 **3.** −5, 0, 6 **5.** $-\sqrt{3}, \sqrt{7}, \pi$ **7.** Commutative property of addition **9.** Associative property of multiplication **11.** Multiplicative identity **13.** Associative property of addition **15.** Closure property of multiplication **17.** Multiplicative inverse **19.** Commutative property of addition **21.** The additive identity is 0 and the multiplicative identity is 1.

Chapter 7 Review

The number in brackets after each answer refers to the section of the chapter that discusses that type of problem.
1. 24 [7-1] **3.** 3 [7-1] **5.** −11 [7-1] **7.** −243 [7-1] **9.** 1 [7-1] **11.** 8 [7-1] **13.** $-6a^4$ [7-1] **15.** $-5xy$ [7-1] **17.** $-10x^3 + 15x^2 - 5x$ [7-1] **19.** $23x + 20$ [7-1] **21.** $4a - 12$ [7-1] **23.** $5x^2 - 18x$ [7-1] **25.** $x^2 + 25x$ [7-1] **27.** −375 [7-1] **29.** 8 [7-1] **31.** 45 [7-1] **33.** −51 [7-1] **35.** $C = -20$ [7-1] **37.** $\frac{1}{16}$ [7-2] **39.** 500 [7-2] **41.** 41,200 [7-2] **43.** 3.76×10^7 [7-2] **45.** 5.12×10^{-5} [7-2] **47.** +13 and −13 [7-3] **49.** 14 [7-3] **51.** $b = \sqrt{27}$ in., $c = 6$ in. [7-4] **53.** 6.71 in. [7-4] **55.** Commutative property of multiplication [7-5] **57.** Additive inverse [7-5]

Chapter 7 Test

The number in brackets after each answer refers to the section of the chapter that discusses that type of problem.
1. −6 [7-1] **2.** 13 [7-1] **3.** −32 [7-1] **4.** −72 [7-1] **5.** $-6x^4 + 4x^3 + 10x^2$ [7-1] **6.** $5xy$ [7-1] **7.** $11x + 36$ [7-1] **8.** 62 [7-1] **9.** 0.00024 [7-2] **10.** 2.9×10^7 [7-2] **11.** 51,300 [7-2] **12.** +7 and −7 [7-3] **13.** 15 [7-1] **14.** 5 ft [7-4] **15.** Commutative property of addition [7-5]

Chapters 1-7 Cumulative Test

The number in brackets after each answer refers to the chapter and section that discusses that type of problem.
1. 225 [1-5] **2.** $(2^2)(3^2)(7)$ [2-7] **3.** $\frac{9}{28}$ [3-3] **4.** $\frac{3}{8} > \frac{4}{11}$ [3-4] **5.** $\frac{1}{4}$ [6-6] **6.** 73° [5-4] **7.** Ten thousands [1-2] **8.** $\frac{x}{5y}$ [3-1] **9.** 3.1 [4-2] **10.** $-5a^2 + 24a$ [7-1] **11.** 12 [6-1] **12.** $17\frac{1}{2}$ ft³ [5-6] **13.** 3.56×10^{-4} [7-2] **14.** −1 [6-4] **15.** 59.496 [4-4] **16.** 70 [2-8] **17.** 84.23 [4-3] **18.** 0 [7-1] **19.** $-2a - 1$ [6-5] **20.** 34.05 [4-6] **21.** $a + 10b$ [1-7] **22.** 78 R18 [2-2] **23.** 5.39 ft [7-4] **24.** $3\frac{3}{8}$ [3-5] **25.** $45.00 [5-7]

Chapter 8 Pretest

The number in brackets after each answer refers to the section of the chapter that discusses that type of problem.
1. A polynomial is a sum or difference of monomials and the term $x^{-1}y^2$ is not a monomial because the exponent −1 is not a whole number. [8-1] **2.** Degree 2 [8-1] **3.** Conditional equation [8-2] **4.** Yes [8-2] **5.** 30 [8-3] **6.** $\frac{2}{3}$ [8-4] **7.** −5 [8-5] **8.** −16 [8-4] **9.** −4 [8-5] **10.** $-\frac{1}{3}$ [8-5] **11.** −22 [8-5] **12.** 8 in. [8-4]

8-1-1

1. Yes **3.** Yes **5.** Yes **7.** No **9.** Yes **11.** 3 **13.** 3 **15.** 4 **17.** 3 **19.** 2

8-2-1

1. Identity **3.** Conditional **5.** Contradiction **7.** Identity **9.** Conditional **11.** 3 **13.** 11 **15.** -4 **17.** 10 **19.** 6 **21.** $\frac{3}{5}$ **23.** $\frac{1}{3}$ **25.** -5 **27.** 7 **29.** 6 **31.** No **33.** One

8-3-1

1. 2 **3.** -5 **5.** 9 **7.** -2 **9.** -9 **11.** -0.8 **13.** 4.2 **15.** $1\frac{1}{10}$ **17.** $\frac{1}{15}$ **19.** 2 **21.** 0 **23.** -4 **25.** 6.3 **27.** -2 **29.** $3\frac{13}{20}$ **31.** 8 **33.** 8 **35.** 23.4 m **37.** $44.40 **39.** $38.75 **41.** Symmetric property

8-4-1

1. 12 **3.** 25 **5.** -12 **7.** -25 **9.** $-10\frac{2}{3}$ **11.** 12 **13.** -5 **15.** $\frac{1}{3}$ **17.** $-\frac{1}{2}$ **19.** -14 **21.** $\frac{15}{16}$ **23.** $-\frac{2}{5}$ **25.** $\frac{4}{5}$ **27.** $-1\frac{1}{3}$ **29.** 6 **31.** 17 m **33.** 31 cm **35.** 69.4 km/hr **37.** 30° C **39.** 32 ft/sec^2 **41.** Multiply each side by $-\frac{1}{5}$ or divide each side by -5.

8-5-1

1. 3 **3.** 1 **5.** -4 **7.** -6 **9.** -3 **11.** 2 **13.** -1 **15.** $-1\frac{2}{3}$ **17.** 7 **19.** $-\frac{7}{11}$ **21.** -5 **23.** -2 **25.** $-1\frac{3}{7}$ **27.** 48 **29.** $1\frac{19}{21}$ **31.** 140 **33.** 36 **35.** $-5\frac{5}{7}$ **37.** $-62\frac{2}{5}$ **39.** $\frac{112}{113}$ **41.** 25° **43.** $1\frac{1}{3}$ ft

Chapter 8 Review

The number in brackets after each answer refers to the section of the chapter that discusses that type of problem.
1. It is the sum and difference of monomials. [8–1] **3.** 3 [8–1] **5.** Conditional [8–2] **7.** Identity [8–2] **9.** Conditional [8–2] **11.** 3 [8–3] **13.** -8 [8–4] **15.** -24 [8–3] **17.** 39 [8–3] **19.** -8 [8–4] **21.** 6 [8–4] **23.** $4\frac{1}{2}$ [8–5] **25.** -9 [8–4] **27.** -3 [8–5] **29.** $2\frac{4}{9}$ [8–5] **31.** -11 [8–5] **33.** $\frac{19}{35}$ [8–5] **35.** $4\frac{1}{5}$ [8–5] **37.** 138 [8–5] **39.** 5 [8–5] **41.** 70 [8–5] **43.** $-15\frac{1}{11}$ [8–4] **45.** $\frac{5}{12}$ [8–5] **47.** 0 [8–3] **49.** $1\frac{5}{21}$ [8–4] **51.** -6 [8–5] **53.** $7\frac{1}{2}$ [8–5] **55.** 18 ft [8–4] **57.** $5.80 per hour [8–4] **59.** 84.2 cm [8–5]

Chapter 8 Test

The number in brackets after each answer refers to the section of the chapter that discusses that type of problem.
1. A polynomial is a sum or difference of monomials and the term $x^{-2}y^5$ is not a monomial because the exponent -2 is not a whole number. [8–1] **2.** Degree 4 [8–1] **3.** Identity [8–2] **4.** No [8–2] **5.** -9 [8–3] **6.** 9 [8–5] **7.** 13 [8–4] **8.** $\frac{1}{6}$ [8–4] **9.** -3 [8–5] **10.** 30 [8–5] **11.** $-\frac{6}{31}$ [8–5] **12.** $7\frac{1}{3}$ ft [8–5]

Chapters 1–8 Cumulative Test

The number in brackets after each answer refers to the chapter and section that discusses that type of problem.

1. 9,859 [1–5] **2.** 151,235 [2–1] **3.** $\frac{5x}{6y^2}$ [3–1] **4.** $\frac{9ab^2}{35c^2}$ [3–3] **5.** -8 [8–3] **6.** -60 [6–6] **7.** -1 [6–4]
8. 214.578 [4–4] **9.** 36° [5–4] **10.** 5.64 [4–6] **11.** Ten thousands [1–2] **12.** 180 [2–8] **13.** 32.00 [4–2]
14. 1.83×10^{-5} [7–2] **15.** 27 [8–4] **16.** $x^2 - 4xy$ [6–5] **17.** $(2^3)(5)(7^2)$ [2–7] **18.** 8 m [7–4] **19.** $2x^2y +$ [illegible]xy [1–7]
20. $9\frac{1}{6}$ [3–5] **21.** $-10x^2 + 5x$ [7–1] **22.** 50 [8–5] **23.** 22 in. [5–5] **24.** 8 ft [8–4] **25.** $810.00 [5–7]

Chapter 9 Pretest

The number in brackets after each answer refers to the section of the chapter that discusses that type of problem.

1. $\frac{2}{7}$ [9–1] **2.** No [9–1] **3.** Yes [9–2] **4.** Yes [9–2] **5.** 4 [9–3] **6.** 9 [9–3] **7.** 10.5 [9–3] **8.** 8 students [9–3]
9. 336 miles [9–3] **10.** 12 teaspoons [9–3] **11.** 104.6 km/hr [9–4] **12.** 1.85 gal [9–4] **13. a.** 37.5% [9–5]
b. 0.2% [9–5] **14.** 42.5% [9–6] **15.** $89.25 [9–6]

9–1–1

1. $\frac{5}{7}$ **3.** $\frac{8}{9}$ **5.** $\frac{2}{5}$ **7.** $\frac{3}{17}$ **9.** $\frac{1}{3}$ **11.** $\frac{6}{5}$ **13.** $\frac{1}{9}$ **15.** $\frac{4}{5}$ **17.** $\frac{7}{3}$ **19.** Yes **21.** Yes **23.** Yes **25.** No
27. Yes **29.** 40 mph **31.** 0.305

9–2–1

1. Yes **3.** Yes **5.** Yes **7.** No **9.** Yes **11.** No **13.** Yes **15.** No **17.** No **19.** Yes **21.** No **23.** Yes
25. Yes **27.** Yes **29.** Yes

9–3–1

1. 4 **3.** 20 **5.** 20 **7.** 30 **9.** 4 **11.** 9 **13.** 20 **15.** 15 **17.** 10.8 **19.** 10 **21.** $56 **23.** 18 cans **25.** 250 miles
27. $6.45 **29.** 440 miles **31.** $3\frac{1}{3}$ ft **33.** $a = 3\sqrt{3}$ in. or $\sqrt{27}$ in., $b = 3$ in. **35.** $a = 2\sqrt{3}$ m or $\sqrt{12}$ m, $b = 2$ m,
$a' = 5\sqrt{3}$ m or $\sqrt{75}$ m, $c' = 10$ m **37.** $605 **39.** 60 bulbs **41.** 7.9 gallons **43.** 219 games **45.** 3,125 black bass

9–4–1

1. 11.8 in. **3.** 36.6 **5.** 137.8 in. **7.** 1.9 liters **9.** 4.6 m **11.** 48.3 km **13.** 10.9 oz **15.** 1,553.5 mi **17.** 109.4 yd
19. 1.9 qt **21.** 590.4 mi **23.** 14.3 qt **25.** Answers will vary. **27.** 11.9 gallons

9–4–2

1. 2.18 m **3.** 460 cm **5.** 1.84 m **7.** 3.5 cm **9.** 4.28 km **11.** 16,500 mg **13.** 5.92 kg **15.** 6,000 ml **17.** 0.0305 ℓ
19. 31.08 m **21.** 3.75 g **23.** 9 ℓ **25.** To change centimeters to meters move the decimal point two places to the left.

9–5–1

1. $\frac{1}{2}$ **3.** $\frac{13}{20}$ **5.** $\frac{49}{200}$ **7.** $\frac{5}{6}$ **9.** $\frac{7}{8}$ **11.** 0.15 **13.** 0.03 **15.** 0.156 **17.** 2.38 **19.** 0.0009 **21.** 92% **23.** 320%

25. 40.3% **27.** 7.2% **29.** 0.4% **31.** 50% **33.** 75% **35.** 7% **37.** 37.5% **39.** 175% **41.** $14\frac{2}{7}$% **43.** $66\frac{2}{3}$%

45. $83\frac{1}{3}$% **47.** $62\frac{1}{2}$% **49.** $233\frac{1}{3}$% **51.** Per hundred

9-6-1

1. 16 **3.** 9.6 **5.** 151.8 **7.** 10 **9.** 124 **11.** 225 **13.** 75% **15.** 20.8% **17.** 89.3% **19.** $700 **21.** 75 members **23.** $24,000 **25.** 3,915 miles **27.** 15% **29.** 43.6% **31.** 5.6%

Chapter 9 Review

The number in brackets after each answer refers to the section of the chapter that discusses that type of problem.

1. $\frac{3}{8}$ [9-1] **3.** $\frac{4}{9}$ [9-1] **5.** Yes [9-1] **7.** No [9-1] **9.** Yes [9-2] **11.** Yes [9-2] **13.** No [9-2] **15.** Yes [9-2] **17.** 20 [9-3] **19.** 14 [9-3] **21.** 3.2 [9-3] **23.** 2 [9-3] **25.** 6.9 m [9-4] **27.** 84.7 in. [9-4] **29.** 12.4 mi [9-4] **31.** 4.162 m [9-4] **33.** 0.37 [9-5] **35.** 0.138 [9-5] **37.** 0.007 [9-5] **39.** 55% [9-5] **41.** 6% [9-5] **43.** 1.6% [9-5] **45.** 40% [9-5] **47.** 70% [9-5] **49.** 250% [9-5] **51.** 36 [9-6] **53.** 134.4 [9-6] **55.** 20.0% [9-6] **57.** 17.5% [9-6] **59.** 5 [9-6] **61.** 1,163.64 [9-6] **63.** 15 [9-3] **65.** Yes [9-2] **67.** 14 teeth [9-3] **69.** 1.85 qt [9-4] **71.** 97 people [9-6] **73.** 21.5% [9-6] **75.** 11,420 [9-6]

Chapter 9 Test

The number in brackets after each answer refers to the section of the chapter that discusses that type of problem.

1. $\frac{5}{6}$ [9-1] **2.** Yes [9-1] **3.** Yes [9-2] **4.** No [9-2] **5.** 10 [9-3] **6.** 35 [9-3] **7.** 4.8 [9-3] **8.** 18 boys [9-3] **9.** 7 gallons [9-3] **10.** 45 people [9-3] **11.** 88.5 kph [9-4] **12.** 53 liters [9-4] **13. a.** 62.5% [9-5] **b.** 130% [9-5] **14.** 41.25% [9-6] **15.** 9.5% [9-6]

Chapters 1-9 Cumulative Test

The number in brackets after each answer refers to the chapter and section that discusses that type of problem.

1. One million, two hundred thousand, three hundred [1-2] **2.** 2,219 [2-2] **3.** 42 [2-8] **4.** 400.03 [4-4] **5.** $\frac{2a}{7c}$ [3-1] **6.** -4 [6-4] **7.** $+16$ [6-1] **8.** 56.25 cm² [5-4] **9.** $1\frac{1}{24}$ [3-5] **10.** $a^2b + 3ab$ [1-7] **11.** 6 [7-1] **12.** 461.5% [9-5] **13.** 5.83 in. [7-4] **14.** $\frac{5}{13} < \frac{2}{5}$ [3-4] **15.** 15.702 [4-2] **16.** 4 [1-6] **17.** 21.61 [4-6] **18.** -2 [8-3] **19.** 0.83 km [9-4] **20.** $(3^2)(7)(13)$ [2-7] **21.** -45 [8-4] **22.** -1 [6-6] **23.** 152° [5-1] **24.** 14 [8-5] **25.** $-\frac{1}{35}$ [6-3] **26.** 182.68 [4-3] **27.** 5.3×10^7 [7-2] **28.** 44 hr [3-5] **29.** $177.45 [5-7] **30.** 4.5 lb [9-3]

Chapter 10 Pretest

The number in brackets after each answer refers to the section of the chapter that discusses that type of problem.
1. 9.7 [10-1] **2.** 10.5 [10-1] **3.** 15 [10-1] **4.** 12 [10-1] **5.** 66 [10-1] **6.** 45% [10-3] **7.** 50% [10-3] **8.** Sports [10-3] **9.** Movies [10-3] **10.** Police/detective shows [10-3] **11.** 24% [10-4] **12.** 25% [10-4] **13.** Ban certain types [10-4] **14.** 72% [10-4] **15.** 60 [10-4]

10-1-1

1. a. 8 **b.** 9 **c.** 5 **d.** 6 **3. a.** 6.8 **b.** 7 **c.** 7 **d.** 6 **5. a.** 6.5 **b.** 5.5 **c.** 5 **d.** 8 **7. a.** 8 **b.** 7 **c.** 3 and 7 **d.** 15 **9. a.** 9.5 **b.** 9.5 **c.** 7 and 11 **d.** 9 **11. a.** 74° **b.** 74° **c.** 74° **d.** 11° **13. a.** $-3°$ **b.** $-5.5°$ **c.** $-10°$ and $-7°$ **d.** 20° **15.** 3.2 **17.** 79 **19.** Median and mode

10-2-1

1. 20,000 **3.** 1990 **5.** 1991 and 1992 **7.** 5,000 **9.** 12.5% increase **11.** 4 **13.** 31 **15.** Warren Moon **17.** 13 **19.** 70% **21.** $2,651 **23.** $3,139 **25.** $2,636 **27.** $712 **29.** $12,247.32 **31.** $11,146.62 **33.** 98,935,331 **35.** 10.2% **37.** 1950–1960

10-3-1

1. 20% **3.** US and France **5.** Sweden **7.** 2 percentage points **9.** 79% **11.** 160,000 sq mi **13.** Idaho **15.** California **17.** 10,000 sq mi **19.** 25% **21.** 22,000 students **23.** 1988 and 1992 **25.** 1990 **27.** 5,000 students **29.** 14% **31.** $15,000 **33.** Store A, April **35.** April and June **37.** $55,000 **39.** $20,000

10-4-1

1. 20% **3.** Stocks **5.** 60% **7.** 25% **9.** $30,000 **11.** 8% **13.** Discount stores **15.** Toy and discount stores **17.** 18° **19.** $2.8 million **21.** 15% **23.** Rent **25.** $\frac{1}{5}$ **27.** 40% **29.** 108° **31.** $450

Chapter 10 Review

The number in brackets after each answer refers to the section of the chapter that discusses that type of problem.
1. a. 21.6 **b.** 18 **c.** 18 **d.** 21 [10–1] **3. a.** 12.5 **b.** 12.5 **c.** 10 and 13 **d.** 14 [10–1] **5.** 2.8 [10–1] **7.** Town B [10–2] **9.** 37,500 [10–2] **11.** 25,000 [10–2] **13.** 95,000 [10–2] **15.** 39% [10–2] **17.** 346 [10–2] **19.** 36 [10–2] **21.** 78% [10–2] **23.** 56% [10–2] **25.** Air-popped popcorn and unsalted pretzels [10–2] **27.** 15,000 [10–3] **29.** College B [10–3] **31.** College C [10–3] **33.** Colleges A, C, and E [10–3] **35.** 5,000 [10–3] **37.** 25,000 [10–3] **39.** 1988, 1989, 1991 [10–3] **41.** 1989–1990 [10–3] **43.** 70,000 [10–3] **45.** 20% [10–3] **47.** 18% [10–4] **49.** 25% [10–4] **51.** Radio, tape and CD players [10–4] **53.** All others [10–4] **55.** 64.8° [10–4] **57.** 1,275 [10–4]

Chapter 10 Test

The number in brackets after each answer refers to the section of the chapter that discusses that type of problem.
1. 8 [10–1] **2.** 8 [10–1] **3.** 4 and 8 [10–1] **4.** 12 [10–1] **5.** 98 [10–1] **6.** Oct [10–3] **7.** $24,000,000,000 [10–3] **8.** Sept and Oct [10–3] **9.** $74,000,000,000 [10–3] **10.** 17% [10–3] **11.** 10% [10–4] **12.** 25% [10–4] **13.** 85% [10–4] **14.** 6 [10–4] **15.** 54° [10–4]

Chapters 1–10 Cumulative Test

The number in brackets after each answer refers to the chapter and section that discusses that type of problem.

1. 3,843 [1–5] **2.** 3,241 R22 [2–2] **3.** $\frac{5b^2}{3a}$ [3–1] **4.** 9.49 [4–4] **5.** 32 [6–1] **6.** 128 [7–1] **7.** -5 [8–3]

8. 60% [9–5] **9.** 210 [2–8] **10.** Two million, five thousand, twenty one [1–2] **11.** $15\frac{5}{24}$ [3–5] **12.** 6.58 [4–6]

13. -8 [6–4] **14.** 0.155 m [9–4] **15.** 30 [10–1] **16.** $(2^2)(3^2)(11)$ [2–7] **17.** $\frac{4}{7} < \frac{3}{5}$ [3–4] **18.** 29° [5–4]

19. 1.04×10^{-6} [7–2] **20.** -20 [8–4] **21.** -49 [6–6] **22.** $\frac{37}{50}$ [4–7] **23.** 61,600 [1–8] **24.** -15 [8–5]

25. 4.24 in. [7–4] **26.** $\frac{3}{4}$ [9–1] **27.** 120 in.^{2} [5–6] **28.** $278.40 [5–7] **29.** 8% [9–6] **30.** March [10–3]

Chapter 11 Pretest

The number in brackets after each answer refers to the section of the chapter that discusses that type of problem.

1. $2x - 9$ [11–1] **2.** $\frac{y}{3} + 1$ [11–1] **3.** $0.173x$ [11–1] **4.** $5n$ [11–1] **5.** $8q$ [11–2]

6. x = width, $x + 4$ = length [11–2] **7.** $p - \$25$ [11–2] **8.** Length of rope A = 19 ft, length of rope B = 8 ft [11–3] **9.** 6 in., 12 in., 8 in. [11–3] **10.** 10 students [11–3]

11-1-1

1. $x + 5$ **3.** $a + 5$ **5.** $N - 9$ **7.** $q + 10$ **9.** $2r$ **11.** $2x + 2$ **13.** $6x - 4$ **15.** $\frac{1}{5}x$ **17.** $x - 3$ **19.** $8a$ **21.** $\frac{5x}{2}$

23. $\frac{b + 7}{2}$ **25.** $3(x - 9)$ **27.** $0.15x$ **29.** $0.1x$ **31.** $0.095x$ **33.** $10d$ **35.** $10d + 25q$ **37.** $12y$ **39.** $\frac{d}{7}$

41. We need a complete equation in order to solve for the variable.

11-2-1

The following represent one way of answering each problem.

1. $x + 8$ = first number
x = second number

3. x = height of World Trade Center
$x + 104$ = height of Sears Tower

5. x = units produced first week
$x + 612$ = units produced second week

7. x = no. students in 7:00 AM
$x + 8$ = no. students in 9:00 AM

9. x = pop. of Long Beach
$2x$ = pop. of San Francisco

11. x = width
$x + 5$ = length

13. x = width
$2x + 3$ = length

15. x = first number
$40 - x$ = second number

17. x = first number
$x + 18$ = second number

19. x = first investment
$10{,}000 - x$ = second investment

21. x = last year's price
$x + 0.05x$ = this year's price

23. $x + 4$ = first number
x = second number
$x + 13$ = third number

25. x = first number
$2x$ = second number
$\frac{x}{3}$ = third number

27. x = first even integer
$x + 2$ = second even integer
$x + 4$ = third even integer

29. x = cost in 1970
$\frac{3}{2}x$ = cost in 1976
$3x$ = cost now

31. $x + 5$ = number dollars Jane has
x = number dollars Bob has
$x + 18$ = number dollars Jim has

33. $x + 4$ = mileage of Buick
x = mileage of Chevrolet
$x + 12$ = mileage of Toyota

35. x = original amount
$0.14x$ = interest
$x + 0.14x$ = amount in the account

37. x = length of first piece
$100 - x$ = length of second piece

39. x = number of nickels
$2x$ = number of dimes
$2x + 2$ = number of quarters

41. Yes, since $x + 0.07x = 1.07x$

11-3-1

1. First number = 15
Second number = 7

3. Sears Tower = 1,454 ft
World Trade Center = 1,350 ft

5. First week = 8,902
Second week = 9,514

7. 7:00 AM class = 38
9:00 AM class = 46

9. 716,000

11. Length = 14 m
Width = 9 m

13. 65 cm

15. First number = 14
Second number = 26

17. First number = 32
Second number = 50

19. First investment = \$6,628
Second investment = \$3,372

21. First number = 26
Second number = −11

23. 60°, 120°

25. 18°, 72°

27. First number = 15
Second number = 11
Third number = 24

29. Coach ticket = \$109
First class ticket = \$218
Economy ticket = \$79

31. Buick = 22 mpg
Chevrolet = 18 mpg
Toyota = 30 mpg

33. First side = 32 cm
Second side = 16 cm
Third side = 36 cm

Chapter 11 Review

The number in brackets after each answer refers to the section of the chapter that discusses that type of problem.

1. $2n - 7$ [11–1] **3.** $r + 5$ [11–1] **5.** $\frac{y}{100}$ [11–1] **7.** $x, 84 - x$ [11–2] **9.** $2x, 181 - 3x, x$ [11–2]
11. $x + (x + 5) = 23$ [11–3] **13.** $3x - (x - 7) = 13$ [11–3] **15.** $0.09x + x = 1.26$ [11–3] **17.** 1, 8 [11–3]
19. 47 cm, 53 cm [11–3] **21.** 28, 35 [11–3] **23.** 4.5 m, 13.5 m, 22.0 m [11–3]

Chapter 11 Test

The number in brackets after each answer refers to the section of the chapter that discusses that type of problem.

1. $3x + 6$ [11–1] **2.** $\frac{y}{2} - 7$ [11–1] **3.** $0.094n$ [11–1] **4.** $\frac{d}{7}$ [11–1] **5.** $r + 8$ [11–2] **6.** $k - 3$ [11–2]
7. $2x - 5$ [11–2] **8.** 11 feet [11–3] **9.** Cost of CD = \$570, cost of VCR = \$320 [11–3] **10.** 12 m, 24 m, 27 m [11–3]

Chapters 1–11 Cumulative Test

The number in brackets after each answer refers to the chapter and section that discusses that type of problem.

1. Hundred thousands [1–2] **2.** 5,975 [1–5] **3.** $(2^2)(3^2)(7)$ [2–7] **4.** $\frac{3}{5} > \frac{7}{12}$ [3–4] **5.** $1\frac{2}{3}$ [3–5] **6.** $14\frac{1}{7}$ m³ [5–6]
7. 18 [8–3] **8.** 28 [6–1] **9.** $9a^2b + 7ab^2 + 7ab$ [1–7] **10.** $\frac{2a^2b}{3}$ [3–3] **11.** $\frac{29}{48}$ [3–4] **12.** -6 [6–6] **13.** 13.969 [4–3]
14. 17 ft [8–4] **15.** -7 [8–4] **16.** 24,000 [1–8] **17.** $8a^7$ [2–6] **18.** 130.000 [4–2] **19.** $\frac{2}{5}$ [9–1] **20.** 33.68 [4–4]
21. -10 [6–4] **22.** 3.57×10^7 [7–2] **23.** $11a^2 - 2ab$ [6–5] **24.** 19,988 [2–1] **25.** 30 [7–1] **26.** 2.41 [4–6]
27. $10x^2 - 19x$ [7–1] **28.** $-\frac{11}{5}$ [8–5] **29.** 8 [10–1] **30.** 37.5% [9–5] **31.** 420 [2–8] **32.** $\frac{1}{2}n - 3$ [11–1]
33. 114.7 cm² [5–4] **34.** $x + 0.055x$ [11–2] **35.** 6.93 ft [7–4] **36.** \$19.20 [9–3] **37.** \$750 [10–4] **38.** \$5.16 [9–6]
39. \$3,150.00 [5–7] **40.** \$43 [11–3]

Index

How much mathematics did you have before this course? Years in high school (circle) 1 2 3 4
Courses in college 1 2 3

If you had mathematics before, how long ago?

Last 2 years _____ 3-5 years ago _____ 5 years or longer _____

What is your major or your career goal? ______________________ Your age? _________ (optional)

--------------------FOLD HERE--------------------

Can we quote you? Yes _____ No _____

If yes, please print your full name here ______________________________

Signature ______________________________

College ______________________________________ State _____________

--------------------FOLD HERE--------------------

Please place stamp here

BUSINESS REPLY MAIL

Hawkes Publishing
A Division of Quant Systems, Inc.
1023 Wappoo Road, Suite 6A
Charleston, SC 29412

STUDENT QUESTIONNAIRE

In order to create textbooks that fit your needs, it will help us to know what you, the student, think of ***Preparation for Algebra*** by Nanney and Cable. We would appreciate it if you would answer the following questions. Then cut out the page, fold, tape to seal, and mail it; no postage required. Thank you for your help.

Which chapters did you cover? (circle) 1 2 3 4 5 6 7 8 9 10 11 All

What did you like the most about ***Preparation for Algebra***?

What changes need to be made to improve ***Preparation for Algebra***?

Does the book have enough worked-out examples? Yes ____ No ____

enough exercises? Yes ____ No ____

Which helped most?

Explanations ___ Examples ___ Exercises ___ All three ___ Other ______________
(fill in)

Were the answers at the back of the book helpful? Yes ____ No ____

Did the answers have any typos or misprints? If so, which exercise answers were inaccurate?

For you, was the course elective? _______ Required? ________

What grade did you receive upon completion of this course? ______

Do you plan to take more mathematics courses? Yes ____ No ____

If yes, which ones?